Compositional Data Analysis

Compositional Data Analysis

Theory and Applications

Edited by

Vera Pawlowsky-Glahn
Department of Computer Science and Applied Mathematics
University of Girona, Spain

Antonella Buccianti
Department of Earth Sciences
University of Florence, Italy

WILEY

A John Wiley & Sons, Ltd., Publication

Library of Congress Cataloging-in-Publication Data

Compositional data analysis : theory and applications / edited by Vera Pawlowsky-Glahn, Antonella Buccianti.
     p. cm.
   Includes bibliographical references and index.
   ISBN 978-0-470-71135-4 (cloth)
  1. Multivariate analysis.   2. Correlation (Statistics)   I. Pawlowsky-Glahn, Vera.   II. Buccianti, Antonella.
   QA278.C62 2011
   519.5′35–dc23

                                                                          2011012322

A catalogue record for this book is available from the British Library.

Print ISBN: 978-0-470-71135-4
ePDF ISBN: 978-1-119-97647-9
oBook ISBN: 978-1-119-97646-2
ePub ISBN: 978-1-119-97761-2
Mobi ISBN: 978-1-119-97762-9

Typeset in 10/12pt Times by Aptara Inc., New Delhi, India

To John

Contents

Preface

The present book reflects the state-of-the-art in compositional data analysis in 2011, the year in which John Aitchison turns 85 years old. With this book, the joint work of 41 authors, we want to leave a testimony to John's role in this field during the several decades of his rich and fruitful research career. We think that the best way to honour his contributions to science is to show not only the progress in this field of knowledge since he presented the log-ratio approach to the Royal Statistical Society in 1982, but also the immense possibilities it has opened up for further study. Today it is difficult for us to imagine that the statistical analysis of compositional data has been a major issue of concern for more than 100 years. It is even more difficult to realize that so many statisticians and users of statistics are unaware of the particular problems affecting compositional data, as well as their solutions. For example, the issue of 'spurious correlation', as the situation was phrased by Karl Pearson back in 1897, affects all data that measure parts of some whole, such as percentages, proportions, ppm, ppb, and the like. Such measurements are present in all fields of science.

The book has five parts devoted to the theory and practice of compositional data analysis. Part I includes a short introduction to the history and development of compositional data analysis and Aitchison's log-ratio approach, followed by a summary of basic concepts and procedures. Part II corresponds to mainly theoretical aspects related to statistical modelling, including a wide spectrum of statistical areas where compositional data analysis is relevant: for example, robustness, geostatistics and time series. The simplex coordinate system is treated in this part as well as connections to correspondence analysis and the Dirichlet distribution. Part III is a first incursion into algebra and calculus on the simplex, a field that is practically new. Part IV is a collection of applications in different fields of science: genomics, ecology, biology, geochemistry, planetology, survey analysis, chemistry and economics, amongst others. Part V concludes with summaries of the three existing software packages for the analysis of compositional data.

It may be of interest to recount how this book came to be written. We, the two editors Vera and Antonella, became friends in 1994 during the conference of the International Association for Mathematical Geology in Mt Tremblant, Canada. In spite of our different backgrounds and expertise in compositional data analysis, we were both intrigued by the new ideas in the analysis of data with a constant-sum constraint. These ideas helped us to modify our way of thinking as well as our approach to describing the reality of the surrounding world. Starting from this point of view, a profitable collaboration started, aimed at disseminating the idea that the geometry of the sample space is fundamental to data analysis and, in particular, the analysis of compositional data. Our experience, matured after several years of investigation, has taught us that many advances in the description of natural phenomena in several fields of research are possible by interacting with researchers who work on real scientific problems. For this reason we think that the realization of this book, apart from paying tribute to John Aitchison, will present a further opportunity to enlarge our horizons on compositional data analysis, both from a theoretical and practical point of view.

We are deeply indebted to John for his seminal contributions to this field and his constant readiness to share his ideas with others. Our special thanks go to all the contributors who have worked hard to have their chapters completed in time, to the reviewers who have contributed to this publication being up-to-date and of a high standard, to Juan José Egozcue for his support in the various stages of the project, to Michael Greenacre for help with proofreading, and to John's wife Muriel, for her very special support all these years.

This book includes an accompanying website. Please visit `www.wiley.com/go/compositional_data_analysis` for more information.

Vera Pawlowsky-Glahn, Girona, Spain
Antonella Buccianti, Florence, Italy
February 2011

List of contributors

Aguilar, Lucía: Department of Mathematics, Technical School, University of Extremadura, Spain; luciaaz@unex.es

Azevedo Rodrigues, Luis: Secondary School Gil Eanes, Lagos, Portugal; laz.rodrigues@gmail.com

Bacon-Shone, John: Social Sciences Research Centre, The University of Hong Kong, Hong Kong; johnbs@hku.hk

Barceló-Vidal, Carles: Department of Computer Science and Applied Mathematics, University of Girona, Spain; carles.barcelo@udg.edu

Buccianti, Antonella: Department of Earth Sciences, University of Florence, Italy; antonella.buccianti@unifi.it

Daunis-i-Estadella, Josep: Department of Computer Science and Applied Mathematics, University of Girona, Spain; pepus@ima.udg.edu

Díaz-Barrero, José Luis: Department of Applied Mathematics III, Technical University of Catalonia, Spain; jose.luis.diaz@upc.edu

Egozcue, Juan José: Department of Applied Mathematics III, Technical University of Catalonia, Spain; juan.jose.egozcue@upc.edu

Filzmoser, Peter: Department of Statistics and Probability Theory, Vienna University of Technology, Austria; P.Filzmoser@tuwien.ac.at

Folch, Albert: Deparment of Geology, Autonomous University of Barcelona, Spain; albert.folch@uab.cat

Fry, Tim: School of Economics, Finance & Marketing, RMIT University, Melbourne, Australia; tim.fry@rmit.edu.au

Graf, Monique: Swiss Federal Statistical Office, Switzerland; monique.graf@bfs.admin.ch

Graffelman, Jan: Department of Statistics and Operations Research, Technical University of Catalonia, Spain; jan.graffelman@upc.edu

Greenacre, Michael: Department of Economics & Business, Pompeu Fabra University, and Barcelona Graduate School of Economics, Barcelona, Spain; michael.greeenacre@upf.edu

Helliwell, Chris: CSIRO Plant Industry, Canberra, Australia; chris.helliwell@csiro.au

Heredia, Javier: Instituto Geológico y Minero de España (IGME), Geological Survey of Spain, Madrid, Spain; j.heredia@igme.es

Hron, Karel: Department of Mathematical Analysis and Applications of Mathematics, Palacký University, Czech Republic; hronk@seznam.cz

Jarauta-Bragulat, Eusebi: Department of Applied Mathematics III, Technical University of Catalonia, Spain; eusebi.jarauta@upc.edu

Lammer, Helmut: Space Research Institute, Austrian Academy of Sciences, Austria; helmut.lammer@oeaw.ac.at

Lichtenegger, Herbert Iwo Maria: Space Research Institute, Austrian Academy of Sciences, Austria; herbert.lichtenegger@oeaw.ac.at

Lloyd, Christopher David: School of Geography, Archaeology and Palaeoecology, Queen's University, Belfast, UK; c.lloyd@qub.ac.uk

Lovell, David: CSIRO Mathematics, Informatics and Statistics, Canberra, Australia; David.Lovell@csiro.au

Martín-Fernández, Josep Antoni: Department of Computer Science and Applied Mathematics, University of Girona, Spain; josepantoni.martin@udg.edu

Mateu-Figueras, Glòria: Department of Computer Science and Applied Mathematics, University of Girona, Spain; gloria.mateu@udg.edu

McKinley, Jennifer: School of Geography, Archaeology and Palaeoecology, Queen's University, Belfast, UK; j.mckinley@qub.ac.uk

Monti, Gianna Serafina: Department of Statistics, University of Milan-Bicocca, Italy; gianna.monti@unimib.it

Müller, Warren: CSIRO Mathematics, Informatics and Statistics, Canberra, Australia; Warren.Muller@csiro.au

Olea, Ricardo Antonio: US Geological Survey, USA; rolea@usgs.gov

Otero, Neus: Faculty of Geology, University of Barcelona (UB), Spain; notero@ub.edu

Palarea-Albaladejo, Javier: Biomathematics and Statistics Scotland, JCMB, Edinburgh, UK; jpalarea@pdi.ucam.edu

Pardo-Igúzquiza, Eulogio: Instituto Geológico y Minero de España (IGME), Geological Survey of Spain, Madrid, Spain; e.pardo@igme.es

Pawlowsky-Glahn, Vera: Department of Computer Science and Applied Mathematics, University of Girona, Spain; vera.pawlowsky@udg.edu

Pierotti, Michele Edoardo Raffaele: Department of Biology, East Carolina University, USA; pierottim@ecu.edu

Puig, Roger: Faculty of Geology, University of Barcelona (UB), Spain; rpuig@ub.edu

Taylor, Jen: CSIRO Plant Industry, Canberra, Australia; Jen.Taylor@csiro.au

Templ, Matthias: Department of Statistics and Probability Theory, Vienna University of Technology, Austria; templ@statistik.tuwien.ac.at

Thió-Henestrosa, Santiago: Department of Computer Science and Applied Mathematics, University of Girona, Spain; santiago.thio@udg.edu

Tolosana-Delgado, Raimon: Maritime Engineering Laboratory, Technical University of Catalonia (LIM-UPC), Spain; raimon.tolosana@upc.edu

van den Boogaart, Karl Gerald: Institute for Stochastic, Technical University Bergakademie Freiberg, Germany; gerald@boogaart.de

Wurz, Peter: Institute of Physics, University of Bern, Switzerland; peter.wurz@space.unibe.ch

Zwart, Alec: CSIRO Mathematics, Informatics and Statistics, Canberra, Australia; Alec.Zwart@csiro.au

Part I

INTRODUCTION

1

A short history of compositional data analysis

John Bacon-Shone
Social Sciences Research Centre, The University of Hong Kong, Hong Kong

1.1 Introduction

Compositional data are data where the elements of the composition are non-negative and sum to unity. While the data can be generated directly (e.g. probabilities), they often arise from non-negative data (such as counts, area, volume, weights, expenditures) that have been scaled by the total of the components. Geometrically, compositional data with D components has a sample space of the regular unit D-simplex, $\mathcal{S}^D$. The key question is whether standard multivariate analysis, which assumes that the sample space is $\mathbb{R}^D$, is appropriate for data from this restricted sample space and if not, what is the appropriate analysis? Ironically, most multivariate data are non-negative and hence already have a sample space with a restriction to $\mathbb{R}^D_+$. This chapter tries to summarize more than a century of progress towards answering this question and draws heavily on the review paper by Aitchison and Egozcue (2005).

1.2 Spurious correlation

The starting point for compositional data analysis is arguably the paper of Pearson (1897), which first identified the problem of '*spurious correlation*' between ratios of variables. It is easy to show that if X, Y and Z are uncorrelated, then X/Z and Y/Z will not be uncorrelated. Pearson then looked at how to adjust the correlations to take into account the '*spurious correlation*' caused by the scaling. However, this ignores the implicit constraint that scaling only makes sense if the scaling variable is either strictly positive or strictly negative. In short, this approach ignores the range of the data and does not assist in understanding the process by which the data are generated. Tanner (1949) made the essential point that a log transform of the data may avoid the problem and that checking whether the original or log transformed data follow a Normal distribution may provide some guidance as to whether a transform is needed.

Chayes (1960) later made the explicit connection between Pearson's work and compositional data and showed that some of the correlations between components of the composition must be negative because of the unit sum constraint. However, he was unable to propose a means to model such data in a way that removed the effect of the constraint.

1.3   Log and log-ratio transforms

The first step towards modern compositional data analysis was arguably the use by McAlister (1879) of Log-Normal distributions to model data that are constrained to lie in positive real space. Interestingly, he proposed this as the law of the geometric mean (versus the Normal distribution as the law of the arithmetic mean) and pointed out the lack of practical value for variance of a variable that must be positive, which can be seen in retrospect as recognition of the need for a different metric for data from restricted sample spaces, that takes constraints into account. Instead, he emphasized the meaning of the cumulative distribution. This is by no means the only way to model data on the positive real line and competes with, for example, the Gamma and Weibull distributions. It is equivalent to taking a log transform of the data, so that the non-negative constraint is removed, and then assuming a Normal distribution. One of the key texts for the Log-Normal distribution is the book by Aitchison and Brown (1969). However, this only addresses the non-negative constraint of compositional data and does not address the unit sum constraint.

The simplest meaningful example of a composition is with just two components, so the unit-sum constraint implies that the second component is just one minus the first component. This is just the situation that arises with probabilities for a binary outcome. Cox and Snell (1989) use the logit or logistic transformation of the probability in this case, which enables the use of regression models for the logit transformed probabilities. However, it appears that nobody saw the potential for a similar approach for the more general case of compositional data until the first known reference to using the log-ratio transform to solve the constraint problem for compositional (or simplicial) data by Obenchain in a personal communication to Johnson and Kotz (Kotz *et al.* 2000). Indeed, Obenchain contributed to the discussion of the Royal Statistical Society paper by Aitchison (1982), where he stated that he became discouraged by the problem of zero components and thus never attempted to publish his simplex work, even though he had derived many properties of the logistic-normal distribution.

The first public introduction of the properties of the logistic-normal distribution can be found in Aitchison and Shen (1980). This distribution is written in terms of log-ratios relative to the last component, so that $\mathbf{y}(\mathbf{x}) = \{\log(x_1/x_D), \ldots, \log(x_{D-1}/x_D)\}$ follows a Multivariate Normal distribution.

Up to that time, the only known tractable distribution on the simplex was the Dirichlet distribution. However, the Dirichlet distribution has some very restrictive properties, such as complete subcompositional independence, i.e. for each possible partition of the composition, the set of all its subcompositions must be independent. This makes it impossible to model any reasonable dependence structure for compositional data using the Dirichlet distribution. In contrast, the logistic-normal distribution yields a distribution on the interior of the simplex that does not require these inflexible properties, but instead they become testable linear hypotheses on the covariance matrix within a broad flexible modelling framework. In addition, the Aitchison and Shen (1980) paper showed that the logistic-normal distribution is close to any Dirichlet distribution in terms of the Kullback–Leibler divergence. Later Aitchison (1985) derived a more general distribution that contains both the Dirichlet and logistic-normal distributions, although the potential for using this distribution for testing Dirichlet against logistic-normal distributions within the same class is diminished as these hypotheses are on the boundary of the parameter space. More recently, the generalization of the logistic-normal distribution to the additive logistic skew-normal distribution on the simplex (Mateu-Figueras *et al.* 2005) applies the skew-normal distribution (Azzalini 2005) to log-ratios on the simplex and offers the useful possibility of modelling data where the distribution of $\mathbf{y}(\mathbf{x})$ is not symmetrical. Use of the logistic-normal distribution opens up the full range of linear modelling available for the multivariate Normal distribution in $\mathbb{R}^D$.

1.4   Subcompositional dependence

As mentioned above, the logistic-normal distribution has the ability to model useful dependence structures. In his seminal book, Aitchison (1986) developed this idea, showing that the covariance structure can be modelled in terms of covariances on the log scale and is completely determined by the $D(D-1)/2$ log-ratio variances $\tau_{ij} = \mathrm{Var}\{\log(x_i/x_j)\}$ (where $i = 1, \ldots, D-1; j = i+1, \ldots, D$).

However, finding a convenient matrix formulation seems tricky, either yielding formulations that require selecting a specific component as divisor [when using $\mathbf{\Sigma}$, which is the log-ratio covariance matrix for the $D-1$ log-ratios relative to one component as divisor (Aitchison 1986, p. 77)], have a zero diagonal [when using $\mathbf{T}$, which is the variation matrix for all pairs of log-ratios (Aitchison 1986, p. 76)] or are singular [when using $\mathbf{\Gamma}$, which is the centred log-ratio covariance matrix (Aitchison 1986, p. 79)]. However, it turns out that there are simple linear relationships between these alternative formulations, so it is feasible to choose whichever formulation is simplest to use in any specific context.

1.5   alr, clr, ilr: which transformation to choose?

One key question for using the log-ratio transformations is choosing the divisor. Most of the literature initially used an arbitrary component as the divisor, known as using alr (additive log-ratio) transformation. This is potentially problematic because the distances between points in the transformed space are not the same for different divisors. However, as shown in Aitchison (1986) and further developed in Aitchison *et al.* (2000), linear statistical methods with compositional data as the dependent variable are invariant to the choice of divisor as the implicit linear transformations between different representations cancel out in any F ratio of quadratic or bilinear forms, so this is a conceptual rather than practical problem.

One way of avoiding this problem of choosing a divisor is to divide by the geometric mean, known as the clr (centred log-ratio) transformation. As noted above, the disadvantage of this is that the clr covariance matrix is singular, making it difficult to use in some standard statistical procedures without adaption.

A key step forward was recognition that compositions can be represented by their co-ordinates in the simplex with a suitable orthonormal basis. This suggests an alternative transformation, known as ilr (isometric log-ratio) transformations (Egozcue *et al.* 2003), which avoids the arbitrariness of alr and the singularity of clr. Thus ilr has significant conceptual advantages, but unfortunately, there is no clear 'simplest' or canonical basis, unlike $\mathbb{R}^D$. One possibility is to use a sequential binary partition of the components (Egozcue and Pawlowsky-Glahn 2005), known as balances, although this alone still does not ensure uniqueness. This approach is explained in detail in Chapter 3. However, despite the mathematical elegance of this approach, it has practical disadvantages in the relative difficulty of choosing the basis when that is not motivated by the statistical question being investigated and also when relating the coordinates back to the original statistical question.

1.6   Principles, perturbations and back to the simplex

At this point, the reader may have concluded that compositional data analysis is entirely a pragmatic approach to avoiding the unit sum constraint that may have mathematical weaknesses. Indeed, mathematical geologists, typified by Rehder and Zier (2001) argued that log-ratio analysis implied an illogical and arbitrary distance metric. In fact, the log-ratio approach can be derived entirely from a few key principles, which enable the derivation of the entire mathematical framework including an appropriate distance metric on the simplex. As explained in Aitchison *et al.* (2000), it should be obvious that compositional data analysis can only make meaningful statements about ratios of components, i.e. the first principle is scale invariance. This should be obvious in that compositional data is unit-free, but some geologists, such as Watson and Philip (1989), did not find this obvious. The second key principle is subcompositional coherence (Aitchison 1992), which states that inferences about subcompositions should be consistent, regardless of whether the inference is based on the subcomposition or the full composition. For $\mathbb{R}^D$, this would translate into the self-evident principle that inference about a subset of variables should be the same regardless of whether we base the inference on the subset of variables or the full set. Any meaningful metric for the simplex should satisfy these two principles and the Euclidean metric for $\mathbb{R}^D$ clearly does not satisfy either for compositional data. Aitchison (1986) introduced the idea of perturbation as the analogue to linear operations in $\mathbb{R}^D$, which was further developed in Aitchison and Ng (2005). A perturbation $\mathbf{p} = (p_1, \ldots, p_D)$ is a differential scaling operator that when applied to the composition $\mathbf{x} = (x_1, \ldots, x_D)$ yields the composition

$$\mathbf{X} = \mathbf{p} \oplus \mathbf{x} = \mathcal{C}(p_1 x_1, \ldots, p_D x_D),$$

where $\mathcal{C}$ is the closure operator that scales elements to ensure that we remain in the unit simplex.

The set of perturbations (if restricted to $\mathcal{S}^D$) form a group with an inverse and an identity perturbation $\mathbf{e} = (1/D, \ldots, 1/D)$. As any composition can be expressed as a result of a perturbation on any other composition, the distance between any two compositions must be expressible in terms of perturbations. Perturbation clearly corresponds to addition in $\mathbb{R}^D$ and we can define powering to correspond to multiplication in $\mathbb{R}^D$ as

$$\mathbf{X} = \mathbf{a} \odot \mathbf{x} = \mathcal{C}(x_1^a, \ldots, x_D^a).$$

The simplicial metric, or Aitchison distance is then given by

$$d_a(\mathbf{x}, \mathbf{y}) = \left\{ \sum_{i=1}^{D} \left[\log \frac{x_i}{g_m(\mathbf{x})} - \log \frac{y_i}{g_m(\mathbf{y})} \right]^2 \right\}^{\frac{1}{2}},$$

where $g_m(\cdot)$ is the geometric mean of the components, which can be shown to satisfy all the usual metric axioms and to depend only on perturbation distance. It is also easy to show that this metric satisfies the two key principles mentioned above.

The centre for a compositional distribution is then

$$\text{Cen}[\mathbf{x}] = \mathcal{C}(\exp\{E[\log(\mathbf{x})]\}),$$

with the variation matrix, $\mathbf{T}$, as the most convenient measure of variability. In summary, this allows us to transfer the analysis back to the simplex, without the asymmetry of using alr.

1.7 Biplots and singular value decompositions

It is essential to have simple ways to summarize and display multivariate data sets. Fortunately, singular value decompositions and the related graphical tool of the biplot (Gabriel 1971), provide precisely the tools we need for compositional data when adapted to the simplex (Aitchison and Greenacre 2002). The biplot for the simplex is based on a singular value decomposition of the row and column centred log-ratio matrix. It enables us to graphically display which combinations of the log-ratios contain large and small amounts of variability. The former provides a useful simplification of the major contributions to total compositional variability, while the latter identifies any likely linear dependencies amongst the log-ratios.

1.8 Mixtures

One important application of compositional data is as the covariate that determines a mixture. This yields log-contrast models for experiments with fixed mixtures where the dependence is only on the composition (Aitchison and Bacon-Shone 1984).

Compositional data can also occur doubly as the mixture of compositions. In this case, the mixed composition does not stay within the class of logistic-normal distributions, but can often be approximated well by a logistic-normal distribution as shown in Aitchison and Bacon-Shone (1999).

One specific problem where the mixture of compositions occurs is what geologists call the end-member problem (Renner 1993; Weltje 1997). In this case, the key question is which of the end members (of usually known compositions) are being mixed to form the outcome composition. A full Bayesian analysis of the end-member problem including spatial dependence is found in Palmer and Douglas (2008).

1.9   Discrete compositions

In the discussion of the Royal Statistical Society paper by Aitchison (1982), R.L. Plackett raised the question about how best to model discrete compositions and whether this might provide a solution to the problem of zeros (see below). The first full analysis of discrete data using compositional data models can be found in Billheimer *et al.* (2001), who use the logistic-normal distribution to model the probabilities for a multinomial distribution to allow much more sophisticated modelling of the occurrence data for species. This can be seen as a more sophisticated approach to the multivariate count modelling of Aitchison and Ho (1989), who use a logistic-normal model for the log means of Poisson data. However, the resultant data in both cases were counts, rather than discrete compositions, although in Billheimer's case, it is the relative occurrence that is of interest. As shown in Bacon-Shone (2008), it may be helpful to model discrete compositions even without knowing the total counts, helped by the knowledge that the original counts are non-negative integers.

1.10   Compositional processes

One of the nice consequences of analysing in the simplex, is that it is easy to investigate compositional processes as in Thomas and Aitchison (2005), who examine dependence on time through

$$Dx(t) = C \left\{ \exp \left[\frac{d}{dt} \log x(t) \right] \right\},$$

where d/dt denotes differentiation with respect to time.

This approach allows easy investigation and modelling of any processes that can be parameterized, including the possibility of change-points (Bacon-Shone 2011).

1.11   Structural, counting and rounded zeros

Aitchison recognized from the start that there was a need to solve the problem of zeros in compositional data as the log-ratio is undefined in this case. He wrote a much earlier paper (Aitchison 1955) which looked at the related problem of zeros for non-negative data, which presents a similar problem when using Log-Normal, Gamma or Weibull distributions, all of which have zero probability of observing zero. This paper used a conditional approach that separates the zeros from the continuous distribution and was applied to household expenditure data, which is a compositional data problem when classified into different categories of expenditure.

The original approach to compositional zeros in Aitchison (1982) was to simply replace all zeros by a small positive amount less than the detection limit, with the closure operator applied to apply the unit sum constraint and then sensitivity analysis to check the impact of the replacement value. However, this positive replacement approach potentially distorts the compositional data.

Proper 'in the simplex' approaches were independently proposed by Martín-Fernández et al. (2000) and Fry et al. (2000). Palarea-Albaladejo et al. (2007) used a parametric model to handle the zeros as missing data using the expectation-maximization (EM) algorithm.

As pointed out by Aitchison and Kay (2003) and by Bacon-Shone (2003), zeros can occur for at least three distinct reasons. First, there may be a structural reason why the component must be zero, such as alcohol expenditure components for household expenditure data in families that do not drink alcohol. This situation is best modelled by the conditional approach. Secondly, there may be a zero because of an underlying discrete process (Bacon-Shone 2008), such as expenditure on white goods (i.e. major household appliances) in household expenditure data, where people may go several years between making purchases and may miss capture in the data collection process. This situation is best modelled by modelling the underlying discrete process. Lastly, there may be a limit in the measurement or recording processes, such that very small components are recorded as zero. For this situation, the approaches of Martín-Fernández et al. (2000) and Fry et al. (2000) mentioned above seem most relevant.

Recently Butler and Glasbey (2008) proposed another modelling approach to compositional data with zeros, using Euclidean projections onto the simplex, with the probability 'outside' the simplex used to model the point probabilities on the boundaries. Unfortunately, this approach does not have a special case of log-ratio analysis and fails the test of the two key principles mentioned above.

A more comprehensive discussion of how to handle zeros in compositions can be found in Chapter 4.

1.12   Conclusion

This brief summary shows how much progress has been made in the last century in finding appropriate analyses for compositional data, much of it in the last 30 years, relying heavily on the insights of John Aitchison.

Acknowledgement

This research has been partially supported by the Research Grants Council, Hong Kong (Grant HKU 700303).

References

Aitchison J 1955 On the distribution of a positive random variable having a discrete probability mass at the origin. *Journal of the American Statistical Association* **50**(271), 901–908.

Aitchison J 1982 The statistical analysis of compositional data (with discussion). *Journal of the Royal Statistical Society, Series B (Statistical Methodology)* **44**(2), 139–177.

Aitchison J 1985 A general class of distributions on the simplex. *Journal of the Royal Statistical Society, Series B (Statistical Methodology)* **47**(1), 136–146.

Aitchison J 1986 *The Statistical Analysis of Compositional Data*. Monographs on Statistics and Applied Probability. Chapman and Hall Ltd (reprinted 2003 with additional material by The Blackburn Press), London (UK).

Aitchison J 1992 On criteria for measures of compositional difference. *Mathematical Geology* **24**(4), 365–379.

Aitchison J and Bacon-Shone J 1984 Log contrast models for experiments with mixtures. *Biometrika* **71**(2), 323–330.

Aitchison J and Bacon-Shone J 1999 Log contrast models for experiments with mixtures. *Biometrika* **86**(2), 351–364.

Aitchison J and Brown JAC 1969 *The Lognormal Distribution with Special Reference to its Uses in Econometrics*. Department of Applied Economics Monograph: 5. Cambridge University Press, Cambridge (UK). 176 p.

Aitchison J and Egozcue JJ 2005 Compositional data analysis: where are we and where should we be heading?. *Mathematical Geology* **37**(7), 829–850.

Aitchison J and Greenacre M 2002 Biplots for compositional data. *Applied Statistics* **51**(4), 375–392.

Aitchison J and Ho C 1989 The multivariate Poisson-log normal distribution. *Biometrika* **76**(4), 643–653.

Aitchison J and Kay J 2003 Possible solution of some essential zero problems in compositional data analysis. In *Proceedings of CoDaWork'03, The 1st Compositional Data Analysis Workshop* (ed. Thió-Henestrosa S and Martín-Fernández JA). http://ima.udg.es/Activitats/CoDaWork03/. University of Girona, Girona (Spain). CD-ROM.

Aitchison J and Ng K 2005 The role of perturbation in compositional data analysis. *Statistical Modelling* **5**(2), 173–185.

Aitchison J and Shen SM 1980 Logistic-normal distributions. Some properties and uses. *Biometrika* **67**(2), 261–272.

Aitchison J, Barceló-Vidal C, Martín-Fernández JA and Pawlowsky-Glahn V 2000 Logratio analysis and compositional distance. *Mathematical Geology* **32**(3), 271–275.

Azzalini A 2005 The skew normal distribution and related multivariate families. *Scandinavian Journal of Statistics* **32**(2), 159–188.

Bacon-Shone J 2003 Modelling structural zeros in compositional data. In *Proceedings of CoDaWork'03, The 1st Compositional Data Analysis Workshop* (ed. Thió-Henestrosa S and Martín-Fernández JA). http://ima.udg.es/Activitats/CoDaWork03/. University of Girona, Girona (Spain). CD-ROM.

Bacon-Shone J 2008 Discrete and continuous compositions. In *Proceedings of CoDaWork'08, The 3rd Compositional Data Analysis Workshop* (ed. Daunis-i Estadella J and Martín-Fernández J), p. http://hdl.handle.net/10256/723. University of Girona, Girona (Spain). 11 p.

Bacon-Shone J 2011 Mixing of compositions at points and along lines. *Computers & Geosciences* **37**(5), 692–695.

Billheimer D, Guttorp P and Fagan W 2001 Statistical interpretation of species composition. *Journal of the American Statistical Association* **96**(456), 1205–1214.

Butler A and Glasbey C 2008 A latent gaussian model for compositional data with zeros. *Journal of the Royal Statistical Society. Series C (Applied Statistics)* **57**(5), 505–520.

Chayes F 1960 On correlation between variables of constant sum. *Journal of Geophysical Research* **65**(12), 4185–4193.

Cox D and Snell E 1989 *Analysis of Binary Data*, 2nd edition. Chapman and Hall/CRC, London (UK). p. 236.

Egozcue JJ and Pawlowsky-Glahn V 2005 Groups of parts and their balances in compositional data analysis. *Mathematical Geology* **37**(7), 795–828.

Egozcue JJ, Pawlowsky-Glahn V, Mateu-Figueras G and Barceló-Vidal C 2003 Isometric logratio transformations for compositional data analysis. *Mathematical Geology* **35**(3), 279–300.

Fry JM, Fry TRL and McLaren KR 2000 Compositional data analysis and zeros in micro data.. *Applied Economics* **32**(8), 953–959.

Gabriel KR 1971 The biplot – graphic display of matrices with application to principal component analysis. *Biometrika* **58**(3), 453–467.

Kotz S, Balakrishnan N and Johnson NL 2000 *Continuous Multivariate Distributions. Volume I, Models and Applications*. Wiley Series in Probability and Statistics. Wiley-Interscience, New York, NY (USA). 730 p.

Martín-Fernández JA, Barceló-Vidal C and Pawlowsky-Glahn V 2000 Zero replacement in compositional data sets. In *Studies in Classification, Data Analysis, and Knowledge Organization. Proceedings of the 7th Conference of the International Federation of Classification Societies (IFCS'2000)* (ed. Kiers H, Rasson J, Groenen P and Shader M). Springer-Verlag, Berlin (Germany) pp. 155–160.

Mateu-Figueras G, Pawlowsky-Glahn V and Barceló-Vidal C 2005 The additive logistic skew-normal distribution on the simplex. *Stochastic Environmental Research and Risk Assessment (SERRA)* **19**(3), 205–214.

McAlister D 1879 The law of the geometric mean. *Proceedings of the Royal Society of London* **29**, 367–376.

Palarea-Albaladejo J, Martín-Fernández JA and Gómez-García JA 2007 Parametric approach for dealing with compositional rounded zeros. *Mathematical Geology* **39**(7), 625–645.

Palmer MJ and Douglas GB 2008 A bayesian statistical model for end member analysis of sediment geochemistry, incorporating spatial dependences. *Journal of the Royal Statistical Society. Series C: Applied Statistics* **57**(3), 313–327.

Pearson K 1897 Mathematical contributions to the theory of evolution. On a form of spurious correlation which may arise when indices are used in the measurement of organs. *Proceedings of the Royal Society of London* **LX**, 489–502.

Rehder S and Zier U 2001 Letter to the Editor: Comment on 'Logratio analysis and compositional distance' by J. Aitchison, C. Barceló-Vidal, J.A. Martín-Fernández and V. Pawlowsky-Glahn. *Mathematical Geology* **33**(7), 845–848.

Renner RM 1993 The resolution of a compositional data set into mixtures of fixed source components. *Journal of the Royal Statistical Society, Series C (Applied Statistics)* **42**(4), 615–631.

Tanner J 1949 Fallacy of per-weight and per-surface area standards, and their relation to spurious correlation. *Journal of Applied Physiology* **2**(1), 1–15.

Thomas CW and Aitchison J 2005 Compositional data analysis of geological variability and process: a case study. *Mathematical Geology* **37**(7), 753–772.

Watson DF and Philip GM 1989 Measures of variability for geological data. *Mathematical Geology* **21**(2), 233–254.

Weltje JG 1997 End-member modeling of compositional data: numerical-statistical algorithms for solving the explicit mixing problem. *Mathematical Geology* **29**(4), 503–549.

2

Basic concepts and procedures

Juan José Egozcue[1] and Vera Pawlowsky-Glahn[2]

[1]*Department of Applied Mathematics III, Technical University of Catalonia, Spain*
[2]*Department of Computer Science and Applied Mathematics, University of Girona, Spain*

2.1 Introduction

In all experimental fields, large amounts of compositional data (CoDa) can be found. They describe quantitatively the parts of some whole. They appear as proportions, percentages, concentrations, absolute and relative frequencies, spreading and distribution functions. Their units are also diverse; they range from percentages, parts per unit, or parts per million, to other non-closed units like molar concentrations or absolute frequencies. Often, the total amount is irrelevant or the analyst is not interested in it. For instance, in order to study the political and sociological framework of an election, the total number of electors per circumscription, representing the size of such a region, is considered external to the study, and only the proportions between the number of votes to candidates are considered. In this case, the proportions of votes to lists can be considered compositional, even if they are presented as absolute number of votes. In geology or biology, the mass of a material sample can be considered irrelevant if analysts are only interested in the geo-/biochemical composition of that sample. The analysis of data expressed as proportions carries a number of problems that have been studied for a long time in fields like geology and biology. One of the first examples comes from the field of biologic morphology and is authored by one of the founders of modern statistics: K. Pearson (1897). In geology, the study of CoDa was particularly intensive in the 1950s and 1960s (Chayes 1960). In biology, some attempts can be found (Mosimann 1962; Connor and Mosimann 1969). But the first consistent methodological proposal to deal with CoDa did not

Compositional Data Analysis: Theory and Applications, First Edition. Edited by Vera Pawlowsky-Glahn and Antonella Buccianti.
© 2011 John Wiley & Sons, Ltd. Published 2011 by John Wiley & Sons, Ltd.

arrive until the 1980s. It was introduced by J. Aitchison (1982, 1986). The main point is the statistical analysis of log-ratios and the statement of the principles of CoDa analysis.

Despite the advantages offered by techniques based on log-ratios, they did not have the success one could expect, and many scientists continued (and continue) applying the traditional statistical methods without taking into account the compositional character of their data. At the beginning of the 2000s, several formal contributions were published (Billheimer *et al.* 2001; Pawlowsky-Glahn and Egozcue 2001; Aitchison *et al.* 2002), which allow a better systematic approach to the methods already proposed in the 1980s (Aitchison and Egozcue 2005) (see Chapter 1). Nowadays, CoDa analysis can be reduced to three steps: the representation of data in log-ratio type coordinates; the (traditional) analysis of the coordinates as real random variables; and the interpretation of resulting models either in coordinates, or expressing the results in terms of the original units. This method, termed *principle of working in coordinates* (see Chapter 3), is based on the invariance of the analysis under change of basis. However, these techniques are not yet widely used, but a growing interest is emerging in several scientific fields.

2.2   Election data and raw analysis

To illustrate the anomalous behaviour of standard statistical methods when applied to raw CoDa (Aitchison 1997) we use the provisional results of the November 2010 elections to the *Parlament de Catalunya* (parliament of Catalonia), an autonomous community of Spain. The votes have been recorded for 41 regions, which are a subdivision of the electoral provinces. The data set (Cat10) contains the number of electors (elect), the number of votes, including *none of the above* (nota), null votes (null) and valid votes to parties and coalitions. The difference of electors minus votes gives the abstention (abst). Table 2.1 shows the first,

Table 2.1   Three records (first, intermediate and last) of Cat10. Number of votes per region and categories: number of electors (elect); abstention (abst); none of the above (nota); null votes (null); parties and coalitions in the outgoing Catalan Parliament; votes to amalgamated lists of candidates (other).

	Alt Camp	Barcelonès	Vallès Oriental
elect	32027	1572425	283230
abst	13418	629865	111884
nota	483	28306	5106
null	173	5722	1009
C's	285	39194	6305
CiU	8183	317695	67887
ERC	1679	56200	11199
ICV	931	85579	13119
PSC	2856	188802	29738
PP	1556	139890	18757
other	2463	81172	18226

an intermediate and the last record of the data set by columns. Several minor parties and coalitions, one of them achieving four representatives in the parliament in these elections but not present in the outgoing parliament, have been amalgamated into a single category, called *other*; major parties or coalitions are *Convergència i Unió* (CiU), virtual winner of the 2010 elections, and *Partit dels Socialistes de Catalunya* (PSC); other parties represented in the outgoing parliament of Catalonia, are (in alphabetical order): *Ciutadans-Partido de la Ciudadanía* (C's), *Esquerra Republicana de Catalunya* (ERC), *Iniciativa per Catalunya Verds-Esquerra Unida i Alternativa* (ICV), and *Partit Popular* (PP).

Attention is focused on the inter-relations between all categories of votes (including abstention) and a possible dependence of the distribution of votes among the different parties on the total population of the region.

A first (fruitless) attempt to analyse the election data consists in a correlation study of the raw number of votes. Correlations obtained in this way are clearly dominated by the size of the region, represented by the number of electors, and all of them attain values larger than 0.9. Some of these correlations are shown in Table 2.2, case **A**. To filter out the influence of size dividing each number of votes by the total number of electors in a region is common practice. Effectively this removes the *size of region* effect. Denote this approach **B**, corresponding to per unit or, equivalently, percentage of vote for each category. The correlation matrix obtained for approach **B** differs completely from the correlation matrix of case **A**. The high positive correlations completely disappear. Also, some significant correlations are observed. But the fact that each record of data adds up to 100 when data are expressed in percentage, induces some negative correlations. This effect, called *negative bias*, appears systematically when analysing the correlation of *closed data*, i.e. of a data set with records (observations) which are vectors with positive components adding up to a constant (in this case 100). Some values are shown in Table 2.2, replacing the large positive correlations in case **A**, supporting the impression that the high correlations were a size effect.

It should be possible to reproduce the conclusions obtained by an analyst from correlations for case **B** – or they should be at least compatible with those obtained by another analyst – examining correlations when proportions of votes are expressed in a different way. For instance, approach **B** considers proportions of votes over the whole census of electors; approach **C** may consider only proportion of votes to parties and coalitions, thus excluding abstention, nota and null votes; approach **D** may define proportions of parties and coalitions

Table 2.2 Spurious correlation: Pearson's correlation coefficients between votes to parties or coalitions and the total number of electors depending on approaches **A**, **B**, **C**, **D**, **F**. See text for an explanation.

		elect C's	elect PSC	C's ERC	C's CiU	PSC C's	PSC PP	PSC ERC	PSC ICV
A	absolute vote	0.995	0.997	0.981	0.985	0.999	0.996	0.986	0.998
B	electors	0.668	0.282	−0.794	−0.792	0.392	0.471	−0.334	0.324
C	candidates	0.646	0.304	−0.797	−0.804	0.521	0.590	−0.456	0.407
D	in parliament	0.642	0.306	−0.805	−0.764	0.509	0.537	−0.525	0.372
E	minor	0.544		−0.784					
F	C's, PSC, PP	0.588	−0.362			−0.712	−0.955		

that were present in the outgoing parliament; option **E** studies the proportion of votes distributed among minor parties (C's, ERC, ICV, other); option **F** may be to study relationships between votes to two large parties (PSC, PP), well represented in the Spanish parliament, with a recently born C's that is supposed to take some votes from the former parties. Correlation matrices for all these situations have been computed. Approach **B** assigns each potential vote to a category and takes the proportions of such votes. **B** considers a complete composition. Situations **C–F**, correspond to analysing some *subcomposition* of the original one. As this is common practice when studying percentage data (closed data), one expects to find coherent results when analysing different subcompositions with common categories. Some results have been given in Table 2.2 for an easy comparison; they show some incoherent results. Even the correlations with an external variable (elect) change incoherently from approach to approach. This phenomenon is called *spurious correlation* (Pearson 1897; Aitchison 1986). It invalidates multivariate techniques based on covariances of the raw composition.

2.3 The compositional alternative

Unsuccessful experiences analysing CoDa, similar to the example presented in Section 2.2, have been accumulated over a century. Based on them, J. Aitchison (1982, 1986) stated some principles for the analysis of CoDa. They have been reformulated several times (Barceló-Vidal *et al.* 2001; Martín-Fernández *et al.* 2003; Aitchison and Egozcue 2005; Egozcue 2009) according to new theoretical developments. The first step is the definition of CoDa, which essentially coincides with that stated in Section 2.1 above: *compositional data quantitatively describe the parts of some whole and they provide only relative information between their components.* Note that CoDa appear as vectors of two or more positive components, although frequently one component is omitted as it is the difference of the shown components to the total. However, only ratios of components carry information. This leads to the following principles.

2.3.1 Scale invariance: vectors with proportional positive components represent the same composition

In other words, if a composition is scaled by a constant, e.g. changing from parts per unit to percentages, the information carried is completely equivalent. Accordingly, vectors of proportional positive components form an equivalence class. Therefore, it is natural to select a representative of the equivalence class to facilitate both the analysis and the interpretation. The traditional way to do that is to normalize the vector in such a way that the components sum to a given constant κ, which can be 1, 100, 1000, 10^6, or any other positive constant. This selection is formalized by the *closure* operation. For $\mathbf{x} = (x_1, x_2, \ldots, x_D)$ a vector with D positive components, its *closure* is defined as

$$\mathcal{C}\mathbf{x} = \left(\frac{\kappa x_1}{\sum_{i=1}^{D} x_i}, \frac{\kappa x_2}{\sum_{i=1}^{D} x_i}, \ldots, \frac{\kappa x_D}{\sum_{i=1}^{D} x_i} \right). \tag{2.1}$$

The components of the closed vector are called *parts*, relative to a *total* κ. The set of vectors with D positive components summing to the constant κ form the D-part simplex, denoted by $\mathcal{S}^D$. The compositions equivalent to $\mathbf{x}$ are represented by $\mathcal{C}\mathbf{x}$.

For instance, the votes in Cat10 add up to the number of electors in each region. If we are not interested in the number of electors, which is a measure of the size of the region, the ratio of the number of votes obtained by each party or coalition over the total of electors gives a per-unit distribution or proportion of votes, and the size has been filtered out. Multiplying these values by 100 we get percentages, and the vector of per-units and the vector of percentages convey exactly the same information.

2.3.2   Subcompositional coherence: analyses concerning a subset of parts must not depend on other non-involved parts

A subcomposition is a subset of components or parts of a composition. The study of a subcomposition, requires that the results are not contradictory with those obtained from the full composition. The principle of coherence can be summarized as two criteria: (a) the principle of scale invariance should hold for any of the possible subcompositions thus implying preservation of ratios of parts; (b) if a distance or divergence is used to compare compositions, this distance or divergence should be grater than or equal to that obtained comparing the corresponding subcompositions (*subcompositional dominance*).

Coherence is more subtle than the principle of scale invariance. In the example of Cat10, it means that the analysis of the complete distribution of votes (situation **B**, including abstention, nota and null votes) should be coherent with situation **C**, where only votes to parties and coalitions are taken into account, or even coherent with situation **F**, where the subcomposition (C's, PSC, PP) is analysed. The reason is that the selection of a subcomposition does not change the ratios between the parts and, since these ratios are the only information considered, the analysis should remain invariant when using the same parts from the composition and the subcomposition. Correlations shown in Table 2.2 depend on the subcomposition considered; they are clear examples of violation of the principle of subcompositional coherence: a measure of association between two parts should not depend on which other parts are in the subcomposition. See extreme cases in Table 2.2 where correlation between PSC and PP takes values as diverse as 0.590 (case **C**) and −0.955 (case **F**).

Subcompositional dominance calls for a way to measure distances between compositions and subcompositions which follows the rule of a projection: distances become smaller in a projection. It is logical to ask if the ordinary Euclidean distance between real vectors can be used. This is not the case, as both the principle of scale invariance and of subcompositional dominance are violated. In fact, if two vectors with positive components are multiplied by a positive constant c, then the Euclidean distance between them is multiplied by c, violating the principle of scale invariance. Subcompositional dominance is also violated by the ordinary Euclidean distance between compositional vectors. This fact is illustrated in Table 2.3. Euclidean distances between compositions of votes in two couples of regions are

Table 2.3   Euclidean distances between regions 1 and 2 measured on subcompositions corresponding to situations **B**, **C**, **D**, **F**.

Region 1	Region 2	**B**	**C**	**D**	**F**
Barcelonès	Alt Camp	0.087	0.155	0.178	0.116
Barcelonès	Vallès Oriental	0.049	0.083	0.093	0.049

considered: Barcelonès-Alt Camp and Barcelonès-Vallès Oriental. The raw number of votes for these regions are shown in Table 2.1. Consider the sequence of subcompositions corresponding to situations **B, C, D, F** defined in Section 2.2. As the corresponding subcompositions are nested from right to left in Table 2.3, subcompositional dominance would expect decreasing distances from left to right. This is not the case and, therefore, Euclidean distance for compositions is inappropriate for CoDa analysis (Palarea-Albaladejo *et al.* 2011).

2.3.3   Permutation invariance: the conclusions of a compositional analysis should not depend on the order of the parts

This is obvious in many cases. For example, in geochemical compositions it is quite frequent to record the parts in alphabetical order. The same happens in the example Cat10. But in some cases the parts can be assumed to be ordered. A typical example is the grain size distribution of a sediment: the particles are classified, after sieving, into size categories. Applying a compositional analysis, the information due to the order of the different classes, plays no role.

2.4   Geometric settings

To satisfy the requirements described in Section 2.3, a geometry of the D-part simplex, $\mathcal{S}^D$, is required. The development of the concepts suggested by Aitchison (1986) has lead to the *Aitchison geometry* of the simplex (Pawlowsky-Glahn and Egozcue 2001). This geometry, being Euclidean, requires specific definitions and a specific metric.

Consider the compositions $\mathbf{x}$, $\mathbf{y} \in \mathcal{S}^D$. The *perturbation* of $\mathbf{x}$ with $\mathbf{y}$ is defined as the composition

$$\mathbf{x} \oplus \mathbf{y} = \mathcal{C}(x_1 y_1, x_2 y_2, \dots, x_D y_D), \tag{2.2}$$

and *powering* of $\mathbf{x}$ by a real number α as the composition

$$\alpha \odot \mathbf{x} = \mathcal{C}(x_1^\alpha, x_2^\alpha, \dots, x_D^\alpha). \tag{2.3}$$

It is easy to show that for $\mathbf{n} = \mathcal{C}(1, 1, \dots, 1)$ it holds $\mathbf{x} \oplus \mathbf{n} = \mathbf{x}$. Thus, the composition with all equal parts is the neutral element of perturbation. *Perturbation* and *powering*, defined in $\mathcal{S}^D$, satisfy the requirements for operations of a vector space. But the main advantage of perturbation is that, in addition to satisfying the principles of compositional analysis, it has usually an interpretation in the field analysed.

The usefulness of perturbation and powering is illustrated using the example Cat10 and the regions Barcelonès, Alt Camp and Vallès Oriental. If interest lies, e.g., on the change of the distribution of votes from one region to another, a simple model is that the composition of votes in any region (Alt Camp or Vallès Oriental), $\mathbf{y}_i$, can be obtained from the composition in Barcelonès, $\mathbf{x}$, perturbed by some composition $\mathbf{p}_i$ that describes the change as a compositional shift in the simplex. Note that perturbation $\mathbf{p}_i$ can also be expressed as a difference perturbation denoted with the symbol $\ominus$:

$$\mathbf{y}_i = \mathbf{x} \oplus \mathbf{p}_i, \, \mathbf{p}_i = \mathbf{y}_i \ominus \mathbf{x} = \mathbf{y}_i \oplus ((-1) \odot \mathbf{x}). \tag{2.4}$$

Table 2.4 Modelling change between regions. See text for details.

	x %	y_1 %	y_2 %	p_1 %	p_2 %	p_1 factor	p_2 factor	p_1 % change	p_2 % change
abst	40.1	41.9	39.5	10.7	10.0	1.046	0.986	4.6	−1.4
nota	1.8	1.5	1.8	8.6	10.1	0.838	1.001	−16.2	0.1
null	0.4	0.5	0.4	15.2	9.9	1.484	0.979	48.4	−2.1
C's	2.5	0.9	2.2	3.7	9.0	0.357	0.893	−64.3	−10.7
CiU	20.2	25.6	24.0	12.9	12.0	1.265	1.186	26.5	18.6
ERC	3.6	5.2	4.0	15.0	11.2	1.467	1.106	46.7	10.6
ICV	5.4	2.9	4.6	5.5	8.6	0.534	0.851	−46.6	−14.9
PSC	12.0	8.9	10.5	7.6	8.9	0.743	0.874	−25.7	−12.6
PP	8.9	4.9	6.6	5.6	7.5	0.546	0.744	−45.4	−25.6
other	5.2	7.7	6.4	15.2	12.6	1.490	1.247	49.0	24.7
A. dist.				1.51	0.48				

Table 2.4 shows the perturbation from Barcelonès to Alt Camp, p_1, and to Vallès Oriental, p_2, in three different versions: as a composition expressed in percentage (%); as multiplicative (non-closed) factors; and as percentage of increase/decrease, which is a traditional form of presenting perturbations. For instance, a factor of 1.046 in abst from Barcelonès to Alt Camp means a 4.6% increment of abstention; a factor of 0.357 for C's, corresponds to a 64.3% decrease from Barcelonès to Alt Camp. The Alt Camp region, a rural area compared with Barcelona and its surroundings, shows substantial increments of votes to Catalan nationalist parties (CiU, ERC, other), a moderate increase of abstention, and a heavy null vote increase with respect to Barcelonès. Vallès Oriental is an industrial and populated area. Accordingly, it may have similarities with Barcelonès, which is mainly occupied by Barcelona, a crowded city, with a lot of services and partially industrial. In fact, although CiU, ERC and other also increase with respect to Barcelonès, the increase is not as large as observed for Alt Camp. A remarkable fact is that the abstention and null votes in Vallès Oriental are less than in Barcelonès, while nota votes are similar. The global size or magnitude of a perturbation is its Aitchison norm which is actually the Aitchison distance between the regions. The last row in Table 2.4 gives Aitchison distances (A. dist.) for both couples of regions, as discussed below.

The change of units in some or all the parts of a composition can also be viewed as a perturbation. Typical examples are found in chemistry, when concentrations in parts per million (ppm) of weight are changed to molar concentrations (Buccianti and Pawlowsky-Glahn 2005). This is done by multiplying each component by the inverse of the molar weight. Closing the resulting composition may be unnecessary in many cases. Still, it retains its compositional character. For the example Cat10, we can imagine some meaningful change of units. For instance, considering the subcomposition of candidate parties and coalitions, each one has invested some money in the campaign and, thus, a per-capita (per-elector) amount can be computed for each list. The votes can then be viewed as a return of the investment. The vector of these returns, expressed in monetary units, is still compositional, although closing it to percentages does not have a clear meaning.

Scale invariance required for a compositional analysis leads to the use of ratios of parts so that scale constants are cancelled. Furthermore, ratios can be considered in a relative scale and taking their logarithms is then a natural choice. The analysis of CoDa is essentially based on the statistical analysis of log-ratios of parts. The simplest log-ratios are those comparing two parts. In Cat10, $\ln(\text{CiU/PSC})$ or $\ln(\text{abst/nota})$ are two of these simple log-ratios. More complex log-ratios can be useful in the analysis, but they must be scale invariant. Scale invariant log-ratios are called log-contrasts (Aitchison 1986) and are defined as

$$\ln\left(\prod_{i=1}^{D} x_i^{\alpha_i}\right) = \sum_{i=1}^{D} \alpha_i \ln(x_i), \quad \sum_{i=1}^{D} \alpha_i = 0, \tag{2.5}$$

where the condition on the coefficients α_i guarantees scale invariance.

Frequently, some questions can be answered analysing an appropriate log-contrast. The choice of the log-contrast depends on the stated problem and the interpretation of the composition. In the Cat10 context we can ask ourselves whether the log-ratio between parties declared as Catalan nationalists (CiU, ERC, other) and parties declared as opposite to Catalan nationalism (C's, PP) depends on the total number of electors of the region. A choice of the log-contrast is

$$z = \sqrt{\frac{6}{5}} \ln \frac{(\text{CiU} \cdot \text{ERC} \cdot \text{other})^{1/3}}{(\text{C's} \cdot \text{PP})^{1/2}},$$

where the coefficient in the square root has been added for normalization, as described below when defining balances. The log-contrast z can be computed for all regions and then it is correlated with the logarithm of the number of electors (elect). After removing one outlier (Vall d'Aran, a valley in the Pyrenees), the estimated correlation coefficient is -0.778 and the p-value for the F-test is less than 10^{-6}. Figure 2.1 shows $\ln(\text{elect})$ and the log-contrast z

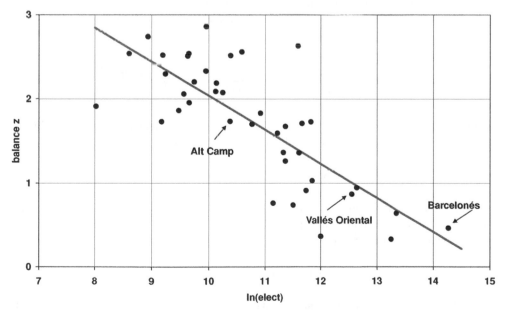

Figure 2.1 Regions considered in Cat10 characterized by the logarithm of the number of electors and the log-contrast z. The correlation coefficient is -0.778.

for each region and the linear model fitted. The significant negative correlation illustrated in Figure 2.1 allows an interpretation. Roughly speaking, the smaller logarithm of the number of electors, the more rural characteristics of the region, and vice versa the more industrial and economically active regions have a larger number of electors. The fitted model reflects the fact that the Catalan nationalist vote attains higher values in rural areas than in urban and industrial areas.

The previous example of regression of a log-contrast on an external variable illustrates the fact that certain aspects of compositions can be analysed using the appropriate log-contrast. Proper and complete representations of a composition using a set of log-contrasts were proposed in the 1980s (Aitchison 1986), so that all the information from the composition is reverted into the set of log-ratios. A first choice of these representations was the *additive-log-ratio* transformation (alr). If $\mathbf{x}$ is a composition in the D-part simplex $\mathcal{S}^D$,

$$\text{alr}(\mathbf{x}) = \ln \left(\frac{x_1}{x_D}, \frac{x_2}{x_D}, \ldots, \frac{x_{D-1}}{x_D} \right), \tag{2.6}$$

where the natural logarithm ln is applied componentwise. Consequently, the ith component is the simple log-ratio $\text{alr}_i(\mathbf{x}) = \ln(x_i/x_D)$. The alr transformation is easily inverted to get the composition from the $D - 1$ alr components and also reduces perturbation and powering to ordinary operations in the $D - 1$ dimensional real space:

$$\text{alr}((\alpha \odot \mathbf{x}) \oplus (\beta \odot \mathbf{y})) = \alpha \cdot \text{alr}(\mathbf{x}) + \beta \cdot \text{alr}(\mathbf{y}),$$

for any compositions $\mathbf{x}, \mathbf{y}$ and any real constants α and β. However, alr has the inconvenient of not being invariant under permutation of components and some statistical procedures may fail. In fact, we favour the Dth part (or any other part selected) which is present in all denominators of the log-ratios. To avoid these problems, Aitchison (1986) introduced the *centered log-ratio* transformation (clr), which represents a D-part composition using D clr coefficients. It is defined as

$$\mathbf{v} = \text{clr}(\mathbf{x}) = \ln \left[\frac{x_1}{g_m(\mathbf{x})}, \frac{x_2}{g_m(\mathbf{x})}, \ldots, \frac{x_D}{g_m(\mathbf{x})} \right], \quad g_m(\mathbf{x}) = \left(\prod_{i=1}^{D} x_i \right)^{1/D}, \tag{2.7}$$

where the D coefficients $\text{clr}_i(\mathbf{x}) = \ln(x_i/g_m(\mathbf{x}))$ are log-contrasts [see Equation (2.5)]. From $\text{clr}(\mathbf{x})$, the composition $\mathbf{x}$ is recovered with the inverse clr transformation

$$\mathbf{x} = \text{clr}^{-1}(\mathbf{v}) = \mathcal{C} \exp(\mathbf{v}), \tag{2.8}$$

where the exponential function is applied componentwise to $\mathbf{v} = \text{clr}(\mathbf{x})$. Similarly to the alr transformation, perturbation and powering in $\mathcal{S}^D$ correspond to sum and product in D-dimensional real space $\mathbb{R}^D$, i.e.

$$\text{clr}((\alpha \odot \mathbf{x}) \oplus (\beta \odot \mathbf{y})) = \alpha \cdot \text{clr}(\mathbf{x}) + \beta \cdot \text{clr}(\mathbf{y}).$$

The drawback of the clr transformation is that it uses D coefficients, adding to zero, to represent a composition which has only $D - 1$ free components, the dimension of $\mathcal{S}^D$. Moreover, the clr components change when working with a subcomposition.

The clr representation of compositions can be used to define a metric structure in the simplex. The Aitchison inner product, norm and distance for compositions in $\mathcal{S}^D$ are

$$\langle \mathbf{x}, \mathbf{y} \rangle_a = \langle \text{clr}(\mathbf{x}), \text{clr}(\mathbf{y}) \rangle, \tag{2.9}$$

$$\|\mathbf{x}\|_a = \|\text{clr}(\mathbf{x})\|, \quad d_a(\mathbf{x}, \mathbf{y}) = d(\text{clr}(\mathbf{x}), \text{clr}(\mathbf{y})), \tag{2.10}$$

where $\langle \cdot, \cdot \rangle$, $\| \cdot \|$, $d(\cdot, \cdot)$, denote the ordinary Euclidean inner product, norm and distance in $\mathbb{R}^D$. For instance, the Aitchison distance is

$$d_a(\mathbf{x}, \mathbf{y}) = \sqrt{\sum_{i=1}^{D} [\text{clr}_i(\mathbf{x}) - \text{clr}_i(\mathbf{y})]^2}.$$

Table 2.4, in its last row, shows the Aitchison distance between Barcelonès and the other two regions (Alt Camp, Vallès Oriental) in example Cat10. The first one is about three times the second, thus reflecting a moderate socio-economical proximity of Barcelonès and Vallès Oriental compared with the large difference between Barcelonès and Alt Camp. Aitchison distances between regions provide a tool for further multivariate analysis, e.g. cluster analysis or multidimensional scaling.

The Aitchison inner product, norm and distance honour the principles of compositional analysis (Section 2.3) and are therefore tools for a compositional analysis free of inconsistencies. Jointly with perturbation and powering, they provide an Euclidean structure to the simplex, called Aitchison simplicial geometry. This suggests to exploit the well-known properties of Euclidean spaces to analyse compositions: orthonormal basis, (orthonormal) coordinate representation, orthogonal projections, definitions of angles, ellipses, etc. An important step to use these concepts is to build orthonormal bases and their corresponding coordinates.

An orthonormal basis of $\mathcal{S}^D$ is a set of compositions $\mathbf{e}_1, \mathbf{e}_2, \ldots, \mathbf{e}_{D-1}$ such that $\langle \mathbf{e}_i, \mathbf{e}_j \rangle_a = 0$ for $i \neq j$, and $\|\mathbf{e}_i\|_a = 1$. For a fixed basis the coordinates of a composition are obtained using the function

$$\mathbf{x}^* = \text{ilr}(\mathbf{x}) = (\langle \mathbf{x}, \mathbf{e}_1 \rangle_a, \langle \mathbf{x}, \mathbf{e}_2 \rangle_a, \ldots, \langle \mathbf{x}, \mathbf{e}_{D-1} \rangle_a), \tag{2.11}$$

with inverse,

$$\mathbf{x} = \text{ilr}^{-1}(\mathbf{x}^*) = \bigoplus_{j=1}^{D-1} x_j^* \odot \mathbf{e}_j. \tag{2.12}$$

The construction of orthonormal coordinates has been called *isometric log-ratio transformation* (ilr) (Egozcue *et al.* 2003) because the coordinates $x_j^* = \text{ilr}_j(\mathbf{x})$ are log-contrasts and are isometric:

$$\text{ilr}((\alpha \odot \mathbf{x}) \oplus (\beta \odot \mathbf{y})) = \alpha \cdot \text{ilr}(\mathbf{x}) + \beta \cdot \text{ilr}(\mathbf{y}), \tag{2.13}$$

$$\langle \mathbf{x}, \mathbf{y} \rangle_a = \langle \text{ilr}(\mathbf{x}), \text{ilr}(\mathbf{y}) \rangle, \tag{2.14}$$

$$\|\mathbf{x}\|_a = \|\text{ilr}(\mathbf{x})\|, \tag{2.15}$$

$$d_a(\mathbf{x}, \mathbf{y}) = d(\text{ilr}(\mathbf{x}), \text{ilr}(\mathbf{y})), \tag{2.16}$$

analogous to the properties given in (2.9) and (2.10) for the clr transformation. The difference is that the inner product, norm and distance between vectors of ilr coordinates correspond to $D - 1$ dimensional real space, which is isomorphic to $\mathcal{S}^D$.

As in any other Euclidean space, an infinite number of orthonormal bases exist in $\mathcal{S}^D$. A simple way to represent compositions in ilr coordinates is to perform a *Singular Value Decomposition* (SVD) of the matrix of a clr transformed sample, an operation which is usually done to obtain a *biplot*, as described in Section 2.5. But the interpretation of the resulting log-contrasts [Equation (2.5)] can be difficult. Therefore, it can be convenient to build log-contrasts which have an interpretation adequate to the problem studied and, in particular, orthonormal coordinates defined by the analyst according to the problem he or she is trying to solve. One technique for doing so is based on a *Sequential Binary Partition* (SBP) of the parts of the composition (Egozcue and Pawlowsky-Glahn 2005a, 2006; Pawlowsky-Glahn and Egozcue 2006; Thió-Henestrosa *et al.* 2008). Each step of the partition, of a total of $D - 1$ steps, gives rise to an ilr coordinate, now called *balance*, which is usually easy to interpret. Table 2.6 illustrates the SBP process. In a first step, SBP consists of dividing the composition into two groups of parts which are indicated by $+1$ and -1, as shown in the first row of Table 2.6. In further steps, each previously obtained group of parts is again subdivided into two groups until all groups are made of a single part. The first step in Table 2.6 consists of separating votes to parties and coalitions (-1) from abstention and nota or null votes $(+1)$. The second step separates abstention $(+1)$ from nota and null votes (-1). The fifth step separates declared Catalan nationalist coalitions $(+1)$ from parties which are present in the whole of Spain (-1), etc. Each step in the SBP is associated with one element of an orthonormal basis and one ilr coordinate. These ilr coordinates are called *balances* due to their peculiar form. For the jth row of the SBP matrix (Table 2.6) denote by $\mathbf{x}_+$ the group of r parts marked with a $+1$ and by $\mathbf{x}_-$ the group of s parts marked with a -1; then, the balance is

$$b_j = \sqrt{\frac{rs}{r+s}} \ln \frac{g_m(\mathbf{x}_+)}{g_m(\mathbf{x}_-)},$$

where $g_m(\cdot)$ is the geometric mean of its arguments. Balances are unit-norm orthogonal log-contrasts (Aitchison 1986) and they have a relatively easy interpretation (Egozcue and Pawlowsky-Glahn 2005b), as they are log-ratios of geometric means of groups of parts.

2.5   Centre and variability

Statistics synthesizes information from a sample using simple descriptors. The mean and variance-covariance are the most popular in multivariate scenarios. When dealing with CoDa, the geometry of their sample space $\mathcal{S}^D$ must be taken into account and, particularly, the Aitchison distance. Following Pawlowsky-Glahn and Egozcue (2001), consider a random composition $\mathbf{X}$ in $\mathcal{S}^D$ and define variability of $\mathbf{X}$ with respect to a composition $\mathbf{z} \in \mathcal{S}^D$ as $\mathrm{Var}(\mathbf{X}, \mathbf{z}) = \mathrm{E}[d_a^2(\mathbf{X}, \mathbf{z})]$, where E and Var denote the ordinary expectation and variance in real space. The composition $\mathbf{z}$ minimizing $\mathrm{Var}(\mathbf{X}, \mathbf{z})$ is called the centre of $\mathbf{X}$, and the minimum variability attained is the total variance. The centre of $\mathbf{X}$ is then expressed as

$$\mathrm{Cen}[\mathbf{X}] = \mathrm{ilr}^{-1}\{\mathrm{E}[\mathrm{ilr}(\mathbf{X})]\} = \mathcal{C} \exp(\mathrm{E}[\ln \mathbf{X}]). \tag{2.17}$$

The right-hand side of the equation can be used to define the centre as a *closed geometric mean* (Aitchison 1997). The centre Cen[**X**] plays the role of the multivariate mean when the sample space is the simplex.

The total variance can be expressed in three different ways, each of them providing a decomposition of total variance:

$$\text{totVar}[\mathbf{X}] = \frac{1}{D} \sum_{i=1}^{D-1} \sum_{j=i+1}^{D} \text{Var}\left[\ln \frac{X_i}{X_j}\right] \tag{2.18}$$

$$= \sum_{i=1}^{D} \text{Var}[\text{clr}_i(\mathbf{X})] \tag{2.19}$$

$$= \sum_{j=1}^{D-1} \text{Var}[\text{ilr}_j(\mathbf{X})]. \tag{2.20}$$

The first one (2.18) proposes the variance of all simple log-ratios as components, with the advantage of easy interpretation of each component. The decomposition into ilr components of variance (2.20) is globally more understandable because of the orthogonality of the ilr coordinates, while each individual coordinate may require an additional effort of interpretation. The centre, as well as the total variance and its components, can be estimated using the ilr components; their properties correspond to those of the estimators of the mean and the variance-covariance in real sample spaces (Pawlowsky-Glahn and Egozcue 2001, 2002).

A typical way to present estimations of the centre and the variability is the variation array (Aitchison 1986). Table 2.5 shows the variation array for Cat10 data. The lower triangle shows the sample mean value of the simple log-ratio of the corresponding two parts (numerator by column, denominator by row). Therefore, the last row of the array is the alr transformation, using *other* in the denominator, of the estimated centre. For instance, the mean log-ratio between CiU and PSC is positive (1.06) and indicates that, in mean, CiU gets more votes than PSC.

The upper triangle contains the sample variances of the same log-ratios. They are the sample version of the terms appearing in the decomposition (2.18) of the total variance in simple log-ratios. The first row of the table shows the sample clr-variances, which are also a decomposition of the total variance (2.19). The lower part of the Table 2.5 compares the sample centre of the composition with the percentages of overall Catalonia (Cat%), i.e. adding all votes across regions and then expressed as percentages. The Cat% row shows some important regional departures from the centre. Specially important is the deviation of C's, whose percentage in Cat% doubles that reported in the centre by regions. This is due to the high variance of this party across regions, as can be seen in the variation array (bold numbers). Also the variance of the log-ratio CiU over ERC is small (0.08) thus suggesting proportionality of these votes across the regions.

The singular value decomposition (SVD) is an important tool in statistics and, in particular, in CoDa analysis (Aitchison 1984). Its normal use is related to reduction of dimensionality, but it also provides an ilr transformation and the corresponding orthonormal basis. It starts with a matrix containing a log-ratio representation of the data set of size n, usually the clr-transformed data. Subtracting the mean value of the clr-components an (n, D)-matrix **M** is obtained. **M** is decomposed using SVD, $\mathbf{M} = \mathbf{U}\boldsymbol{\Lambda}\mathbf{V}^{\top}$, where $(\cdot)^{\top}$ stands for transpose. The

Table 2.5 Variation array for Cat10. Upper triangle: simple log-ratio means; lower triangle: simple log-ratio variances. Total variance is 0.9952. The first row contains clr-variances. The centre of the regions is reported as cen%; the last row is the percentage computed from the total votes in Catalonia.

	abst	nota	null	C's	CiU	ERC	ICV	PSC	PP	other
clrVar	0.06	0.07	0.10	**0.22**	0.07	0.12	0.08	0.08	0.11	0.10
abst		0.067	0.15	0.29	0.07	0.19	0.08	0.05	0.10	0.14
nota	3.01		0.08	**0.53**	0.03	0.09	0.13	0.10	0.24	0.11
null	4.22	1.21		**0.77**	0.07	0.05	0.25	0.21	0.36	0.13
C's	3.76	0.75	−0.46		**0.55**	**0.85**	0.26	0.31	0.16	**0.59**
CiU	0.34	−2.67	−3.88	−3.42		0.08	0.15	0.14	0.26	0.07
ERC	1.95	−1.07	−2.28	−1.82	1.60		0.24	0.26	0.45	0.10
ICV	2.55	−0.46	−1.67	−1.25	2.21	0.61		0.10	0.18	0.17
PSC	1.40	−1.61	−2.82	−2.36	1.06	−0.54	−1.15		0.12	0.26
PP	2.02	−0.99	−2.20	−1.74	1.68	0.07	−0.54	0.62		0.38
other	1.75	−1.27	−2.48	−2.01	1.40	−0.20	−0.81	0.34	−0.27	
Cen%	38.9	1.9	0.57	0.91	27.6	5.6	3.0	9.6	5.2	6.8
Cat%	40.1	1.8	0.4	2.0	22.9	4.2	4.4	10.9	7.3	6.0

diagonal (D, D)-matrix Λ contains the singular values. Starting with clr-transformed data, the last of these singular values must be null. The squares of the singular values add up to the total variance and are variances of ilr-components (2.20). The V term is a (D, D)-matrix which reduces to a $(D, D − 1)$-matrix after removal of the column corresponding to the null singular value. Its $D − 1$ columns are the clr transformation of the elements of an orthonormal basis of the simplex. After removal of the column which corresponds to the null singular value, the rows of the $(n, D − 1)$ matrix $U\Lambda$ are the ilr coordinates of the centred data with respect to the orthonormal basis of the simplex defined by V (see Chapter 11). The biplot of this SVD (Gabriel 1971; Aitchison and Greenacre 2002) consists in representing a projection (usually bidimensional) of the first orthonormal components, called principal components, in the same plot as the projection of the centred clr variables. The centred clr-variables are scaled by the singular values so that the rays from the origin have length proportional to the standard deviation of the variables. The links between the vertices of rays are proportional to the standard deviation of the simple log-ratio of the parts corresponding to the rays. The cosine of the angle between two links approaches the correlation coefficient between the corresponding simple log-ratios; orthogonality of links suggests uncorrelation of the simple log-ratios. Figure 2.2 shows the CoDa-biplot for Cat10 explaining 83.5% of the total variance. The first principal component is clearly related to the links between C's and ERC, CiU, null, nota. This corresponds to the variances of simple log-ratios involving C's (Table 2.5). A political interpretation of this first principal component may be: positive values correspond to a larger proportion of non supporters of Catalan nationalism, while negative values correspond to a larger proportion of different versions of Catalan nationalism. The second principal component can also be interpreted as a kind of non-conservative parties or coalitions (positive values) against conservative parties (including the socialist PSC). Abstention is badly represented in the biplot and corresponds to a low variability principal

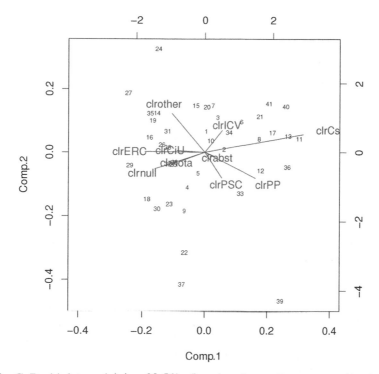

Figure 2.2 CoDa-biplot explaining 83.5% of total variance. Rays proportional to the standard deviation of clr variables. Regions: 1, Alt Camp; 13, Barcelonès; 41, Vallès Oriental; 39, Vall D'Aran (outlier).

component. Regions shown in Table 2.1 are numbered as 1 for Alt Camp, 13 for Barcelonès and 41 for Vallès Oriental. Vall D'Aran, treated as an outlier in the regression of Section 2.4, appears far apart from the centre in the lower part of the biplot (number 39).

Singular value decomposition and the CoDa-biplot determine an ilr transformation of the data, but the coordinates corresponding to principal axes are log-contrasts usually involving all parts with rather different coefficients. Interpretation of such principal components may be difficult and/or vague, although they correspond to the maximum variance explained. In many situations, the questions put forward by analysts suggest grouping parts of the composition and contrasting such groups. In those cases, a user designed basis of the simplex may be more understandable than the principal axes obtained using SVD. The technique is based on an SBP of the compositional vector.

The structure of the SBP, the ilr decomposition of the total variance (2.20), and the mean and dispersion of the sample coordinates can be summarized in a so called CoDa-dendrogram. Figure 2.3 shows the CoDa-dendrogram for Cat10 data following the SBP shown in Table 2.6. The vertical bars connect two groups of parts. The length of these bars is only that required to link the labels; but all of them, whatever their length, are equally scaled, in this case as a segment $(-4, +4)$. On each vertical bar a box-plot of quantiles (0.05, 0.25, 0.50, 0.75, 0.95) of the corresponding balance is represented to visualize sample dispersion. For instance the location of the box-plot of balance between C's and the rest of the parties shows that C's gets few votes in comparison with the other parties and coalitions. Anchored on each

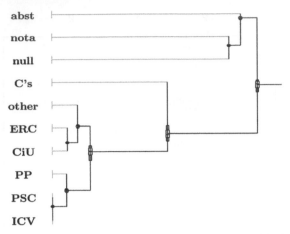

Figure 2.3 CoDa-dendrogram for Cat10 data. SBP as given in Table 2.6. Total variance is 0.9952. Scale of vertical bars ranges from -4 to $+4$.

vertical bar and pointing to the right, there is a horizontal bar. Its length is proportional to the variance of the balance, the ilr-variance in Equation (2.20). Contact of the vertical bar with the left end point of the horizontal bar is the mean balance (coordinate of the sample centre). Comparison of the mean balance and its median (0.50 quantile) in the box-plot gives an idea of the symmetry of a sample balance.

Figure 2.3 shows the relatively small variance of balances involving abstention, null and nota votes. The largest variance corresponds to the partition separating C's from the rest of the parties and coalitions (b_4), followed by b_5, which groups Catalan-only parties and coalitions (CiU, ERC, other) from lists of candidates with a presence in the whole of Spain (ICV, PSC, PP). Further partitions within these groups have smaller variances, thus indicating that the most variable balances across regions are precisely b_4 and b_5. Both balances are quite correlated (-0.835), thus showing that regions with a large relative presence of ICV, PSC, and/or PP in front of CiU, ERC and other are also those with a larger presence of C's.

Table 2.6 Sequential binary partition of Cat10, represented in Figure 2.3 as a CoDa-dendrogram.

	abst	nota	null	C's	CiU	ERC	ICV	PSC	PP	other
b_1	1	1	1	-1	-1	-1	-1	-1	-1	-1
b_2	1	-1	-1	0	0	0	0	0	0	0
b_3	0	1	-1	0	0	0	0	0	0	0
b_4	0	0	0	1	-1	-1	-1	-1	-1	-1
b_5	0	0	0	0	1	1	-1	-1	-1	1
b_6	0	0	0	0	-1	-1	0	0	0	1
b_7	0	0	0	0	-1	1	0	0	0	0
b_8	0	0	0	0	0	0	-1	-1	1	0
b_9	0	0	0	0	0	0	-1	1	0	0

2.6   Conclusion

Compositional data appear frequently in most experimental scientific fields. The statistical analysis of compositional data requires attention to their characteristics in order to avoid misleading results and conclusions. Spurious correlation is one of the most striking dangers when analysing compositional data; it was detected by K. Pearson at the end of the nineteenth century. The principles of compositional data analysis, established by John Aitchison in the 1980s, allow the development of a methodology free of all these problems. It relies on the study of log-ratios that transform the parts of a composition (e.g. concentrations in percentage, parts per one, parts per million) into real random variables that can be studied with conventional statistical methods. The developments related to the geometry of the simplex, called Aitchison geometry, allows the representation of the compositions in real orthogonal coordinates in such a way that usual statistical methods can be applied systematically. The use of coordinates in the simplex facilitates the computations and the analysis, but requires interpretation of (log)-ratios to obtain conclusions. That way, one will not be tempted to interpret each element individually, without acknowledgement of their relative character.

Acknowledgements

This research has been supported by the Spanish Ministry of Science and Innovation (projects CSD2006-00032 and MTM2009-13272) and by the Agència de Gestió d'Ajuts Universitaris i de Recerca of the Generalitat de Catalunya (Ref. 2009SGR424).

References

Aitchison J 1982 The statistical analysis of compositional data (with discussion). *Journal of the Royal Statistical Society, Series B (Statistical Methodology)* **44**(2), 139–177.

Aitchison J 1984 Reducing the dimensionality of compositional data sets. *Mathematical Geology* **16**(6), 617–636.

Aitchison J 1986 *The Statistical Analysis of Compositional Data*. Monographs on Statistics and Applied Probability. Chapman and Hall Ltd (reprinted 2003 with additional material by The Blackburn Press), London (UK). 416 p.

Aitchison J 1997 The one-hour course in compositional data analysis or compositional data analysis is simple. In *Proceedings of IAMG'97 – The III Annual Conference of the International Association for Mathematical Geology* (ed. Pawlowsky-Glahn V), vol. I, II and addendum. International Center for Numerical Methods in Engineering (CIMNE), Barcelona (Spain). pp. 3–35.

Aitchison J and Egozcue JJ 2005 Compositional data analysis: where are we and where should we be heading? *Mathematical Geology* **37**(7), 829–850.

Aitchison J and Greenacre M 2002 Biplots for compositional data. *Applied Statistics* **51**(4), 375–392.

Aitchison J, Barceló-Vidal C, Egozcue JJ and Pawlowsky-Glahn V 2002 A concise guide for the algebraic-geometric structure of the simplex, the sample space for compositional data analysis. In *Proceedings of IAMG'02 – The VIII Annual Conference of the International Association for Mathematical Geology* (ed. Bayer U, Burger H and Skala W), vol. I and II. Selbstverlag der Alfred-Wegener-Stiftung, Berlin (Germany). pp. 387–392.

Barceló-Vidal C, Martín-Fernández JA and Pawlowsky-Glahn V 2001 Mathematical foundations of compositional data analysis. In *Proceedings of IAMG' 01 – The VII Annual Conference of the International Association for Mathematical Geology* (ed. Ross G). Kansas Geological Survey, Cancun (Mexico). 20 p.

Billheimer D, Guttorp P and Fagan W 2001 Statistical interpretation of species composition *Journal of the American Statistical Association* **96**(456), 1205–1214.

Buccianti A and Pawlowsky-Glahn V 2005 New perspectives on water chemistry and compositional data analysis. *Mathematical Geology* **37**(7), 703–727.

Chayes F 1960 On correlation between variables of constant sum. *Journal of Geophysical Research* **65**(12), 4185–4193.

Connor RJ and Mosimann JE 1969 Concepts of independence for proportions with a generalization of the Dirichlet distribution. *Journal of the American Statistical Association* **64**(325), 194–206.

Egozcue JJ 2009 Reply to 'On the Harker variation diagrams; . . .' by J. A. Cortés.. *Mathematical Geosciences* **41**(7), 829–834.

Egozcue JJ and Pawlowsky-Glahn V 2005a Coda-dendrogram: a new exploratory tool. In *Proceedings of CoDaWork' 05, The 2nd Compositional Data Analysis Workshop* (ed. Mateu-Figueras G and Barceló-Vidal C). http://ima.udg.es/Activitats/CoDaWork05/. University of Girona, Girona (Spain).

Egozcue JJ and Pawlowsky-Glahn V 2005b Groups of parts and their balances in compositional data analysis. *Mathematical Geology* **37**(7), 795–828.

Egozcue JJ and Pawlowsky-Glahn V 2006 Exploring compositional data with the coda-dendrogram. In *Proceedings of IAMG' 06 – The XI Annual Conference of the International Association for Mathematical Geology* (ed. Pirard E, Dassargues A and Havenith HB). University of Liège, Liège (Belgium). CD-ROM.

Egozcue JJ, Pawlowsky-Glahn V, Mateu-Figueras G and Barceló-Vidal C 2003 Isometric logratio transformations for compositional data analysis. *Mathematical Geology* **35**(3), 279–300.

Gabriel KR 1971 The biplot – graphic display of matrices with application to principal component analysis. *Biometrika* **58**(3), 453–467.

Martín-Fernández JA, Barceló-Vidal C and Pawlowsky-Glahn V 2003 Dealing with zeros and missing values in compositional data sets using nonparametric imputation. *Mathematical Geology* **35**(3), 253–278.

Mosimann JE 1962 On the compound multinomial distribution, the multivariate β-distribution and correlations among proportions. *Biometrika* **49**(1–2), 65–82.

Palarea-Albaladejo J, Martín-Fernández JA and Soto JA 2011 C-means clustering of compositional data. *Journal of Classification* (in press).

Pawlowsky-Glahn V and Egozcue JJ 2001 Geometric approach to statistical analysis on the simplex. *Stochastic Environmental Research and Risk Assessment (SERRA)* **15**(5), 384–398.

Pawlowsky-Glahn V and Egozcue JJ 2002 BLU estimators and compositional data. *Mathematical Geology* **34**(3), 259–274.

Pawlowsky-Glahn V and Egozcue JJ 2006 Análisis de datos composicionales con el coda-dendrograma. In *Actas del XXIX Congreso de la Sociedad de Estadística e Investigación Operativa (SEIO' 06)* (ed. Sicilia-Rodríguez J, González-Martín C, González-Sierra MA and Alcaide D). Sociedad de Estadística e Investigación Operativa, Tenerife (Spain). CD-ROM.

Pearson K 1897 Mathematical contributions to the theory of evolution. On a form of spurious correlation which may arise when indices are used in the measurement of organs. *Proceedings of the Royal Society of London* **LX**, 489–502.

Thió-Henestrosa S, Egozcue JJ, Pawlowsky-Glahn V, Kovács LO and Kovács G 2008 Balance-dendrogram a new routine of codapack. *Computer and Geosciences* **34**(12), 1682–1696.

Part II

THEORY – STATISTICAL MODELLING

3

The principle of working on coordinates

Glòria Mateu-Figueras[1], Vera Pawlowsky-Glahn[1] and Juan José Egozcue[2]

[1]*Department of Computer Science and Applied Mathematics, University of Girona, Spain*
[2]*Department of Applied Mathematics III, Technical University of Catalonia, Spain*

3.1 Introduction

Given a k-dimensional Euclidean vector space E, with the internal operation $\oplus$, the external operation $\odot$, and inner product $\langle \cdot, \cdot \rangle_E$, linear algebra theory assures the existence of orthonormal bases, i.e. of bases with orthogonal and unit norm elements. Infinitely many bases exist. But once one has been chosen, any element $\mathbf{x} \in E$ can be uniquely expressed as a linear combination of the basis elements, denoted here as $\{\mathbf{e}_1, \mathbf{e}_2, \ldots, \mathbf{e}_k\}$. It follows that there exists a vector $\mathbf{y} = (y_1, y_2, \ldots, y_k)$, with components denoted as coordinates or coefficients with respect to the basis, such that

$$\mathbf{x} = (y_1 \odot \mathbf{e}_1) \oplus (y_2 \odot \mathbf{e}_2) \oplus \cdots \oplus (y_k \odot \mathbf{e}_k), \quad y_i = \langle \mathbf{x}, \mathbf{e}_i \rangle_E, \quad i = 1, 2, \ldots, k.$$

Any element of this space can be represented by its vector of coordinates, which is an element of $\mathbb{R}^k$. Mathematical statistics relies on real analysis, and real analysis is commonly performed using coordinates with respect to an orthonormal basis. Therefore, statistical analysis on E can be conducted in a straightforward way, using standard methods and techniques, on orthonormal coordinates.

Compositional Data Analysis: Theory and Applications, First Edition. Edited by Vera Pawlowsky-Glahn and Antonella Buccianti.
© 2011 John Wiley & Sons, Ltd. Published 2011 by John Wiley & Sons, Ltd.

In statistics, the real space, $\mathbb{R}^k$, is usually assumed to be the natural sample space for a given set of observations. For this reason standard statistics has been developed in $\mathbb{R}^k$ using its particular algebraic-geometric structure, which is commonly known as Euclidean geometry. When the sample space is different from real space, two strategies can be used: (a) directly working in the sample space based on its particular structure and operations; or (b) using coordinates with respect to an orthonormal basis. Both strategies are equivalent. However, a word of caution is necessary when the sample space E is a subset of real space and when E entails a different algebraic-geometric structure. This is the case of the simplex, which is the natural sample space of compositional data in general, or of the positive real line. In these cases, observations are expressed as real vectors. It is tempting to apply standard statistics directly, instead of using coordinates with respect to an orthonormal basis. Caution must be taken, however, as incompatibilities or incoherencies can result from this method. Regarding the simplex, the most famous example is spurious correlation. It was first mentioned by Karl Pearson (1897), and then studied in detail by John Aitchison (1986, chapter 3), who provides a list of additional difficulties, including those appearing in parametric models.

The main objective of this paper is to establish the role of coordinates in the simplex. In Section 3.3 we analyze the simplex as the sample space of compositions. Particularly, we revise its vector space structure and we focus on coordinates with respect to an orthonormal basis using simple examples. Finally, in Section 3.4, we compare the proposed methodology with the traditional log-ratio approach developed by Aitchison in the early 1980s. Although the two approaches are essentially the same, we note some differences. To distinguish the two approaches, the log-ratio approach is termed the *move* approach, while our coordinate-focused approach is termed the *stay-in-the-simplex* approach, or *stay* approach for short (Mateu-Figueras 2003).

3.2   The role of coordinates in statistics

Standard or classical statistics has been traditionally used in real space by taking into account the particular structure of Euclidean vector space. The sum is the internal operation, the scalar product is the external operation, and the ordinary inner product, which induces the Euclidean norm and distance. This algebraic-geometric structure has been considered an implicit or explicit assumption in the development of virtually any statistical methodology, as well as in the interpretation of statistical results. In $\mathbb{R}^k$, as in any Euclidean vector space, linear algebra theory assures the existence of an orthonormal basis, but it is not unique. Once a basis has been chosen, it is referred to as the *canonical basis*, and is denoted by $\{(1, 0, \ldots, 0), (0, 1, 0, \ldots, 0), \ldots, (0, \ldots, 0, 1)\}$. Any vector $\mathbf{x} = (x_1, x_2, \ldots, x_k) \in \mathbb{R}^k$ can be written as

$$\mathbf{x} = x_1 \cdot (1, 0, \ldots, 0) + x_2 \cdot (0, 1, 0, \ldots, 0) + \cdots + x_k \cdot (0, \ldots, 0, 1).$$

The vector of coordinates is identical to the original element in $\mathbb{R}^k$.

Given a sample space E with an Euclidean vector space structure different from the usual one in real space, our proposal is to apply standard statistics to the vectors of coordinates with respect to an orthonormal basis. As mentioned in Section 3.1, these vectors of coordinates are elements of real space; thus, real analysis in general and standard statistics in particular can be applied. The coordinates of any element of E, and vice versa, can be obtained by means of simple linear combinations. Consequently, linear algebra allows us to translate standard statistics into any sample space, other than $\mathbb{R}^k$, with an Euclidean vector space structure. We only have to ensure that we are working with coordinates with respect to an orthonormal basis. This is what we call *the principle of working on coordinates*.

Some references to this principle can be found in the literature. The foundations of modern probability were derived by Kolmogorov (1933). Though not explicitly stated, he identifies any random vector by its coordinates, as he refers to observations interchangeably by the terms *random vector* and *coordinates* (Kolmogorov 1933, p. 27). Later, Kolmogorov and Fomin (1957) work with coordinates in a functional analysis context. Additionally, Eaton (1983) provides a geometric approach to multivariate statistics by importantly emphasising finite dimensional inner product spaces. In particular, Eaton (1983, chapter 2) uses the principle of working on coordinates to define density functions and to compute corresponding mean values.

3.3   The simplex

Most notation used in this section is standard in compositional data analysis (see Chapters 2 and 11). The vector $\mathbf{x} = (x_1, x_2, \ldots, x_D)$ denotes a D-part composition with sample space the simplex

$$\mathcal{S}^D = \left\{ \mathbf{x} = (x_1, x_2, \ldots, x_D) : x_i > 0 \ (i = 1, 2, \ldots, D), \sum_{i=1}^{D} x_i = \kappa \right\},$$

where κ is a given positive constant, which is usually 1 or 100, depending on whether the parts are measured in per unit or as percentages, respectively. Aitchison (1986) introduces the following two basic operations on the simplex, namely, perturbation and powering, as the respective internal and external operations,

$$\mathbf{x} \oplus \mathbf{x}^* = \mathcal{C}(x_1 x_1^*, \ldots, x_D x_D^*), \tag{3.1}$$

$$\alpha \odot \mathbf{x} = \mathcal{C}(x_1^\alpha, \ldots, x_D^\alpha), \tag{3.2}$$

where $\mathcal{C}$ denotes the closure operation that normalises any vector $\mathbf{x}$ to a constant sum (Aitchison 1986), $\mathbf{x}, \mathbf{x}^* \in \mathcal{S}^D$ and $\alpha \in \mathbb{R}$. As stated in Aitchison (2001), perturbation and powering induce a real vector space structure on the simplex. Furthermore, the following inner product, with its associated norm and distance, can be used to obtain a finite $(D-1)$-dimensional Hilbert space structure; note that this is an Euclidean vector space structure on the simplex (Billheimer *et al.* 2001; Pawlowsky-Glahn and Egozcue 2001):

$$\langle \mathbf{x}, \mathbf{x}^* \rangle_a = \frac{1}{D} \sum_{i<j} \ln \frac{x_i}{x_j} \ln \frac{x_i^*}{x_j^*} \, ; \tag{3.3}$$

$$\|\mathbf{x}\|_a = \sqrt{\frac{1}{D} \sum_{i<j} \left(\ln \frac{x_i}{x_j} \right)^2} \, ; \tag{3.4}$$

$$d_a(\mathbf{x}, \mathbf{x}^*) = \sqrt{\frac{1}{D} \sum_{i<j} \left(\ln \frac{x_i}{x_j} - \ln \frac{x_i^*}{x_j^*} \right)^2} \, . \tag{3.5}$$

Note that operations (3.1–3.5) are defined in terms of the original parts of the compositions, not in terms of coordinates. When referring to the properties of $(\mathcal{S}^D, \oplus, \odot)$ as an Euclidean space, it is usual to refer to the Aitchison geometry on the simplex, particularly the Aitchison distance, norm and inner product. Hence the subscript a is used in Equations (3.3–3.5).

3.3.1 Basis of the simplex

The simplex $\mathcal{S}^D$ is a $(D-1)$-dimensional subset of D-dimensional real space. Thus, compositions are usually expressed in terms of the canonical basis of $\mathbb{R}^D$. That is any composition $\mathbf{x} \in \mathcal{S}^D$ can be written as follows:

$$\mathbf{x} = x_1 \cdot (1, 0, \ldots, 0) + x_2 \cdot (0, 1, 0, \ldots, 0) + \cdots + x_D \cdot (0, \ldots, 0, 1). \qquad (3.6)$$

Because the sum of x_i's is 1 and all x_i are positive, Equation (3.6) is a convex linear combination in $\mathbb{R}^D$. Thus, $\mathbf{x}$ is expressed as a mixture, or, put differently, it is a composition, with the x_i's its parts. The problem is, that the set of vectors $\{(1, 0, \ldots, 0), (0, 1, 0, \ldots, 0), \ldots, (0, \ldots, 0, 1)\}$ is neither a basis nor a generating system on $\mathcal{S}^D$ with respect to the vector space structure defined above; these vectors do not even belong to the simplex. Moreover, Equation (3.6) is not a linear combination on $\mathcal{S}^D$ because neither the sum nor the product is a closed operation on the simplex. Nevertheless, observations are usually registered in this way.

Within Aitchison geometry, a generating system of $\mathcal{S}^D$ can be obtained as $\{\mathbf{w}_1, \mathbf{w}_2, \ldots, \mathbf{w}_D\}$ with $\mathbf{w}_i = \mathcal{C}(1, 1, \ldots, e, \ldots, 1)$ (where e is the ith component) for $i = 1, 2, \ldots, D$. In fact, any vector $\mathbf{x} \in \mathcal{S}^D$ can be written as follows:

$$\mathbf{x} = (\ln x_1 \odot \mathbf{w}_1) \oplus (\ln x_2 \odot \mathbf{w}_2) \oplus \cdots \oplus (\ln x_D \odot \mathbf{w}_D).$$

As perturbation and powering implicitly include closure, they are scale invariant. Adding any constant to all perturbed terms leaves $\mathbf{x}$ unaltered. The following equivalent expression can be employed.

$$\mathbf{x} = \left(\ln \frac{x_1}{g_m(\mathbf{x})} \odot \mathbf{w_1} \right) \oplus \left(\ln \frac{x_2}{g_m(\mathbf{x})} \odot \mathbf{w_2} \right) \oplus \cdots \oplus \left(\ln \frac{x_D}{g_m(\mathbf{x})} \odot \mathbf{w_D} \right), \qquad (3.7)$$

where $g_m(\mathbf{x})$ denotes the geometric mean across the components of the argument. The coefficients in Equation (3.7) are the centred log-ratio transformation (clr) defined in Aitchison (1986) (see also Chapters 2 and 11). Note that non-uniqueness is consistent with treating a composition as an equivalence class as given in Barceló-Vidal et al. (2001). Moreover, the clr coefficients $\ln[x_i/g_m(\mathbf{x})]$ have been selected so that their sum is null.

From the previous generating system, a basis can be obtained by taking any $(D-1)$ vectors, such as $\{\mathbf{w}_1, \mathbf{w}_2, \ldots, \mathbf{w}_{D-1}\}$. Any vector $\mathbf{x} \in \mathcal{S}^D$ can be written

$$\mathbf{x} = \left(\ln \left(\frac{x_1}{x_D} \right) \odot \mathbf{w}_1 \right) \oplus \left(\ln \left(\frac{x_2}{x_D} \right) \odot \mathbf{w}_2 \right) \oplus \cdots \oplus \left(\ln \left(\frac{x_{D-1}}{x_D} \right) \odot \mathbf{w}_{D-1} \right). \qquad (3.8)$$

The coefficients correspond to the well-known additive log-ratio transformation (alr) introduced by Aitchison (1986). The basis $\{\mathbf{w}_1, \mathbf{w}_2, \ldots, \mathbf{w}_{D-1}\}$ is not orthogonal, nor do the vectors have unit norm, as can be shown with the inner product (3.3) and the norm (3.4) (Egozcue and Pawlowsky-Glahn 2005) (see also Chapter 11).

The Euclidean space structure of the simplex guarantees the existence of an orthonormal basis, which can be readily obtained by applying the usual Gram-Schmidt orthonormalisation process to any basis. This in turn can be obtained from any given generating system. As is the case for any inner product, an orthonormal basis is not unique. Moreover, it is not obvious how to determine which basis is the most appropriate for any given problem. Egozcue *et al.* (2003) give a particular orthonormal basis $\{e_1, e_2, \ldots, e_{D-1}\}$ with

$$\mathbf{e}_i = \mathcal{C}\left(\exp\left(\frac{1}{\sqrt{i(i+1)}} \right), \ldots, \exp\left(\frac{1}{\sqrt{i(i+1)}} \right), \exp\left(-\sqrt{\frac{i}{i+1}} \right), 1, \ldots, 1 \right) \quad (3.9)$$

and coordinates

$$\mathbf{x} = (y_1 \odot \mathbf{e}_1) \oplus (y_2 \odot \mathbf{e}_2) \oplus \cdots \oplus (y_D \odot \mathbf{e}_{D-1}),$$

with

$$y_i = \frac{1}{\sqrt{i(i+1)}} \ln\left[\frac{x_1 x_2 \cdots x_i}{(x_{i+1})^i} \right], \qquad i = 1, 2, \ldots, D-1.$$

These coordinates correspond to a particular case of the isometric log-ratio transformation (ilr) (Egozcue *et al.* 2003). Another orthonormal basis of the simplex and their respective ilr coordinates can be obtained using a strategy based on a sequential binary partition of the composition (Egozcue and Pawlowsky-Glahn 2005). This allows us to design a particular orthonormal basis with interpretable coordinates called *balances*. Another possibility is to use principal component analysis (Aitchison 1983; Aitchison and Greenacre 2002), although the basis obtained can be difficult to interpret. The important point here is that, once an orthonormal basis has been chosen, all standard statistical methods can be applied to the coordinates and transferred to the simplex, thereby preserving their properties.

3.3.2   Working on orthonormal coordinates

The principle of working on coordinates described above implies that all standard statistical methods can be applied to coordinates of any composition with respect to an orthonormal basis. In this section, some simple examples are used to illustrate this methodology. The specific orthonormal basis (3.9) is used, but any other orthonormal basis could be taken as well. In some cases, the result will be directly interpretable, but in other cases, it will be better to re-express the result as a composition instead of a vector of coordinates.

The simplest cases are the basic operations on the simplex, namely, the perturbation and the powering. Let $\mathbf{x} = (0.6, 0.3, 0.1)$ and $\mathbf{x}^* = (0.3, 0.3, 0.4)$ be two compositions in $\mathcal{S}^3$. To compute $\mathbf{x} \oplus \mathbf{x}^*$ and $2 \odot \mathbf{x}$, we can directly apply operations (3.1) and (3.2) with the original parts of the compositions. Alternatively, we can work with vectors $\mathbf{y} = (0.490, 1.180)$ and $\mathbf{y}^* = (0, -0.235)$, which are corresponding orthonormal coordinates with respect to (3.9). As the standard operations in real space are the sum and the scalar product, we need only compute $\mathbf{y} + \mathbf{y}^* = (0.490, 0.945)$ and $2\mathbf{y} = (0.980, 2.360)$. The results themselves are also coordinates, and we obtain the corresponding compositions by means of the linear combinations $\mathbf{x} \oplus \mathbf{x}^* = (0.490 \odot \mathbf{e}_1) \oplus (0.945 \odot \mathbf{e}_2) = (0.581, 0.290, 0.129)$ and $2 \odot \mathbf{x} = (0.980 \odot \mathbf{e}_1) \oplus (2.360 \odot \mathbf{e}_2) = (0.782, 0.196, 0.022)$. Straightforward expressions of the back-transformation (ilr^{-1}) for higher dimensions are found in Chapter 11. These simple computations can be carried out using clr and alr (non-orthogonal) coordinates as well.

Other simple examples include the computation of the Aitchison inner product $\langle \mathbf{x}, \mathbf{x}^* \rangle_a$, norm $\|\mathbf{x}\|_a$, and distance $d_a(\mathbf{x}, \mathbf{x}^*)$. Expressions (3.3), (3.4) and (3.5) could be used, but if we work with orthonormal coordinates, we can apply the ordinary inner product, norm and Euclidean distance to coordinates. Denoting these ordinary metrics without any subscript, we obtain $\langle \mathbf{x}, \mathbf{x}^* \rangle_a = \langle \mathbf{y}, \mathbf{y}^* \rangle = -0.277$, $\|\mathbf{x}\|_a = \|\mathbf{y}\| = 1.278$ and $d_a(\mathbf{x}, \mathbf{x}^*) = d(\mathbf{y}, \mathbf{y}^*) = 1.497$. These results are not elements of the support space, as they are only numerical values. As such, back-transformation is nonsensical. When these computations are done using alr coordinates, they fail because they are coordinates with respect to an oblique basis. Using the clr representation, the correct answer is still obtained, although the operations involved correspond to $\mathbb{R}^D$.

When non-orthonormal coordinates (e.g. alr, clr, or simple log-ratio transformations) are used, some problems can appear. For instance, the basis $\{\mathbf{w}_1, \mathbf{w}_2, \ldots, \mathbf{w}_{D-1}\}$ used in (3.8) is neither unitary nor orthogonal; rather, it is an oblique basis (Egozcue *et al.* 2003). Therefore, if the alr vectors are mapped onto orthogonal axes, the distances and the angles between the alr coordinates are deformed. Figure 3.1 shows the axes corresponding to an orthonormal basis (3.9) and the axes corresponding to the alr basis $\{\mathbf{w}_1, \ldots, \mathbf{w}_{D-1}\}$, represented on the ternary diagram. Moreover, a square is plotted that is defined by vertices $A = (0.436, 0.436, 0.128)$, $B = (0.576, 0.140, 0.284)$, $C = (0.185, 0.185, 0.630)$ and $D = (0.140, 0.576, 0.284)$. This figure may not appear to be a square, but under the Aitchison geometry and, using (3.3–3.5), we can check that vectors AB, AD, BC and DC have equal length and that vectors AC and BD are orthogonal. Figure 3.2(a) shows an alr representation of the square. The distances and the angles have been clearly deformed. Figure 3.2(b) contains the ilr representation of the square.

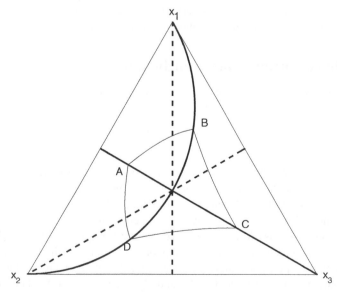

Figure 3.1   ilr axes (solid line), alr axes (dashed line) and a square in $\mathcal{S}^3$.

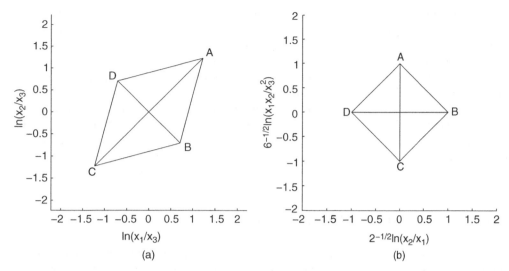

Figure 3.2 (a) alr transformed square represented in alr axes; (b) ilr transformed square represented in ilr axes.

In this case, there is no deformation, as the basis is orthonormal. Additionally, the area of the square in Figure 3.2(b) equals the Aitchison area of the region (ABCD) in Figure 3.1.

In the previous cases, working with orthogonal coordinates does not represent a great simplification, although some advantages are evident when plotting data or curves. Real differences are obtained during statistical analysis. Few models work directly on compositions. Some examples can be found in Martín-Fernández *et al.* (1998) regarding cluster analysis and in Barceló-Vidal (1996) regarding discriminant analysis. Regression of compositions using the Dirichlet distribution can be considered in this context as well (Gueorguieva *et al.* 2008; Hijazi and Jernigan 2009). The reason for it is that the probability distribution should be given as a probability density defined on $\mathcal{S}^D$, and the Dirichlet distribution is the only well-known probability density model defined directly on $\mathcal{S}^D$. In contrast, the prototype of a distribution in the simplex is the logistic-normal distribution (Aitchison and Shen 1980; Aitchison 1982), which is defined as a normal distribution on the alr coordinates. More recently (Mateu-Figueras 2003), the logistic-normal distribution has been redefined on ilr coordinates using a re-parameterisation of the logistic-normal model. Working on coordinates greatly simplifies the computation of probabilities and likelihood functions compared with using the expression of the compositions in the simplex.

An important advantage of working on coordinates is that linear models become tractable and easy to understand. Assume, for instance, that we are trying to predict a random composition based on the values of some covariates or factors (e.g. time, space or social parameters). Typically, a linear combination of the covariates ranges over the entire real line. Therefore, the linear combination fails as a predictor of a part of a composition (e.g. percentage, proportion or concentration). Alternatively, if the composition is represented by its coordinates, the model is feasible. Here, the transformation into coordinates plays the role of a *link function* in a generalised linear model. In this kind of scenario, the sample covariance matrix of the coordinates plays a major role, and it is normally assumed to be non-singular. Thus, it excludes the use of clr transformed data, where the covariance matrix is singular by construction. Additionally, linear models may use distances or norms during estimation (e.g. least squares), the analysis of residuals and model-checking techniques. This prevents the use of alr coordinates that, being oblique, burden the computation of Aitchison metrics. The advantage of ilr coordinates (that is, orthonormal coordinates) is that the analyst can be assured that all requirements related to the transformation of the composition are fulfilled.

In summary, the principle of working on (orthonormal, ilr) coordinates is easy to apply on the simplex; simply work with coordinates, apply standard statistical methodology, and, if necessary, return to the simplex by means of a back-transformation. Although this proposal is very simple, in some cases greater care is necessary to correctly interpret the results. This is due to the fact that coordinates are expressed in terms of the log-ratios of groups of parts.

3.4 *Move* or *stay* in the simplex

In the statistical literature, there is a long history of searching for a proper approach to the statistical analysis of compositional data (Aitchison and Egozcue 2005) (see also Chapter 1). The main contribution to a solution was made by John Aitchison, in the early 1980s (Aitchison 1982) when he introduced the log-ratio approach, using the intuitive concept of difference. The idea behind this approach was first to transform compositions into real space using a log-ratio transformation and, then to apply standard statistical methodology, finally returning to the simplex using the inverse log-ratio transformation. The only distinction between the log-ratio approach and the principle of working on coordinates is the recommended use of orthonormal or ilr coordinates in the latter approach. However, both approaches are essentially the same, because difficulties due to singular covariance matrices (clr) or to the distortion of the Aitchison metrics (alr) can be overcome using adequate algebraic techniques. The aim of this section is to compare both approaches with respect to the representation of probabilistic models and corresponding expectations.

The difference between the two approaches is more psychological than mathematical. The log-ratio approach has long been thought to be a transformation technique similar to that of a log-normal model; a log-normal random variable, when log-transformed, becomes a normal random variable (Aitchison and Brown 1957). Both random variables assume $\mathbb{R}$ as the sample space, although the former is restricted to positive values. No attention is paid to the fact that $\mathbb{R}_+$ is itself an Euclidean space, with its own operations, metrics and reference measure (Pawlowsky-Glahn and Egozcue 2001). This kind of transformation mentality is associated with the traditional Aitchison log-ratio approach. The alternative is to consider $\mathbb{R}_+$ as a space different from $\mathbb{R}$ and then pay special attention to the reference measure. In $\mathbb{R}$, the Lebesgue measure λ assigns the length $\lambda\{(a, b)\} = |b - a|$ to the interval (a, b), whereas the reference measure in $\mathbb{R}_+$ is $\lambda_+\{(a, b)\} = |\ln b - \ln a|$. This alternative has at least two consequences: (a) probability densities are expressed with respect to the reference measure; and (b) the computation of expectation changes according to the reference measure and the operations of the sample space. When we assume that the log-normal random variable exists in $\mathbb{R}$, we are *moving* from $\mathbb{R}_+$ to $\mathbb{R}$, thereby using a *move*-procedure. When we consider that the random variable has the sample space $\mathbb{R}_+$ and its logarithm is just its coordinate, than we are said to *stay* in $\mathbb{R}_+$.

Similar concepts can be used to address random compositions. The sample space is $\mathcal{S}^D$, and we can think of $\mathcal{S}^D$ as equipped with Aitchison geometry. This allows a coordinate representation of the random composition (that is, the *stay* approach). Alternatively, we can transform the random composition into a real random vector (that is, the *move* approach). One of the major implications of the *stay* approach is the use of the natural reference measure; as such, it deserves a brief explanation. The natural reference measure in $\mathcal{S}^D$ is the Aitchison measure (Mateu-Figueras and Pawlowsky-Glahn 2005, 2007). The Aitchison measure in $\mathcal{S}^D$ is easily defined using orthonormal coordinates. Consider a hyper-parallelepiped V in the space of ilr coordinates $\mathbb{R}^{D-1}$. Its Lebesgue measure or hyper-volume $\lambda(V)$ is computed as the product of the lengths of its edges (Pawlowsky-Glahn 2003). Then, $U = \mathrm{ilr}^{-1}(V) \subseteq \mathcal{S}^{D-1}$ has Aitchison measure $\lambda_a(U) = \lambda(V)$. For instance, the region within the contour ABCD in Figure 3.1 has Aitchison measure equal to the area (i.e. the Lebesgue measure) of the square ABCD in Figure 3.2(b).

As implicitly suggested by Eaton (1983), the principle of working on coordinates together with the *stay* approach can be used to define parametric models on the simplex using the density function of orthonormal coordinates. In coordinate space, this density is a classical density, as it is a density with respect to the Lebesgue measure, but on the simplex, it is a density with respect to the Aitchison measure. The normal on $\mathcal{S}^D$ (Mateu-Figueras and Pawlowsky-Glahn 2008) and the skew-normal on $\mathcal{S}^D$ (Mateu-Figueras *et al.* 2005; Mateu-Figueras and Pawlowsky-Glahn 2007) are defined using this strategy as probability densities with respect to the Aitchison measure. Figure 3.3(a) shows the isodensity curves of a normal on $\mathcal{S}^3$ with respect to the Aitchison measure. It corresponds to a zero-mean, unitary covariance matrix, binormal model for the two ilr coordinates. When the density is translated back into the simplex, the reference measure is the Aitchison measure. This means that an element of probability should be computed as the height of the density times the Aitchison-area element under the surface element. In the plain log-ratio approach (i.e. the *move* approach), the random composition is transformed into the real space of coordinates (typically alr coordinates). The back-transformation from the log-ratio coordinates to the simplex is viewed as a simple transformation of a real vector of coordinates into a real vector in the simplex (i.e. a random composition). To obtain the density of the random composition, a Jacobian is required to maintain the Lebesgue reference measure assumed for the coordinates. This Jacobian expresses the ratio of the Lebesgue and the Aitchison measures over the simplex, and it has an important role in computing moments and plotting isodensity contours. Figure 3.3(b) contains

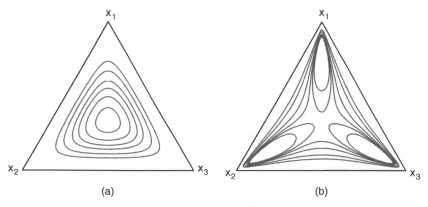

Figure 3.3 Isodensity plots of (a) a normal on $\mathcal{S}^3$ and (b) a logistic normal both with $\boldsymbol{\mu} = (0, 0)$ and $\boldsymbol{\Sigma} = Id$.

the isodensity curves of a logistic normal model. The differences are obvious, particularly regarding the mode(s). The multi-mode exhibited in Figure 3.3(b) is only due to the Jacobian. The two plotted densities correspond to the same probability law, but in Figure 3.3(a) we have a density with respect to the Aitchison measure, while in Figure 3.3(b) we have a density with respect to the Lebesgue measure on the simplex. Although, the normal in $\mathcal{S}^3$ in Figure 3.3(a) seems to be simpler than the logistic-normal density in Figure 3.3(b), none of them allows a direct computation of probabilities. Coordinate expression (that is, the normal model) is much more useful, which explains why probabilistic models are commonly expressed in terms of a coordinate system (Eaton 1983).

Defining expected values (or any moment) directly on the simplex without using coordinates is not easy. The reason is simple; the classical definition uses standard operations and Lebesgue integration in real space, which must be adapted to the simplex. Aitchison (2001) adapts the expected value concept and defines the center of a random composition using the geometric interpretation of the expected value. Applying the principle of working on coordinates, we can compute the expected value using the classical definition applied to the coordinates. In studies by Mateu-Figueras and Pawlowsky-Glahn (2007, 2008), the normal and the skew-normal models on the simplex are defined using coordinates, and the expected value is computed as usual. This is clear in Chapter 2 (Section 2.5) where the center of a random composition $\mathbf{X}$ is expressed as:

$$\text{Cen}[X] = \mathcal{C} \exp(\text{E}[\ln X]) = \text{ilr}^{-1}(\text{E}[\text{ilr}(X)]) \,. \tag{3.10}$$

It is remarkable that the expectation of $\ln X$ is based on the probability density of $X \in \mathcal{S}^D$ with the Lebesgue measure as a reference (i.e. the *move* approach). The expectation of $\text{ilr}(X)$ is based on the probability density of the ilr coordinates using the Lebesgue measure as reference. That is, it is the density in $\mathcal{S}^D$ using the Aitchison measure as a reference (i.e. the *stay* approach). For the right-hand side of Equation (3.10), the following procedure is used: compute the ilr coordinates of X, take the standard expectation of coordinates, and then back-transform the result to the simplex (Pawlowsky-Glahn and Egozcue 2001, 2002), thus using the principle of working on coordinates. The Dirichlet distribution is an exception to these coordinate procedures because the distribution of its coordinates cannot be easily applied. However, the computation of the center with respect to the Aitchison measure is still possible (see Chapter 10).

3.5   Conclusions

Standard statistical methodology was traditionally developed on real space. When the sample space has a different algebraic-geometric structure, standard statistical methodology is not adequate. Sometimes, the sample space is a subset of real space (e.g. positive orthants, cones, multivariate bounded intervals of real space and the simplex). In those cases, there is the possibility of using an ordinary sum and distance restricted to the subset. If operations and distances in these restricted sample spaces differ from the real ones, this practice may produce misleading – or even wrong – conclusions and thus should be avoided. In this chapter, the role of coordinates in statistics has been revisited. Using linear algebra theory, the principle of working on coordinates with respect to an orthonormal basis can be applied to any sample space with an Euclidean vector space structure. The simplex is one such particular case.

The notion of working on coordinates of the simplex is generally the same as that of working with the log-ratio approach. Nevertheless, when we work on coordinates, we emphasise orthonormal coordinates, such as ilr coordinates, moreover, the analyst is encouraged to choose the most interpretable coordinates for the given problem. In the log-ratio approach, the alr, clr and ilr transformations are available such that the analyst must choose the most suitable transformation for each particular analysis by taking into account their respective advantages and limitations. The principle of working on coordinates in the simplex is easy to use: apply standard statistics to (orthonormal) coordinates and, if necessary, return to the simplex. Basically, the log-ratio approach on the simplex and the proposal to work on coordinates do not differ substantially. However, the additive (alr) and centred log-ratio (clr) vectors are not coordinates with respect to an orthonormal basis, and consequently, each has its own limitations. Additionally, the principle of working on coordinates is naturally associated with *stay-in-the-simplex* strategies, which use the entire space structure, including the Aitchison measure, to define probability densities and moments.

Acknowledgements

This research has been supported by the Spanish Ministry of Science and Innovation (projects CSD2006-00032 and MTM2009-13272) and by the Agència de Gestió d'Ajuts Universitaris i de Recerca of the Generalitat de Catalunya (Ref. 2009SGR424).

References

Aitchison J 1982 The statistical analysis of compositional data (with discussion). *Journal of the Royal Statistical Society, Series B (Statistical Methodology)* **44**(2), 139–177.

Aitchison J 1983 Principal component analysis of compositional data. *Biometrika* **70**(1), 57–65.

Aitchison J 1986 *The Statistical Analysis of Compositional Data*. Monographs on Statistics and Applied Probability. Chapman and Hall Ltd (reprinted 2003 with additional material by The Blackburn Press), London (UK). 416 p.

Aitchison J 2001 Simplicial inference. In *Algebraic Methods in Statistics and Probability* (ed. Viana MA and Richards DS), vol. 287. American Mathematical Society, Providence, RI (USA). pp. 1–22.

Aitchison J and Brown JAC 1957 *The Lognormal Distribution*. Cambridge University Press, Cambridge (UK). 176 p.

Aitchison J and Egozcue JJ 2005 Compositional data analysis: where are we and where should we be heading?. *Mathematical Geology* **37**(7), 829–850.

Aitchison J and Greenacre M 2002 Biplots for compositional data. *Applied Statistics* **51**(4), 375–392.

Aitchison J and Shen SM 1980 Logistic-normal distributions. Some properties and uses. *Biometrika* **67**(2), 261–272.

Barceló-Vidal C 1996 *Mixturas de Datos Composicionales*. PhD thesis, Universitat Politècnica de Catalunya, Barcelona (Spain).

Barceló-Vidal C, Martín-Fernández JA and Pawlowsky-Glahn V 2001 Mathematical foundations of compositional data analysis. In *Proceedings of IAMG'01 – The VII Annual Conference of the International Association for Mathematical Geology* (ed. Ross G). Kansas Geological Survey, Cancun (Mexico). 20 p.

Billheimer D, Guttorp P and Fagan W 2001 Statistical interpretation of species composition. *Journal of the American Statistical Association* **96**(456), 1205–1214.

Eaton ML 1983 *Multivariate Statistics. A Vector Space Approach*. John Wiley & Sons, Ltd, New York, NY (USA).

Egozcue JJ and Pawlowsky-Glahn V 2005 Groups of parts and their balances in compositional data analysis. *Mathematical Geology* **37**(7), 795–828.

Egozcue JJ, Pawlowsky-Glahn V, Mateu-Figueras G and Barceló-Vidal C 2003 Isometric logratio transformations for compositional data analysis. *Mathematical Geology* **35**(3), 279–300.

Gueorguieva R, Rosenheck R and Zelterman D 2008 Dirichlet component regression and its applications to psychiatric data. *Computational Statistics and Data Analysis* **52**, 5344–5355.

Hijazi RH and Jernigan RW 2009 Modelling compositional data using Dirichlet regression models. *Journal of Applied Probability and Statistics* **4**(1), 77–91.

Kolmogorov 1933 *Grundbegriffe der Wahrscheinlichkeitsrechnung*. Springer, Berlin (Germany).

Kolmogorov AN and Fomin SV 1957 *Elements of the Theory of Functions and Functional Analysis*, vols. 1 and 2. Dover Publications, Inc., Mineola, NY (USA).

Martín-Fernández J, Barceló-Vidal C and Pawlowsky-Glahn V 1998 A critical approach to non-parametric classification of compositional data. In *Advances in Data Science and Classification*. In *Proceedings of the 6th Conference of the International Federation of Classification Societies (IFCS-98)* (ed. Rizzi A, Vichi M and Bock H). Springer-Verlag, Berlin (Germany). pp. 49–56.

Mateu-Figueras G 2003 *Models de distribució sobre el símplex*. PhD thesis, Universitat Politècnica de Catalunya, Barcelona (Spain).

Mateu-Figueras G and Pawlowsky-Glahn V 2005 The Dirichlet distribution with respect to the Aitchison measure on the simplex – a first approach. In *Proceedings of CoDaWork'05, The 2nd Compositional Data Analysis Workshop* (ed. Mateu-Figueras G and Barceló-Vidal C). http://ima.udg.es/Activitats/CoDaWork05/. University of Girona, Girona (Spain).

Mateu-Figueras G and Pawlowsky-Glahn V 2007 The skew-normal distribution on the simplex. *Communications in Statistics – Theory and Methods* **36**(9), 1787–1802.

Mateu-Figueras G and Pawlowsky-Glahn V 2008 A critical approach to probability laws in geochemistry. *Mathematical Geosciences* **40**(5), 489–502.

Mateu-Figueras G, Pawlowsky-Glahn V and Barceló-Vidal C 2005 The additive logistic skew-normal distribution on the simplex. *Stochastic Environmental Research and Risk Assessment (SERRA)* **19**(3), 205–214.

Pawlowsky-Glahn V 2003 Statistical modelling on coordinates. In *Proceedings of CoDaWork'03, The 1st Compositional Data Analysis Workshop* (ed. Thió-Henestrosa S and Martín-Fernández JA). http://ima.udg.es/Activitats/CoDaWork03/. University of Girona, Girona (Spain). CD-ROM.

Pawlowsky-Glahn V and Egozcue JJ 2001 Geometric approach to statistical analysis on the simplex. *Stochastic Environmental Research and Risk Assessment (SERRA)* **15**(5), 384–398.

Pawlowsky-Glahn V and Egozcue JJ 2002 BLU estimators and compositional data. *Mathematical Geology* **34**(3), 259–274.

Pearson K 1897 Mathematical contributions to the theory of evolution. On a form of spurious correlation which may arise when indices are used in the measurement of organs. *Proceedings of the Royal Society of London* **LX**, 489–502.

4

Dealing with zeros

Josep Antoni Martín-Fernández[1], Javier Palarea-Albaladejo[2] and Ricardo Antonio Olea[3]

[1] *Department of Computer Science and Applied Mathematics, University of Girona, Spain*
[2] *Biomathematics and Statistics Scotland, JCMB, Edinburgh, UK*
[3] *US Geological Survey, USA*

4.1 Introduction

Since the 1980s, progress in compositional data analysis – hereafter CODA – has been based on the log-ratio methodology. Both ratios as well as logarithms are operations that require non-zero elements in the data matrix. As a consequence, any analysis of a vector of components must be preceded by a treatment of the zeros. In the past, some authors have tried to escape this requirement by exploring different ways to deal with zeros. Recently, Wang *et al.* (2007) proposed a hyperspherical transformation instead of the ordinary log-ratio family. The hyperspherical approach has been the subject of controversy in the CODA literature – see e.g. discussion in Aitchison (1982, p. 175) or in Aitchison (1986, p. 417). Basically, the hyperspherical transformation moves the data into another constrained space which is topologically different to the simplex. Once there, it is not possible to find a counterpart in this hypersphere – actually the positive orthant of the unit hypersphere – to the well-known simplicial operations. In addition, the method fails when one part equals 1, and all others equal 0. Butler and Glasbey (2008) introduced a treatment of zeros that models compositional data as censored observations from a latent multivariate normal distribution. Unfortunately, the principles – scale invariance and subcompositional coherence – that characterize CODA, as well as the vector space structure of the simplex, are violated by these strategies. Consequently, in some cases, the results can be misleading (Aitchison 1986; Martín-Fernández *et al.* 2003).

Compositional Data Analysis: Theory and Applications, First Edition. Edited by Vera Pawlowsky-Glahn and Antonella Buccianti.
© 2011 John Wiley & Sons, Ltd. Published 2011 by John Wiley & Sons, Ltd.

After the initial approaches to zeros proposed by Aitchison's monograph (Aitchison 1986), several authors have contributed to solve the 'zero problem' according to CODA principles. New ideas and strategies to deal with zeros have been presented at the three CoDaWork meetings (2003, 2005, 2008). The recent papers which exclusively focused on zeros are by Martín-Fernández *et al.* (2003), Martín-Fernández and Thió-Henestrosa (2006), Palarea-Albaladejo *et al.* (2007) and Palarea-Albaladejo and Martín-Fernández (2008). Nevertheless this topic is far from being exhausted. Probably there is not a general approach to the 'zero problem' because the presence of zero values in a compositional data set can be due to multiple and different reasons. The underlying reason itself is informative and determines the approach to be applied. In this work, no novel techniques are given to the reader; only some details are new contributions. Nevertheless, this chapter offers an updated account of recent techniques to deal with each particular kind of zero: rounded, count, and essential. The following sections describe these methods separately. Finally, in the discussion we outline some of the theoretical limitations of the approaches and discuss some possible ways in which these techniques could be improved in the future.

4.2   Rounded zeros

This kind of zero mostly appears in those studies where the variables are continuous (e.g. percentages by weight, time, expenditure, or length). Sometimes the number of significant digits means that very small observed percentages are rounded to zero. That is, actually a zero value in the data matrix is not a true zero, but rather represents an observed value below a particular maximum possible rounding-off error ($\varepsilon = 10^{-d}$). This kind of zero is called a *rounded zero*. In other situations there are very small values that cannot be recorded. Usually, this happens because the true value cannot be measured or observed due to the low concentration of a substance or element. In these cases, the value included in the data matrix is a zero or an annotation like '$< \varepsilon_{ij}$', where ε_{ij} is the threshold or detection limit of the measuring process applied to variable j in composition i. In this case, the unknown value is called a *below-detection value*. Note that, again, the true value is unknown but the information about its possible maximum value is available. Since essentially the framework and the treatment are the same for both rounding zeros and below-detection values then, in this work, we use indistinguishably both names.

Most of the approaches for dealing with zeros that have been proposed in the past have been inspired by methods for dealing with missing data. Actually, in some cases, there is a strong connection between both concepts. Missing data are usually classified into three types (Little and Rubin 2002): MCAR, MAR and NMAR. One type is Missing Completely at Random (MCAR), when the reason why a value is missing is independent of all other variables and parameters of interest in the study. A MAR (Missing At Random) value assumes that the probability to be missing depends on the observed components of the vector but not on the missing information. The NMAR (Not Missing At Random) class appears when the probability that one value is missing may depend on the missing part itself. This case implies that the mechanism behind the missing value cannot be ignored and special models and methods are usually required. In CODA, the rounded zeros are a particular NMAR case: data cannot be observed because their values are below a known value ε_{ij}.

In a broad sense, techniques for dealing with missing data can be classified into non-parametric and parametric methods. This classification, also applicable to the zeros case, assumes that the first type of techniques, the group of non-parametric techniques for rounded zeros, consists essentially of a family of imputation strategies. Here imputation means a replacement strategy that inserts a small quantity for each zero and completes in some way the incomplete data set. On the other hand, the group of parametric techniques for rounded zeros [expectation-maximization (EM) algorithm], rely on fully parametric models for multivariate data, usually the normal distribution on the simplex (Mateu-Figueras and Pawlowsky-Glahn 2008).

4.2.1   Non-parametric replacement of rounded zeros

One should be careful in using a replacement strategy because the covariance structure and the metric properties of the data set could be seriously distorted. Note that this distortion may result in misleading further analyses. Both from theoretical and practical viewpoints, multiplicative replacement shows better behaviour (Martín-Fernández *et al.* 2003) than other previously proposed imputation techniques: additive replacement and simple substitution. The interested reader may refer to Martín-Fernández and Thió-Henestrosa (2006) for further details on imputation techniques.

The multiplicative strategy simply consists of replacing each rounded zero in the composition by an appropriate small value δ_{ij} and, then, modifying the non-zero values in a multiplicative way. This modification is forced by the sum-constraint requirement of a composition. In more precise terms, let $\mathbf{x}_i$ be a D-composition $\mathbf{x}_i = (x_{i1}, \ldots, x_{iD})$ containing rounded zeros. The composition $\mathbf{x}_i$ is replaced by a new composition $\mathbf{xr}_i = (xr_{i1}, \ldots, xr_{iD})$ according to the formula

$$
xr_{ij} = \begin{cases} \delta_{ij} & \text{if } x_{ij} = 0, \\[2ex] x_{ij}\left(1 - \dfrac{\sum\limits_{k|x_{ik}=0} \delta_{ik}}{c_i}\right) & \text{if } x_{ij} > 0, \end{cases} \tag{4.1}
$$

where c_i is usually the constant of the sum-constraint; for example, $c_i = 100$ when data are percentages. For those cases that the replacement (4.1) is applied to the raw data, the constant c_i is also equal to the sum of the observed components $c_i = \sum_{j|x_{ij}\neq 0} x_{ij}$.

In Martín-Fernández and Thió-Henestrosa (2006, p. 200), the authors conclude 'The multiplicative replacement appears to be the easier, faster, more coherent and natural formula for substitution of the rounded zeros by appropriate small values'. Obviously, one important question immediately arises: how to select an appropriate δ_{ij} small value? When the zero is a small observed value rounded to zero, this δ_{ij} value must be lower than the maximum rounding-off error ε_{ij}. Because in practice this error is usually the same for all the variables and compositions in the study, then a single common δ value is selected. On the other hand, for the below-detection values case this δ_{ij} value must be lower than the threshold or detection limit ε_{ij}. Usually the variables have different detection limit ε_j but all compositions $\mathbf{x}_i$ have the same threshold therefore a common δ_j is selected for each component. But sometimes, it happens that different compositions were measured by different observers (laboratories) whose detection limits were different. In these situations a particular δ_{ij} must be selected for each variable and composition. Taking into account all these considerations, and after the analysis of an extensive case study, Martín-Fernández *et al.* (2003) found that when the proportion of these zeros is not large (less than 10% of the values in the data matrix) a replacement which uses an imputation value equal to 65% of the threshold value minimizes the distortion of the covariance structure. Note that the analyst applying non-parametric replacement decides which small value δ_{ij} is used. In below-detection values contexts, the analyst could have external information on unobserved data. But this is not a common situation in the rounded zeros case, thus a sensitivity analysis of δ_{ij} is typically needed (Martín-Fernández and Thió-Henestrosa 2006).

Let us use an example to illustrate this replacement and to show its coherence with one crucial operation – subcomposition. Butler and Glasbey (2008) analysed the subcompositon (protein, fat, carbohydrate) from eight food groups. The subcomposition of the swordfish is **sx** = (83.16, 16.84, 0). Note that the per cent of carbohydrate is entered as zero. These data come from the Release 19 of the USDA National Nutrient Database for Standard Reference (Agricultural Research Service 2010). In the data documentation files, it is specified that the nutrient value is rounded using two decimal places. Furthermore, the documentation details that, actually, the carbohydrate, when present, was determined as the difference between 100 and the sum of the percentages of water, protein, total lipid (fat), ash, and also, when present, alcohol. Since alcohol is not present in the swordfish case, then the full composition (water, protein, fat, ash, carbohydrate) is **x** = (75.62, 19.8, 4.01, 1.48, 0). Note that, as it is frequent in practical cases, the rounding-off error prevents the sum of the full composition from being exactly equal to 100. In all adequate CODA techniques, however, the multiplicative replacement is a scale invariant method and, therefore, this error does not have an effect on the results.

Suppose there are two researchers: J_1 and J_2. Researcher J_1 applies the replacement to the full composition **x** and then makes the subcomposition. On the other hand, researcher J_2 first makes the subcomposition **sx** and then replaces the zeros. J_1 considers that the zero value in the swordfish is a rounded zero and its associated maximum rounding-off error is $\varepsilon = 10^{-2}$%. In consequence J_1 takes $\delta_1 = 0.65 \times 10^{-2}$ in Equation (4.1). After this, J_1 obtains the full composition $\mathbf{xr}_1 = (75.615, 19.799, 4.01, 1.48, 6.5 \times 10^{-3})$ and its subcomposition $\mathbf{sxr}_1 = (83.136, 16.837, 2.729 \times 10^{-2})$, both without zeros. Researcher J_2 initially makes the subcomposition **sx** = (83.158, 16.842, 0). J_2 should be careful selecting the small value δ_2 for the imputation in Equation (4.1) because the rounding-off error produced in the full composition is propagated to the subcomposition. In our case, J_2 should use

$$\delta_2 = \frac{\delta_1}{(\text{Protein+Fat})(1 - \delta_1) + \delta_1}. \tag{4.2}$$

Note that the value of δ_2 may vary between foodstuffs even if δ_1 does not. In our example, applying this formula the result is $\delta_2 = 2.729 \times 10^{-2}$. After the multiplicative replacement, the corresponding subcomposition without zeros $\mathbf{sxr}_2$ is exactly equal to $\mathbf{sxr}_1$, the subcomposition obtained by researcher J_1. Note that in Equation (4.2), when the components protein and fat are major elements, i.e. (Protein+Fat) is close to 1, then δ_2 tends toward δ_1. On the other hand, the lower the value of (Protein+Fat), the higher the value of δ_2. Equation (4.2) is easily generalized to higher dimensionality where samples could have a larger number of zeros. Note that the coherence with the operation subcomposition is achieved because the basis of the method is multiplicative. Neither the additive method nor the simple substitution method exhibit this behaviour (Martín-Fernández et al. 2003).

In the above replacement, which uses an imputation value equal to 65% of the threshold value, some kind of sensitivity analysis (Martín-Fernández and Thió-Henestrosa 2006) is required to analyse to what extent conclusions depend on the imputation value introduced by (4.1). Furthermore, in those cases where the multiplicative replacement (4.1) imputes exactly the same value in all the zeros of the component, this replacement introduces artificial correlation between components which have zeros in the same rows of the data set. This undesirable effect might distort the results of the subsequent multivariate analysis as the number of zeros in the data set gets larger (Palarea-Albaladejo *et al.* 2007). These drawbacks motivate the development of parametric methods for rounded zeros.

4.2.2   Parametric modified EM algorithm for rounded zeros

In Palarea-Albaladejo *et al.* (2007), the main strategies for rounded compositional zeros were reviewed and compared. The EM algorithm is a well-known tool for maximum likelihood problems involving unobserved data in a real space. However, in its standard formulation, it is not able to deal with compositional rounded zeros. For this reason, Palarea-Albaladejo *et al.* (2007) introduced a modification of the common EM algorithm which, in combination with the additive log-ratio (alr) transformation, generates suitable estimates for the values below the detection limit. This method, which is independent of the selected divisor in the alr transformation, replaces unobserved values by small values that are imputed conditionally on the information included in the observed data.

Consider a compositional data set $\mathbf{X} = [x_{ij}]$ with n observed samples (rows) and D components (columns) that includes rounded zeros or below-detection values. Initially, the data set $\mathbf{X}$ must be alr-transformed into an unconstrained real data set $\mathbf{Y} = [y_{ij}]$. Note that if a below-detection values occur when $x_{ij} < \varepsilon_{ij}$, where ε_{ij} is the detection limit, its alr-transformed value y_{ij} is missing data in $\mathbf{Y}$. The unknown value y_{ij} then verifies $y_{ij} < ln(\varepsilon_{ij}/x_{iD}) = \psi_{ij}$. The procedure assumes that random compositions $\mathbf{x}$ are distributed according to an additive logistic normal (aln) model (Aitchison 1986), i.e. real random vectors $\mathbf{y} = \mathrm{alr}(\mathbf{x})$ are distributed according to a $(D-1)$-dimensional normal distribution with mean $\boldsymbol{\mu}$ and covariance $\boldsymbol{\Sigma}$. Then, on the tth iteration, the modified E step replaces the values y_{ij} in data set $\mathbf{Y}$ by

$$
y_{ij}^{(t)} = \begin{cases} y_{ij} & \text{if } y_{ij} \geq \psi_{ij}, \\ \mathrm{E}[y_{ij}|\mathbf{y}_{i,-j}, y_{ij} < \psi_{ij}, \boldsymbol{\mu}^{(t)}, \boldsymbol{\Sigma}^{(t)}] & \text{if } y_{ij} < \psi_{ij}, \end{cases} \tag{4.3}
$$

where $\mathbf{y}_{i,-j}$ denotes the set of observed variables for the row i of the data matrix $\mathbf{Y}$. Respectively denoting by ϕ and Φ the normal density and distribution functions, the expectation in (4.3) is equal to

$$
\mathrm{E}[y_{ij}|\mathbf{y}_{i,-j}, y_{ij} < \psi_{ij}] = \mathbf{y}_{i,-j}^{\mathsf{T}}\boldsymbol{\beta}_j - \sigma_j \frac{\phi\left(\frac{\psi_{ij} - \mathbf{y}_{i,-j}^{\mathsf{T}}\boldsymbol{\beta}_j}{\sigma_j}\right)}{\Phi\left(\frac{\psi_{ij} - \mathbf{y}_{i,-j}^{\mathsf{T}}\boldsymbol{\beta}_j}{\sigma_j}\right)}, \tag{4.4}
$$

where σ_j^2 is the conditional variance of variable $\mathbf{y}_j$, and $\boldsymbol{\beta}_j$ denotes the vector of coefficients of the linear regression of y_{ij} on $\mathbf{y}_{i,-j}$.

Palarea-Albaladejo and Martín-Fernández (2008) showed that the modified EM algorithm improves the behaviour of the non-parametric multiplicative replacement as the proportion of zeros grows. In this case the modified EM algorithm significantly introduces lower bias and better estimations of the variability of the data set. The method imputes values only based on the observed data information and on the threshold values. In addition, it deals with different detection limits for different samples and/or components in a straightforward manner. Even so, the multiplicative replacement could produce similar results when dealing with small samples or with a low number of rounded zeros.

We illustrate the functioning of the modified EM algorithm in a geochemical setting dealing with data from Montero-Serrano *et al.* (2010). In the study, CODA is applied to introduce and test a statistical protocol for the identification of chemofacies in the studied geological sections. The data consist of 96 samples of a 23-component geochemical composition coming from La Paloma stream which traverses the Cerro Pelado Fm., NW Venezuela. The observed composition, measured in µg/g, is:

(Cr, Zn, B, P, Mn, V, Cu, Ti, Ni, Y, Sr, La, Ce, Ba, Li, K, Rb, Fe, Mg, Ca, S, TOC, $CaCO_3$).

Due to the limitations of the measurement equipment, a number of values below the different detection limits could not be reported. Table 4.1 summarizes the different patterns of below-detection values found in the data set (16 patterns in total).

Non-observed values are mainly concentrated in the components Cu (28.12%) and Ni (51.04%), although the overall percentage of below-detection values is not high (4.48%). On the other hand, we can find several components – P, V, Ti, and others – that are free of below-detection values in all the patterns. These components are ideal for use as denominators in the alr transformation within the modified EM algorithm. More than half of the samples (56.25%) contain at least one below-detection value. The patterns containing a higher number of components with below-detection values include numbers 8, 10 and 13 – four out of 23 components each – which represent 3.12% of the samples.

After the modified EM algorithm is applied, the resulting replaced data set is displayed by means of a biplot of the clr-transformed data. Namely, they are represented in the space of the two first axes [Figure 4.1(a)] and in the space of the first and third axes [Figure 4.1(b)] as well. These three axes retain 75% of the original data variability. Note that, due to the high sample variability observed in the data, the principal components from the biplot were obtained using robust methods (Filzmoser *et al.* 2009). As it can be seen, the imputed samples (○) are well blended with the complete samples (●), not exhibiting any remarkable trend or bias. Logically, compositions containing small values are closer to the limits of the simplex space. Hence, most of the samples located at the boundaries of the data in Figure 4.1 correspond to replaced samples.

Two main drawbacks are inherent to the modified EM algorithm. First, the alr transformation cannot be applied when every component contains zeros. For this case, Palarea-Albaladejo and Martín-Fernández (2008) suggested a sequential strategy for applying the algorithm to subsets of samples with some common component free of zeros. If there is some component containing very few zeros, another alternative could be to apply a multiplicative replacement (4.1) to such a component and then use it as alr divisor. Secondly, we have to assume multivariate alr normality, but violations of this assumption may arise from different reasons. For example, it is possible to have a strong asymmetry introduced by the simultaneous presence of major and minor elements; or there could be clusters in the data set. Another situation where the alr normal model could not be appropriate is in studies where the data are discrete

Table 4.1 Pattern of values below detection limit (1 indicates observed values, 0 indicates non-observed values). The rightmost column refers to the percentage of samples exhibiting each pattern. The bottom two rows refer, respectively, to the percentage of below-detection values by component and to the vector of threshold values provided by the laboratory (in $\mu g/g$). In the lower right corner, the overall percentage of below-detection values is reported.

Pattern	Cr	Zn	B	P	Mn	V	Cu	Ti	Ni	Y	Sr	La	Ce	Ba	Li	K	Rb	Fe	Mg	Ca	S	TOC	CaCO3	% sam/pat
1	1	1	1	1	1	1	1	1	1	1	1	1	1	1	1	1	1	1	1	1	1	1	1	43.75
2	1	1	1	1	1	1	1	1	1	1	1	1	1	1	1	1	1	1	1	1	0	1	1	1.04
3	1	1	1	1	1	1	1	1	0	1	1	1	1	1	1	1	1	1	1	1	1	1	1	20.83
4	1	1	1	1	1	1	0	1	1	1	1	1	0	1	1	1	1	1	1	1	1	1	1	1.04
5	1	1	1	1	1	1	1	1	1	1	1	1	1	1	1	0	1	1	0	1	1	1	1	1.04
6	1	1	1	1	1	1	0	1	0	1	1	1	1	1	1	1	1	1	1	1	1	1	1	3.13
7	1	1	1	1	1	1	0	1	0	1	1	1	1	1	1	1	1	1	1	1	1	1	1	14.58
8	1	1	1	1	1	1	0	1	0	1	1	1	1	1	1	1	1	1	1	1	0	1	1	1.04
9	1	1	1	1	1	1	0	1	0	1	1	1	1	1	1	0	0	1	1	1	0	0	1	1.04
10	1	1	1	1	1	1	0	1	0	1	1	1	1	1	1	1	0	1	1	1	1	1	1	1.04
11	1	1	1	1	1	1	0	1	0	1	1	0	1	1	1	1	1	1	1	1	1	1	1	3.13
12	1	1	1	1	0	1	1	1	0	1	1	1	1	1	1	1	1	1	1	1	1	1	1	3.13
13	1	1	1	1	0	1	0	1	0	0	0	1	1	1	1	1	1	1	1	1	1	1	1	1.04
14	1	1	1	1	0	1	0	1	0	1	1	1	1	1	1	1	1	1	1	1	1	1	1	2.08
15	1	0	1	1	1	1	1	1	1	1	1	1	1	1	1	1	1	1	1	1	1	1	1	1.04
16	0	1	0	1	1	1	1	1	1	1	1	1	1	1	1	1	1	1	1	1	1	1	1	1.04
% vbdl	1.04	1.04	1.04	0	6.25	0	28.12	0	51.04	1.04	1.04	3.12	1.04	0	0	2.08	2.08	0	1.04	0	2.08	1.04	0	% tot. vbdl
dl	1.96	3.61	1.22	4.6	1	1	2.21	32	6.34	1	0.6	1	1.12	3	2	632	9.96	9	1	2	100	1	100	4.48

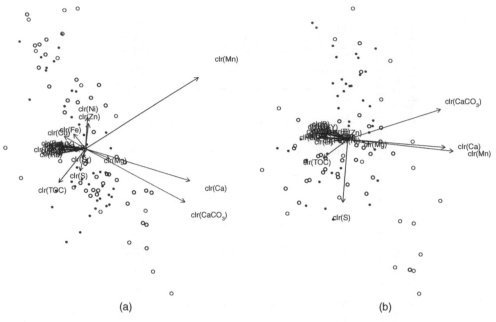

Figure 4.1   Biplot representation of the modified EM-replaced data set: first and second axes (a) and first and third axes (b). Filled circles (•) represent complete samples; empty circles (○) represent imputed samples.

'count data', that is, those cases where the components do not come from measures of continuous components (e.g. weight or time).

4.3   Count zeros

A count is defined as the number of times an event occurs. Vectors of counts are categorical data in which the counts represent the numbers of items falling into each of several categories. For problems where the analyst is not interested in the total sum of the vector, the study should be based on compositional techniques. For example, Pierotti *et al.* (2009) deal with compositional count data when they analyse individual male preferences in female colour polymorphic species of East African cichlids of three colour morphs: P, plain type; WB, white background; and OB, orange background. The experiment consisted of letting one male court one female of each morph inside a transparent, cylindrical aquarium for 15 min. A male courts females by displaying his fins laterally in front of a female within a short distance. The metric being investigated was the proportion of courtships by the males for each colour morph. For each of the 26 males, 6–12 trials were performed with a different set of three females each from a colour morph. Over 204 successful trials were obtained, forming a compositional data set of 204 3-compositions (P, WB, OB). *Count zeros* are present because the male, within a certain trial, did not court a particular female. But in another trial, the same male may exhibit courting, suggesting that in trials with no courtship the cause may be the limit on time (15 min).

Then, a new type of zero related to a sampling problem has arisen, because components may be unobserved due to the limited size of the sample. It is not understood as an essential zero – see next section – because in another situation that component may be greater than zero. Pierotti *et al.* (2009) applied a treatment – first introduced in Daunis-i-Estadella *et al.* (2008) – based on a *Bayesian-multiplicative* approach. This name is due to the two methodologies involved in the treatment, which, respectively, come from Walley (1996) and Martín-Fernández *et al.* (2003).

Let c_i be a counts vector – row – with D categories in a data set $\mathbf{C}$. Let T_i be the total count in c_i and θ_i its associated parameter vector of probabilities from a multinomial distribution. The prior distribution for θ_i is the conjugate distribution of the multinomial: a Dirichlet distribution with parameter vector α_i, where $\alpha_{ij} = s_i p_{ij}$, $j = 1, \ldots, D$. The vector p_i is the a priori expectation for θ_i and the scalar s_i is known as the *strength* of that prior.

From Bayes theorem, after one sample vector of counts c_i is collected, the posterior Dirichlet distribution for θ_i takes a new parameter vector α^*_i, where $\alpha^*_{ij} = c_{ij} + s_i p_{ij} = c_{ij} + \alpha_{ij}$ and the posterior estimation for θ_{ij} is

$$\widehat{\theta}_{ij} = \frac{c_{ij} + s_i p_{ij}}{\sum_{k=1}^{D}(c_{ik} + s_i p_{ik})} = \frac{c_{ij} + \alpha_{ij}}{T_i + s_i}. \tag{4.5}$$

Almost all proposed priors (Bernard 2005) for a fixed T_i are symmetric Dirichlet priors with p_i the uniform vector, i.e. the centre of the simplex $p_i = (1/D, \ldots, 1/D)$, and different possibilities for the 'strength'.

There are other priors, like the Berger–Bernardo reference priors, or those based on the imprecise Dirichlet model, but none of them satisfies all desirable principles: symmetry, embedding, representation invariance, stopping rule, and likelihood – see more details in Bernard (2005). Although Table 4.2 shows the most commonly used priors, the final selection is a decision of the analyst, who must combine knowledge about the priori probabilities of components (vector p) with the strength of this knowledge. Note that subscripts in Equation (4.5) indicate that different prior probabilities and strengths for each component and for each row in the data set can be applied.

When the Haldane prior is selected, an estimation based exclusively on the observed counts is assumed. The other priors consist of adding an amount to each count of the vector. Perks, Jeffreys and Bayes–Laplace, respectively, add $1/D$, $1/2$ and 1 to each count. Therefore, the

Table 4.2 Proposed Dirichlet priors and corresponding posterior estimation $\widehat{\theta}_{ij}$.

Prior	s_i	α_{ij}	$\widehat{\theta}_{ij}$
Haldane	0	0	$\frac{c_{ij}}{T_i}$
Perks	1	$\frac{1}{D}$	$\frac{c_{ij}+1/D}{T_i+1}$
Jeffreys	$\frac{D}{2}$	$\frac{1}{2}$	$\frac{c_{ij}+1/2}{T_i+D/2}$
Bayes–Laplace	D	1	$\frac{c_{ij}+1}{T_i+D}$

total count increases by 1, $D/2$ and D, respectively. This additive modification of the vector of counts does not preserve ratios between non-zero components, distorting the covariance structure (Martín-Fernández et al. 2003). When the counts involved are small, the changes in its ratios may be large. For example, consider that $T = 6$ with $D = 3$ categories and an observed vector of counts $\mathbf{c} = (4, 0, 2)$, where the ratio between the first and third components equals 2. The Bayesian estimate of the composition $\mathbf{x} = (2/3, 0, 1/3)$ is

$$\mathbf{xr} = \left(\frac{4 + \alpha_1}{6 + s}, \frac{0 + \alpha_2}{6 + s}, \frac{2 + \alpha_3}{6 + s} \right).$$

Observe that the ratio between the first and the third elements $\frac{4+\alpha_1}{2+\alpha_3}$ may be lower than 2 (e.g. 5/3 with the Bayes–Laplace prior) and only remains equal to 2 for the Haldane prior. Unfortunately, with that prior, the posterior estimate of a zero count is still zero.

The Bayesian prior technique is improved in order to both estimate the zero and achieve the ratio preservation. This improvement is based on the multiplicative replacement (4.1) and consists on estimating the count composition $\mathbf{x}_i$ by $\mathbf{xr}_i$ applying the expression

$$xr_{ij} = \begin{cases} \frac{\alpha_{ij}}{T_i + s_i} & \text{if } x_{ij} = 0, \\ x_{ij}\left(1 - \sum_{k | x_{ik} = 0} \frac{\alpha_{ik}}{T_i + s_i}\right) & \text{if } x_{ij} > 0. \end{cases} \tag{4.6}$$

From the comparison of Equations (4.6) and (4.1), we conclude that the Bayesian-multiplicative treatment for count zeros coincides with the multiplicative replacement for $\delta_{ij} = \frac{\alpha_{ij}}{T_i + s_i}$. This fact suggests that all desirable properties and undesirable limitations (Martín-Fernández et al. 2003) achieved by replacement (4.1) are also satisfied by (4.6).

In the example from Pierotti et al. (2009), let $\mathbf{c} = (4, 0, 2)$ be a vector of, respectively, courtships from a particular male to the females (P, WB, OB). Applying (4.6) with Perks prior ($s = 1, \alpha_j = 1/3$) the posterior estimate is

$$\mathbf{xr} = \left(\frac{4}{6}(1 - \frac{1/3}{7}), \frac{1/3}{7}, \frac{2}{6}(1 - \frac{1/3}{7}) \right) = \left(\frac{40}{63}, \frac{1}{21}, \frac{20}{63} \right),$$

where the ratio between the first and third elements is preserved: $\frac{40}{63} / \frac{20}{63} = 2$. In terms of counts, one could interpret that the male has got enough time to perform $N^* = 63$ courtships and its count composition is $\mathbf{c}^* = (40, 3, 20)$.

In this way the ratios between non-zero components in count data set $\mathbf{C}$ and the unit-sum constraint in raw data set $\mathbf{X}$ are preserved. After the replacement, we have a new data set $\mathbf{XR}$ without zeros. We can also make a data set of counts without zeros, $\mathbf{CR}$, multiplying each row of $\mathbf{XR}$ by the total count of each row in $\mathbf{C}$. This strategy has the advantage of preserving the initial total count. Nevertheless, as the total count of the vector is not informative from a compositional point of view, both data sets $\mathbf{XR}$ and $\mathbf{CR}$ give the same information. Figure 4.2 shows the distribution of the initial proportions $\mathbf{X}$ and of the adjusted proportions $\mathbf{XR}$ in a ternary diagram. By comparing their univariate distribution and the ternary distribution (Figure 4.2), Pierotti et al. (2009) showed that no relevant differences arise between the data sets with zeros ($\mathbf{C}, \mathbf{X}$) and without zeros ($\mathbf{CR}, \mathbf{XR}$).

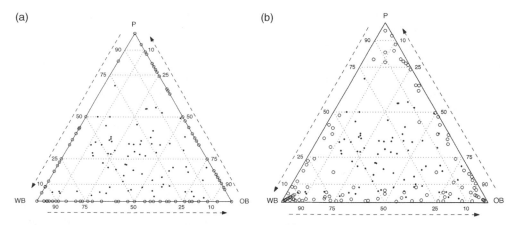

Figure 4.2   Distributions of initial **X** (a) and adjusted **XR** (b) data sets on ternary diagrams. Empty circles (o) represent compositions with count zeros in **C**; dots (●) represent compositions without zeros.

4.4   Essential zeros

Recent relevant contributions specifically focused on the treatment of this kind of zero are by Aitchison and Kay (2003) and Bacon-Shone (2003 2008). All of them consider that (Aitchison and Kay 2003, p. 1) 'by an essential zero we mean a component which is truly zero, not something recorded as zero simply because the experimental design or the measuring instrument has not been sufficiently sensitive to detect a trace of the component'. Other names for this kind of zero value are *absolute* or *structural* zeros. In any case, it is clear that treatments that try to replace the zero by a small value are not appropriate. Nevertheless, in most cases, the decision on whether a zero is essential or not is the responsibility of the CODA analyst. Once one has decided on this, then an appropriate treatment must be selected. For example, in the *household budget pattern* problem, some householders may report a zero in the 'alcohol' commodity group because they spend nothing during the period of observation. In this case, the CODA analyst could assume that the households are teetotal – essential zeros – or could decide that those households allow the drinking of alcohol, and then that the experimental design is the cause of the zeros. In this latter case, the zero in the raw data would be due to the 'short' period of observation or to a rounding-off error. Consequently, a treatment based on imputation as shown in the previous sections could be sensible. In Bacon-Shone (2003) and Fry and Chong (2005) the authors provide interesting discussions on this crucial decision about the kind of zeros we are dealing with.

In spite of the fact that, at the present moment, there is not a general methodology for dealing with essential zeros, the early approaches mainly suggest two different possible situations. On the one hand, essential zeros can be present in data sets where the components are continuous variables or percentages, for example, household or time budgets (Aitchison and Kay 2003). On the other hand, essential zeros can appear in discrete compositions of count data (Bacon-Shone 2008). A typical example of such data is a satisfaction survey, where the questions are the rows or samples and the number of responses in each item – typically from completely agree to completely disagree – are the components. In this case, a zero may represent, for example, a question about which none of the interviewees are in complete disagreement. Nowadays, in the first situation, the more promising approach is based on a binomial conditional logistic normal model (Aitchison and Kay 2003). For the second scenario, an approach based on the Poisson-Log Normal distribution may be more appropriate in some cases (Bacon-Shone 2008). Note that in both situations the analyst assumes that the zero is truly a zero, and is not a consequence of the experimental design or of the sample size. Then, the zero values are incorporated into the modelling process itself.

In some contexts, it is possible to interpret essential zeros in a certain component as indicators of two different subgroups of interest: observations with a value of zero in that component versus observation taking a positive value instead. To illustrate this situation we consider the results by municipalities of the 2006 parliamentary elections in Catalonia (Statistical Institute of Catalonia 2010). Among the 946 municipalities distributed through four electoral districts or provinces, we focus on the results from the province of Girona (221 municipalities). Because the number of seats in the Parliament of Catalonia assigned to each party is related to its percentage of valid votes, we consider the data as percentages where the components are the parties and the rows are the municipalities. In Catalonia the electoral law specifies that any party that obtains less than 3% of votes cannot have representation in Parliament. As a consequence, we focus our analysis on the subcomposition formed by the six parties with seats: CiU, PSC, PP, IC, ERC, and C's. The C's party participated for the first time in 2006 and there were 69 municipalities in Girona where the party obtained no votes. An interesting aspect is to analyse whether or not the distribution of the votes between the five main parties differs depending on the absence or not of party C's. In other words, to analyse whether the votes to C's affect the ratios between the main parties. The data set considered had 221 rows – municipalities – and six components – parties – where the component of C's includes 69 zeros. We consider that these zeros are essential because they reflect the fact that no elector in the municipality gave their vote to the party. The biplots in Figure 4.3 show that the relative relationship between the main parties depends on whether the C's party obtains votes or not. For example, the rays for the CiU and ERC parties completely change relative

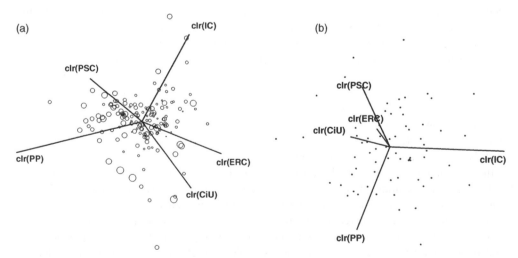

Figure 4.3   Biplots of the subcomposition of votes from the five main parties in municipalities where C's obtained votes (a) and where they did not (b).

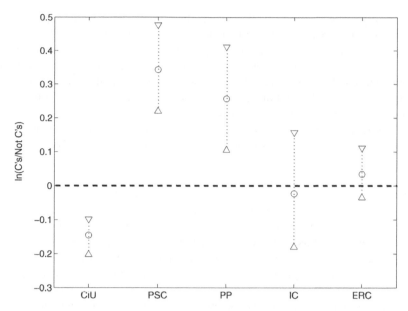

Figure 4.4   Bootstrap percentile-type confidence intervals of the log-ratios between the compositional geometric means of the municipalities with and without votes from the C's party: mean (○); 2.5% percentile (△); 97.5% percentile (▽).

to the ray of PSC. Since the percentage of variance explained by the two first axes is about 70%, one must be careful with the interpretations from these figures. In Figure 4.3(a) the size of the dots is proportional to the percentage of votes obtained by the C's party. No clear trend is present in the distribution of these dots, suggesting that this does not particularly affect any of the other parties.

We focus our analysis on the effect of the absence or presence of votes for the C's party. The compositional geometric mean of the subcomposition (*CiU*, *PSC*, *PP*, *IC*, *ERC*) in the subset of municipalities with votes in C's component is $g_{m1} = (45.63, 18.35, 5.34, 7.66, 23.02)$ whereas in the municipalities with essential zero in C's is $g_{m2} = (52.76, 13.02, 4.13, 7.83, 22.25)$. When log-ratios between the centres are formed, the resulting $\ln(g_{m1}/g_{m2}) = (-0.15, 0.34, 0.26, -0.02, 0.03)$ shows that in some parties the effect is bigger. In order to obtain a measure of variability, a bootstrap experiment was carried out. Figure 4.4 shows the result of the 1000 times resampling by means of the 2.5% and 97.5% percentiles of the log-ratios between the centres. The calculations suggest that the effect in the IC and ERC parties could be considered as not significant and that, on average, the CiU party gained the most in those municipalities where the C's obtained zero votes.

4.5   Difficulties, troubles and challenges

The essential zero is by far the more complicated case because the zero is informative by itself and determines all posterior analysis. Specific models combining zero and non-zero components in a convenient way are required, but a well-founded general approach to this problem has not yet been developed. Perhaps, if the parametric scenario appears to be too complicated, a non-parametric approach to address the treatment of essential zeros could be an alternative. In any case, from the work done up to now, it appears to be absolutely necessary to make a clear differentiation between essential zeros for both continuous compositions and discrete compositions. Although this differentiation is left to the analyst, some guidelines are advisable and more examples and discussion are welcomed. For example, in this paper we have assumed that the zeros in the 'votes' example were essential. However, the assumption of count zero might also be considered. For example, one could consider that the zero in the component is an effect of the abstention in the elections, that is, an effect of the sample size. Such kind of controversy suggests the need for theoretical studies to better define the hypotheses that are assumed in each case, as well as empirical studies to evaluate how sensitive the results actually are to those assumptions. The latter would be specially useful for practitioners in terms of assessing the qualitative conclusions drawn from a data analysis.

The treatment of the count zeros has found an appropriate procedure through the Bayesian-multiplicative strategy. In this scenario, it is necessary to provide a more detailed study in which the different choices for the hyper-parameters in the prior distribution are completely analysed. The analyst needs guidance about values that are more advisable for different situations. Despite the fact that almost all proposed priors are based on the 'uniform' vector, it is clear that analysts may incorporate their knowledge about the data through the prior vector. Simple examples might be those studies where the analyst has historical information about the total maximum per cent observed for a component. In these studies, the vector of priori parameters should be adapted to incorporate this information. Another aspect to improve within the Bayesian-multiplicative treatment is related to the observed values of the data set. In some situations, it might happen that after the treatment of a zero value, the transformed value may be a per cent greater than the minimum observed per cent in such component. Under the assumption that the data set does not include different subpopulations, this fact would be considered as misleading since one would expect a zero to be transformed into a small value. This possibility suggests that the procedure must be improved in order to cover all these situations.

At this time, most of the efforts on the treatment of zeros are concentrated on rounded zeros. Surely, this fact is due to the frequency with which rounded-off values and values below a detection limit arise in fields such as geochemistry, environmental sciences, and earth sciences in general, which are the traditional fields for the application of compositional methods. Both the multiplicative replacement method and the modified EM algorithm allow the analyst to incorporate information about detection limits. Although based on different philosophies, both methods are competent, well-founded methods for undertaking the task of replacing unobserved compositional values. Nevertheless they both, certainly with different intensity, tend to underestimate the data variability. This underestimation is a result of prefixed imputed values, in the case of multiplicative replacement, and the use of an expected value, for the modified EM method. Thus, improvements to minimize this effect are required. This issue may be specially important when, for example, the analysis is focused on inferring characteristics of distributional parameters. In this case, the estimated variability could be seriously distorted. Multiple imputation and resampling methods are well-known statistical proposals for accounting for the added uncertainly due to the presence of non-observed values (Little and Rubin 2002, section 10.2). A first approach to multiple imputation in the rounded zeros context was presented in Martín-Fernández *et al.* (2003). In addition, Olea (2008) recently introduced a univariate approach using bootstrap inference from samples containing below-detection values in which uncorrelated and spatially correlated resamples are used to numerically model the parameters of the distribution of geochemical elements, such as the median or the 75th percentile. Results reveal that the proposed strategy may be a promising way to better reproduce sample variability. However, further research is needed on this method, including extension of the approach to a multivariate realm.

Acknowledgements

This research has been supported by the Spanish Ministry of Science and Innovation (projects CSD2006-00032 and MTM2009-13272) and by the Agència de Gestió d'Ajuts Universitaris i de Recerca of the Generalitat de Catalunya (Ref. 2009SGR424). We would like to thank Michele Pierotti and Jean Carlos Montero-Serrano for kindly providing some of the data sets used in this work. We are indebted to Carles Barceló-Vidal, Adam Butler, Mark Engle, Glòria Mateu-Figueras and L. Andries van der Ark for comments and helpful suggestions.

References

Agricultural Research Service 2010 *USDA Nutrient Database for Standard Reference, Release 19.* US Department of Agriculture (Available from www.ars.usda.gov/nutrientdata), Washington DC (USA).

Aitchison J 1982 The statistical analysis of compositional data (with discussion). *Journal of the Royal Statistical Society, Series B (Statistical Methodology)* **44**(2), 139–177.

Aitchison J 1986 *The Statistical Analysis of Compositional Data.* Monographs on Statistics and Applied Probability. Chapman and Hall Ltd (reprinted 2003 with additional material by The Blackburn Press), London (UK). 416 p.

Aitchison J and Kay J 2003 Possible solution of some essential zero problems in compositional data analysis. In *Proceedings of CoDaWork'03, The 1st Compositional Data Analysis Workshop* (ed. Thió Henestrosa S and Martín-Fernández JA). http://ima.ud.es/Activitats/CoDaWork03/. University of Girona, Girona (Spain). CD-ROM.

Bacon-Shone J 2003 Modelling structural zeros in compositional data. In *Proceedings of CoDaWork'03, The 1st Compositional Data Analysis Workshop* (ed. Thió-Henestrosa S and Martín-Fernández JA). http://ima.ud.es/Activitats/CoDaWork03/. University of Girona, Girona (Spain). CD-ROM.

Bacon-Shone J 2008 Discrete and continuous compositions. In *Proceedings of CoDaWork'08, The 3rd Compositional Data Analysis Workshop* (ed. Daunis-i-Estadella J and Martín-Fernández JA). University of Girona, Girona (Spain). CD-ROM.

Bernard J 2005 An introduction to the imprecise Dirichlet model for multinomial data. *International Journal of Approximate Reasoning* **35**(2–3) 123–150.

Butler A and Glasbey C 2008 A latent Gaussian model for compositional data with zeros. *Journal of the Royal Statistical Society. Series C (Applied Statistics)* **57**, 505–520.

Daunis-i-Estadella J, Martín-Fernández JA and Palarea-Albaladejo J 2008 Bayesian tools for count zeros in compositional data. In *Proceedings of CoDaWork'08, The 3rd Compositional Data Analysis Workshop* (ed. Daunis-i-Estadella J and Martín-Fernández JA). University of Girona, Girona (Spain). CD-ROM.

Filzmoser P, Hron K and Reimann C 2009 Principal component analysis for compositional data with outliers. *Environmetrics* **20**, 621–632.

Fry TRL and Chong D 2005 A tale of two logits, compositional data analysis and zero observations In *Proceedings of CoDaWork'05, The 2nd Compositional Data Analysis Workshop* (ed. Mateu-Figueras G and Barceló-Vidal C). http://ima.udg.es/Activitats/CoDaWork05/. University of Girona, Girona (Spain).

Little RJA and Rubin DB 2002 *Statistical Analysis with Missing Data*, 2nd edition. John Wiley & Sons, Ltd, New York (USA). 381 p.

Martín-Fernández JA and Thió-Henestrosa S 2006 Rounded zeros: some practical aspects for compositional data. In *Compositional Data Analysis in the Geosciences: From Theory to Practice*, vol. 264. Geological Society, London, pp. 191–201.

Martín-Fernández JA, Barceló-Vidal C and Pawlowsky-Glahn V 2003 Dealing with zeros and missing values in compositional data sets using nonparametric imputation. *Mathematical Geology* **35**(3), 253–278.

Mateu-Figueras G and Pawlowsky-Glahn V 2008 A critical approach to probability laws in geochemistry. *Mathematical Geosciences* **40**(5), 489–502.

Montero-Serrano JC, Palarea-Albaladejo J, Martín-Fernández JA, Martínez-Santana M and Gutiérrez-Martín JV 2010 Multivariate analysis applied to chemostratigraphic data: identification of chemofacies and stratigraphic correlation. *Sedimentary Geology* **228**(3–4) 218–228.

Olea RA 2008 Inference of distributional parameters from compositional samples containing nondetects. In *Proceedings of CoDaWork'08, The 3rd Compositional Data Analysis Workshop* (ed. Daunis-i-Estadella J and Martín-Fernández JA). University of Girona, Girona (Spain). CD-ROM.

Palarea-Albaladejo J and Martín-Fernández JA 2008 A modified EM alr-algorithm for replacing rounded zeros in compositional data sets. *Computers and Geosciences* **34**(8), 902–917.

Palarea-Albaladejo J, Martín-Fernández JA and Gómez-García JA 2007 Parametric approach for dealing with compositional rounded zeros. *Mathematical Geology* **39**(7), 625–645.

Pierotti MER, Martín-Fernández JA and Seehausen O 2009 Mapping individual variation in male mating preference space: multiple choice in a colour polymorphic cichlid fish. *Evolution* **63**(9), 2372–2388.

Statistical Institute of Catalonia 2010 *Electoral Statistics: Catalonia, Counties and Municipalities*. Idescat (Available from www.idescat.cat), Barcelona (Spain).

Walley P 1996 Inferences from multinomial data: learning about a bag of marbles (with discussion). *Journal of the Royal Statistical Society, Series B (Statistical Methodology)* **58**, 3–57.

Wang H, Liu Q, Mok HMK, Fu L and Tse WM 2007 A hyperspherical transformation forecasting model for compositional data. *European Journal of Operational Research* **179**(2), 459–468.

5

Robust statistical analysis

Peter Filzmoser[1] and Karel Hron[2]

[1]*Department of Statistics and Probability Theory, Vienna University of Technology, Austria*
[2]*Department of Mathematical Analysis and Applications of Mathematics, Palacký University, Czech Republic*

5.1 Introduction

Robust statistical methods are nowadays widely used when one has to deal with data that are affected by artifacts such that classical methods fail. Robust methods allow for certain deviations from a predefined statistical model. Although a lot of work has been devoted to the development of robust statistical methods, their development for compositional data is still at the early stages. One of the reasons was the lack of a regular and isometric transformation from the simplex, the sample space of compositions, to the Euclidean real space. This problem was solved by introducing the isometric log-ratio (ilr) transformations (Egozcue *et al.* 2003), together with the concept of balances (Egozcue and Pawlowsky-Glahn 2005a 2005b), that are closely related to the concept of orthonormal bases on the simplex (with respect to the Aitchison geometry) (see Chapter 3). In fact, the ilr variables $\mathbf{z} = (z_1, \ldots, z_{D-1})^\top$ of a D-part composition $\mathbf{x} = (x_1, \ldots, x_D)^\top$ represent coefficients of an orthonormal basis on the simplex (in the sense of the Aitchison geometry). Both the theoretical and the practical advantages of ilr transformations overcame the original additive log-ratio (alr) and centred log-ratio (clr) transformations, defined in Aitchison (1986), as they correspond to an oblique basis and a generating system of compositions, respectively. Nowadays, it seems that the clr transformation is only important for the construction of the compositional biplot (Aitchison and Greenacre 2002), because it makes its interpretation possible in terms of the original compositional parts.

Compositional Data Analysis: Theory and Applications, First Edition. Edited by Vera Pawlowsky-Glahn and Antonella Buccianti.
© 2011 John Wiley & Sons, Ltd. Published 2011 by John Wiley & Sons, Ltd.

Since the time when the ilr transformations were proposed, several papers have been published that describe robustness aspects for the most frequently used multivariate statistical methods, like outlier detection (Filzmoser and Hron 2008), principal component analysis (Filzmoser *et al.* 2009a), factor analysis (Filzmoser *et al.* 2009c), discriminant analysis (Filzmoser *et al.* 2009b), or the estimation of missing values (Hron *et al.* 2010). Although each of these methods represents a specific use of the robust approach, all of them follow the idea of providing a statistical tool that is able to deal with compositional data sets including outliers.

The main task of this contribution is to provide a comprehensive view of robustness for compositional data. The chapter is organised as follows. Section 5.2 introduces basic concepts of robust statistics with focus on compositional aspects to robustness and outliers. Explanations are provided using real compositional data sets. Section 5.3 gives an overview of three robust methods, probably the most used by exploratory compositional data analysis (ECDA): outlier detection, principal component analysis (together with the concept of the compositional biplot) and discriminant analysis. In Section 5.4 the theoretical concepts are applied to real data sets from geochemistry. The chapter concludes with a summary in Section 5.5.

5.2   Elements of robust statistics from a compositional point of view

In general, robust statistics provides methods and estimators that emulate the classical ones, but that are less affected by outlying observations or other small departures from model assumptions. Classical statistical methods usually heavily rely on the underlying assumptions, like the underlying probability distribution, which are often not met in practice. When outliers occur in the data, the behaviour of such classical methods may be very poor. In order to quantify the robustness of a method, one usually uses the concepts of breakdown point and influence function. The breakdown point of an estimator is the proportion of arbitrary observations (outliers, other deviating points) that an estimator can handle before giving a nonsensical result. The influence function gives an idea of how an estimator behaves under small amounts of data contamination. The influence of such an infinitesimal contamination on a robust estimator is bounded, while it becomes unbounded for nonrobust estimators (see, e.g. Maronna *et al.* 2006).

An important task for multivariate statistics is the estimation of location and covariance. Consider a $(D - 1)$-dimensional sample of order n from the distribution of a composition $\mathbf{x}$ in coordinates $\mathbf{z}_1, \ldots, \mathbf{z}_n$. The usual way of estimation is to compute the arithmetic mean $\bar{\mathbf{z}} = \frac{1}{n} \sum_{i=1}^{n} \mathbf{z}_i$ and the sample covariance matrix $\mathbf{S} = \frac{1}{n} \sum_{i=1}^{n} (\mathbf{z}_i - \bar{\mathbf{z}})(\mathbf{z}_i - \bar{\mathbf{z}})^{\top}$. However, both estimators are very sensitive to outliers, and only small departures from the model assumptions can completely change their behaviour and consequently also the interpretation of the results. Several robust counterparts have been proposed in the literature, like the MCD or S estimator (Maronna *et al.* 2006). Concretely, the MCD (Minimum Covariance Determinant) estimator yields robust estimates of location and covariance with maximum breakdown point 50%. It is defined by those h observations that result in the smallest determinant of their sample covariance matrix, and taking $h \approx n/2$ yields the maximum breakdown point. The location estimator $\mathbf{t} = \mathbf{t}(\mathbf{z}_1, \ldots, \mathbf{z}_n)$ is the centre of these h observations, and the covariance estimator $\mathbf{C} = \mathcal{C}(\mathbf{z}_1, \ldots, \mathbf{z}_n)$ is given by their sample covariance matrix, multiplied by a factor for consistency at normal distributions.

An important property of the MCD estimator is its affine equivariance. In general, a location estimator $\mathbf{t}$ and a covariance estimator $\mathbf{C}$ are called affine equivariant, if for any nonsingular $(D - 1) \times (D - 1)$ matrix $\mathbf{A}$ and for any vector $\mathbf{b} \in \mathbf{R}^{D-1}$ the conditions

$$\mathbf{t}(\mathbf{Az}_1 + \mathbf{b}, \ldots, \mathbf{Az}_n + \mathbf{b}) = \mathbf{At}(\mathbf{z}_1, \ldots, \mathbf{z}_n) + \mathbf{b},$$

$$\mathbf{C}(\mathbf{Az}_1 + \mathbf{b}, \ldots, \mathbf{Az}_n + \mathbf{b}) = \mathbf{AC}(\mathbf{z}_1, \ldots, \mathbf{z}_n)\mathbf{A}^\top$$

are fulfilled. This implies that both location and covariance estimators obtained from the MCD estimator behave like the classical estimators arithmetic mean and the sample covariance matrix under affine transformations. The property of affine equivariance is of special importance for the robustification of multivariate methods for compositional data.

For almost all traditional statistical methods it is necessary to work in the proper geometry, i.e. in the Euclidean geometry. This is also the case for their robust counterparts. A robust treatment of the data cannot *repair* the inappropriate geometry that is caused by working with the raw or even log-transformed compositional data (i.e. the logarithm is applied to each of the compositional parts) (see, e.g. Aitchison 1986; Filzmoser and Hron (2008); Filzmoser *et al.*, 2009a). For example, the distances between the data points could be interpreted in a wrong way, just because some data points might appear far away from the data majority and thus appear as outliers. The picture could be completely different in the Aitchison geometry.

The idea of the relative scale is quite an intuitive concept of differences for this particular type of data. While the difference between 5% and 10% is the same as between 45% and 50%, the proportions show a quite contrasting relation, because 5% is half of 10%, while 45% is 0.9 of 50%. Thinking in terms of differences in ratios is natural for this kind of data, and the Aitchison distance works exactly in this way (Buccianti *et al.* 2006). A consequence of this concept is that for compositions near the boundary of the simplex the distances are in general higher than for compositions near the centre, although their Euclidean distances are the same. A more detailed discussion is provided for instance in Aitchison *et al.* (2000) and Hron *et al.* (2010).

This concept of distances is illustrated in the following on a real data example. The data set `haplogroups` from the R library `robCompositions` contains the distribution of the European Y-chromosome DNA (Y-DNA) haplogroups in 38 countries and regions in percentages. The human Y-chromosome DNA can be divided in genealogical groups sharing a common ancestor, and these are called haplogroups. The compositional parts x_1, x_2, x_3 contain the three main haplogroups corresponding to Mesolithic Europeans (Nordic, Celtic, Danubian, Basque, etc.), Mesolithic Eurasians (Uralo-Finnic, Balto-Slavic, Germanic, Italic, etc.) and Neolithic, Bronze and Iron Age immigrants (North and Eastern African, Near Eastern, Jewish, Arabic, etc.). Three-part compositional data like here can be displayed using the ternary diagram [Figure 5.1,(a)] . In most of the countries the Mesolithic Eurasians group dominates (the compositions are close to the corresponding vertex), however also some deviating data points occur, namely the observations 1 (Albania), 5 (Bulgaria), 6 (Croatia), 17 (Greece), 23–25 (South Italy, Sicily, Sardinia) and 34 (Serbia).

In Figure 5.1(b) the data are expressed in coordinates, chosen as

$$z_1 = \frac{1}{\sqrt{2}} \ln \frac{x_1}{x_2}, \qquad z_2 = \frac{\sqrt{2}}{\sqrt{3}} \ln \frac{\sqrt{x_1 x_2}}{x_3}. \tag{5.1}$$

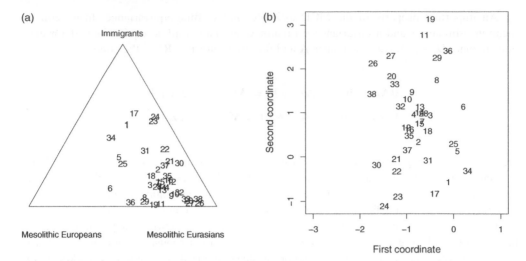

Figure 5.1 Distribution of European Y-DNA haplogroups in 38 countries and regions shown in a ternary diagram (a) and in coordinates (b).

The first coordinate thus explains the ratio between the first and the second haplogroup, while the second coordinate represents the remaining two ratios in the three-part composition. The group of observations that was deviating in the ternary diagram is also slightly deviating from the main data cloud in this presentation. However, now an additional group of observations forming its own cluster is visible, namely the countries 8 (Denmark), 11 (Finland), 19 (Iceland), 29 (Norway) and 36 (Sweden). This group was not identifiable as deviating in the ternary diagram. These observations represent countries (regions) with the negligible role of the Neolithic, Bronze and Iron Age immigrants haplogroup.

Although compositional data are mostly characterized by a constant sum constraint, like 100 in the case of percentages in the previous example, the sum of parts can in general be different, since by definition of compositions the relevant information is contained only in the ratios between them. For example, when sample materials are only analysed for some chemical elements but not analysed completely, the sum of the element concentrations of the different samples will in general not be the same. As an example we consider a real data set, available as `expendituresEU` in the R package `robCompositions` that contains mean consumption expenditures (in Euro) of households on 12 domestic year costs in all 27 member states of the European Union (2005). To see the effect, we focus on two selected costs, clothing and footwear versus expenditures on furnishings, household equipment and routine maintenance of the house. When the costs are displayed [see Figure 5.2(a)], there are some observations that clearly seem to be (Euclidean) outliers due to their high expenditures on both type of costs: L (Luxembourg), M (Malta), ILR (Ireland), and CY (Cyprus). However, due to the relative character of the data, we are rather interested in the variability of the ratio between both expenditures. Because all the data points follow approximately a straight line from the origin, we can expect quite a stable proportion. Indeed, this is clearly visible when the data are expressed in the corresponding coordinate [Figure 5.2(b)] and confirmed with the value $\text{var}(\frac{1}{\sqrt{2}}\ln\frac{x_1}{x_2}) = 0.034$. Moreover, in this presentation no real outliers are visible, only LT (Lithuania) is deviating slightly. The reason is that not the absolute values, but the ratios are

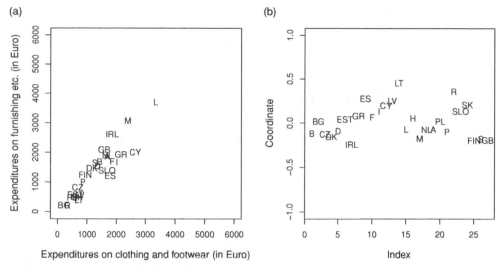

Figure 5.2 Expenditures on clothing and footwear versus furnishings, household equipment and routine maintenance of the house, displayed in the scatterplot (a) and in the corresponding coordinate (b).

of interest. Thus, consequently, if all the data would be shifted along the line from the origin, their ratio would not change and thus also the variance (and the Aitchison distance) of such compositions would be zero. More formally, here all the observed compositions are members of the corresponding equivalence class of a composition $\mathbf{x}$ (in general a D-part composition), $\mathbf{x} = \{c\mathbf{x},\ c \in \mathbf{R}^+\}$, which contain the same information. They are also called *compositionally equivalent*, see, e.g. of Hron *et al.* (2010) for further discussion.

Both concepts of the relative scale and the equivalence classes for compositions should be taken into account before robust statistical analyses, like the methods presented in the following, are carried out.

5.3 Robust methods for compositional data

When applying exploratory methods to multivariate data, one is usually interested in (a) the data quality, i.e. presence of outliers or of any other deviating observations; (b) relations between the variables and their relations to the observations; (c) the grouping structure of the data set, i.e. whether there are any significant patterns. The answers can be obtained by using appropriate statistical tools: outlier detection, principal component analysis (resulting in a biplot), and discriminant analysis. All these tools can also be applied to compositional data because of the orthonormal coordinates. However, for their robust versions there are some specifics that need to be considered. The first and crucial one, that is common to all the methods described in the following, is the use of affine equivariant estimators like the mentioned MCD estimator. Different orthonormal bases on the simplex result in orthonormal transformations of the corresponding coordinates, thus it is important that the robust methods are invariant to the choice of the balances that provide a reasonable interpretation of the coordinates. Another problem that needs to be solved is related to the clr coordinates (in fact coordinates to a generating system on the simplex), necessary to construct the compositional biplot according to Aitchison and Greenacre (2002). Because of the singularity of the resulting data, robust methods that rely on regular observations cannot be applied.

5.3.1   Multivariate outlier detection

A search for possible outliers should be the first step in each ECDA. Identified outliers are candidates for aberrant data that may otherwise adversely lead to model misspecification, biased parameter estimation, and incorrect results. It is therefore important to identify them prior to modelling and analysis. Outlier detection for compositional data is usually based on robust Mahalanobis distances, defined for regular $(D - 1)$-dimensional data $\mathbf{z}_i$ as

$$\mathrm{MD}(\mathbf{z}_i) = [(\mathbf{z}_i - \mathbf{t})^\top \mathbf{C}^{-1}(\mathbf{z}_i - \mathbf{t})]^{1/2}, \ i = 1, \ldots, n,$$

with robust estimators $\mathbf{t}$ of location and $\mathbf{C}$ of covariance, respectively. Here, the estimated covariance structure is used to assign a distance to each observation indicating how far the observation is from the centre of the data cloud with respect to the covariance structure. For the computed squared robust Mahalanobis distances it is common to use a certain quantile (like the quantile 0.975) of the χ^2 distribution with $D - 1$ degrees of freedom as a cut-off value for outlier identification (Filzmoser and Hron 2008); observations with larger squared robust Mahalanobis distance are considered as potential outliers. However, compositional data first need to be moved to the real space using a suitable transformation. A natural way is to express the data in orthonormal coordinates. Nevertheless, if equivariant estimators, like the results of MCD, are taken for $\mathbf{t}$ and $\mathbf{C}$, it is also possible to use the alr coordinates to obtain invariant results (Filzmoser and Hron 2008). However, in the latter case any interpretation of the data structure could be misleading.

5.3.2   Principal component analysis

This prominent statistical method is used for dimension reduction of a multivariate data set, and the results are usually presented in biplots that display both samples and variables of a data matrix graphically in terms of the resulting scores and loadings (Gabriel 1971). For compositional data one would intuitively construct the biplot in orthonormal coordinates, however, due to the complex interpretation of the new variables it is still more common to construct the compositional biplot for compositions in clr coordinates as proposed in Aitchison and Greenacre (2002). The scores represent the structure of the compositional data set in the Euclidean space, so they can be used to see patterns and groups in the data. The loadings (rays) represent the corresponding clr variables, forming a vector $\mathbf{y} = (y_1, \ldots, y_D)^\top$ with the constraint $y_1 + \ldots + y_D = 0$. Accordingly, their interpretation is different from the usual case: the main interest is in the links which are the distances between the vertices of the rays. In more detail, for the rays i and j $(i, j = 1, \ldots, D)$, the link approximates, up to a constant, the (usual) variance $\mathrm{var}(\frac{1}{\sqrt{2}} \ln \frac{x_i}{x_j})$ of the corresponding coordinate for the compositional parts x_i and x_j. Hence, when the vertices coincide, or nearly so, then the ratio between x_i and x_j is constant, or nearly so. In addition, directions of the rays signal where observations with dominance of the corresponding compositional part are located. Again, outliers can substantially affect results of the underlying principal component analysis and depreciate the predicative value of the biplot. For this reason, again the robust version of the biplot is needed. However, as the robust methods cannot work with singular data, the robust scores and loadings must be computed in orthonormal coordinates and the result needs to be back-transformed. Here the linear relation between clr and orthonormal coordinates (Egozcue *et al.* 2003) is used, together with the advantageous properties of the MCD estimator, to express the resulting scores and loadings in the clr plane [see Filzmoser *et al.* (2009a) for details]. Afterwards, the robust compositional biplot (with the above interpretation) can be constructed.

5.3.3   Discriminant analysis

Discriminant analysis is a statistical technique which allows the study of the differences between two or more groups of objects with respect to considered variables simultaneously. Discriminant analysis assumes that the group membership for data from the training set is known. Using certain statistical assumptions, this knowledge leads to an assignment of the group membership of new observations, coming from the test data set. Sometimes, discriminant analysis is also used as a confirmatory tool, by verifying the group structure resulting from the visualization of a (robust) compositional biplot.

The assignment of an observation from the test set to a group is based on the discriminant rules. Usually either the *Bayesian* or the *Fisher* discriminant rules are used. For both types a comprehensive literature is available (see, e.g. Johnson and Wichern 2007); Filzmoser *et al.* 2009b. For this reason only a brief overview will be given here. Using the Bayesian rule for compositions in orthonormal coordinates, multivariate normal distribution with mean vector $\boldsymbol{\mu}_j$ and covariance matrix $\boldsymbol{\Sigma}_j$ for each of the $j = 1, \ldots, g$ groups is assumed. Moreover, the groups are supposed to follow certain prior probabilities p_j which sum up to one. Then a new observation $\mathbf{z}$ is assigned to that group $k \in \{1, \ldots, g\}$ for which

$$d_j^Q(\mathbf{z}) = -\frac{1}{2} \ln[\det(\boldsymbol{\Sigma}_j)] - \frac{1}{2}(\mathbf{z} - \boldsymbol{\mu}_j)^\top \boldsymbol{\Sigma}_j^{-1}(\mathbf{z} - \boldsymbol{\mu}_j) + \ln(p_j) \qquad (5.2)$$

is the largest for $j = 1, \ldots, g$. The corresponding method is known by the name *quadratic discriminant analysis* (QDA). With the additional assumption of equal group covariance matrices $\boldsymbol{\Sigma}_1 = \cdots = \boldsymbol{\Sigma}_g := \boldsymbol{\Sigma}$, the rule (5.2) can be simplified to

$$d_j^L(\mathbf{z}) = \boldsymbol{\mu}_j^\top \boldsymbol{\Sigma}^{-1} \mathbf{z} - \frac{1}{2} \boldsymbol{\mu}_j^\top \boldsymbol{\Sigma}^{-1} \boldsymbol{\mu}_j + \ln(p_j), \qquad (5.3)$$

resulting in *linear discriminant analysis* (LDA). The use of linear or quadratic discriminant rules depends on the underlying data structure. In spite of the more restrictive assumptions of LDA, this method requires less parameters to estimate. As a result, overfitting of the training data can be avoided, and often smaller misclassification errors can be achieved compared with QDA. Also the *Fisher discriminant rule* leads to linear classification boundaries, and this approach is mainly used for dimension reduction. In contrast to principal component analysis, that searches for directions in the data having largest variance, here such directions are used that best capture the differences among the groups. Like in the case of biplots, the resulting Fisher discriminant scores are usually displayed in a planar graph picturing the grouping structure in the data.

There are many papers devoted to practical aspects of discriminant analysis for compositional data (von Eynatten *et al.* 2003; Kovács *et al.* 2006; Thomas and Aitchison 2006). Here the classical linear discriminant analysis is favoured as it is invariant with respect to the choice of all alr, clr or orthonormal coordinates. However, as the discrimination is from the principle intended to objects and not to variables, any reasonable invariance is necessary only with respect to the choice of orthonormal coordinates. In addition, the mentioned discriminant rules may lead to completely useless results whenever outliers are present in the data. Since the discriminant rules lose their optimality (with respect to the misclassification rate) if the assumptions are violated, the influence of outlying observations needs to be reduced. This can again be achieved by robust estimates of location and covariance, forming the main ingredients to the discriminant rules. It is easy to show (Filzmoser *et al.* 2009b) that LDA, QDA as well as the Fisher discriminant rule are invariant to the choice of balances if affine equivariant estimators are used.

5.4   Case studies

The concepts discussed in Section 5.3 are applied here to real data from geochemistry. In this field of science one usually deals with compositional data, because the available data are typically concentrations of chemical elements measured in soil samples. The data values are thus proportions in a whole, and if all the elements would be measured, they would sum up to 100%, or to 1 million mg/kg. All data sets used in this section are available in R (R Development Core Team 2009), and the exact references will be provided in the corresponding subsections. The purpose, however, is not a comprehensive statistical analysis of the data sets, but they should rather be used as an illustration of the different methods.

5.4.1   Multivariate outlier detection

From 1992 to 1998, the Geological Surveys of Finland and Norway, and the Central Kola Expedition, Russia, carried out a large geochemical mapping project in an area covering $188000 \, km^2$ at the peninsula Kola in Northern Europe. This area was sampled in four different layers (moss, humus, B-horizon, C-horizon), and in each layer around 600 samples were taken. The soil samples were analysed by a number of different techniques for more than 50 chemical elements. The project was primarily designed to reveal the environmental conditions in the area, which consists of very pristine parts but also of regions with serious pollution caused by the Russian industry, especially by the smelters in Nikel/Zapoljarnij and Monchegorsk. More details can be found in Reimann *et al.* (1998) which also includes maps of the single element distributions, and in Reimann *et al.* (2008). The data are available in the package StatDA of the statistical software R.

Here a subset of the moss data is used, available in R as data frame moss. This layer is especially interesting because the resulting element concentrations originate from the atmosphere and not from the soil. We focus on Ag, As, Co, Cu and Ni, which are typical smelter-related elements. Multivariate outlier detection might thus be helpful in identifying locations with atypical behaviour of the multivariate information contained in the distribution of the five elements. Following an ilr transformation, the MCD estimator is applied to the four-dimensional data, and the resulting robust estimators of location and covariance are plugged into the formula for the Mahalanobis distance, see Section 5.3. The result is presented on the

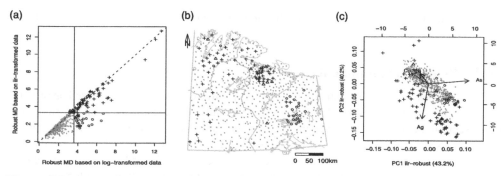

Figure 5.3 Multivariate outlier detection of a subcomposition of the Kola moss data based on the ilr- and log-transformed data; distance–distance plot (a), points in the map (b), and compositional biplot (c) to support the interpretation.

vertical axis of Figure 5.3(a). The horizontal axis is for the robust Mahalanobis distances of the log-transformed data. A log transformation of the variables is frequently carried out in practice because it symmetrizes the data distributions. However, the log transformation ignores the compositional nature of the data, and thus it is inappropriate. Nevertheless, the resulting plot is interesting, because it reveals the effects of data transformation on multivariate outlier detection. The horizontal and vertical lines in Figure 5.3(a) indicate the 0.975 quantiles of the corresponding χ^2 distributions, which are used as cut-off values for outlier identification. Accordingly, the lower left quadrant (grey dots) contains regular observations, and the upper right quadrant (+ symbols) includes observations that would be identified as multivariate outliers by both types of transformations. The lower right quadrant (o symbols) points at observations that would erroneously be declared as outliers, while the upper left quadrant ($\triangle$ symbols) includes samples that are only identified as outliers when using the appropriate transformation.

The same symbols are used in Figure 5.3(b), which represents the map of the investigated project area, and the symbols are plotted at the locations of the sample origins. Most of the circles plot around the smelter Monchegorsk, and these samples are highlighted as outliers only by the incorrect procedure. The multivariate outliers identified by both methods are concentrated around Nikel/Zapoljarnij, but they are also found in the northern (pristine) part of the area, and in the south-west, and some triangles support the abnormal behaviour in these regions. A wrong transformation would thus lead to a different picture and therefore to wrong conclusions. On the other hand, the reason for outlyingness cannot be answered by these plots. Multivariate outliers are not characterized by just abnormally high concentrations in all elements. Specifically for compositional data they point at abnormal compositions, being characterized by the ratios between the parts. The compositional biplot can be used to get more insight, see Figure 5.3(c). Again the same plot symbols as in the other two plots are used. The different behaviour of the potential outliers becomes immediately clear: The incorrectly identified points o are more dominated by As than the correctly identified outliers + and $\triangle$. Thus the observations around Monchegorsk are not exceptional in their compositions of the considered elements. They could be exceptional if additional elements would be considered, because the ratios to other elements could reveal the rather atypical behaviour of this region. Moreover, although the outliers + are correctly identified, they differ with respect to their composition. This example demonstrates that an inappropriate procedure may indeed lead to erroneous conclusions.

5.4.2   Principal component analysis

Figure 5.3(c) has already demonstrated that the compositional biplot, which is based on principal component analysis, is helpful in interpreting the multivariate compositional information. In this section the focus is on comparing the classical and robust version of principal component analysis for ilr-transformed and log-transformed data, respectively. These methods are applied to all variables with reasonable data quality of the data set `OsloTransect`, included in the R package `rrcov`. For 360 observations, the concentrations of chemical elements have been determined. The samples originate from nine different plant species collected along a 120 km transect running through the city of Oslo, Norway (Reimann *et al.* 2007, and references therein). The species are birch bark (1), birch leaves (2), birch wood (3), fern (4), moss (5), European mountain ash leaves (6), spruce needles (7), spruce tree wood (8), and spruce tree twigs (9). For each species 40 samples are available.

The aim of principal component analysis is to find the relationships between the 16 compositional parts used in this study, but also to characterize the chemical structure of the nine plant species, or at least to reveal their major differences. Since the data consist of different groups, the plant species, a robust analysis will focus on the homogeneous data majority, and it is likely that single groups rather than single observations are treated as *outliers*. In contrast, classical principal component analysis can be driven mainly by the *outlying groups*, and thus the focus is less on the homogeneous majority.

Similar to the Kola data set, the distribution of the single elements is skewed, and a log-transformation leads to better symmetry. Although this approach is common practice, it is not useful in terms of transforming the simplex sample space to the Euclidean space. Since this mistake is frequently made, a comparison of the results will be of interest. Figure 5.4 shows the resulting biplots constructed with the first two principal components, for the classical (left column) and robust (right column) analysis, using log-transformed (upper row) and ilr-transformed (lower row) data. The results change quite dramatically, and the degenerated solution of the log transformation is clearly visible in the biplots where almost all the variables are arranged in one half-plane. This would indicate mainly positive intercorrelations, a conclusion which is misleading. The biplots based on the ilr-transformed data reveal the real associations, and in particular the robust version (Figure 5.4, lower right) leads to a reliable interpretation since it is not affected by outlying groups. The plant species turn out to be relatively homogeneous with respect to their chemical composition, and the biplot is indeed helpful for their characterization.

5.4.3   Discriminant analysis

The data set from Oslo (see previous subsection) is again used to illustrate discriminant analysis for compositional data. Here we focus on the elements Cr, P, and Pb, and on the groups 1, 6, 7, and 8 (see previous subsection). Since only three compositional parts are used, the data can be visualized in the ternary diagram in Figure 5.5(a). The groups 6 and 7 show a certain overlap, but also groups 1 and 8 appear close together. The representation with ilr-coordinates in Figure 5.5(b) shows that the groups 1 and 8 are clearly separated. This was hidden in the ternary diagram where both groups went close to the

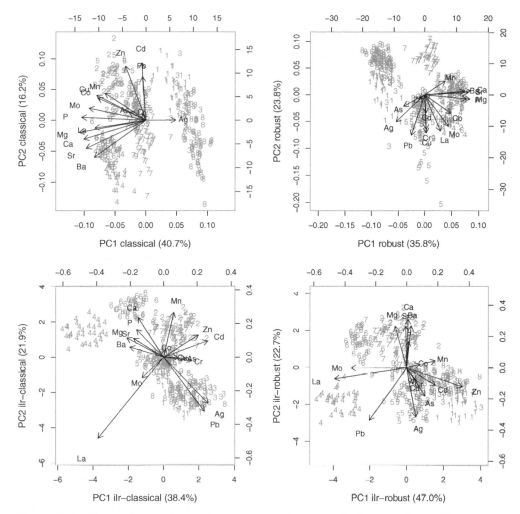

Figure 5.4 Biplots for the data `OsloTransect` based on the log-transformed (upper row) and ilr-transformed (lower row) data, for classical (left column) and robust (right column) principal component analysis.

boundary. Finally, Figure 5.5(c) visualizes a scatterplot of the log-transformed elements P and Pb which are mainly responsible for the group separation. Although the logarithm is an inappropriate transformation of compositional data, it still allows the group structure to be revealed.

Applying linear discriminant analysis to this data subset allows the estimation of the misclassification rate, and this information will be useful for the practitioner who wants to classify a new observation (originating from one of the four groups) at the basis of the three measured parts (this is only illustrative, and using all the parts and groups might be more relevant to practitioners). The resulting misclassification rates are shown in Table 5.1 for the original, ilr-transformed, and log-transformed data, using classical and robust (MCD-based)

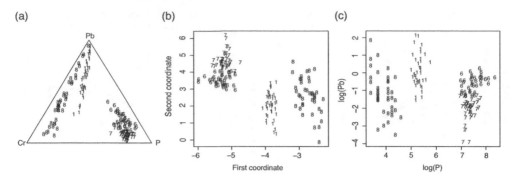

Figure 5.5 Visualization of three selected parts and four groups of the data `OsloTransect` by the ternary diagram (a), by ilr coordinates (b), and by two log-transformed variables (c).

Table 5.1 Misclassification rates of LDA and QDA applied to a subset of the `Oslotransect` data.

	Original data	ilr-transformed data	Log-transformed data
Classical LDA	0.247	0.051	0.063
Robust LDA	0.190	0.025	0.063
Classical QDA	0.044	0.032	0.038
Robust QDA	0.120	0.044	0.051

LDA and QDA. These reported misclassification rates are derived from five-fold cross-validation, which means that consecutively four parts of the data are used for establishing the discrimination rules, while the fifth part is predicted. The resulting error measure is known to be more realistic than the misclassification rate resulting from using all data for model building, or from leave-one-out cross-validation. Table 5.1 shows a strong impact of the transformation, and also robustness can improve the misclassification rate. This is not generally true for every data set, because the data configuration and position of outliers can change the situation (Croux *et al.* 2008). The same statement can be made for the transformation: there is no guarantee that ilr transformation automatically leads the smallest misclassification rate. However, in terms of interpretation, and specifically for a graphical presentation in a lower-dimensional space obtained from the Fisher rule, an appropriate (ilr) transformation is necessary.

5.5   Summary

As any real data set, compositional data also usually contain atypical observations that deviate from the main data structure. The influence of these observations on a statistical analysis can be reduced by using robust methods. For multivariate statistical methods this usually means that the covariance matrix needs to be estimated in a robust way, because this matrix forms the basis for methods like principal component analysis, discriminant analysis, or multivariate outlier detection. Robust covariance estimation, however, can only be done with regular data, and thus the ilr transformations are the preferred transformations in case of compositional data. Since the ilr coordinates are often not straightforward to interpret, a back-transformation to the clrspace is recommended whenever one has to analyse the structure of a compositional data set using a compositional biplot. Data outliers may lead to erroneous conclusions if they are not treated appropriately by robust methods. On the other hand, robust methods cannot *repair* an inappropriate transformation of compositional data, as it results from a log transformation.

Acknowledgement

This work was supported by the Council of the Czech Government under project MSM 6198959214.

References

Aitchison J 1986 *The Statistical Analysis of Compositional Data.* Monographs on Statistics and Applied Probability. Chapman and Hall Ltd (reprinted 2003 with additional material by The Blackburn Press), London (UK). 416 p.

Aitchison J and Greenacre M 2002 Biplots for compositional data. *Applied Statistics* **51**(4), 375–392.

Aitchison J, Barceló-Vidal C, Martín-Fernández JA and Pawlowsky-Glahn V 2000 Logratio analysis and compositional distance. *Mathematical Geology* **32**(3), 271–275.

Buccianti A, Mateu-Figueras G and Pawlowsky-Glahn V (ed.) 2006 *Compositional Data Analysis in the Geosciences: From Theory to Practice.* Geological Society, London (UK). 212 p.

Croux C, Filzmoser P and Joossens K 2008 Classification efficiencies for robust linear discriminant analysis. *Statistica Sinica* **18**, 581–599.

Egozcue JJ and Pawlowsky-Glahn V 2005a Coda-dendrogram: a new exploratory tool. In *Proceedings of CoDaWork'05, The 2nd Compositional Data Analysis Workshop* (ed. Mateu-Figueras G and Barceló-Vidal C). University of Girona, http://ima.udg.es/Activitats/CoDaWork05/, Girona (Spain).

Egozcue JJ and Pawlowsky-Glahn V 2005b Groups of parts and their balances in compositional data analysis. *Mathematical Geology* **37**(7), 795–828.

Egozcue JJ, Pawlowsky-Glahn V, Mateu-Figueras G and Barceló-Vidal C 2003 Isometric logratio transformations for compositional data analysis. *Mathematical Geology* **35**(3), 279–300.

Filzmoser P and Hron K 2008 Outlier detection for compositional data using robust methods. *Mathematical Geosciences* **40**(3), 233–248.

Filzmoser P, Hron K and Reimann C 2009a Principal component analysis for compositional data with outliers. *Environmetrics* **20**, 621–632.

Filzmoser P, Hron K and Templ M 2009b Discriminant analysis for compositional data and robust estimation. Technical Report SM-2009-3, Department of Statistics and Probability Theory, Vienna University of Technology, Austria. 27 p.

Filzmoser P, Hron K, Reimann C and Garrett R 2009c Robust factor analysis for compositional data. *Computers and Geosciences* **35**, 1854–1861.

Gabriel KR 1971 The biplot – graphic display of matrices with application to principal component analysis. *Biometrika* **58**(3), 453–467.

Hron K, Templ M and Filzmoser P 2010 Imputation of missing values for compositional data using classical and robust methods. *Computational Statistics and Data Analysis* **54**(12), 3095–3107.

Johnson R and Wichern D 2007 *Applied Multivariate Statistical Analysis*, 6th Edition. Prentice Hall, New York (USA). 800 p.

Kovács L, Kovács G, Martín-Fernández J and Barceló-Vidal C 2006 Major-oxide compositional discrimination in Cenozoic volcanites of Hungary. In *Compositional Data Analysis in the Geosciences: From Theory to Practice* (ed. Buccianti A, Mateu-Figueras G and Pawlowsky-Glahn V). Geological Society, London (UK). pp. 11–23.

Maronna R, Martin R and Yohai V 2006 *Robust Statistics: Theory and Methods*. John Wiley & Sons, Ltd, New York (USA). 436 p.

R Development Core Team 2009 *R: A Language and Environment for Statistical Computing*. R Foundation for Statistical Computing, Vienna (Austria).

Reimann C, Arnoldussen A, Boyd R, Finne TE, Koller F, Nordgullen O and Englmair P 2007 Element contents in leaves of four plant species (birch, mountain ash, fern and spruce) along anthropogenic and geogenic concentration gradients. *Science of the Total Environment* 377, 416–433.

Reimann C, Äyräs M, Chekushin V, Bogatyrev I, Boyd R, de Caritat P, Dutter R, Finne T, Halleraker J, Jæger Ø, Kashulina G, Lehto O, Niskavaara H, Pavlov V, Räisänen M, T. Strand T and Volden T 1998 *Environmental Geochemical Atlas of the Central Barents Region*. NGU-GTK-CKE Special Publication, Geological Survey of Norway, Trondheim (Norway). 745 p.

Reimann C, Filzmoser P, Garrett R and Dutter R 2008 *Statistical Data Analysis Explained. Applied Environmental Statistics with R*. John Wiley & Sons, Ltd, Chichester (UK). 343 p.

Thomas C and Aitchison J 2006 Log-ratios and geochemical discrimination of Scottish Dalradian limestones: a case study. In *Compositional Data Analysis in the Geosciences: From Theory to Practice* (ed. Bucciants A, matue - figures G and Pawlaowsky-Glahn v). Geological Society, London (UK). pp. 25–41.

von Eynatten H, Barceló-Vidal C and Pawlowsky-Glahn V 2003 Composition and discrimination of sandstones: a statistical evaluation of different analytical methods. *Journal of Sedimentary Research* 73(1), 47–57.

6

Geostatistics for compositions

Raimon Tolosana-Delgado[1], Karl Gerald van den Boogaart[2] and Vera Pawlowsky-Glahn[3]

[1]*Maritime Engineering Laboratory, Technical University of Catalonia, Spain*
[2]*Institute for Stochastic, Technical University Bergakademie Freiberg, Germany*
[3]*Department of Computer Science and Applied Mathematics, University of Girona, Spain*

6.1   Introduction

Geostatistics is a collection of techniques and tools to analyse regionalised data sets. A *regionalised data set* is one where each sample has been collected in a known location of space (i.e. it includes geographic coordinates for each sample), and where samples show a mutual dependence coming from the spatial proximity between their sampling locations.

Virtually any multivariate geostatistical data set is compositional in nature: the most common regionalised sets of variables are geochemical concentrations (in rocks, ores, soils, water, air or environmental proxies, like moss), followed by species partitions (of animals or trees, of grain sizes, of causes of death, etc.) and multinomial probabilities of belonging to a set of classes. All these sets of variables are formed by vectors of positive components showing the *relative importance of a set of components on a whole*. The statistical treatment of these data sets is based on Aitchison's (1986) idea that they convey only *relative* information, and that we must therefore point our statistical machinery towards ratios of components. Any question and answer regarding absolute increments or decrements in compositions is damned to be as spurious as the classical correlation (Chayes 1960), randomly changing sign, sense and meaning. Spatial correlation is not an exception (Pawlowsky 1984).

Compositional Data Analysis: Theory and Applications, First Edition. Edited by Vera Pawlowsky-Glahn and Antonella Buccianti.
© 2011 John Wiley & Sons, Ltd. Published 2011 by John Wiley & Sons, Ltd.

Because the problems of geostatistics with compositions are the same as the problems of classical statistics, the solutions to the former are the same as those found for the latter. To avoid interpreting spurious covariances and dealing with singular matrices, we must build our methods on log-ratios of parts.

The first step of a statistical analysis is the characterisation of the spread of the data set. This task is fulfilled in classical statistics with the study of the variation matrix, the elements of which are the variances of each possible pairwise log-ratio. In the same way, the spatial correlation structure can be unveiled by using *variation-variograms*: variograms of each possible pairwise log-ratio. This topic is covered in Section 6.4. Before, Section 6.2 explains briefly the basics of multivariate geostatistics, and Section 6.3 introduces cokriging for compositions. Section 6.5 briefly discusses what to do in the presence of zero values. Finally, the real case study is treated in Section 6.6, and some conclusions given in Section 6.7.

6.2   A brief summary of geostatistics

This section presents a brief overview of a subject with many good textbooks, including Chilès and Delfiner (1999), Clark and Harper (2000), Isaaks and Srivastava (1989) and Journel and Huijbregts (1978). It is devised to introduce the most basic concepts and notation later used in the chapter, and follows the approach of Myers (1984). Regionalised data sets are modelled with the concept of a *vector random function* $\mathbf{Z}(\vec{x})$, an infinite collection of random vectors indexed with $\vec{x} \in \mathbb{R}^p$ a spatial location (typically $p = 2$, easting and northing). Usually *second-order stationarity* is assumed,

$$E[\mathbf{Z}(\vec{x})] = \boldsymbol{\mu}(\vec{x}) = \boldsymbol{\mu}, \quad \text{and} \quad \text{Cov}[\mathbf{Z}(\vec{x}), \mathbf{Z}(\vec{y})] = \mathbf{C}(\vec{x} - \vec{y}),$$

i.e. the expected value is constant and independent of location, and the covariance matrix between the random function at any two locations only depends on the lag between them. Typically, one relaxes this hypothesis to *intrinsic stationarity*,

$$E[\mathbf{Z}(\vec{x}) - \mathbf{Z}(\vec{y})] = \mathbf{0}, \quad \text{and} \quad \text{Var}[\mathbf{Z}(\vec{x}) - \mathbf{Z}(\vec{y})] = \boldsymbol{\Gamma}(\vec{x} - \vec{y}),$$

i.e. the mean increment is zero and the variance of the increment of the random function between any two locations only depends on the lag between them. The functions $\mathbf{C}(\cdot)$ and $\boldsymbol{\Gamma}(\cdot)$ are, respectively, called the (multivariate) *covariance function* and *variogram*, and in cases where both exist, they are related through

$$\boldsymbol{\Gamma}(\vec{h}) = 2\mathbf{C}(\vec{0}) - [\mathbf{C}(\vec{h}) + \mathbf{C}(-\vec{h})].$$

Diagonal terms are called auto-covariances or direct variograms, and show the spatial continuity of a given variable. Off-diagonal terms are called cross-covariance or -variograms, and explain how two different variables taken at two different locations are related. The typical way to work with these functions is to estimate empirical versions of the direct and cross-variograms, and somehow fit to them a *linear model of coregionalisation* (LMC, e.g. Wackernagel 1998),

$$\Gamma(\vec{h}) = \sum_{k=0}^{K} [1 - \rho^k(\vec{h})] \cdot \mathbf{C}_k, \tag{6.1}$$

where $\Gamma(\vec{h})$ represents the variogram, matrix-valued function (with direct and cross-variograms), $\rho^k(\vec{h})$ are valid correlograms (positive definite scalar-valued functions; any geostatistical textbook includes examples), and $\mathbf{C}_k$ are positive definite matrices. It is typical to set the first correlogram as $\rho^0(\vec{h}) = \delta(\vec{h})$, called the *nugget effect*. This variogram estimation and modelling step is known as *structural analysis*.

Once variograms are modelled, one can use them to interpolate the available samples to estimate the random function at an unsampled location. This is called *cokriging*. The *ordinary cokriging* estimator $\hat{\mathbf{z}}_0$ at an unsampled location $\vec{x}_0$ is a linear combination of the data, $\{\mathbf{z}_i = \mathbf{z}(\vec{x}_i), i = 1, \ldots, n\}$,

$$\hat{\mathbf{z}}_0 = \sum_{i=1}^{n} \Lambda_i \cdot \mathbf{z}_i, \tag{6.2}$$

subject to the unbiasedness condition $\sum_{i=1}^{n} \Lambda_i = \mathbf{I}$. The weights Λ_i are matrices of the same dimension as $\mathbf{C}(\cdot)$ or $\Gamma(\cdot)$, which show the influence of sample $\mathbf{z}_i$ on the predicted $\mathbf{Z}(\vec{x}_0)$. These weights are found solving the ordinary cokriging system,

$$\Lambda = \mathbf{S}^{-1} \cdot \mathbf{S}_0,$$

where all matrices are defined by blocks as follows, taking $\Gamma_{ij} = \Gamma(\vec{x}_i - \vec{x}_j)$:

$$\lambda = \begin{bmatrix} \lambda_1 \\ \vdots \\ \lambda_n \\ \nu \end{bmatrix}, \quad \mathbf{S} = \begin{bmatrix} \Gamma_{11} & \cdots & \Gamma_{1n} & \mathbf{I} \\ \vdots & \ddots & \vdots & \vdots \\ \Gamma_{n1} & \cdots & \Gamma_{nn} & \mathbf{I} \\ \mathbf{I} & \cdots & \mathbf{I} & \mathbf{0} \end{bmatrix}, \quad \mathbf{S}_0 = \begin{bmatrix} \Gamma_{10} \\ \vdots \\ \Gamma_{n0} \\ \mathbf{I} \end{bmatrix}.$$

The same results are obtained if one takes the blocks $\Gamma_{ij} = \mathbf{C}(\vec{x}_i - \vec{x}_j)$, using covariances instead of variograms.

As mentioned in Section 6.1, these techniques cannot be directly applied to regionalised compositional data (i.e. percentages, concentrations, proportions, probabilities). Multivariate covariance functions and variograms of raw compositions are as spurious as the correlation matrix. This is due to the fact that the sum of proportions or percentages is fixed and, thus, $\Gamma(\vec{h})$ or $\mathbf{C}(\vec{h})$ are singular *at any lag* $\vec{h}$; each of their rows and columns must sum up to zero: the presence of positive variances in the diagonal terms of these matrices forces some of the cross-covariances or cross-variograms to be negative. Moreover, kriging presents severe problems. The separate kriging of each component does not ensure that the interpolated results are positive, or sum up to 100% (or 1 when dealing with proportions), and often delivers sums larger than the maximum allowed. Cokriging preserves the total sum, but does not avoid negative interpolations and presents a numerical problem: the inversion of a singular matrix (with a high multiplicity of the null singular value), something that most software is not prepared to deal with. An extended presentation of these problems can be found in Pawlowsky-Glahn and Olea (2004) and Tolosana-Delgado (2006).

6.3   Cokriging of regionalised compositions

A straightforward way of using geostatistics with compositional data sets is to transform the data with any black-box isometric log-ratio transformation ilr (Egozcue *et al.* 2003; Egozcue and Pawlowsky-Glahn 2005) (see also Chapter 11), apply standard techniques to model variograms and cokriging to the obtained scores, and back-transform the interpolated scores with the inverse ilr transformation. This is the application of the *principle of working on coordinates* (Pawlowsky-Glahn 2003). This way valid interpolations are obtained, i.e. vectors of positive components summing up to a fixed total amount (as expected for compositions), and as sensible as implied by the variogram models used. The same procedure applies to simulation and the construction of confidence regions: one applies the classical tools (simulation, 2σ intervals, ...) to the coordinates, and back-transforms the final results to express them in proportions or percentages.

It is easy to show that, as long as the systems of direct and cross-variograms are consistently modelled, the final interpolated compositions do not depend on the actual ilr/basis used for the computations. Take two different ilr matrices $\mathbf{V}_A$ and $\mathbf{V}_B$, each one defining a different ilr transformation:

$$\mathrm{ilr}_A(\mathbf{x}) = \ln(\mathbf{x}) \cdot \mathbf{V}_A^\top =: \boldsymbol{\xi}_A, \quad \mathrm{ilr}_B(\mathbf{x}) = \ln(\mathbf{x}) \cdot \mathbf{V}_B^\top =: \boldsymbol{\xi}_B$$

Denote by $\mathbf{B}_{AB} = \mathbf{V}_A \cdot \mathbf{V}_B^\top$ the matrix of transformation from coordinates of the first basis to the second one. The experimental variograms $\hat{\boldsymbol{\Gamma}}_A(\vec{h})$ and $\hat{\boldsymbol{\Gamma}}_B(\vec{h})$ necessarily satisfy that

$$\hat{\boldsymbol{\Gamma}}_A(\vec{h}) = \mathbf{B}_{AB}^\top \cdot \hat{\boldsymbol{\Gamma}}_B(\vec{h}) \cdot \mathbf{B}_{AB},$$

for any $\vec{h}$. If the variogram models satisfy also this relation, then we say that the models are *mutually consistent*. Mutual consistency is a sufficient condition to ensure that the resulting cokriged compositions are the same, once the respective ilr's are back-transformed. This is a direct implication of the linearity of the kriging predictor [Equation (6.2)], and the fact that $\mathbf{B}_{AB}$ is always invertible. A formal proof can be found in Tolosana-Delgado (2006).

6.4   Structural analysis of regionalised composition

The structural analysis of regionalised compositional data sets can also be done basis-independent, by using the *variation-variograms* (Pawlowsky-Glahn and Olea 2004). The variation matrix (Aitchison 1986) (see also Chapter 24) of a random composition $\mathbf{X}$ is a matrix $\mathbf{T} = [t_{ij}]$, with elements $t_{ij} = \mathrm{Var}[\ln(X_i/X_j)]$. Variation matrices and coordinate covariance matrices $\boldsymbol{\Sigma} = \mathrm{Var}[\mathrm{ilr}(\mathbf{X})]$ satisfy the relation

$$\boldsymbol{\Sigma} = -\frac{1}{2}\mathbf{V} \cdot \mathbf{T} \cdot \mathbf{V}^\top. \tag{6.3}$$

Interestingly, the eigenvectors of both matrices $\mathbf{T}$ and $\mathbf{\Sigma}$ are the same, because $\mathbf{V}$ is an orthonormal matrix, i.e. a *rotation* of the space. On the other hand, the eigenvalues of the former are (-2)-times those of the latter. This implies that $\mathbf{T}$ is a negative semi-definite $D \times D$ matrix, because $\mathbf{\Sigma}$ must be a positive definite $(D-1) \times (D-1)$ matrix.

By analogy, variation-variograms of a D-part compositional random function $\mathbf{Z}(\vec{x})$ are defined as

$$\tau_{ij}(\vec{h}) = \mathrm{Var}\left[\ln\frac{Z_i(\mathbf{x}+\vec{h})}{Z_j(\mathbf{x}+\vec{h})} - \ln\frac{Z_i(\mathbf{x})}{Z_j(\mathbf{x})}\right],$$

and form a matrix of $D \times D$ *direct variograms*. No cross-variogram is thus used, which has several advantages.

First, the resulting functions are strictly positive (except at the origin), and we can thus use an automatic fitting procedure in a log-scale. This enhances the fit of the small values, around the origin, while downweighting the fit at the sill. Because the interpolation itself depends mostly on the behaviour of the variogram around the origin, using this logarithmic fit will improve the kriging results quality. However, this happens at the expense of being less reliable for the assessment of the kriging variance, which is rather controlled by the sill (Chilès and Delfiner 1999).

Secondly, variation-variograms are a valid set if they form a positive semi-definite matrix-function. Therefore, standard multivariate fitting algorithms are still valid: they must just be fed with the variogram system instead of the covariance. To prove that, assume a linear model of coregionalisation [Equation (6.1)], introduce Equation (6.3), and thanks to the linearity of the LMC, one finds that the variation-variogram system must satisfy

$$\mathbf{T}(\vec{h}) = \sum_{i=0}^{K}[1 - \rho^k(\vec{h})]\mathbf{B}_k, \tag{6.4}$$

with $\mathbf{B}_k$ negative semi-definite $D \times D$ matrices.

Thirdly, this equivalence allows us to interpret each structure of the LMC of a variation-variogram system, exactly as they are interpreted in multivariate geostatistical applications. In particular, we may use factor modelling procedures such as the minimax-autocorrelation factors (MAFs; Switzer and Green 1984), common principal components (CPCs; Flury 1988; Xie and Myers 1995a 1995b), or the analysis of the multivariable scalar variogram (Bourgault and Marcotte 1991). Moreover, these techniques all turn out to be just *procedures of selecting a basis* which coordinates are *easy* to model, i.e. which are more or less spatially uncorrelated. This allows to interpolate each coordinate separately, avoiding the use of cross-variograms and reducing the size of the cokriging matrices. This property is called *autokrigeability*.

Fourthly, empirical variation-variograms can be computed even when some missing values occur, or even when data sets are *heterotopic* (i.e. at each location, a different subset of variables can be available, and all of them may or may not be closed to sum up to 100%). To estimate τ_{ij} at lag $\vec{h}$ we only need pairs of observations of variable i and j:

$$\hat{\tau}_{ij}(\vec{h}) = \frac{1}{2N(\vec{h}, i, j)}\sum_{\substack{\vec{x}_n - \vec{x}_m \simeq \vec{h}}}^{N(\vec{h}, i, j)}\left[\ln\frac{z_i(\vec{x}_m)}{z_j(\vec{x}_m)} - \ln\frac{z_i(\vec{x}_n)}{z_j(\vec{x}_n)}\right]^2.$$

It does not matter, whether the composition has been closed at $\vec{x}_m$ but not at $\vec{x}_n$, or we missed a third variable in one of the two locations: $\hat{t}_{ij}(\vec{h})$ can be computed without harm. For each lag and pair of variables we will use a different subset of observations, and there is no guarantee that the resulting system forms a positive semi-definite matrix function. This is nevertheless irrelevant, as the empirical variograms only guide the fit of a model. It only matters that the final, fitted model is a valid one, e.g. an LMC [Equation (6.4)] with negative semi-definite matrices.

Finally, from a purely theoretical point of view, it is worth adding that variation-variograms imply the existence of some sort of abstract spatial covariance description, totally independent of the basis. We can define the spatial covariance as an endomorphism between (the spaces of the compositions at) two different locations (sampled or unsampled). Then, we see any system of multivariate variograms of the observations with respect to a particular orthonormal basis as the coordinate representation of this endomorphism on that basis (Tolosana-Delgado 2006). The same reasoning can be applied to the cokriging weights and cokriging covariance matrix. In particular, cokriging weights are endomorphisms from (the spaces of the compositions at) the observed locations to (the space of the compositions at) an interpolated location.

6.5   Dealing with zeros: replacement strategies and simplicial indicator cokriging

The presence of zero values poses a handicap to the statistical treatment of compositional data with log-ratios. This is particularly common when dealing with percentages of trace elements (and similar small components), and when working with count compositions and probabilities. In most of these cases, zero values indicate that its presence is too low to be detected, or the counting was not extensive enough to catch it, but *not* the absolute absence of the offending part. Therefore, the bottom line of an analysis can be either: (1) to replace the zeros by some suitable value and go ahead with an unmodified analysis (Martín-Fernández *et al.* 2003), (2) to make the analysis independent of the missing value (Bren *et al.* 2008), or (3) to build a hierarchical model taking into account the way the values were missed. Strategies on the first line are reviewed in Chapter 4. The estimation of variation-variograms in the presence of missing values outlined in the preceding section is an example of the second strategy.

The treatment of data sets with zeros from count compositions and probabilities typically fall into the third strategy. One can build a model where such zeros naturally occur, without implying that the actual composition has a real zero, i.e. that the part is utterly absent. Unfortunately, most of these models cannot be analytically solved, and one must seek approximate solutions with simulation techniques, mostly Markov Chain Monte Carlo methods. Tjelmeland and Lund (2003) presented a case study where a sand–silt–clay composition of the sediments of an arctic lake was analysed under the hypothesis of having a second-order stationary, Gaussian RF, and assuming that the zeros observed were actually values below a known detection limit. Diggle *et al.* (1998) present another case in which the observations follow a binomial distribution with unknown probability of success: the log-odds (i.e. the ilr coordinate) of success against failure are assumed a second-order stationary Gaussian RF, and this RF is mapped using information from the binomial observations. This model is further studied by Roeder *et al.* (2011), and extended by Tolosana-Delgado *et al.* (2010) to a multinomial experiment, i.e. when the classification can be done in several classes.

The simplest model of treatment of zeros when working with probabilities (or numbers to be interpreted as such) is *indicator kriging* (IK) (Journel 1983). This is typically applied to define the limits of a body in the geographic space, denoted as $\mathcal{F}$. For this purpose, we define the indicator function

$$I_i = I(\vec{x}_i) = I(\vec{x}_i \in \mathcal{F}) = \begin{cases} 1, & \vec{x}_i \in \mathcal{F}, \\ 0, & \vec{x}_i \notin \mathcal{F}. \end{cases} \tag{6.5}$$

Classically, IK interprets the resulting numbers as the (rather trivial) probabilities of belonging to the body $\mathcal{F}$. One then computes the variogram of the resulting 'observations' and interpolates them with kriging. This produces quite often negative interpolations, that cannot be further interpreted as probabilities. Simplicial IK (Egozcue *et al.* 2007) proposes a log-ratio approach to this problem. This is justified by the sensible assumption of a relative scale for vectors of probabilities $\mathbf{p}(\vec{x}) = [p(\vec{x}), q(\vec{x})]$. Here $p(\vec{x})$ is the probability that $\vec{x}$ belongs to $\mathcal{F}$, and $q(\vec{x}) = 1 - p(\vec{x})$ is the probability that is does not. The idea is simply to take log-ratios $\pi(\vec{x}) = \ln[p(\vec{x})/q(\vec{x})]$ and treat the resulting scores with classical tools. But to be able to do so, we must somehow treat the zeros from the indicators of Equation (6.5), for instance estimating $p(\vec{x})$ with the generalised indicator

$$G_i = G(\vec{x}_i) = G(\vec{x}_i \in \mathcal{F}) = \begin{cases} 1 - \alpha, & \vec{x}_i \in \mathcal{F}, \\ \alpha, & \vec{x}_i \notin \mathcal{F}. \end{cases} \tag{6.6}$$

where $\alpha < 0.5$ is interpreted as the subjective probability that a datum is wrongly classified as being within (outside) the body when it is actual outside (within) it. Egozcue *et al.* (2007) showed that the variograms of $\pi(\vec{x})$ and $I(\vec{x})$ are related through $\gamma_\pi(\vec{h}) = \ln^2[\alpha/(1 - \alpha)]\gamma_I(\vec{h})$. We can therefore study the experimental variance of the increments of $I(\vec{x})$, and obtain an estimator $\hat{\gamma}_I(\vec{h})$. Then we recast these values to estimations of the variogram $\gamma_\pi(\vec{h})$, and use it to interpolate the log-ratios of the G-transformed observations. The resulting kriging estimates [Equation (6.2)] are then back-transformed from log-ratios $\hat{\pi}(\vec{x}_0)$ to probabilities $\hat{p}_0$ of belonging to the body, by

$$\hat{p}_0 = \frac{\exp[\hat{\pi}(\vec{x}_0)]}{1 + \exp[\hat{\pi}(\vec{x}_0)]}. \tag{6.7}$$

Thanks to involving sums of exponentials, results of Equation (6.7) are always positive, bounded to the $(0, 1)$-interval, and consistent with a relative scale.

6.6 Application

The Lyons West oil field in Kansas, USA (Ehm 1965) was characterised by analysing their total porosity (denoted as ϕ) and the total oil/brine saturation (S_w) of its pores along 76 boreholes. These parameters were recast as proportions of solid matter (rock, denoted as s), water brines (w) and oil (o), using the relations

$$w + o = 1 - s = \phi, \quad S_w = \frac{w}{\phi} = \frac{w}{w + o}.$$

Additionally, 34 other boreholes around the field happened to be oil-free, i.e. with $S_w = 1$. Both data sets are displayed in Figure 6.1. The goal is to delimit the oil field and map its resources.

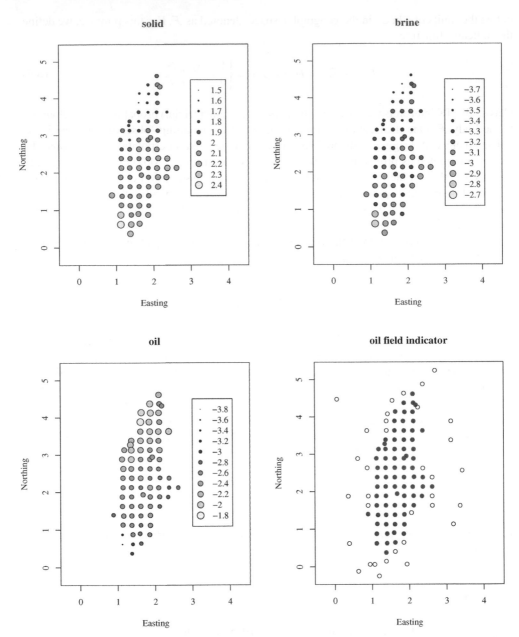

Figure 6.1 Lyons West oil field data set. The lower right diagram shows the boreholes that hit the field (dots) and those which fell outside it (circles). The other three plots show the proportion of each of the three phases (solid, brine, oil) in logistic scale:

Percentage	1	2.5	5	10	20	50	80	90	95
Logit	−4.6	−3.7	−2.9	−2.2	−1.4	0.0	1.4	2.2	2.9

6.6.1   Delimiting the body: simplicial indicator kriging

The first step of the problem is to delimit the oil field, using the information that 76 wells are inside it and 34 are outside. Figure 6.1 shows these data, those in the field as dots and those outside as circles. Let us denote the oil body as $\mathcal{F}$, and the spatial coordinates of each sample as $\vec{x}_1, \ldots, \vec{x}_N$. The analysis follows these steps:

1. *Estimate and model the indicator variogram.* We took $\vec{h}$ at intervals of $\sim$0.25 km, with tolerance half an interval. Then we computed the variance of the increments of all pairs of locations which spatial distance fell within this tolerance. Figure 6.2 shows the resulting empirical variogram values, as circles. This set of values was fitted with variogram model $\gamma_I(h) = c_0 + c \cdot \rho(h)$, with nugget $c_0 = 1/16$ and a Matérn correlogram

$$\rho(h) = \frac{2^{1-\nu}}{\Gamma(\nu)} \left(\frac{\sqrt{2\nu}h}{a} \right)^{\nu} K_{\nu}\left(\frac{\sqrt{2\nu}h}{a} \right),\tag{6.8}$$

of sill $c = 3/16$, range $a = 0.75$ and smoothness $\nu = 2$. (Figure 6.2 also displays this model.)

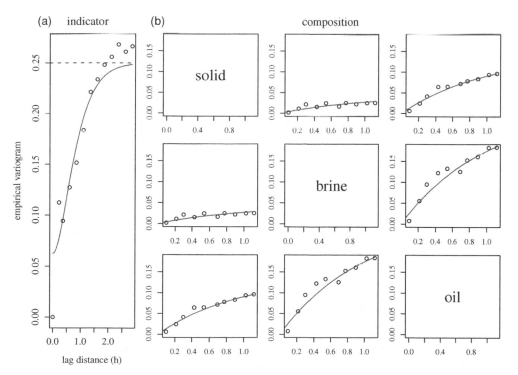

Figure 6.2   (a) Indicator variogram: empirical version (circles) and fitted model (dotted line). Indicator variograms cannot have a sill greater than 1/4. (b) Variation variograms of the solid–brine–oil system: empirical variograms (circles), LMC model (dotted line).

2. *Convert indicator values to probabilities.* We attached to each observation within the body a value of $G(\vec{x}_i) = 0.99 = 1 - \alpha$ and to each outside the body $G(\vec{x}_i) = 0.01 = \alpha$, as explained by Equation (6.6).

3. *Compute the simplicial coordinates of these probabilities.* In this case, that meant taking the log-odds of being within the body. This gives $\pi(\vec{x}_i) = \ln(0.99/0.01) = 4.59$ inside the body and $\pi(\vec{x}_i) = -4.59$ outside it.

4. *Recast the indicator variogram to a coordinate variogram.*

5. *Interpolate the log-odds* using Equation (6.2) or the alternative kriging method desired.

6. *Back-transform the log-odds to probabilities* with Equation (6.7). Final probabilities are displayed in Figure 6.3.

6.6.2 Interpolating the oil–brine–solid content

The second step of the problem is to map the solid, brine and oil content on the Lyons West field, using the 76 samples for which we have the $\{s, w, o\}$-composition available. This was done as follows, using some commands from the R package `compositions`:

1. *Compute and model the empirical variation-variograms.* These are displayed in Figure 6.2. We fitted a variation-LMC [Equation (6.4)] with a correlogram $\rho_1(h)$ following a Matérn model [Equation (6.8)], of parameters smoothness $\nu = 0.5$ (equivalent to the Exponential model), range $a = 2.2$, zero nugget, and a sill variation matrix

$$\mathbf{B}_1 = \begin{pmatrix} 0.00 & 4.21 & 13.96 \\ 4.21 & 0.00 & 26.92 \\ 13.96 & 26.92 & 0.00 \end{pmatrix} \cdot 10^{-2}.$$

The empirical variograms were obtained with command `logratioVariogram`, the LMC was defined with `CompLinModCoReg`, and the automatic fitting was done in a log-scale with command `vgmFit2lrv`.

2. *Choose a basis.* In this case, an arbitrary one was chosen. The observed compositions were expressed in that basis, following Egozcue and Pawlowsky-Glahn (2005). With the corresponding matrix $\mathbf{V}$ of basis definition, the variation-variograms were recasted to coordinate variograms [Equation (6.3)].

3. *Interpolate the coordinates.* Using these variograms, ilr coordinates were interpolated with Equation (6.2). In the case that the coordinates themselves were interpretable, the obtained maps could be useful. This is not our case, having used an arbitrary basis.

4. *Back transform interpolated coordinates.* Direct results are displayed in Figure 6.3. These last three steps are all computed with command `compOKriging`.

However, the interpolated oil content does not take into account that outside the Lyons West field the oil content should be zero. In fact, the obtained interpolations are only valid

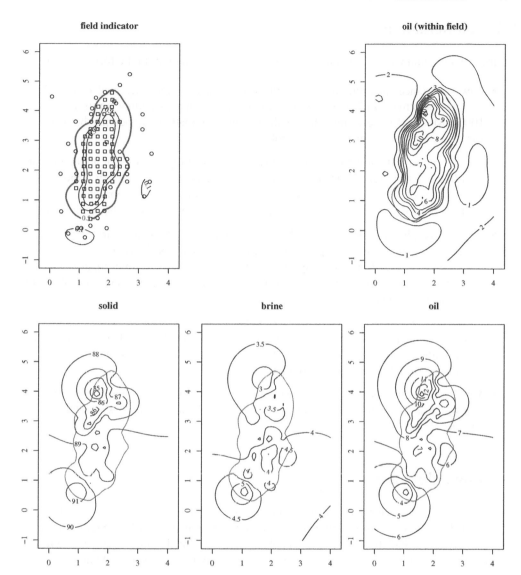

Figure 6.3   Interpolation results. (Upper left) Interpolated probabilities of belonging to the oil field. (Lower plots) Interpolated proportions of solid, brine and oil. (Upper right) Expected oil content, taking into account the estimated probabilities of being inside the field.

conditional to being inside the oil body. We can then weight the kriged oil content by the kriged probability of being inside the body

$$\hat{o}^*(\vec{x}_0) = \hat{p}_0 \cdot \hat{o}(\vec{x}_0) + (1 - \hat{p}_0) \cdot 0 = \hat{p}_0 \cdot \hat{o}(\vec{x}_0),$$

which produces final oil content maps which adapt to the contours of the oil body (Figure 6.3, upper right).

6.7   Conclusions

To finish, let us summarise the main conclusions of this contribution, in the form of a list:

- Raw covariance functions and variograms are spurious, and may lead to meaningless interpolations, with negative proportions or percentages adding up to more than 100%. Most software cannot deal with the singular matrices appearing both in the variogram/covariance modelling and cokriging.

- Existing geostatistical methods and software tools are fully valid when applied to the coordinates of the regionalised composition with respect to an arbitrary orthonormal basis. Interpolations, simulations or confidence regions for the coordinates can be applied to the basis in use to obtain these results as compositions, in the original units.

- The variographic structure may be adequately studied using variation-variograms, without choosing the basis of computation. The factorial analysis (MAF, CPC, etc.) of the estimated structural functions may finally suggest a basis where interpolation will be autokrigeable (i.e. kriging each coordinate independently gives the same as cokriging the whole coordinate vector).

- Variation-variograms can be automatically fitted with a log-scale, enhancing goodness of fit on the small values (around the origin). That requires small modifications of existing algorithms, that should check for negative semidefiniteness of the coefficient matrices in the linear model of coregionalisation.

- Variation-variograms allow a small amount of missing or noisy values in the data set, as estimation can be done using only the subset of variables observed with reliability at each location.

Acknowledgements

This research has been supported by the Spanish Ministry of Science and Innovation through a 'Juan de la Cierva' subprogramme of the European Social Fund (ESF-FSE), as well as through projects CSD2006-00032 and MTM2009-13272, and by the Agència de Gestió d'Ajuts Universitaris i de Recerca of the Generalitat de Catalunya (Ref. 2009SGR424).

References

Aitchison J 1986 *The Statistical Analysis of Compositional Data*. Monographs on Statistics and Applied Probability. Chapman and Hall Ltd (reprinted 2003 with additional material by The Blackburn Press), London (UK). 416 p.

Bourgault G and Marcotte D 1991 Multivariable variogram and its application to the linear model of coregionalization. *Mathematical Geology* **23**(7), 899–928.

Bren M, Tolosana-Delgado R and van den Boogaart KG 2008 News from compositions, the r package. In *Proceedings of CoDaWork'08, The 3rd Compositional Data Analysis Workshop* (ed. Daunis-i Estadella J and Martín-Fernández J), http://hdl.handle.net/10256/723. University of Girona, Girona (Spain). CD-ROM.

Chayes F 1960 On correlation between variables of constant sum. *Journal of Geophysical Research* **65**(12), 4185–4193.

Chilès JP and Delfiner P 1999 *Geostatistics – Modeling Spatial Uncertainty*. Series in Probability and Statistics. John Wiley & Sons, Ltd, New York, NY (USA). 695 p.

Clark I and Harper WV 2000 *Practical Geostatistics 2000*. Ecosse North America Llc, Columbus, OH (USA). 342 p.

Diggle PJ, Tawn JA and Moyeed RA 1998 Model-based geostatistics (with discussion). *Journal of the Royal Statistical Society, Series C (Applied Statistics)* **47**(3), 299–350.

Egozcue JJ and Pawlowsky-Glahn V 2005 Groups of parts and their balances in compositional data analysis. *Mathematical Geology* **37**(7), 795–828.

Egozcue JJ, Pawlowsky-Glahn V, Mateu-Figueras G and Barceló-Vidal C 2003 Isometric logratio transformations for compositional data analysis. *Mathematical Geology* **35**(3), 279–300.

Egozcue JJ, Tolosana-Delgado R and Pawlowsky-Glahn V 2007 Simplicial indicator kriging for boundary assessment. In *Proceedings of IAMG'07 – The XII Annual Conference of the International Association for Mathematical Geology* (ed. Cheng Q, Agterberg F and Pengda Z). China University of Geosciences, Beijing (China). pp. 36–38.

Ehm AE 1965 Lyons West Field *Kansas Oil and Gas Fields*, vol. 4. Kansas Geological Society, Wichita, KS (USA). pp. 861–874.

Flury B 1988 *Common Principal Components and Related Multivariate Models*. John Wiley & Sons, Ltd, New York, NY (USA). 258 p.

Isaaks EH and Srivastava RM 1989 *An Introduction to Applied Geostatistics*. Oxford University Press, New York, NY (USA). 561 p.

Journel AG 1983 Nonparametric estimation of spatial distributions *Mathematical Geology* **15**(3), 445–468.

Journel AG and Huijbregts CJ 1978 *Mining Geostatistics*. Academic Press, London (UK). 600 p.

Martín-Fernández JA, Barceló-Vidal C and Pawlowsky-Glahn V 2003 Dealing with zeros and missing values in compositional data sets using nonparametric imputation. *Mathematical Geology* **35**(3), 253–278.

Myers DE 1984 Co-kriging: New developments. In *Geostatistics for Natural Resources Characterization* (ed. Verly G, David M, Journel A and Marechal A). D. Reidel Publishing Co., Dordrecht (The Netherlands). pp. 295–305.

Pawlowsky V 1984 On spurious spatial covariance between variables of constant sum. *Science de la Terre, Série Informatique* **21**, 107–113.

Pawlowsky-Glahn V 2003 Statistical modelling on coordinates. In *Proceedings of CoDaWork'03, The 1st Compositional Data Analysis Workshop* (ed. Thió-Henestrosa S and Martín-Fernández JA). http://ima.udg.es/Activitats/CoDaWork03/. University of Girona, Girona (Spain). CD-ROM.

Pawlowsky-Glahn V and Olea RA 2004 *Geostatistical Analysis of Compositional Data*. Oxford University Press, New York, NY (USA). 304 p.

Roeder J, Tolosana-Delgado R and Hamprecht F 2011 Indicator kriging and gaussian process classification: A comparison of several prediction methods. *Stochastic Environmental Research and Risk Assessment* (submitted).

Switzer P and Green AA 1984 Min/max autocorrelation factors for multivariate spatial imaging. Technical Report 6, Department of Statistics, Stanford University (USA).

Tjelmeland H and Lund KV 2003 Bayesian modelling of spatial compositional data. *Journal of Applied Statistics* **30**(1), 87–100.

Tolosana-Delgado R 2006 *Geostatistics for constrained variables: positive data, compositions and probabilities. Application to environmental hazard monitoring*. PhD thesis, University of Girona, Girona (Spain).

Tolosana-Delgado R, Sánchez-Arcilla A and Gómez J 2010 Classifying wave forecasts with model-based geostatistics and the aitchison distribution. In *8th International Conference on Geostatistics for Environmental Applications* (ed. Cockx L, van Meirvenne M, Bogaert P and D'Or D). Ghent University, Ghent (Belgium). pp. 25–27.

Wackernagel H 1998 *Multivariate Geostatistics, An Introduction With Applications*, 2nd edition. Springer Verlag, Berlin (Germany). 291 p.

Xie T and Myers DE 1995a Fitting matrix-valued variogram models by simultaneous diagonalization (Part I: Theory). *Mathematical Geology* **27**(7), 867–875.

Xie T and Myers DE 1995b Fitting matrix-valued variogram models by simultaneous diagonalization (Part II: Application). *Mathematical Geology* **27**(7), 877–888.

7

Compositional VARIMA time series

Carles Barceló-Vidal[1], Lucía Aguilar[2] and Josep Antoni Martín-Fernández[1]
[1]*Department of Computer Science and Applied Mathematics, University of Girona, Spain*
[2]*Department of Mathematics, Technical School, University of Extremadura, Spain*

7.1   Introduction

Multivariate time series of proportions, or compositions, arise in many areas of application. Such series are characterized by D non-negative components $x_{1t}, \ldots, x_{Dt}$ which sum to a constant at each time t. Without loss of generality, it can be assumed that the constant in question is 1. It is usual to refer to the series $\{\mathbf{x}_t : t = 1, \ldots, n\}$, where $\mathbf{x}_t = (x_{1t}, \ldots, x_{Dt})^\top$, as a *compositional time series*, or CTS for short. The $\mathbf{x}_t$ are elements of the simplex $\mathcal{S}^D$. Data of this kind frequently arise in disciplines as disparate as biology, demography, ecology, economics, geology and politics. Although a CTS constitutes a multivariate time series, standard techniques such as those in multivariate autoregressive integrated moving average (VARIMA) modelling are not applicable because of the constant sum constraint (Barceló-Vidal *et al.* 2007).

Historically, CTS modelling has been based almost exclusively on the transformation approach, which consists of the application of an initial transformation to break the unit sum constraint followed by the use of standard techniques to model the transformed time series. Then, the possibility of VARIMA modelling is available. In this context, one of the most frequently employed transformations has been the *additive log-ratio* (alr) transformation. This transformation depends on the choice of the component used as the common denominator in the log-ratios, so there are as many possible alr transformations as the number of parts, D, of the compositional data. The VARIMA based modelling approach of alr-transformed compositional time series has been employed by Brunsdon (1987), Smith and Brunsdon (1989) and Brunsdon and Smith (1998). Ravishanker *et al.* (2001) generalized the approach of Brunsdon and Smith (1998) to an extension of VARMA models incorporating covariates. Recent applications of this approach appear in Mills (2009 2010). Most of these contributions conclude that the final model obtained using the approach is invariant to the choice of the component used in the common denominator of the alr transformation. This is true if, as is the case in the cited publications, only full VARIMA models are contemplated. However, none of these contributions studies the dependence of the final model on the alr transformation used when, in the model building process, the possibility of post-estimation simplification of the model using restricted models is contemplated. Other contributions to CTS analysis using the alr transformation are Silva (1996) and Silva and Smith (2001), who employ a state-space modelling approach for the transformed time series.

The *centred* (or *symmetric*) *log-ratio* transformation (clr) was used by Quintana and West (1988) to analyse CTS data using a type of dynamic regression model which allowed for subjective as well as exogenous interventions. They handled singularities of the covariances matrices associated with the clr-transformed time series $\{z_t : t = 1, \ldots, n\}$ ignoring the zero-sum constraint on the z_t. They modelled the z_t series assuming non-singularity of the covariances matrices and by imposing *a posteriori* the zero-sum constraint on the estimated model. Direct modelling of the clr-transformed series is impossible due to the singularity of the covariance matrices, and the strategies used to date to circumvent this problem are simply incorrect.

Bergman (2008) used the *isometric log-ratio* (ilr) to fit a VAR model to monthly compositional time series from the Swedish Labour Force Survey. The ilr transformation depends on the orthonormal basis of $\mathcal{S}^D$ chosen in its definition. Bergman (2008) employed only full models for the ilr-transformed data, so that the final models do not depend on the orthonormal basis used in the ilr transformation.

Bhaumik *et al.* (2003) used the well-known Box–Cox transformation applied to ratios of components of a CTS as an alternative to the alr-transformation. These authors modelled the Box–Cox transformed ratios using dynamic linear models incorporating a rich class of distributions for the errors based on scale mixtures of multivariate normal distributions. The Box–Cox transformation has the advantage of including the alr transformation as a special case but has the disadvantage of increasing the number of parameters to be estimated. As with the alr transformation, the Box–Cox transformation also depends on the chosen denominator in the ratios.

If the x_t series contains components that are exactly zero, none of the log-ratio transformations (alr, clr nor ilr) can be applied directly. In an attempt to resolve this problem, some authors (Bell *et al.* 1986) follow the suggestions of Aitchison (1986) and Martín-Fernández *et al.* (2003) and propose the replacement of any zeros by 'very small' positive values. However, if x_t contains many zeros, the replaced values can have a very strong effect on the estimation procedure. These problems have led some authors (Nolan and Smith 1995; Wang *et al.* 2007) to propose the hyperspherical transformation, based on the arccos function, as a means of circumventing them. However, the fact that this transformation fails if one of the components of x_t equals 1, and the resulting models are very difficult to interpret, provide strong arguments against the use of this transformation.

Direct modelling on the simplex of CTS has been used by Grunwald *et al.* (1993). These authors developed state–space models in which the distribution of the CTS conditioned on the unobserved state is assumed to be Dirichlet distributed. In turn, the state distribution is assumed to be conjugate Dirichlet.

In Section 7.2 we provide a brief summary of the basic elements underpinning Compositional Data Analysis (CODA), placing special emphasis on the fact that the use of any one of the log-ratio transformations is equivalent to the selection of a particular base to represent vectors in $\mathcal{S}^D$ space. In Section 7.3 we define the basic elements associated with any CTS that are independent of the representation chosen. We also define $\mathcal{C}$-VARIMA models, by so doing completing the work initiated by Barceló-Vidal *et al.* (2007). Although the different representations in $\mathcal{S}^D$ are mathematically equivalent, in Section 7.4 we illustrate, using a CTS data set, how, in the application of the usual model building procedure, the chosen representation influences the final model and its interpretation. In Section 7.5, we present some final conclusions and provide some practical advice regarding the modelling of compositional time series. The basic time series analysis concepts we make use of in this chapter are described in standard textbooks on multivariate time series, such as Peña *et al.* (2001) and Lütkepohl (2005). For those concepts associated with compositional data analysis, see Aitchison (1986) and also Chapter 11.

7.2 The simplex $\mathcal{S}^D$ as a compositional space

7.2.1 Basic concepts and notation

The *centred* (or *symmetric*) *log-ratio transformation* (clr) is the mapping from the compositional space $\mathcal{S}^D$ to $\mathbb{R}^D$, defined by $\mathbf{x} \longrightarrow \mathbf{z} = \text{clr}(\mathbf{x}) = \log\left[\mathbf{x}/g_{\text{m}}(\mathbf{x})\right]$, where $g_{\text{m}}(\mathbf{x})$ is the geometric mean of the components of $\mathbf{x}$, i.e., $g_{\text{m}}(\mathbf{x}) = (x_1 x_2 \dots x_D)^{1/D}$. This transformation maps $\mathcal{S}^D$ in the subspace $V = \{\mathbf{z} \in \mathbb{R}^D : z_1 + \dots + z_D = 0\}$ of $\mathbb{R}^D$, which can be seen to be a hyperplane through the origin of $\mathbb{R}^D$, orthogonal to $\mathbf{1}_D$ (vector of units). The one-to-one linear transformation clr permits the translation of the real Euclidean structure defined in $\mathbb{R}^D$ to $\mathcal{S}^D$. In this way, $\mathcal{S}^D$ becomes an Euclidean space and the transformation clr is the natural isometry between $\mathcal{S}^D$ and the subspace V of $\mathbb{R}^D$.

Let $\{\mathbf{v}_1, \dots, \mathbf{v}_{D-1}\}$ be any orthonormal basis of V, and $\mathbf{V}$ be the $D \times (D-1)$ matrix $[\mathbf{v}_1 : \dots : \mathbf{v}_{D-1}]$. Then the *isometric log-ratio transformation* associated with the matrix $\mathbf{V}$ (ilr$_{\text{V}}$) is the one-to-one linear transformation from $\mathcal{S}^D$ to $\mathbb{R}^{D-1}$ which assigns to each composition $\mathbf{x}$ the coordinates of clr$(\mathbf{x})$ with respect to the orthonormal basis $\{\mathbf{v}_1, \dots, \mathbf{v}_{D-1}\}$ of V. Therefore, the components of ilr$_{\text{V}}(\mathbf{x})$ can also be viewed as the coordinates of the composition $\mathbf{x}$ with respect to the $\mathcal{C}$-orthonormal basis $\{\text{clr}^{-1}(\mathbf{v}_1), \dots, \text{clr}^{-1}(\mathbf{v}_{D-1})\}$ of the Euclidean vector space $\mathcal{S}^D$. It can be proved that ilr$_{\text{V}}(\mathbf{x}) = (\mathbf{FV})^{-1}\mathbf{F}\log(\mathbf{x})$, for any $\mathbf{x} \in \mathcal{S}^D$, where $\mathbf{F}$ is the $(D-1) \times D$ matrix $[\mathbf{I}_{D-1} : -\mathbf{1}_{D-1}]$.

Similarly, the *additive log-ratio transformation* of index k ($k = 1, \dots, D$) (alr$_k$) is the one-to-one linear transformation from $\mathcal{S}^D$ to $\mathbb{R}^{D-1}$ defined as $\mathbf{x} \longrightarrow \mathbf{y} = \text{alr}_k(\mathbf{x}) = \log(\mathbf{x}_{-k}/x_k)$, where $\mathbf{x}_{-k}$ denotes the vector $\mathbf{x}$ with the component x_k deleted. In particular, we use alr –without any subscript– to denote the transformation alr$_D$. The components of the vector alr$_k(\mathbf{x})$ are the coordinates of this vector with respect to the canonical basis $\mathbf{e}_1 = (1, 0, \dots, 0)^\top, \mathbf{e}_2 = (0, 1, 0, \dots, 0)^\top, \dots, \mathbf{e}_{D-1} = (0, 0, \dots, 0, 1)^\top$ of $\mathbb{R}^{D-1}$. Therefore, the components of the vector alr$_k(\mathbf{x})$ can also be viewed as the coordinates of the composition $\mathbf{x}$ in the $\mathcal{C}$-basis $\{\text{alr}_k^{-1}(\mathbf{e}_1), \dots, \text{alr}_k^{-1}(\mathbf{e}_{D-1})\}$ of the vector space $\mathcal{S}^D$. It is easy to prove that $\mathrm{alr}_k(\mathbf{x}) = \mathbf{F}_k \log(\mathbf{x})$, for any $\mathbf{x} \in \mathcal{S}^D$, where $\mathbf{F}_k$ is the $(D-1) \times D$ matrix $[\mathbf{e}_1 : \ldots : \mathbf{e}_{k-1} : -\mathbf{1}_{D-1} : \mathbf{e}_k : \ldots : \mathbf{e}_{D-1}]$. In particular, $\mathrm{alr}(\mathbf{x}) = \mathbf{F} \log(\mathbf{x})$. It is important to note that the $\mathcal{C}$-basis $\{\mathrm{alr}_k^{-1}(\mathbf{e}_1), \ldots, \mathrm{alr}_k^{-1}(\mathbf{e}_{D-1})\}$ is not an orthonormal basis of the Euclidean vector space $\mathcal{S}^D$.

In summary, we can interpret the components of the transformed vectors $\mathrm{alr}_k(\mathbf{x})$ $(k = 1, \ldots, D)$ and ilr_V –for any orthonormal basis $\mathbf{V}$ of V– as the coordinates of the same composition $\mathbf{x}$ with respect to different bases of the Euclidean vector space $(\mathcal{S}^D, \oplus, \odot)$. Therefore, the vectors $\mathbf{u} = \mathrm{ilr}_V(\mathbf{x})$, $\mathbf{y} = \mathrm{alr}(\mathbf{x})$ and $\mathbf{z} = \mathrm{clr}(\mathbf{x})$ associated with the same composition $\mathbf{x} \in \mathcal{S}^D$ will be related by linear relationships (see Appendix).

Let $\mathbf{x}$ be a random D-part composition defined in $\mathcal{S}^D$. The $\mathcal{C}$-mean of $\mathbf{x}$, symbolized by $\boldsymbol{\xi}$ or $\mathrm{E}_\mathcal{C}\{\mathbf{x}\}$, has to be defined according to the metric structure of $\mathcal{S}^D$. Pawlowsky-Glahn and Egozcue (2002) prove that $\mathrm{E}_\mathcal{C}\{\mathbf{x}\}$ is nothing more than the closure of the geometric mean of $\mathbf{x}$, i.e.

$$\boldsymbol{\xi} = \mathrm{E}_\mathcal{C}\{\mathbf{x}\} = \mathcal{C}\left(\exp\left(\mathrm{E}\{\log x_1\}\right), \ldots, \exp\left(\mathrm{E}\{\log x_D\}\right)\right)^\top,$$

where $\mathcal{C}$ denotes the *closure* operator defined for any $\mathbf{x} \in \mathbb{R}_+^D$ as $\mathcal{C}(\mathbf{x}) = \mathbf{x}/\sum_{i=1}^D x_i$. It holds that $\mathrm{alr}(\boldsymbol{\xi}) = \boldsymbol{\mu}_Y$, $\mathrm{clr}(\boldsymbol{\xi}) = \boldsymbol{\mu}_Z$, and $\mathrm{ilr}(\boldsymbol{\xi}) = \boldsymbol{\mu}_U$, where $\boldsymbol{\mu}_Y = \mathrm{E}\{\mathrm{alr}(\mathbf{x})\}$, $\boldsymbol{\mu}_Z = \mathrm{E}\{\mathrm{clr}(\mathbf{x})\}$ and $\boldsymbol{\mu}_U = \mathrm{E}\{\mathrm{ilr}(\mathbf{x})\}$. The same linear relation will exist between $\boldsymbol{\mu}_Y$, $\boldsymbol{\mu}_Z$ and $\boldsymbol{\mu}_U$ as exists between their respective random vectors $\mathbf{y}$, $\mathbf{z}$ and $\mathbf{u}$.

Finally, we use $\mathcal{A}_{D \times D}$ to denote the family of all real $D \times D$ matrices such that $\mathbf{A}\mathbf{1}_D = \mathbf{A}^\top \mathbf{1}_D = \mathbf{0}_D$. By definition, these matrices are singular. If $\mathbf{x} \in \mathcal{S}^D$ and $\mathbf{A} \in \mathcal{A}_{D \times D}$, we define the *product* $\mathbf{A} \odot \mathbf{x}$ as

$$\mathbf{A} \odot \mathbf{x} = \mathcal{C}\left(\prod_{j=1}^D x_j^{a_{1j}}, \ldots, \prod_{j=1}^D x_j^{a_{Dj}}\right)^\top.$$

Thus, the function $\mathbf{x} \to \mathbf{A} \odot \mathbf{x}$ is an endomorphism of the vector space $(\mathcal{S}^D, \oplus, \odot)$. Moreover, any endomorphism of $\mathcal{S}^D$ can be written in this form. The matrix associated with the identity endomorphism is the well-known *centring matrix* $\mathbf{G}_D = \mathbf{I}_D - D^{-1}\mathbf{J}_D$ of order $D \times D$, where $\mathbf{J}_D$ is the $D \times D$ unity matrix $\mathbf{1}_D \mathbf{1}_D^\top$.

7.2.2   The covariance structure on the simplex

The existence of three types of log-ratio transformation (alr, ilr and clr) allows us to define the covariance structure of a random D-part composition $\mathbf{x}$ defined in $\mathcal{S}^D$ in three different, although equivalent, ways via the covariance matrices of $\mathrm{alr}(\mathbf{x})$, $\mathrm{clr}(\mathbf{x})$ and $\mathrm{ilr}(\mathbf{x})$. The symmetry of the clr transformation and its independence of the denominators and bases used leads us to the choice of the covariance matrix of $\mathrm{clr}(\mathbf{x})$ as the *compositional covariance matrix*.

Let $\mathbf{x}$ be a random D-part composition defined in $\mathcal{S}^D$. We define the $\mathcal{C}$-covariance matrix of $\mathbf{x}$ as

$$\boldsymbol{\Sigma}_\mathcal{C} = \left[\mathrm{Cov}\left\{\log \frac{x_i}{g_m(\mathbf{x})}, \log \frac{x_j}{g_m(\mathbf{x})}\right\}\right]_{i,j=1}^D,$$

i.e. as the covariance matrix of the random vector $\mathbf{z} = \text{clr}(\mathbf{x})$, known as the *centred log-ratio matrix*. The singularity of the distribution of $\mathbf{z} = \text{clr}(\mathbf{x})$ is reflected in the singularity of $\mathbf{\Sigma}_C$ since this covariance matrix belongs to the family of $\mathcal{A}_{D \times D}$ matrices.

The *log-ratio covariance matrix* of $\mathbf{x}$ is defined as

$$\mathbf{\Sigma}_Y = \left[\text{Cov}\left\{ \log \frac{x_i}{x_D}, \log \frac{x_j}{x_D} \right\} \right]_{i,j=1}^{D-1},$$

i.e. by the covariance matrix of the random vector $\mathbf{y} = \text{alr}(\mathbf{x})$ in $\mathbb{R}^{D-1}$. It is clear that $\mathbf{\Sigma}_Y$ will depend on the denominator used in the alr transformation. We will use the notation $\mathbf{\Sigma}_{Y_k}$ for the covariance matrix of the random vector $\text{alr}_k(\mathbf{x})$ when we wish to stress the denominator used in the alr transformation.

Finally, the covariance matrix of the random vector $\mathbf{u} = \text{ilr}(\mathbf{x})$ in $\mathbb{R}^{D-1}$ will be denoted by $\mathbf{\Sigma}_U$. This covariance matrix will depend on the matrix $\mathbf{V}$ used in the ilr transformation. It will be convenient to write $\mathbf{\Sigma}_{U_V}$ when it is of interest to highlight the orthonormal base, $\mathbf{V}$, used.

Although the C-covariance structure of $\mathbf{x}$ is given by $\mathbf{\Sigma}_C$, the linear relationships between the vectors $\mathbf{z}, \mathbf{y}$ and $\mathbf{u}$ (see Appendix) give rise to the well known relations between the corresponding covariance matrices $\mathbf{\Sigma}_C, \mathbf{\Sigma}_{Y_k}$ and $\mathbf{\Sigma}_{U_V}$.

In the same way as we have done with the C-covariance matrix of a random compositional vector in $\mathcal{S}^D$, we will define the covariance structure of the joint distribution of a bivariate random compositional vector $(\mathbf{x}_1, \mathbf{x}_2)$ defined in $\mathcal{S}^D \times \mathcal{S}^D$ via the clr-transformed bivariate random vector.

Let $(\mathbf{x}_1, \mathbf{x}_2)$ be a bivariate random compositional vector defined in $\mathcal{S}^D \times \mathcal{S}^D$. If $\boldsymbol{\xi}_i = E_C\{\mathbf{x}_i\}$ (where $i = 1, 2$), the C-covariance matrix of $(\mathbf{x}_1, \mathbf{x}_2)$ is defined as

$$\mathbf{\Gamma}_C(\mathbf{x}_1, \mathbf{x}_2) = \left[E\left\{ \left(\log \frac{x_{1i}}{g(\mathbf{x}_1)} - \log \frac{\xi_{1i}}{g(\boldsymbol{\xi}_1)} \right) \left(\log \frac{x_{2j}}{g(\mathbf{x}_2)} - \log \frac{\xi_{2j}}{g(\boldsymbol{\xi}_2)} \right) \right\} \right]_{i,j=1}^{D}.$$

Therefore, $\mathbf{\Gamma}_C(\mathbf{x}_1, \mathbf{x}_2)$ coincides with the covariance matrix of the bivariate random vector $(\mathbf{z}_1, \mathbf{z}_2) = (\text{clr}(\mathbf{x}_1), \text{clr}(\mathbf{x}_2))$ defined in $V \times V \subset \mathbb{R}^D \times \mathbb{R}^D$. The matrix $\mathbf{\Gamma}_C(\mathbf{x}_1, \mathbf{x}_2)$ is not symmetric but is singular because it belongs to the family of $\mathcal{A}_{D \times D}$ matrices.

We denote by $\mathbf{\Gamma}_Y(\mathbf{y}_1, \mathbf{y}_2)$ the covariance matrix of $(\mathbf{y}_1, \mathbf{y}_2) = (\text{alr}(\mathbf{x}_1), \text{alr}(\mathbf{x}_2))$, and by $\mathbf{\Gamma}_U(\mathbf{u}_1, \mathbf{u}_2)$ the covariance matrix of $(\mathbf{u}_1, \mathbf{u}_2) = (\text{ilr}(\mathbf{x}_1), \text{ilr}(\mathbf{x}_2))$. As before, it will be convenient to denote the covariance matrix of $(\text{alr}_k(\mathbf{x}_1), \text{alr}_k(\mathbf{x}_2))$ by $\mathbf{\Gamma}_{Y_k}$ when we want to identify the denominator used in the alr transformation. Similarly, we write $\mathbf{\Gamma}_{U_V}$ when it is relevant to identify the orthonormal base, $\mathbf{V}$, used in the ilr transformation. As before, matrix relationships exist between the covariance matrices $\mathbf{\Gamma}_C, \mathbf{\Gamma}_{Y_k}$ and $\mathbf{\Gamma}_{U_V}$.

7.3   Compositional time series models

Let $\{\mathbf{x}_t : t = 0, \pm 1, \pm 2, \ldots\}$ be a compositional time series formed by random variables of the form $\mathbf{x}_t = (x_{1t}, \ldots, x_{Dt})^\top$ defined in $\mathcal{S}^D$ (i.e. a process). The second-order properties of $\{\mathbf{x}_t\}$ are then specified by the C-mean vectors, $\boldsymbol{\xi}_t = E_C\{\mathbf{x}_t\} = (\xi_{t1}, \ldots, \xi_{tD})^\top$, and the

$\mathcal{C}$-autocovariance matrices,

$$\boldsymbol{\Gamma}_C(t+h,t) = \mathrm{E}\left\{(\mathrm{clr}(\mathbf{x}_{t+h}) - \mathrm{clr}\,(\boldsymbol{\xi}_{t+h}))(\mathrm{clr}(\mathbf{x}_t) - \mathrm{clr}\,(\boldsymbol{\xi}_t))^\top\right\} = \left[\gamma_{C,ij}(t+h,t)\right]_{i,j=1}^D.$$

Notice that, in the compositional context, given a compositional time series $\{\mathbf{x}_t\}$ it makes no sense to analyse any of the individual parts $\{x_{it}\}$ as univariate time series. However, in some cases one might be interested in analysing the relative behaviour of two parts i and j ($i \neq j$) or, in general, of a subcompositional time series $\{\mathbf{x}_{St}\}$, where S symbolizes a subset of two or more of the parts $1, \ldots, D$ of $\mathbf{x}_t$.

When applied to a compositional process $\{\mathbf{x}_t\}$, the clr, alr$_k$ and ilr$_V$ transformations induce the processes $\{\mathbf{z}_t\}$, $\{\mathbf{y}_t\}$ and $\{\mathbf{u}_t\}$, respectively. The former, $\{\mathbf{z}_t\}$, defined in $\mathbb{R}^D$, is restricted to the hyperplane V because $\mathbf{z}_t^\top \mathbf{1}_D = 0$. The other two processes are defined in $\mathbb{R}^{D-1}$ but $\{\mathbf{y}_t\}$ depends on the denominator used in the alr$_k$ transformation and $\{\mathbf{u}_t\}$ on the matrix $\mathbf{V}$ used in the ilr$_V$ transformation. We denote by $\boldsymbol{\mu}_{Z,t}$, $\boldsymbol{\mu}_{Y,t}$ and $\boldsymbol{\mu}_{U,t}$ the mean vectors of $\{\mathbf{z}_t\}$, $\{\mathbf{y}_t\}$ and $\{\mathbf{u}_t\}$, respectively, and by $\boldsymbol{\Gamma}_Z(t+h,t)$, $\boldsymbol{\Gamma}_Y(t+h,t)$ and $\boldsymbol{\Gamma}_U(t+h,t)$ the autocovariance matrices of these processes. From Section 7.2, all of these elements are related through matrix formulae. Observe that $\boldsymbol{\mu}_{Z,t} = \mathrm{clr}\,(\boldsymbol{\xi}_t)$ and, by definition, $\boldsymbol{\Gamma}_Z(t+h,t) = \boldsymbol{\Gamma}_C(t+h,t)$.

7.3.1 $\mathcal{C}$-stationary processes

By analogy with the standard definition, a compositional process $\{\mathbf{x}_t\}$ is said to be (weakly) $\mathcal{C}$-*stationary* if $\boldsymbol{\xi}_t$ and $\boldsymbol{\Gamma}_C(t+h,t)$, $h = 0, \pm 1, \ldots$ are independent of t. For a $\mathcal{C}$-stationary process we use the notation

$$\boldsymbol{\xi} = \mathrm{E}_C\{\mathbf{x}_t\}\,;\, \boldsymbol{\Gamma}_C(h) = \mathrm{E}\left\{(\mathrm{clr}(\mathbf{x}_{t+h}) - \mathrm{clr}\,(\boldsymbol{\xi}))(\mathrm{clr}(\mathbf{x}_t) - \mathrm{clr}\,(\boldsymbol{\xi}))^\top\right\} = \left[\gamma_{C,ij}(h)\right]_{i,j=1}^D.$$

So we shall refer to $\boldsymbol{\xi}$ as the $\mathcal{C}$-mean of $\{\mathbf{x}_t\}$, to $\boldsymbol{\Gamma}_C(h)$ as the $\mathcal{C}$-autocovariance at lag h, and to $\boldsymbol{\Gamma}_C(h)_{h=0,1,\ldots}$ as the $\mathcal{C}$-autocovariance function. The $\mathcal{C}$-autocorrelation function $\mathbf{R}_C(h)_{h=0,1,\ldots}$ is defined by

$$\mathbf{R}_C(h) = \left[\gamma_{C,ij}(h)/\sqrt{\gamma_{C,ii}(0)\gamma_{C,jj}(0)}\right]_{i,j=1}^D = \left[\rho_{C,ij}(h)\right]_{i,j=1}^D.$$

In consequence, the $\mathcal{C}$-stationary property of $\{\mathbf{x}_t\}$ is equivalent to the stationary property of any of the transformed processes $\{\mathbf{z}_t\}$, $\{\mathbf{y}_t\}$ and $\{\mathbf{u}_t\}$.

A compositional time series process $\{\mathbf{w}_t\}$ is said to be $\mathcal{C}$-*white noise* with $\mathcal{C}$-covariance matrix $\boldsymbol{\Sigma}_C$, denoted by $\{\mathbf{w}_t\} \sim \mathrm{WN}_C(\mathbf{1}_C, \boldsymbol{\Sigma}_C)$, if it has $\mathcal{C}$-mean vector $\mathbf{1}_C = (1/D, \ldots, 1/D)^\top$ and $\mathcal{C}$-autocovariance function

$$\boldsymbol{\Gamma}_C(0) = \boldsymbol{\Sigma}_C; \quad \boldsymbol{\Gamma}_C(h) = \mathbf{0}_{D \times D}, \text{ for } h \neq 0.$$

Obviously $\mathcal{C}$-*white noise* is $\mathcal{C}$-stationary. Note that $\{\mathbf{x}_t\}$ is $\mathcal{C}$-white noise if and only if $\{\mathbf{z}_t\}$, or $\{\mathbf{y}_t\}$, or $\{\mathbf{u}_t\}$, is white noise. If the elements of $\{\mathbf{y}_t\}$ (or $\{\mathbf{u}_t\}$) are independent and identically normally distributed, $\{\mathbf{x}_t\}$ is called $\mathcal{C}$-Gaussian white noise. In this case the elements of $\{\mathbf{z}_t\}$ are also independent and identically distributed as a degenerate (singular) normal distribution.

7.3.2 $\mathcal{C}$-VARIMA processes

Combining standard definitions and operations in the simplex space, for a D-part variate compositional time series $\{\mathbf{x}_t\}$, the class of multivariate autoregressive moving average, $\mathcal{C}$-VARMA(p, q), models takes the form,

$$(\mathbf{x}_t \ominus \boldsymbol{\xi}) \ominus \left(\boldsymbol{\Phi}_{\mathcal{C},1} \odot (\mathbf{x}_{t-1} \ominus \boldsymbol{\xi})\right) \ominus \ldots \ominus \left(\boldsymbol{\Phi}_{\mathcal{C},p} \odot (\mathbf{x}_{t-p} \ominus \boldsymbol{\xi})\right) =$$
$$\mathbf{w}_t \ominus \left(\boldsymbol{\Theta}_{\mathcal{C},1} \odot \mathbf{w}_{t-1}\right) \ominus \ldots \ominus \left(\boldsymbol{\Theta}_{\mathcal{C},q} \odot \mathbf{w}_{t-q}\right),$$

where $\boldsymbol{\Phi}_{\mathcal{C},1}, \ldots, \boldsymbol{\Phi}_{\mathcal{C},p}, \boldsymbol{\Theta}_{\mathcal{C},1}, \ldots, \boldsymbol{\Theta}_{\mathcal{C},q}$ are $\mathcal{A}_{D \times D}$-matrices and $\mathbf{w}_t \sim \mathrm{WN}_{\mathcal{C}}(\mathbf{1}_{\mathcal{C}}, \boldsymbol{\Sigma}_{\mathcal{C}})$. These equations can be written in the more concise form

$$\boldsymbol{\Phi}_{\mathcal{C}}(L_{\mathcal{C}})(\mathbf{x}_t \ominus \boldsymbol{\xi}) = \boldsymbol{\Theta}_{\mathcal{C}}(L_{\mathcal{C}})\mathbf{w}_t, \quad \{\mathbf{w}_t\} \sim \mathrm{WN}_{\mathcal{C}}(\mathbf{1}_{\mathcal{C}}, \boldsymbol{\Sigma}_{\mathcal{C}}),$$

where $\boldsymbol{\Phi}_{\mathcal{C}}(L_{\mathcal{C}}) = \mathbf{G}_D \ominus (\boldsymbol{\Phi}_{\mathcal{C},1} \odot L_{\mathcal{C}}) \ominus \ldots \ominus (\boldsymbol{\Phi}_{\mathcal{C},p} \odot L_{\mathcal{C}}^p)$ and $\boldsymbol{\Theta}_{\mathcal{C}}(L_{\mathcal{C}}) = \mathbf{G}_D \ominus (\boldsymbol{\Theta}_{\mathcal{C},1} \odot L_{\mathcal{C}}) \ominus \ldots \ominus (\boldsymbol{\Theta}_{\mathcal{C},q} \odot L_{\mathcal{C}}^q)$ are $\mathcal{A}_{D \times D}$-matrix-valued polynomials, $\mathbf{G}_D$ is the centring matrix and $L_{\mathcal{C}}$ is the $\mathcal{C}$-backshift operator. In the compositional context, the operator $1 - L_{\mathcal{C}}$ represents the $\mathcal{C}$-difference operator, i.e., $(1 - L_{\mathcal{C}})\mathbf{x}_t = \mathbf{x}_t \ominus \mathbf{x}_{t-1}$. Applying $1 - L_{\mathcal{C}}$ to $\{\mathbf{x}_t\}$ is equivalent to applying $1 - L$ to the transformed processes $\{\mathbf{z}_t\}$, $\{\mathbf{y}_t\}$ and $\{\mathbf{u}_t\}$.

If $\{\mathbf{x}_t\}$ is a $\mathcal{C}$-VARMA(p, q) process then $\{\mathbf{z}_t\}$ is a VARMA(p, q) process because

$$\boldsymbol{\Phi}_{\mathrm{Z}}(L)(\mathbf{z}_t - \boldsymbol{\mu}_{\mathrm{Z}}) = \boldsymbol{\Theta}_{\mathrm{Z}}(L)\mathbf{w}_{\mathrm{Z},t}, \quad \{\mathbf{w}_{\mathrm{Z},t}\} \sim \mathrm{WN}(\mathbf{0}_D, \boldsymbol{\Sigma}_{\mathcal{C}}),$$

where $\boldsymbol{\Phi}_{\mathrm{Z}}(L) = \mathbf{G}_D - \sum_{i=1}^p \boldsymbol{\Phi}_{\mathcal{C},i} L^i$, $\boldsymbol{\Theta}_{\mathrm{Z}}(L) = \mathbf{G}_D - \left(\sum_{i=1}^q \boldsymbol{\Theta}_{\mathcal{C},i} L^i\right)$ and $\boldsymbol{\mu}_{\mathrm{Z}} = \mathrm{clr}(\boldsymbol{\xi})$. From the above relations between the transformed processes, it follows that $\{\mathbf{y}_t\}$ is a VARMA(p, q) process because

$$\boldsymbol{\Phi}_{\mathrm{Y}}(L)(\mathbf{y}_t - \boldsymbol{\mu}_{\mathrm{Y}}) = \boldsymbol{\Theta}_{\mathrm{Y}}(L)\mathbf{w}_{\mathrm{Y},t}, \quad \{\mathbf{w}_{\mathrm{Y},t}\} \sim \mathrm{WN}(\mathbf{0}_{D-1}, \boldsymbol{\Sigma}_{\mathrm{Y}}),$$

where

$$\boldsymbol{\Phi}_{\mathrm{Y}}(L) = \mathbf{I}_{D-1} - \left(\sum_{i=1}^p \mathbf{F}\boldsymbol{\Phi}_{\mathcal{C},i}\mathbf{F}^\top \mathbf{H}^{-1} L^i\right),$$

$$\boldsymbol{\Theta}_{\mathrm{Y}}(L) = \mathbf{I}_{D-1} - \left(\sum_{i=1}^q \mathbf{F}\boldsymbol{\Theta}_{\mathcal{C},i}\mathbf{F}^\top \mathbf{H}^{-1} L^i\right),$$

$\boldsymbol{\mu}_{\mathrm{Y}} = \mathrm{alr}(\boldsymbol{\xi})$ and $\boldsymbol{\Sigma}_{\mathrm{Y}} = \mathbf{F}\boldsymbol{\Sigma}_{\mathcal{C}}\mathbf{F}^\top$. Also, $\{\mathbf{u}_t\}$ is a VARMA(p, q) process because

$$\boldsymbol{\Phi}_{\mathrm{U}}(L)(\mathbf{u}_t - \boldsymbol{\mu}_{\mathrm{U}}) = \boldsymbol{\Theta}_{\mathrm{U}}(L)\mathbf{w}_{\mathrm{U},t}, \quad \{\mathbf{w}_{\mathrm{U},t}\} \sim \mathrm{WN}(\mathbf{0}_{D-1}, \boldsymbol{\Sigma}_{\mathrm{U}}),$$

where

$$\boldsymbol{\Phi}_{\mathrm{U}}(L) = \mathbf{I}_{D-1} - \left(\sum_{i=1}^p \mathbf{V}^\top \boldsymbol{\Phi}_{\mathcal{C},i}\mathbf{V} L^i\right), \quad \boldsymbol{\Theta}_{\mathrm{U}}(L) = \mathbf{I}_{D-1} - \left(\sum_{i=1}^q \mathbf{V}^\top \boldsymbol{\Theta}_{\mathcal{C},i}\mathbf{V} L^i\right),$$

$\mu_U = \text{ilr}(\xi)$ and $\Sigma_U = V^\top \Sigma_C V$. Conversely, if $\{y_t\}$ or $\{u_t\}$ or $\{z_t\}$ is a VARMA(p, q) process, then $\{x_t\}$ is a C-VARMA(p, q) process. The C-VARMA(p, q) model equation of $\{x_t\}$ can be deduced from the VARMA(p, q) model of $\{y_t\}$ or $\{u_t\}$.

In the context of standard VARMA models, a VARMA(p, q) process is stationary when all the roots of $|\Phi(L)| = 0$ are greater than one in absolute value. In this standard context, the VARIMA(p, d, q) class of models takes the form

$$\Phi(L)(1 - L)^d (z_t - \mu) = \Theta(L)w_t, \quad \{w_t\} \sim \text{WN}(0_D, \Sigma),$$

where $\Phi(L) = I_D - \sum_{i=1}^{p} \Phi_i L^i$ and $\Theta(L) = I_D - \left(\sum_{i=1}^{q} \Theta_i L^i\right)$.

Thus, this model simply states that the process $\{z_t\}$ is nonstationary but the process $(1 - L)^d (z_t - \mu)$ is a stationary VARMA(p, q) process. In a natural way, then, for a compositional time series $\{x_t\}$, the C-VARIMA(p, d, q) class of models has the form

$$\Phi_C(L_C)(1 - L_C)^d (x_t - \xi) = \Theta_C(L_C)w_t, \quad \{w_t\} \sim \text{WN}_C(1_C, \Sigma_C),$$

where $\Phi_C(z) = G_D \ominus (\Phi_{C,1} \odot z) \ominus \ldots \ominus (\Phi_{C,p} \odot z^p)$ and $\Theta_C(z) = G_D \ominus (\Theta_{C,1} \odot z) \ominus \ldots \ominus (\Theta_{C,q} \odot z^q)$ are $\mathcal{A}_{D \times D}$-matrix-valued polynomials.

7.4   CTS modelling: an example

From a compositional data analysis perspective, in practice the modelling of a CTS, $\{x_t : t = 1, \ldots, n\}$ defined in $\mathcal{S}^D$, should be conducted by modelling one of the different coordinate representations of the CTS which, as we have seen in Section 7.2, are directly associated with the different transformations one can apply to compositional data. It would appear logical to ask if the resulting final model will depend on the representation used; be it that which results from the application of one of the transformations $\text{alr}_1, \ldots, \text{alr}_D$, or be it that based on one of the infinite possible transformations ilr_V that might be applied on varying the orthonormal base V of the subspace V of $\mathbb{R}^D$. If the modelling of CTS depends on the representation used then objective criteria for choosing the best transformation to apply will clearly be of interest. In this section we base the discussion of these issues around the analysis of an example.

7.4.1   Expenditure shares in the UK

We consider the CTS of quarterly consumption (x_1), investment (x_2), government expenditure (x_3) and exports (x_4) shares of UK gross final expenditure from the first quarter of 1955 to the final quarter of 2005. This data set was used by Mills (2010) to obtain forecasts out to the final quarter of 2008. Mills transformed the initial CTS by means of the alr transformation using x_4 (exports) as the denominator of the log-ratio transformation. A VARIMA$(1,1,0)$ model was then fitted to the log-ratios. After imposing some reasonable restrictions to enhance interpretability, the model proposed by Mills (2010) was

$$(1 - L)(\text{alr}_4(\mathbf{x}_t)) = \begin{bmatrix} 0 & 0 & 0 \\ 0 & 0 & \begin{matrix} -0.10 \\ (0.04) \end{matrix} \\ \begin{matrix} 0.33 \\ (0.07) \end{matrix} & \begin{matrix} -0.42 \\ (0.10) \end{matrix} & \begin{matrix} -0.33 \\ (0.07) \end{matrix} \end{bmatrix} (1 - L)(\text{alr}_4(\mathbf{x}_{t-1})) + \boldsymbol{\epsilon}_t,$$

where $\{\boldsymbol{\epsilon}_t\}$ is Gaussian white noise in $\mathbb{R}^3$. The standard errors of the nonzero parameter estimates are shown in parentheses.

The question of interest here is whether the final model would have been equivalent to that identified above if, rather than applying the alr$_4$ transformation (which uses x_4 as the denominator of the alr transformation), another of the transformations alr$_1$, alr$_2$ or alr$_3$ had been used, or, indeed, an ilr transformation associated with any particular base. Here we compare the results obtained from applying all four alr transforms and two ilr transforms associated with two different bases, assuming membership of the VARI(p, d) class of models for the transformed time series. The bases $\mathbf{V}_1$ and $\mathbf{V}_2$ used in the ilr transformations are arbitrary and are identified in the Appendix.

It is important to stress that, although we have introduced C-VARIMA models through the covariance and autocovariance structures of the clr transformed series, the modelling of a CTS should be carried out on the alr- or ilr-transformed series due to the singularity of the clr-transformed series.

7.4.2 Model selection

When modelling any given transformed CTS, we employed an iterative model building process consisting of specification, estimation and diagnostic checking for an assumed VARI(p, d) model. With regard to the tentative specification of a model for the time series $\{\text{alr}_1\mathbf{x}_t\}$, the proximity to unity of the moduli of the three eigenvalues of the matrix $\mathbf{\Phi}_{Y_1,1}$ for a fitted VARI(1, 0) model suggested the series should be differenced in order to achieve stationarity. As the eigenvalues of the matrices $\mathbf{\Phi}_{Y_k,1}$ and $\mathbf{\Phi}_{U_V,1}$ do not depend on the alr$_k$ or ilr$_V$ transformation used, the decision to difference the CTS would have been taken whichever of the two transformations had been used.

Assuming that the process $\{\mathbf{x}_t\}$ is C-Gaussian, we used likelihood ratio testing to identify the order p of a VARI($p, 1$) model. It is known that, when a nonsingular linear transformation is applied, the difference between the log-likelihood of the original data and that of the transformed data (for a specific model), is proportional to the sample size and the logarithm of the determinant (Jacobian) of the linear transformation. As the determinants of the matrices which allow one to move between the alr$_j$-transformed data and the alr$_k$-transformed data ($j \neq k$) equal ± 1, the log-likelihood of the alr-transformed data (for a specific model) does not depend on the denominator used in the transformation. For this reason, the identification of the order p of a VARI($p, 1$) model based on the use of the likelihood ratio test (LRT) does not depend on the alr transformation used. The same is true in relation to the orthonormal base, $\mathbf{V}$, used in the definition of the ilr-transformed data. Then again, given that the difference between the log-likelihood of the ilr- and the alr-transformed data remains constant for the different VARI($p, 1$) models explored, the successive values of the LRT statistic obtained from exploring the different VARI(1,1), VARI(2,1),... models for the alr-transformed data will coincide with those obtained for the ilr-transformed data. Thus, the selection of the order p of a VARI(p, 1) model based on the use of the LRT is independent of the alr or ilr transformation used. The same can be said in relation to the use of other selection criteria for the order p of a VARI(p, 1) model – such as Final Prediction Error (FPE), Akaike Information (AIC) or Hannan–Quinn (HQ) – given that they too are based on the log-likelihood of the data. For the data under consideration, the LRT as well as the FPE, AIC and HQ criteria identify a model with $p = 1$. Thus, finally, a VARI(1,1) model was chosen for the transformed CTS of expenditure shares in the UK.

7.4.3   Estimation of parameters

A VARI(1,1) model was fitted to each one of the four possible alr transformed time series and the two ilr transformed series. In Table 7.1 are listed the equations of the full models fitted to the alr_3- and ilr_1-transformed data, where the standard errors of the parameter estimates are shown in parentheses.

All six of the equations of the fitted models correspond to the same $\mathcal{C}$-VARI(1,1) model

$$(1 - L_{\mathcal{C}})\mathbf{x}_t = \begin{bmatrix} 0.250 \\ 0.250 \\ 0.250 \\ 0.250 \end{bmatrix} \oplus \begin{bmatrix} -0.053 & -0.031 & 0.066 & 0.018 \\ -0.097 & -0.183 & 0.118 & 0.163 \\ 0.202 & 0.264 & -0.207 & -0.259 \\ -0.052 & -0.050 & 0.024 & 0.077 \end{bmatrix} \odot (1 - L_{\mathcal{C}})\mathbf{x}_{t-1} \oplus \mathbf{w}_t,$$

where

$$\hat{\mathbf{\Sigma}}_{\mathcal{C}} = \begin{bmatrix} 2.978 & -0.742 & -4.455 & 2.218 \\ -0.742 & 8.137 & -6.275 & -1.120 \\ -4.445 & -6.275 & 16.491 & -5.761 \\ 2.218 & -1.120 & -5.761 & 4.663 \end{bmatrix} \times 10^{-4}$$

is the estimated $\mathcal{C}$-covariance matrix of the $\mathcal{C}$-Gaussian white noise $\{\mathbf{w}_t\}$.

As is usual, once the parameters of the full VARI(1,1) model had been estimated, we considered its potential simplification identifying those coefficients that could be omitted without reducing the log-likelihood significantly. In Table 7.1 those parameter estimates for the full model that are significantly different from zero ($\alpha = 0.05$) have been identified in boldface. For each one of the transformed series, restricted VARI(1,1) models were then fitted by sequentially forcing those coefficients that were not significantly different from zero to be equal to zero, starting with the least significant coefficient. In none of the cases was the reduction in the log-likelihood for the reduced model significantly different from that for the full model. Table 7.1 presents the final reduced models obtained from the full models fitted to the alr_3- and ilr_1-transformed data. Table A7.1 presents all six final reduced models expressed in compositional form, that is as $\mathcal{C}$-VARI(1,1) models.

7.4.4   Interpretation and comparison

A comparison of the different reduced $\mathcal{C}$-VARI(1,1) models in Table A7.1 immediately highlights the fact that the final fitted model depends on the transformation used in the modelling

Table 7.1 Fitted full and reduced VARI(1,1) models for the alr_3 and ilr_1 transformations of the UK expenditure shares series. In the full models, the parameters that are significantly different from zero ($\alpha = 0.05$) appear in bold. The standard errors of the nonzero parameter estimates appear in parentheses.

Full models

$$(1 - L)(\text{alr}_3(\mathbf{x}_t)) =$$

$$= \begin{bmatrix} -0.001 \\ (0.004) \\ 0.001 \\ (0.004) \\ 0.000 \\ (0.004) \end{bmatrix} + \begin{bmatrix} -0.255 & \mathbf{-0.295} & 0.277 \\ (0.232) & (0.123) & (0.206) \\ -0.299 & \mathbf{-0.447} & 0.422 \\ (0.265) & (0.118) & (0.236) \\ -0.254 & \mathbf{-0.313} & 0.337 \\ (0.248) & (0.110) & (0.221) \end{bmatrix} (1 - L)(\text{alr}_3(\mathbf{x}_{t-1})) + \boldsymbol{\epsilon}_t$$

$$(1 - L)(\text{ilr}_1(\mathbf{x}_t)) =$$

$$= \begin{bmatrix} -0.001 \\ (0.002) \\ 0.000 \\ (0.003) \\ -0.000 \\ (0.002) \end{bmatrix} + \begin{bmatrix} -0.054 & 0.086 & 0.118 \\ (0.095) & (0.055) & (0.112) \\ 0.054 & \mathbf{-0.415} & -0.329 \\ (0.168) & (0.098) & (0.199) \\ 0.002 & 0.070 & 0.104 \\ (0.094) & (0.055) & (0.111) \end{bmatrix} (1 - L)(\text{ilr}_1(\mathbf{x}_{t-1})) + \boldsymbol{\epsilon}_t$$

Reduced models

$$(1 - L)(\text{alr}_3(\mathbf{x}_t)) = \begin{bmatrix} 0 & -0.275 & 0 \\ & (0.057) & \\ 0 & -0.352 & 0 \\ & (0.006) & \\ 0 & -0.248 & 0 \\ & (0.062) & \end{bmatrix} (1 - L)(\text{alr}_3(\mathbf{x}_{t-1})) + \boldsymbol{\epsilon}_t$$

$$(1 - L)(\text{ilr}_1(\mathbf{x}_t)) = \begin{bmatrix} 0 & 0.136 & 0.207 \\ & (0.040) & (0.075) \\ 0 & -0.368 & -0.274 \\ & (0.066) & (0.124) \\ 0 & 0 & 0 \end{bmatrix} (1 - L)(\text{ilr}_1(\mathbf{x}_{t-1})) + \boldsymbol{\epsilon}_t$$

---

of the CTS. However, it is important to stress that the differences between the different $\boldsymbol{\Phi}$ matrices corresponding to the different models do not necessarily imply differences in terms of prediction. Thus, for example, the prediction of $\mathbf{x}_t$ for the first quarter of 2006 is, to three decimal places, $(0.492, 0.207, 0.132, 0.169)^\top$, whatever the transformation used. Any differences between the different predictions only manifest themselves in the fourth decimal place. Nevertheless, the differences in terms of the interpretation of the fitted models associated with the different transformations are certainly stark. Thus, for example, the reduced model obtained using the alr_3-transformed data would appear to be the most easily interpreted. From its representation in Table A7.1, this model can be written as

$$\frac{x_{1t}}{x_{1,t-1}} = k \left(\frac{x_{2,t-1}}{x_{2,t-2}}\right)^{-0.056} \times \left(\frac{x_{3,t-1}}{x_{3,t-2}}\right)^{0.056} \times w_{1t},$$

$$\frac{x_{2t}}{x_{2,t-1}} = k \left(\frac{x_{2,t-1}}{x_{2,t-2}}\right)^{-0.133} \times \left(\frac{x_{3,t-1}}{x_{3,t-2}}\right)^{0.133} \times w_{2t},$$

$$\frac{x_{3t}}{x_{3,t-1}} = k \left(\frac{x_{2,t-1}}{x_{2,t-2}}\right)^{0.219} \times \left(\frac{x_{3,t-1}}{x_{3,t-2}}\right)^{-0.219} \times w_{3t},$$

$$\frac{x_{4t}}{x_{4,t-1}} = k \left(\frac{x_{2,t-1}}{x_{2,t-2}}\right)^{-0.030} \times \left(\frac{x_{3,t-1}}{x_{3,t-2}}\right)^{0.030} \times w_{4t},$$

where $\{\mathbf{w}_t\}$ is C-white noise and k represents any positive constant.

From this model we see that the relative changes (increases or decreases) manifested by the four components x_1, x_2, x_3 and x_4 of $\mathbf{x}_t$ between two consecutive periods, $t-1$ and t, depend exclusively on the relative changes manifested by the components x_2 and x_3 in the previous transition from $t-2$ to $t-1$. According to the magnitudes of the exponents of the ratios that appear in these equations, the relative changes in the transition from $t-1$ to t most influenced by the relative changes in the immediately preceding transition are those corresponding to the components x_2 and x_3, above all the latter. Thus, the relative increase of x_3 in the transition from $t-1$ to t will be great if the relative increase of x_2 in the preceding transition was great and the relative decrease in x_3 was also great. Regarding x_2, given the transposition of the signs of the exponents, its relative changes behave in exactly the opposite directions to those for x_3.

The exponents of the ratios that appear in the equation for x_4 are all close to zero and hence the ratios $x_{4,t}/x_{4,t-1}$ are close to unity. Thus the relative changes in x_4 between consecutive periods are very weak.

It is evident that this simple interpretation arising from the use of the alr$_3$ transformation would have been difficult to draw on the basis of any of the other reduced models. This is not to say, however, that the different interpretations which can be drawn contradict one another.

We can also choose to interpret the reduced models via the equations of the log-ratios. Thus, regarding the reduced model associated with the alr$_3$-transformed data in Table 7.1, the equations are

$$\log \frac{x_{1t}}{x_{3t}} - \log \frac{x_{1,t-1}}{x_{3,t-1}} = -0.275 \left(\log \frac{x_{1,t-1}}{x_{3,t-1}} - \log \frac{x_{1,t-2}}{x_{3,t-2}}\right) + \epsilon_{1t},$$

$$\log \frac{x_{2t}}{x_{3t}} - \log \frac{x_{2,t-1}}{x_{3,t-1}} = -0.352 \left(\log \frac{x_{2,t-1}}{x_{3,t-1}} - \log \frac{x_{2,t-2}}{x_{3,t-2}}\right) + \epsilon_{2t},$$

$$\log \frac{x_{4t}}{x_{3t}} - \log \frac{x_{4,t-1}}{x_{3,t-1}} = -0.248 \left(\log \frac{x_{4,t-1}}{x_{3,t-1}} - \log \frac{x_{4,t-2}}{x_{3,t-2}}\right) + \epsilon_{3t},$$

where $\{\epsilon_t\}$ is white noise. From these equations, one deduces that the change (additive increase or decrease) of any one of the log-ratios $\log(x_{jt}/x_{3t})$ (where $j = 1, 2, 4$) between two consecutive periods $t-1$ to t tends to be in the opposite direction to that of the same log-ratio in the immediately preceding transition. The absolute magnitude of the change between $t - 1$ and t reduces to approximately a quarter of that between $t - 2$ and $t - 1$ for the log-ratios $\log(x_{1t}/x_{3t})$ and $\log(x_{4t}/x_{3t})$, and a third for $\log(x_{2t}/x_{3t})$. As before, we will draw other interpretations if we use the reduced models derived using the other alr transformations.

The arbitrariness in the choice of the orthonormal bases used to define the ilr_1 and ilr_2 transformations does not permit a simple interpretation of the reduced models associated with these transformations. If possible, the interpretation should be drawn using the corresponding models in Table A7.1. The only outstanding observation we would like to make is that, according to the reduced model associated with the ilr_1-transformed data, the relative change $x_{4t}/x_{4,t-1}$ in x_4 between two consecutive periods is considered to be random.

7.5   Discussion

We have shown that, thanks to the Euclidean vector space of the simplex $\mathcal{S}^D$, it is possible to define the concept of a compositional process and $\mathcal{C}$-VARIMA models completely independently of the log-ratio transformations used to model compositional time series, although effective modelling of CTS data should be conducted using alr- or ilr-transformed data.

In the standard iterative model building process, the specification of a full $\mathcal{C}$-VARIMA model and the estimation of its parameters do not depend on the alr or ilr transformations applied to the data. However, the simplification of full models, leading to restricted models, results in different $\mathcal{C}$-ARIMA models depending on the transformation applied. If the applied selection criteria are the same, one should expect the reduced $\mathcal{C}$-ARIMA models to produce similar short term forecasts, despite potential differences in their compositional formulations. Such differences can, however, lead to very different model interpretations. Neither would we expect between the restricted models significant differences in terms of goodness of fit. In fact, using the total $\mathcal{C}$-variability of the data explained by the model (Daunis-i-Estadella *et al.* 2002) as a global measure of goodness of fit, we found that it varies from 8.7% for the restricted model for the ilr_1-transformed data to 10.2% for the restricted model for the alr_4-transformed data. It should be noted, however, that the full model only managed to explain 10.6% of the total $\mathcal{C}$-variability of the data.

One possible alternative approach to overcoming these inconveniences would be to choose an orthonormal base, $\mathbf{V}$, coinciding with the principal components of the residuals for the originally estimated full model. If this were done, the simplification of the model should then be performed using the estimates of the parameters of the full model fitted to the ilr-transformed data using the previous orthonormal base. Finally, interpretations can be drawn by considering the $\mathcal{C}$-ARIMA formulation of the model or, alternatively, the alr-transformed model employing that component which best aids the interpretation of the final model as the denominator.

Acknowledgements

This research has been supported by the Spanish Ministry of Science and Innovation (projects CSD2006-00032 and MTM2009-13272) and by the Agència de Gestió d'Ajuts Universitaris i de Recerca of the Generalitat de Catalunya (Ref. 2009SGR424).

References

Aitchison J 1986 *The Statistical Analysis of Compositional Data*. Monographs on Statistics and Applied Probability. Chapman and Hall Ltd (reprinted 2003 with additional material by The Blackburn Press), London (UK). 416 p.

Barceló-Vidal C, Aguilar L and Martín-Fernández J 2007 Time series of compositional data: A first approach. In *Proceedings of the 22nd International Workshop of Statistical Modelling (IWSM 2007)*, (ed. del Castillo J, Espinal A and Puig P). Institut d'Estadística de Catalunya (IDESCAT), Barcelona (Spain). pp. 81–86.

Bell W, Bozik J, McKenzie S and Shulman H 1986 Time series analysis of household headship proportions: 1959–1985. Technical Report 86/01, Statistical Research Division, Bureau of the Census, Washington (USA).

Bergman J 2008 Compositional time series: An application. In *Proceedings of CoDaWork'08, The 3rd Compositional Data Analysis Workshop* (ed. Daunis-i Estadella J and Martín-Fernández J), p. http://hdl.handle.net/10256/723. University of Girona, Girona (Spain). CD-ROM.

Bhaumik A, Dey DK and Ravishanker N 2003 A dynamic linear model approach for compositional time series analysis. Technical Report, University of Connecticut (USA).

Brunsdon TM 1987 *Time series of compositional data*. PhD thesis, University of Southampton, Southampton (UK).

Brunsdon TM and Smith TMF 1998 The time series analysis of compositional data. *Journal of Official Statistics* **14**(3), 237–253.

Daunis-i-Estadella J, Egozcue JJ and Pawlowsky-Glahn V 2002 Least squares regression in the simplex. In *Proceedings of IAMG'02 – The VIII Annual Conference of the International Association for Mathematical Geology* (ed. Bayer U, Burger H and Skala W), vol. I and II. Selbstverlag der Alfred-Wegener-Stiftung, Berlin (Germany). pp. 411–416.

Grunwald GK, Raftery AE and Guttorp P 1993 Time series of continuous proportions *Journal of the Royal Statistical Society, Series B (Statistical Methodology)* **55**(1), 103–116.

Lütkepohl L 2005 *New Introduction to Multiple Time Series Analysis*. Springer, Berlin (Germany). 764 p.

Martín-Fernández JA, Barceló-Vidal C and Pawlowsky-Glahn V 2003 Dealing with zeros and missing values in compositional data sets using nonparametric imputation *Mathematical Geology* **35**(3), 253–278.

Mills T 2009 Forecasting obesity trends in England. *Journal of the Royal Statistics Society, Series A* **172**(1), 107–17.

Mills T 2010 Forecasting compositional time series. *Journal Quality and Quantity* **44**, 673–690.

Nolan T and Smith G 1995 Time series analysis of the prevalence of endoparasitic infections in cats and dogs presented to a veterinary teaching hospital. *Veterinary Parasitology* **59**(2), 87–96.

Pawlowsky-Glahn V and Egozcue JJ 2002 BLU estimators and compositional data. *Mathematical Geology* **34**(3), 259–274.

Peña D, Tiao G and Tsay R 2001 *A Course in Time Series Analysis*. John Wiley & Sons, Ltd, New York, NY (USA). 456 p.

Quintana JM and West M 1988 Time series analysis of compositional data. In *Bayesian Statistics 3* (ed. Bernardo JM, DeGroot MH, Lindley DV and Smith AFM). Oxford University Press, New York, NY (USA). pp. 747–756.

Ravishanker N, Dey D and Iyengar M 2001 Compositional time series analysis of mortality proportions. *Communications in Statistics - Theory and Methods* **30**(11), 2281–2291.

Silva D 1996 *Modelling compositional time series from repeated surveys.* PhD thesis University of Southampton, Southampton (UK).

Silva D and Smith T 2001 Modelling compositional time series from repeated surveys. *Survey Methodology* **27**, 205–215.

Smith T and Brunsdon T 1989 The time series analysis of compositional data. In *Proceedings of the Survey Research Methods Section.* American Statistical Association, Alexandria, VA (USA). pp. 26–32.

Wang H, Liu Q, Mok H, Fu L and Tse W 2007 A hyperspherical transformation forecasting model for compositional data. *European Journal of Operational Research* **179**(2), 459–468.

Appendix

Let $\mathbf{x}$ be any compositional vector in $\mathcal{S}^D$. Let

$$\mathbf{y}_k = \mathrm{alr}_k(\mathbf{x}), \qquad \mathbf{z} = \mathrm{clr}(\mathbf{x}), \qquad \mathbf{u_V} = \mathrm{ilr_V}(\mathbf{x}),$$

be the corresponding log-ratio transformed vectors, where $k = 1, \ldots, D$, and $\mathbf{V}$ is a $D \times (D-1)$ matrix $[\mathbf{v}_1 : \ldots : \mathbf{v}_{D-1}]$ whose columns are the vectors of any orthonormal basis of the subspace $V = \{\mathbf{z} \in \mathbb{R}^D : z_1 + \ldots + z_D = 0\}$ of $\mathbb{R}^D$. Table A7.2 gives the relationships between the transformed vectors $\mathbf{y}_k$, $\mathbf{z}$ and $\mathbf{u_V}$ from the elementary matrices in Table A7.3 and the matrices $\mathbf{V}_1$ and $\mathbf{V}_2$ associated with the ilr$_1$ and ilr$_2$ transformations in Table A7.4.

Table A7.1 Fitted restricted VARI(1,1) models, expressed in compositional form, for different transformations of the UK expenditure shares series.

alr$_1$ transformation

$$(1 - L_C)\mathbf{x}_t = \left(\begin{bmatrix} -0.045 & -0.036 & 0.081 & 0 \\ 0.093 & -0.174 & 0.081 & 0 \\ -0.049 & 0.246 & -0.197 & 0 \\ 0.000 & -0.036 & 0.036 & 0 \end{bmatrix} \odot (1 - L_C)\mathbf{x}_{t-1} \right) \oplus \mathbf{w}_t$$

alr$_2$ transformation

$$(1 - L_C)\mathbf{x}_t = \left(\begin{bmatrix} 0 & -0.037 & 0.087 & -0.050 \\ 0 & -0.103 & 0.087 & 0.017 \\ 0 & 0.198 & -0.214 & 0.017 \\ 0 & -0.057 & 0.041 & 0.017 \end{bmatrix} \odot (1 - L_C)\mathbf{x}_{t-1} \right) \oplus \mathbf{w}_t$$

alr$_3$ transformation

$$(1 - L_C)\mathbf{x}_t = \left(\begin{bmatrix} 0 & -0.056 & 0.056 & 0 \\ 0 & -0.133 & 0.133 & 0 \\ 0 & 0.219 & -0.219 & 0 \\ 0 & -0.030 & 0.030 & 0 \end{bmatrix} \odot (1 - L_C)\mathbf{x}_{t-1} \right) \oplus \mathbf{w}_t$$

alr$_4$ transformation

$$(1 - L_C)\mathbf{x}_t = \left(\begin{bmatrix} 0 & -0.044 & 0.063 & -0.019 \\ 0 & -0.201 & 0.110 & 0.090 \\ 0 & 0.288 & -0.190 & -0.098 \\ 0 & -0.044 & 0.018 & 0.026 \end{bmatrix} \odot (1 - L_C)\mathbf{x}_{t-1} \right) \oplus \mathbf{w}_t$$

ilr$_1$ transformation

$$(1 - L_C)\mathbf{x}_t = \left(\begin{bmatrix} -0.012 & -0.012 & 0.054 & -0.030 \\ -0.175 & -0.175 & 0.127 & 0.224 \\ 0.187 & 0.187 & -0.181 & -0.194 \\ 0 & 0 & 0 & 0 \end{bmatrix} \odot (1 - L_C)\mathbf{x}_{t-1} \right) \oplus \mathbf{w}_t,$$

Table A7.1   *(Continued)*

$$\text{ilr}_2 \text{ transformation}$$

$$(1 - L_{\mathcal{C}})\mathbf{x}_t = \left(\begin{bmatrix} -0.011 & -0.036 & 0.062 & -0.015 \\ 0.025 & -0.143 & 0.121 & -0.003 \\ -0.052 & 0.261 & -0.208 & 0 \\ 0.038 & -0.082 & 0.025 & 0.018 \end{bmatrix} \odot (1 - L_{\mathcal{C}})\mathbf{x}_{t-1} \right) \oplus \mathbf{w}_t.$$

Table A7.2   Relationships between $\mathbf{y}_k$, $\mathbf{z}$ and $\mathbf{u}_V$.

Transformation	Relationship	Transformation	Relationship
$\mathbf{y}_k \to \mathbf{z}$:	$\mathbf{z} = \mathbf{F}_k^\top \mathbf{H}^{-1} \mathbf{y}_k$	$\mathbf{z} \to \mathbf{y}_k$:	$\mathbf{y}_k = \mathbf{F}_k \mathbf{z}$
$\mathbf{y}_k \to \mathbf{u}_V$:	$\mathbf{u}_V = (\mathbf{F}_k \mathbf{V})^{-1} \mathbf{y}_k$	$\mathbf{u}_V \to \mathbf{y}_k$:	$\mathbf{y}_k = \mathbf{F}_k \mathbf{V} \mathbf{u}_V$
$\mathbf{u}_V \to \mathbf{z}$:	$\mathbf{z} = \mathbf{V} \mathbf{u}_V$	$\mathbf{z} \to \mathbf{u}_V$:	$\mathbf{u}_V = \mathbf{V}^\top \mathbf{z}$
$\mathbf{y}_k \to \mathbf{y}_j$:	$\mathbf{y}_j = \mathbf{F}_j \mathbf{F}_k^\top \mathbf{H}^{-1} \mathbf{y}_k$	$\mathbf{u}_{V*} \to \mathbf{u}_V$:	$\mathbf{u}_V = \mathbf{V}^\top \mathbf{V}^* \mathbf{u}_{V*}$

Table A7.3   Elementary matrices.

Notation	Definition	Order
$\mathbf{I}_{D-1}$	identity matrix	$(D-1) \times (D-1)$
$\mathbf{1}_{D-1}$	column vector of ones	$(D-1) \times 1$
$\mathbf{J}_{D-1} = \mathbf{1}_{D-1}\mathbf{1}_{D-1}^\top$	matrix of ones	$(D-1) \times (D-1)$
$\mathbf{F}$	$[\mathbf{I}_{D-1} : -\mathbf{1}_{D-1}]$	$(D-1) \times D$
$\mathbf{F}_k$	identity matrix $\mathbf{I}_{D-1}$ with extra column of -1's inserted between columns $k-1$ and k	$(D-1) \times D$
$\mathbf{G}_D$	$\mathbf{I}_D - D^{-1}\mathbf{J}_D$	$D \times D$
$\mathbf{H}_{D-1}$	$\mathbf{I}_{D-1} + \mathbf{J}_{D-1}$	$(D-1) \times (D-1)$

Table A7.4   Matrices associated with the isometric log-ratio transformations.

Matrix associated with the ilr_1 transformation

$$\mathbf{V}_1 = \begin{bmatrix} 0.707107 & 0.408248 & 0.288675 \\ -0.707107 & 0.408248 & 0.288675 \\ 0 & -0.816497 & 0.288675 \\ 0 & 0 & -0.866025 \end{bmatrix}$$

Matrix associated with the ilr_2 transformation

$$\mathbf{V}_2 = \begin{bmatrix} 0.154303 & 0.635001 & -0.568300 \\ -0.771517 & 0.127000 & 0.372334 \\ 0.617213 & 0 & 0.607493 \\ 0 & -0.762001 & -0.411527 \end{bmatrix}$$

8

Compositional data and correspondence analysis

Michael Greenacre

Department of Economics & Business, Pompeu Fabra University, and Barcelona Graduate School of Economics, Barcelona, Spain

8.1 Introduction

In this chapter we consider two alternative approaches to the reduction of dimensionality of a table of positive data: log-ratio analysis and correspondence analysis, referred to hereafter as LRA and CA, respectively. These two approaches have several aspects in common: (i) they are both special cases of a generalized definition of principal component analysis (PCA), relying on the singular value decomposition (SVD) for the computation of their solutions; (ii) they both treat the rows and columns of the table in a symmetric fashion; that is, there is no difference in the results if the transposed version of the table is analysed; and (iii) they both consider the relative values of the data rather than their absolute values. With respect to this last aspect the main distinguishing feature of the two approaches is that LRA considers each data element relative to the others, while CA considers each data element relative to the corresponding margins of the table. This means that all the data have to be strictly positive for LRA while for CA only the margins have to be strictly positive.

Although both methods apply to the general case of positive data (or nonnegative data, in the case of CA) we restrict our attention here to the special case of compositional data, when the rows (say) of the table have a constant sum, usually 1 or 100%. This property is sometimes called one of *closure*: that is, the row elements, for example, of the table are closed. Since CA considers the rows or columns relative to their margins, this means that CA is inherently analysing the data compositionally, whether the original data are closed or not.

Compositional Data Analysis: Theory and Applications, First Edition. Edited by Vera Pawlowsky-Glahn and Antonella Buccianti.
© 2011 John Wiley & Sons, Ltd. Published 2011 by John Wiley & Sons, Ltd.

   The LRA approach to compositional data originates in papers by Aitchison (1983, 1990). A weighted form of LRA for positive data appeared at more or less the same time in work by Lewi (1976, 1980), who called this method *spectral mapping*. CA has a long history, since the 1930s, as a method for analysing count data, and its geometric properties as a PCA-style method for categorical data were recognized by Benzécri (1973). Recent treatments of these different historical tracks of research are by Aitchison and Greenacre (2002), Greenacre (2009) and Greenacre and Lewi (2009). In particular, Greenacre (2009) has shown that CA and LRA, in both its weighted and unweighted forms, can be embedded in a common family of methods, thanks to the Box–Cox power transformation – in fact, LRA turns out to be a limiting case of CA as the power parameter tends to zero.

   After a comparative technical description of the two methods, a summary of their properties and some recent results, we give an application to a set of data on fatty acid compositions in marine biology. This data set includes zeros, which is a problem for LRA but not for CA. We shall use the close connection between CA and LRA to analyse the data including the zero values, without replacing them by small positive numbers which is the usual approach in the LRA framework.

8.2   Comparative technical definitions

We first give the general definitions of LRA and CA, and afterwards describe the special case of compositional data. Suppose that the original $I \times J$ data matrix is denoted by $\mathbf{N}$, and let $\log(\mathbf{N})$ denote the matrix of logarithms of $\mathbf{N}$. Suppose that $\mathbf{P}$ is $\mathbf{N}$ divided by its grand total $n : \mathbf{P} = (1/n)\mathbf{N}$ and let the marginal sums of $\mathbf{P}$ be denoted by $\mathbf{r}$ and $\mathbf{c}$, respectively – these are the weights, or *masses*, associated with the rows and columns. [Notice that the definition of LRA given here coincides with what Greenacre and Lewi (2009) call weighted LRA, to distinguish it from unweighted LRA where there are constant weights: $r_i = 1/I$ and $c_j = 1/J$ – unweighted LRA has been more commonly used in compositional data analysis up to now.] Let $\mathbf{D}_r$ and $\mathbf{D}_c$ be the diagonal matrices of these masses. The computations of LRA and CA differ only slightly in their first steps, but then follow identical paths:

Step 1: Define a weighted double-centred matrix $\mathbf{S}$:

   LRA: $\mathbf{S} = \mathbf{D}_r^{1/2}(\mathbf{I} - \mathbf{1r}^\top)\log(\mathbf{N})(\mathbf{I} - \mathbf{1c}^\top)^\top\mathbf{D}_c^{1/2}$;

   CA: $\mathbf{S} = \mathbf{D}_r^{1/2}(\mathbf{I} - \mathbf{1r}^\top)(\mathbf{D}_r^{-1}\mathbf{PD}_c^{-1})(\mathbf{I} - \mathbf{1c}^\top)^\top\mathbf{D}_c^{1/2}$.

Step 2: Perform the SVD of $\mathbf{S}$: $\mathbf{S} = \mathbf{UD}_\gamma\mathbf{V}^\top$, where the singular vectors are orthonormal: $\mathbf{U}^\top\mathbf{U} = \mathbf{V}^\top\mathbf{V} = \mathbf{I}$, and the singular values in the diagonal matrix $\mathbf{D}_\gamma$ are positive and in descending order: $\gamma_1 \geq \gamma_2 \geq \cdots > 0$.

Step 3: Calculate the standard and principal coordinates of the rows and columns:

$$
\begin{aligned}
\text{(row standard)} \quad & \mathbf{X} &=& \ \mathbf{D}_r^{-1/2}\mathbf{U}; \\
\text{(column standard)} \quad & \mathbf{Y} &=& \ \mathbf{D}_c^{-1/2}\mathbf{V}; \\
\text{(row principal)} \quad & \mathbf{F} &=& \ \mathbf{XD}_\gamma = \mathbf{D}_r^{-1/2}\mathbf{UD}_\gamma; \\
\text{(column principal)} \quad & \mathbf{G} &=& \ \mathbf{YD}_\gamma = \mathbf{D}_c^{-1/2}\mathbf{VD}_\gamma.
\end{aligned}
$$

It is common to construct planar maps and biplots from these results, using the first two columns of these coordinate matrices to position the row and column points. The most common way to make biplots is to plot one set (usually the samples in the rows) in principal coordinates and the other set in standard coordinates. The set in principal coordinates is approximating interpoint distances in each case – see distance Equations (8.2) and (8.3) – while the set in standard coordinates defines biplot axes for the interpretation of the plot. When there are many columns (variables or components) a convenient alternative is the *contribution biplot* [called the *standard biplot* in Greenacre (2007, chapter 15)], using the matrix $\mathbf{V}$ in step 3 above to display the columns, that is $\mathbf{D}_c^{1/2}$ times the standard coordinate matrix $\mathbf{Y}$ (Greenacre 2010a) – the elements of $\mathbf{D}_c^{1/2}\mathbf{Y} = \mathbf{V}$ are called *contribution coordinates*. This option shows the variables according to their contributions to the principal axes of the solution, allowing a visual separation of the determinant variables from those that could be effectively glossed over in the interpretation. Greenacre (1993, 2007, chapters 6 and 15) gives more details about the different plotting options, which are all available in the **ca**-package by Nenadić and Greenacre (2007) in the R project (R Development Core Team 2009).

As shown by Greenacre (2009), both methods can be defined in terms of the matrix $\mathbf{D}_r^{-1}\mathbf{PD}_c^{-1}$ of the so-called *contingency ratios* $p_{ij}/(r_i c_j)$, since in step 1 of the LRA definition above the matrix $\log(\mathbf{N})$ can be replaced by $\log(\mathbf{D}_r^{-1}\mathbf{PD}_c^{-1})$ without changing $\mathbf{S}$. This is because $\log[p_{ij}/(r_i c_j)] = \log[n_{ij}/(n r_i c_j)] = \log(n_{ij}) - \log(n) - \log(r_i) - \log(c_j)$ and the double-centring removes all the additive terms except $\log(n_{ij})$. Hence the matrix $\mathbf{S}$ in step 1 can be written equivalently as:

$$\mathbf{S} = \mathbf{D}_r^{1/2}(\mathbf{I} - \mathbf{1r}^\top)\log(\mathbf{D}_r^{-1}\mathbf{PD}_c^{-1})(\mathbf{I} - \mathbf{1c}^\top)^\top \mathbf{D}_c^{1/2}. \tag{8.1}$$

Comparing this with the matrix $\mathbf{S}$ for CA, it is clear that the only essential difference between the two methods is the log-transformation.

As a result of their definitions above, LRA and CA have their corresponding inherent distance functions, given here as squared distances between rows i and i':

LRA:

$$d_{ii'}^2 = \sum_{j<j'}\sum c_j c_{j'} \left(\log\frac{n_{ij}}{n_{ij'}} - \log\frac{n_{i'j}}{n_{i'j'}}\right)^2 = \sum_j c_j \left[\log\frac{n_{ij}}{\overline{\mathrm{g_m}}(\mathbf{n}_i)} - \log\frac{n_{i'j}}{\overline{\mathrm{g_m}}(\mathbf{n}_{i'})}\right]^2, \tag{8.2}$$

where $\overline{\mathrm{g_m}}(\mathbf{n}_i) = n_{i1}^{c_1} n_{i2}^{c_2}, \ldots, n_{iJ}^{c_J}$ is the weighted geometric mean of the ith row.

CA:

$$d_{ii'}^2 = \sum_j \left(\frac{p_{ij}}{r_i} - \frac{p_{i'j}}{r_{i'}}\right)^2 \Big/ c_j = \sum_j c_j \left(\frac{p_{ij}}{r_i c_j} - \frac{p_{i'j}}{r_{i'} c_j}\right)^2. \tag{8.3}$$

For distances between columns, simply interchange the indices i and j, and the masses r_i and c_j.

Notice the standardizing effect of the column masses c_j in the log-ratio distance (8.2) where the high variance ratios resulting from small values of n_{ij} are downweighted through multiplication by small masses. In CA, where the distance is called the chi-square distance, the smaller profile elements p_{ij}/r_i have lower variances and are upweighted by the division by the corresponding small masses – this can also be written as multiplying the squared differences in the contingency ratios by the masses.

When the table is one of compositional data, with the rows having constant sums, the row weights are constant: $r_i = 1/I$, so the differential weighting in distances (8.2) and (8.3) only applies to the columns, or components of the composition. Since the rarer components usually have higher relative error as well as high variance in their log-ratios, the weighting in (8.2) serves to diminish their influence on the low-dimensional solution. In (8.3) the profile values are equal to the original data (assuming constant sum of 1), and the smaller variances in the lower compositional values are compensated for through the division by their corresponding averages c_j. If columns are equally weighted: $c_j = 1/J$, the distances in (8.2) are, apart from an overall scale factor, equal to the so-called *Aitchison distances*, introduced by Aitchison (1983) (see also Chapter 1, Section 1.6).

8.3 Properties and interpretation of LRA and CA

Although in many applications the results of LRA and CA might appear to be quite similar, their properties and interpretations are quite different. In CA the points lie inside a bounded simplex, whereas the logarithmic transformation in LRA takes the points into unbounded real vector space. This means that CA has an upper bound on the total variance for any given problem, equal to the dimensionality of the data set minus 1, whereas the total variance in LRA is unbounded. If there is very little variance in the compositional vectors and thus the profiles in CA are very close to their average and far from the outer limits of the simplex, then the two methods give practically the same result. Both methods obey the principal of distributional equivalence, which is one of the foundational principles of CA, but only LRA is subcompositionally coherent (Greenacre and Lewi 2009) (see also Chapter 2, Section 2.3). LRA plots have the special property that vector links between pairs of variables (for example, pairs of components for compositional data) represent the corresponding log-ratio vectors. This property is useful for diagnosing power-law relationships between subsets of components.

Greenacre (2009) showed that LRA is a limiting case of CA if one considers the Box–Cox family of power transformations of the data and then lets the power parameter tend to zero. Specifically, consider the following transformation of the contingency ratios:

$$q_{ij}(\alpha) = (1/\alpha)[p_{ij}/(r_i c_j)]^{\alpha}. \tag{8.4}$$

Note that there is a minus 1 in the Box–Cox transformation, but this is superfluous here because it will be eliminated by the double-centring to come.

If these transformed contingency ratios (8.4) are used in the definition of matrix **S** in the first step of the CA instead of the original one, where $\alpha = 1$, then as α tends to 0, so the CA solution tends to the LRA solution. We will use this result in the application presented in the next section.

8.4 Application to fatty acid compositional data

We use LRA and CA to study a compositional data set consisting of the 40-part fatty acid compositions measured on a sample of 42 samples of a marine organism, *Calanus glacialis*, collected in a Norwegian fjord as part of a larger research project on the food chain in Arctic ecosystems. This species is one of the primary elements of the marine food chain. A small

Table 8.1 A part of the 42 × 40 table of fatty acid compositions.

ID	14:0	14:1(n-5)	i-15:0	a-15:0	15:0	15:1(n-6)
B5	13.8541	0.2025	1.1903	0.4464	0.8473	0.1033
B6	11.8266	0.1480	1.2359	0.4599	1.0559	0.0840
B7	6.4571	0.0000	0.7680	0.2481	0.5140	0.0000
B8	12.0115	0.1484	1.1297	0.4114	0.7958	0.1214
H5	6.5795	0.1933	0.3287	0.0000	0.2642	0.0000
H6	6.7042	0.1935	0.3153	0.0000	0.2544	0.0000
H7	6.2799	0.2009	0.3426	0.0000	0.2658	0.0000
H8	5.2029	0.0000	0.3919	0.1805	0.2630	0.0000
D11	8.9500	0.1570	0.7913	0.2839	0.5892	0.1914

part of the data matrix is shown in Table 8.1. Just over 11% of the data values are zeros, which is a problem for LRA but not for CA. Figure 8.1 is the CA biplot of these data, with columns shown in contribution coordinates (i.e. standard coordinates multiplied by the square roots of the respective masses).

There are three clusters of samples apparent in the data, and the most contributing fatty acids stand out in the map, showing that 16:1(n-7) is chiefly responsible for the cluster at the bottom left, 18:4(n-3) for the cluster at the bottom right, and 18:0 for the cluster stretching out at the top. In order to compare these results with those of LRA, we need to do something

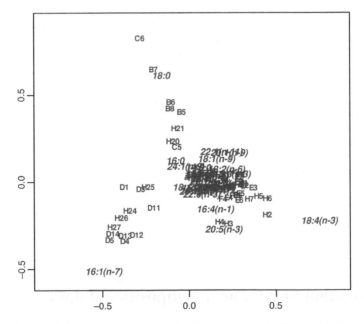

Figure 8.1 Correspondence analysis contribution biplot of the fatty acid data, with samples in principal coordinates and fatty acids in contribution coordinates. This scaling option shows how each fatty acid contributes to the principal axes of the two-dimensional solution. The total variance (inertia) is 0.2058 and the percentage explained in this two-dimensional solution is 42.7 + 31.5 = 74.2%.

about the zeros. When the zeros are not structural zeros but are due to the imprecision of measurement, there are several strategies for coping with them: the simplest is just to replace them with some reasonable value such as half (or another fraction) of the detection limit of the measuring instrument – see Martín-Fernández *et al.* (2003) for a more complete discussion of the zero replacement problem, as well as Chapter 4. An alternative approach is to study how the results are affected by ever-decreasing small amounts being substituted for the zeros and then choose the smallest value before the results become unstable. Using this latter approach we found that replacing the zero percentages by 0.01 was suitable, and was considerably less than the smallest nonzero percentage (0.0255) in the data. Nonzero data were proportionally reduced after the zeros were replaced. Figure 8.2 is the weighted LRA biplot of the data set with zero percentages replaced by 0.01, again using contribution coordinates for the component points. Again three clusters of samples are visible: compared with Figure 8.1, the main difference is that 18:0 is not so clearly pulling out the cluster at the top left.

The fact that power-transformed CA converges to LRA can be used as a way to combine the two approaches. As stated before, CA has no problems with zeros, so we can apply the power transformation with ever-diminishing values of the power parameter $\alpha \in (1, 0.99, 0.98,$ etc.) and monitor the solutions at each step. A point will be reached when the solution becomes unstable as logarithmic convergence is reached when $\alpha = 0$, since there are zeros in the data. An interesting statistic to monitor as the power is lowered is what Greenacre (2011) has defined as the *subcompositional incoherence*, which measures how close the method is to being subcompositionally coherent. This measure is based on the given distance function between pairs of components in the full composition and the distances computed on the corresponding two-part compositions, which appear to be the most affected by re-closing. Specifically, the measure is one of the classic measures of stress in multidimensional scaling

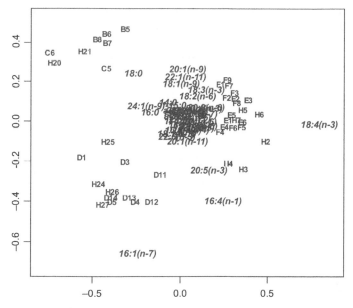

Figure 8.2 Weighted log-ratio biplot of fatty acid data, where zero percentages have been replaced by 0.01, with samples in principal coordinates and fatty acids in contribution coordinates. The total (weighted) log-ratio variance is 0.2836, and the percentage explained in this two-dimensional solution is $51.8 + 22.7 = 74.5\%$.

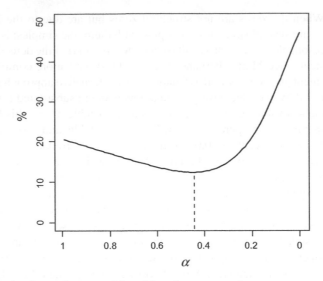

Figure 8.3 Monitoring subcompositional incoherence as data are powered by decreasing values of α.

(Börg and Groenen 2007), used to measure overall similarity between two sets of distances. Figure 8.3 shows the sequence of subcompositional incoherence measures as a function of descending α.

The value of the power $\alpha = 0.46$ shown in Figure 8.3 gives us the least incoherence for this transformed data set. Figure 8.4 shows the power-transformed CA biplot using this value of α and the result indeed looks like an intermediate analysis between the regular CA (Figure 8.1) and the LRA with the zero-substituted value (Figure 8.2). In practice, the

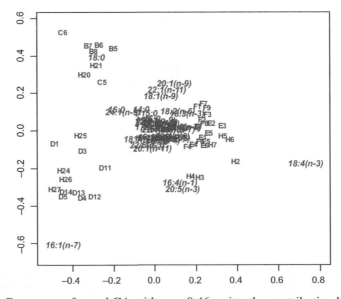

Figure 8.4 Power-transformed CA with $\alpha = 0.46$, using the contribution biplot scaling.

data analyst might choose 0.5, the square root transformation, which gives a minimum in Figure 8.3 only slightly higher than that for 0.46, thus partially justifying a square root transformation of the fatty acid data.

8.5 Discussion and conclusions

We have reviewed some results that show how CA and LRA are intimately related technically and how CA can be used to come as close as one likes to a LRA, thanks to Box–Cox power transformations. The fact that CA can handle data zeros explains in part why it has become popular in the fields of linguistics, archaeology and ecology, where large sparse tables of counts are encountered, often with many more observed zeros than positive values.

By LRA we mean the weighted form of LRA described by Greenacre and Lewi (2009). Actually, Lewi defined the weighted form of LRA in early papers (Lewi 1976, 1980), where the method was applied to biological activity spectra and so was named *spectral mapping* – see also Wouters *et al.* (2003). Greenacre and Lewi (2009) demonstrated that the use of differential weights significantly improves the properties of the method.

The unweighted form of LRA can also be shown to be a limiting case of CA if the original data themselves, not the contingency ratios, are power-transformed (Greenacre 2009, 2010b). Hence, a power transformation of the original compositional data, followed by reclosure and using the chi-square distance can be made to come arbitrarily close to the Aitchison distance which is the unweighted form of the log-ratio distance. To illustrate this fact, we consider the small artificial data set used by Martín-Fernández and Thió-Henestrosa (2006) as a counter-example against the use of Euclidean-type distances. This example contains four samples of three-part compositions:

	1	2	3
A	0.1	10.0	89.9
B	10.0	0.1	89.9
C	30.1	40.0	29.9
D	40.0	30.1	29.9

The point made by these authors was that the Euclidean distance between A and B is equal to that between C and D, whereas the Aitchison distance gives a relatively much higher difference between the A and B samples. The chi-square distance also calculates these distances to be equal, but as soon as a power transformation is applied the distance between A and B increases relative to that between C and D. For example, taking a fourth-root transformation of the data, followed by reclosure, the chi-square distances are equal to $d_{AB} = 3.763$ and $d_{CD} = 0.173$. [Notice that here the original data are transformed, not the contingency ratios; in addition, as shown by Greenacre (2010b), the chi-square distances on power-transformed distances, using a power α, are multiplied by $1/\alpha$, as in the Box–Cox transformation, and finally, to find distances comparable to the Aitchison distance used by Martín-Fernández and Thió-Henestrosa (2006), the chi-square distances need to be multiplied by a scale factor equal to the square root of the number of components, i.e. $\sqrt{3}$ in this case.] Going further to a tenth-root transformation the chi-square distances are $d_{AB} = 6.437$ and $d_{CD} = 0.404$, which are almost equal to the Aitchison distances of 6.513 and 0.402. In the limit the chi-square distances tend to the exact Aitchison distances.

Hence the power-transformation approach can be used to avoid zero-replacement, whether the zeros be actual structural zeros or so-called *rounded zeros* (Martín-Fernández *et al.* 2003) (see also Chapter 1, Section 1.11 and Chapter 4, Section 4.2). Since reducing the power parameter, in the presence of zeros, will eventually lead to an unstable result as we approach the log-transformation, the use of the measure of subcompositional incoherence gives a method for choosing an appropriate value of the power. So, in summary, a Euclidean-type distance measure such as the chi-square statistic can give substantively justifiable results if the data are suitably transformed, and the family of power-transformed correspondence analyses linking regular correspondence analysis and the weighted and unweighted forms of log-ratio analysis can be profitably exploited for the analysis of compositional data.

Acknowledgements

This research has been supported by the Fundación BBVA, Madrid, Spain, and the author expresses his gratitude to the Foundation's director, Prof. Rafael Pardo. Partial support of Spanish Ministry of Science grants MTM2008-00642 and MTM2009-09063 is also acknowledged. The fatty acid data on *Calanus glacialis* were kindly provided by a research team at the Norwegian Polar Institute in Tromsø, Norway – thanks to Eva Leu, Janne Søreide and Stig Falk-Petersen.

References

Aitchison J 1983 Principal component analysis of compositional data. *Biometrika* **70**(1), 57–65.

Aitchison J 1990 Relative variation diagrams for describing patterns of compositional variability. *Mathematical Geology* **22**(4), 487–511.

Aitchison J and Greenacre M 2002 Biplots for compositional data. *Applied Statistics* **51**(4), 375–392.

Benzécri JP 1973 *L'Analyse des Données. Tôme 1: L'Analyse des Correspondances.* Dunod, Paris (France).

Börg I and Groenen P 2007 *Modern Multidimensional Scaling*, 2nd edition. Springer, New York, NY (USA).

Greenacre M 1993 Biplots in correspondence analysis. *Journal of Applied Statistics* **20**, 251–269.

Greenacre M 2007 *Correspondence Analysis in Practice*, 2nd edition. Chapman and Hall/CRC Press, London (UK).

Greenacre M 2009 Power transformations in correspondence analysis. *Computational Statistics and Data Analysis* **53**, 3107–3116.

Greenacre M 2010a Contribution biplots. Technical Report, Department of Economics and Business, Universitat Pompeu Fabra, Barcelona (Spain). http://www.econ.upf.edu/en/research/onepaper.php?id=1162.

Greenacre M 2010b Log-ratio analysis is a limiting case of correspondence analysis. *Mathematical Geosciences* **42**, 129–134.

Greenacre M 2011 Measuring subcompositional incoherence. *Mathematical Geosciences*. http://dx.doi.org/10.1007/s11004-011-9338-5.

Greenacre M and Lewi P 2009 Distributional equivalence and subcompositional coherence in the analysis of compositional data, contingency tables and ratio-scale measurements. *Journal of Classification* **26**(1), 29–54.

Lewi P 1976 Spectral mapping, a technique for classifying biological activity profiles of chemical compounds. *Arzneimittelforschung (Drug Research)* **26**, 1295–1300.

Lewi P 1980 Multivariate data analysis in apl. In *Proceedings of APL-80 Conference* (ed. van der Linden G). North-Holland, Amsterdam (The Netherlands). pp. 267–271.

Martín-Fernández J and Thió-Henestrosa S 2006 Rounded zeros: some practical aspects for compositional data. In *Compositional Data Analysis: From Theory to Practice* (ed. Buccianti A, Mateu-Figueras G and Pawlowsky-Glahn V). The Geological Society, London (UK). pp. 191–201.

Martín-Fernández JA, Barceló-Vidal C and Pawlowsky-Glahn V 2003 Dealing with zeros and missing values in compositional data sets using nonparametric imputation. *Mathematical Geology* **35**(3), 253–278.

Nenadić O and Greenacre M 2007 Correspondence analysis in R, with two- and three-dimensional graphics: the **ca** package. *Journal of Statistical Software*. http://www.jstatsoft.org/v20/i03/paper.

R Development Core Team 2009 *R: A Language and Environment for Statistical Computing*. R Foundation for Statistical Computing, Vienna (Austria).

Wouters L, Göhlmann H, Bijnens L, Kass S, Molenberghs G and Lewi P 2003 Graphical exploration of gene expression data: a comparative study of three multivariate methods. *Biometrics* **59**, 1131–1139.

9

Use of survey weights for the analysis of compositional data[*]

Monique Graf
Swiss Federal Statistical Office, Switzerland

9.1 Introduction

Surveys are essentially built to optimize the estimation of totals in population subgroups for a number of variables. Practically, a key variable is chosen and the design is optimized for this variable, the trade-off being between cost and precision. Totals are estimated by weighted sums of the sampled values. The weights are extrapolation factors that depend on the survey design. It is an important aspect of the data quality to inform the user on the measurement error of the published figures.

In a survey context, analysis of compositions is a byproduct, that is the design is never optimized for compositions, moreover sums have no meaning in the context of compositions. Yet the problem of measurement error remains and it is legitimate to estimate a covariance matrix related to the composition estimators for subgroups in the population.

Let us consider the two well known operations on compositions (Chapter 1): perturbation and powering (Aitchison 1986). Are they a kind of weighting operation? Let $\left[p_i = (p_{i1}, \ldots, p_{iD}) \right]$ be a compositional data set consisting of N D-part compositions p_i, $i = 1, \ldots, N$. Because perturbation acts differently on the parts of p_i, perturbation is not analogous to weighting the compositions. On the other hand powering acts multiplicatively

---

[*]The views expressed in this chapter are the responsibility of the author, they do not necessarily reflect the policy of the Swiss Federal Statistical Office.

Compositional Data Analysis: Theory and Applications, First Edition. Edited by Vera Pawlowsky-Glahn and Antonella Buccianti.
© 2011 John Wiley & Sons, Ltd. Published 2011 by John Wiley & Sons, Ltd.

on the logarithm of parts. Thus when the compositions are log transformed, the power b_i acts on e.g. clr(p_i) as a multiplicative factor,

$$\mathrm{clr}\left(b_i \odot p_i\right) = b_i\mathrm{clr}\left(p_i\right).$$

This is also true for other transformations like alr(p_i) (Aitchison 1986) or ilr(p_i) (Egozcue *et al.* 2003). Thus weighting a composition at the log scale is a special case of powering, because weights must be positive and powers are unrestricted.

Section 9.2 contains a brief review of survey methodology. Section 9.3 applies the design-based principles to the estimation of compositions and their covariance matrix at a population sub-group level. Section 9.4 discusses the concepts.

9.2   Elements of survey design

The peculiarity of data collection in finite population surveys is the existence of registers, that is of lists of statistical units (persons, households, companies). The access to the registers for statistical reasons is ruled by a Statistics Act. In some countries, there exist persons registers that record many demographic variables (age, gender, education, employment status, etc.). In other countries these variables are traditionally observed only when censuses are conducted. Censuses themselves tend to disappear, because most of the needed information exists in official registers at the community level. Thus, e.g. in Switzerland, the effort is presently directed to the harmonization of these numerous registers. The result is a more efficient use of the available information together with a reduction of burden for the surveyed individuals.

Coverage An ideal register is exhaustive. Of course in reality, there are coverage problems, i.e. not all units appear in the register. For instance a phonebook only lists households with a registered phone number. Thus some households would be excluded from a phonebook-based survey. An important aspect of the quality description of the survey is the measurement of the coverage error.

Actuality An ideal register contains for each unit a number of characteristics that are recorded without errors. One problem is that the information may be outdated, e.g. the employment status. A specific register may be updated only once a year, but the information should be available on a quarterly basis for the needs of a particular survey. In this case, the register information will be considered as auxiliary information.

Thematic surveys Not all the needed statistical information is recorded in registers. Surveys are thus conducted on specific themes. The concomitant information contained in the registers will be used at the survey design stage and at the estimation stage.

An authoritative reference on the design and analysis of surveys based on register information is Särndal *et al.* (1992). A succinct review is given in the following paragraphs.

9.2.1   Randomization

In most surveys access to individual elements of the population is provided by a *sampling frame* that is obtained from one or several registers. To simplify the theory, we shall admit that there is a one to one relationship between the elements of the population and the units in the frame.

Definition 9.2.1 (Sampling frame)   *A sampling frame is a device that associates the elements of the population U with the sampling units in the frame.*

A sample is selected from the frame. Sound scientific rules require that the selection should include a chance mechanism.

Definition 9.2.2 (Probability sample)   *Consider a finite population whose elements are listed in a frame. A probability sample S is an element taken randomly out of a set $\mathbf{S}$ of possible samples with a known probability of selection p(.). The procedure defining*

$$\mathrm{P}(S = s) = p(s), s \in \mathbf{S}$$

must give every element in the frame a nonzero probability of selection.

Thus the probability π_k of selection of unit k in the population is given by the probability that one of the samples containing k ($S \ni k$) is selected. (The symbol $\ni$ is used in order to emphasize the fact that S is variable and k is fixed.)

$$\pi_k = \mathrm{P}(S \ni k) = \sum_{\{s \in \mathbf{S}\,:\,s \ni k\}} p(s). \tag{9.1}$$

In the same way, the joint probability of selection π_{ik} of units i and k is given by the probability that one of the samples containing both i and k is selected.

$$\pi_{ik} = \sum_{\{s \in \mathbf{S}\,:\,s \ni i \text{ and } k\}} p(s) \qquad \pi_{kk} = \pi_k. \tag{9.2}$$

9.2.1.1   Examples of survey designs

Example 9.2.3 (Bernoulli design)   In the Bernoulli design, each unit is sampled with a fixed probability π and independently. We have

$$\pi_k = \pi_{kk} = \pi \quad \text{and} \quad \pi_{ik} = \pi^2, i \neq k.$$

The sample size is random in this case with expectation $N\pi$, where N is the population size.

Example 9.2.4 (Simple random sampling without replacement)   Here is the simplest example of a survey design with fixed sample size. Let N be the number of units in the population; N is known from the frame. In simple random sampling without replacement, possible samples have the same probability of selection and a fixed size n, $n < N$. Thus $\mathbf{S}$ is the set of all subsets of the population U that have size n and the number of possible samples is given by $\binom{N}{n}$. So, for every $s \in \mathbf{S}$, $p(s) = 1 \big/ \binom{N}{n}$. In the same way, the number of possible samples containing unit k is given by $\binom{N-1}{n-1}$. It is easy to see that, if $i \neq k$,

$$P(S \ni i | S \ni k) = \binom{N-2}{n-2} \Big/ \binom{N-1}{n-1}. \text{ Thus}$$

$$\pi_k = P(k \in s) = \binom{N-1}{n-1} \Big/ \binom{N}{n} = \frac{n}{N} \quad \text{and} \quad \pi_{ik} = P(i, k \in s) = \frac{n(n-1)}{N(N-1)}.$$

Example 9.2.5 (Stratified design)   Sometimes it is of advantage to partition the population into more homogeneous subpopulations called strata. In each stratum, independently of the other strata, a simple random sample is extracted.

Let H be the number of strata. Every sample s can be decomposed into $s = s_1 \cup s_2 \cup, \ldots, \cup s_H$ and, by the independence between strata, $p(s) = p_1(s_1)p_2(s_2), \ldots, p_H(s_H)$. The inclusion probabilities are entirely determined by $p_h(.)$ and follow along the lines of Example 9.2.4. With unit k in stratum h,

$$\pi_k = \pi_{k|h} \qquad \pi_{ik} = \begin{cases} \pi_i \pi_k & \text{if } i, k \in \text{different strata,} \\ \pi_{ik|h} & \text{if } i, k \in \text{stratum } h. \end{cases}$$

Example 9.2.6 (Two-stage design)   Another widely used instance of survey design is the two-stage design. The population is formed of primary units (e.g. households). Primary units are selected randomly according to a first stage design, and then in each selected primary unit, individuals are randomly chosen according to a second stage design.

Let $p_I(.)$ define the selection probability of the primary sample s_I. For every primary unit $k_p \in s_I$, let $p_{k_p}(.|s_I)$ define the selection probability of the secondary sample s_{k_p}. The sample s is given by $s = \bigcup_{k_p \in s_I} s_{k_p}$. One generally requires

- Invariance: $p_{k_p}(.|s_I) = p_{k_p}(.)$, the second stage sampling design within primary unit k_p does not depend on the other selected primary units.

- Independence: $P(\bigcup_{k_p \in s_I} s_{k_p} | s_I) = \prod_{k_p \in s_I} p_{k_p}(s_{k_p} | s_I)$

Under these conditions, the inclusion probabilities of secondary units i and k belonging respectively to primary units i_p and k_p are given by

$$\pi_k = \pi_{Ik_p} \pi_{k|k_p} \qquad \pi_{ik} = \begin{cases} \pi_{Ii_p k_p} \pi_{i|i_p} \pi_{k|k_p} & \text{if } i \in i_p \text{ and } k \in k_p, i_p \neq k_p, \\ \pi_{Ii_p} \pi_{ik|i_p} & \text{if } i, k \in i_p. \end{cases}$$

9.2.1.2   Sample membership indicator

Let us denote as before by S a random sample with distribution $p(.)$. To each unit in the population an indicator variable of sample inclusion is attached:

$$z_k(S) = \begin{cases} 1 & \text{if } k \in S, \\ 0 & \text{otherwise.} \end{cases}$$

The (randomization) distribution of the binary random variables $z_k(S)$, $k = 1, \ldots, N$ gives the inclusion probability of element k and the joint inclusion probability of elements i and k:

$$P(z_k(S) = 1) = \pi_k, \qquad P(z_i(S)z_k(S) = 1) = \pi_{ik}.$$

The first and second moments of $z_k(S)$ are easily obtained. Using Equations (9.1) and (9.2):

$$E(z_k(S)) = \sum_{s \in S} p(s)z_k(s) = \sum_{\{s \in S : s \ni k\}} p(s) = \pi_k,$$

$$\mathrm{Var}(z_k(S)) = E(z_k^2(S)) - E(z_k(S))^2 = \pi_k - \pi_k^2, \tag{9.3}$$

$$\mathrm{Cov}(z_i(S), z_k(S)) = \pi_{ik} - \pi_i \pi_k. \tag{9.4}$$

Note that the sample size is given by $n_S = \sum_{k \in U} z_k(S)$ and, depending on the sampling design, it can be fixed or random.

9.2.2 Design-based estimation

The purpose of the survey is the estimation of some statistic $Q(S)$, viewed as a function of the random sample S. Its design-based distribution is derived from $p(.)$.

$$E(Q(S)) = \sum_{s \in S} p(s)Q(s),$$

$$\mathrm{Var}(Q(S)) = \sum_{s \in S} p(s)Q^2(s) - [E(Q(s))]^2,$$

$$\mathrm{Cov}(Q_1(S), Q_2(S)) = \sum_{s \in S} p(s)Q_1(s)Q_2(s) - E(Q_1(s))E(Q_2(s)).$$

The sample membership estimator is a fundamental example of a statistic $Q(S)$.

9.2.2.1 Linear statistic

Let y_k be some quantity of interest attached to individual k, for instance an age class indicator or the income. The quantity y_k is observed if k is in the sample and unknown otherwise. The aim is to estimate the total of y_k, $k \in U$, for instance the total number of persons in a given age class or the total income in the population. The parameter of interest is the sum

$$\theta = \sum_{k \in U} y_k.$$

Definition 9.2.7 (Horvitz–Thompson or π estimator)   *The estimator*

$$\hat{\theta}_1(S) = \sum_{k \in U} \frac{y_k z_k(S)}{E(z_k(S))} = \sum_{k \in U} \frac{y_k z_k(S)}{\pi_k} = \sum_{k \in S} \frac{y_k}{\pi_k} \tag{9.5}$$

is called the Horvitz–Thompson or π estimator.

It is unbiased for θ:

$$E\left(\hat{\theta}_1(S)\right) = \sum_{k \in U} \frac{y_k E\left(z_k(S)\right)}{E\left(z_k(S)\right)} = \sum_{k \in U} y_k = \theta,$$

and its variance is given by:

$$\text{Var}\left(\hat{\theta}_1(S)\right) = \sum_{i,k \in U} \frac{y_i y_k}{\pi_i \pi_k} \text{Cov}\left(z_i(S), z_k(S)\right) = \sum_{i,k \in U} \frac{y_i y_k}{\pi_i \pi_k} \left(\pi_{ik} - \pi_i \pi_k\right), \quad (9.6)$$

where $\text{Cov}\left(z_i(S), z_k(S)\right)$ is given by Equation (9.3) if $i = k$ and by Equation (9.4) if $i \neq k$.

Notice that y_k is not considered as a random variable, but rather as a multiplicative constant applied to the random sample inclusion indicator.

If the sample size is random, the π estimator is inefficient. It is better to estimate the population size $\hat{N}$ on the sample [setting $y_k = 1$ in Equation (9.5)], then deduce an estimated mean and multiply that mean by the actual population size N,

$$\hat{\theta}_2(S) = N \frac{\sum_{k \in U} y_k z_k(S)/\pi_k}{\sum_{k \in U} z_k(S)/\pi_k} = N \frac{\sum_{k \in S} y_k/\pi_k}{\sum_{k \in S} 1/\pi_k}. \quad (9.7)$$

We see that $\hat{\theta}_2(S)$ is a nonlinear function of $z_k(S)$ (where $k = 1, \ldots, N$).

9.2.2.2   Nonlinear statistic

Variance estimation through linearization is a widely used method in the context of surveys. For differentiable statistics, see Särndal *et al.* (1989, 1992), Binder (1996) and Demnati and Rao (2004). For nondifferentiable statistics (e.g. quantiles) (Deville and Särndal 1999; Tillé 2001), apply influence functions.

Unlike the above authors who take the derivatives with respect to the variable of interest (Särndal *et al.* 1989, 1992; Binder 1996) or to the survey weights (Demnati and Rao 2004), here the classical theory of Taylor linearization for twice differentiable statistics is applied (Stuart and Ord 1994, chapter 10), that is we take the derivatives with respect to the variables associated with the inclusion indicators. Suppose that the statistic $Q(S)$ is twice differentiable with respect to $z_k(S)$ (where $k = 1, \ldots, N$). Let us write for simplicity $z_k(S) = z_k$ and $Q(S) = Q(z_1, z_2, \ldots, z_N)$.

Recall that $E(z_k) = \pi_k$. Let us denote the first partial derivative of Q with respect to z_k evaluated at $(\pi_1, \pi_2, \ldots, \pi_N)$ by Q'_k and expand Q up to the order 1:

$$Q(z_1, z_2, \ldots, z_N) \approx Q(\pi_1, \pi_2, \ldots, \pi_N) + \sum_{k \in U} Q'_k(z_k - \pi_k).$$

Replacing $E\left(Q(z_1, z_2, \ldots, z_N)\right)$ by its first order approximation $Q(\pi_1, \pi_2, \ldots, \pi_N)$, we have that the variance is evaluated by the variance of the linear statistic $\sum_{k \in U} Q'_k z_k$:

$$\text{Var}\left(Q(z_1, z_2, \ldots, z_N)\right) \approx \sum_{i,k \in U} Q'_i Q'_k \text{Cov}\left(z_i(S), z_k(S)\right) = \sum_{i,k \in U} Q'_i Q'_k \left(\pi_{ik} - \pi_i \pi_k\right). \quad (9.8)$$

If, for example $Q = \hat{\theta}_2$ [see Equation (9.7)], it is found that

$$Q'_k = N \frac{(y_k - \theta/N)/\pi_k}{\sum_{i \in U} \pi_i/\pi_i} = (y_k - \theta/N)/\pi_k, \quad k \in U,$$

where θ/N is the population mean and $k \in U$. If the design is of fixed sample size, like in Examples 9.2.4 and 9.2.5, $\hat{\theta}_1$ and $\hat{\theta}_2$ are equivalent and one can show that the linearized variance of $\hat{\theta}_2$ obtained from Equation (9.8) gives the same result as the variance of $\hat{\theta}_1$ in Equation (9.6).

Example 9.2.8 (Bernoulli design)   In the Bernoulli design of Example 9.2.3, the sample size is random. We have

$$\text{Var}\left(\hat{\theta}_1(S)\right) = \sum_{k \in U} \frac{y_k^2}{\pi^2}\left(\pi - \pi^2\right) > \sum_{k \in U} \frac{(y_k - \theta/N)^2}{\pi^2}\left(\pi - \pi^2\right) \approx \text{Var}\left(\hat{\theta}_2(S)\right).$$

9.2.2.3   Variance estimator

The variance of $\hat{\theta}_1(S)$ in Equation (9.6) is given by a sum over the whole population, but y_i and y_k are only known on the sample, so we need an estimator. Its Horvitz–Thompson estimator is given by

$$\widehat{\text{Var}}\left(\hat{\theta}_1(S)\right) = \sum_{i,k \in U} \frac{y_i y_k}{\pi_i \pi_k}(\pi_{ik} - \pi_i \pi_k)\frac{z_i(S)z_k(S)}{\pi_{ik}} = \sum_{i,k \in S} \frac{y_i y_k}{\pi_i \pi_k}\frac{\pi_{ik} - \pi_i \pi_k}{\pi_{ik}}. \tag{9.9}$$

This estimator is unbiased for $\text{Var}\left(\hat{\theta}_1(S)\right)$ in Equation (9.6), because $\text{E}\left(z_i(S)z_k(S)\right) = \pi_{ik}$.

In the case of $\hat{\theta}_1$ and $\hat{\theta}_2$, Q'_k is known for $k \in S$. It is not always the case and an alternative method is to expand Q around the actually observed sample. Let us denote by $\hat{z}_1, \hat{z}_2, \hat{z}_N$ the indicators of the actual sample and let

$$\widehat{Q'_k} = \frac{\partial Q(S)}{\partial z_k}|_{\hat{z}_1, \hat{z}_2, \hat{z}_N}.$$

The linearized statistics is then

$$Q(S) - \text{E}(Q(S)) \approx \sum_{k \in U} \widehat{Q'_k}(z_k - \pi_k).$$

By this expansion, the estimator of the linearized variance is

$$\widehat{\text{Var}}\left(Q(S)\right) \approx \sum_{i,k \in U} \widehat{Q'_i}\widehat{Q'_k}(\pi_{ik} - \pi_i \pi_k)\frac{z_i(S)z_k(S)}{\pi_{ik}} = \sum_{i,k \in S} \widehat{Q'_i}\widehat{Q'_k}\frac{\pi_{ik} - \pi_i \pi_k}{\pi_{ik}}. \tag{9.10}$$

For example, if $Q = \hat{\theta}_2$ and $\hat{N} = \sum_{i \in S} 1/\pi_i$,

$$\widehat{Q'_k} = \frac{N}{\hat{N}} \frac{(y_k - \hat{\theta}_2/N)}{\pi_k}, \quad k \in S,$$

and substitution into Equation (9.10) leads to

$$\widehat{Var}\left(\hat{\theta}_2(S)\right) \approx \left(\frac{N}{\hat{N}}\right)^2 \sum_{i,k \in S} \frac{y_i - \hat{\theta}_2/N}{\pi_i} \frac{y_k - \hat{\theta}_2/N}{\pi_k} \frac{\pi_{ik} - \pi_i \pi_k}{\pi_{ik}}.$$

The difference between the two linearizations of $\hat{\theta}_2$ is the term $N/\hat{N}$ which is asymptotically equal to 1.

Remark: The usual way of presenting the Taylor linearization in the context of surveys is via the derivation with respect to the variable of interest y_k, $k \in S$ and gives rise to the same estimators. The advantage of the above presentation is that it generalizes immediately to vector statistics.

The design-based covariance matrix estimator of a vector statistic $\mathbf{Q}(S)$ follows along the same lines:

$$\widehat{\mathbf{Q}'_k} = \frac{\partial \mathbf{Q}(S)}{\partial z_k}|_{\hat{z}_1, \hat{z}_2, \hat{z}_N}$$

$$\widehat{Var}(\mathbf{Q}(S)) \approx \sum_{i,k \in S} \widehat{\mathbf{Q}'_i}(\widehat{\mathbf{Q}'_k})^\top \frac{\pi_{ik} - \pi_i \pi_k}{\pi_{ik}}. \tag{9.11}$$

9.2.2.4 Nonresponse

Missing data in surveys are due to nonresponse. The causes of nonresponse may be of different nature: impossibility to contact, refusal, unreturned questionnaires, etc. Surveys of course record many variables and the nonresponse pattern may depend on the variable. The nonresponse rate can reach a high level on sensitive questions and just ignoring the fact may lead to large biases in the estimations. There are no perfect solutions to the problem of missing data, but successful approaches to reduce the impact have been developed.

We sketch one possible solution to unit nonresponse, the so-called 'response homogeneity group model'. Suppose the units in the population can be classified into response homogeneity groups, such that the response probability of units is constant within the group, and units answer independently. Let s_g be the sampled units within the gth homogeneity group. s_g can be partitioned into

$$s_g = r_g \cup \bar{r}_g,$$

where r_g is the respondents subset and $\bar{r}_g$ the nonrespondents subset. If the response probability τ_g is interpreted as an inclusion probability into the respondents set, then the response model can be seen as a stratified Bernoulli sampling, conditionally on the realized sample s. The overall inclusion probabilities are given by

$$\pi_k^* = \pi_{kk}^* = \pi_k \tau_{g_k} \qquad \pi_{ik}^* = \pi_{ik} \tau_{g_i} \tau_{g_k}, \ i \neq k,$$

with π_k and π_{ik} the inclusion probabilities in the original sampling design, g_i and g_k being the response homogeneity groups of units i and k, respectively. The difference with a pure stratified sample is that the groups s_g can only be determined on the realized sample, when the strata are fixed prior to sampling. Moreover, τ_g must be estimated.

Under these conditions, the π-estimator of the total $\hat{\theta}_1^*(S)$ is obtained from Equation (9.5) upon substituting π_k^* to π_k and summing on the overall respondents set $r(S) = \bigcup_g r_g$ instead of S. The variance is derived from Equation (9.9) in a similar way.

$$\hat{\theta}_1^*(S) = \sum_{k \in r(S)} \frac{y_k}{\pi_k^*} \qquad \widehat{\mathrm{Var}}\left(\hat{\theta}_1^*(S)\right) = \sum_{i,k \in r(S)} \frac{y_i y_k}{\pi_i^* \pi_k^*} \frac{\pi_{ik}^* - \pi_i^* \pi_k^*}{\pi_{ik}^*}. \qquad (9.12)$$

Recently, Hron *et al.* (2010) proposed several imputation procedures for compositions based on k-nearest neighbour followed by iterative regressions that could be applied also in surveys.

9.2.2.5   Calibration

Recall that auxiliary information on the population U can be present in the register used to build the sampling frame. The design weights $1/\pi_k$ or $1/\pi_k^*$ are usually modified in such a way that the weighted total over key auxiliary variables, measured on the sample, equals the total of the auxiliary variable over the population (e.g. totals by age group, marital status for individuals or number of employees for companies). Calibration is a very efficient way to correct bias in the sample when auxiliary information is related to the study variable.

Let $w_k = g_k/\pi_k^*$ be the final weight, where the calibration correction $g_k = g_k(r(S))$ is a nonlinear function of the respondent set. The final calibrated estimator is thus a nonlinear statistics of the type $Q(S)$, see Section 9.2.2.2. It can be linearized by the present method and the same results are obtained as by Demnati and Rao (2004).

Algorithms for simultaneous calibration on several variables exist.

9.3   Application to compositional data

The design-based theory sketched in Section 9.2 is the basic estimation approach used by the data producers. Thus the incorporation of design weights is unavoidable if one wants to get results that are comparable with those officially published. In model-based estimation on the other hand, model assumptions are posed on the study variable y_k and weights have another meaning. For example in regression, weights are used to model heteroscedasticity. It would be pure chance if the survey weights w_k were inversely proportional to the residual variance in a regression model. Thus model-based variances may differ considerably from design-based variance. A good overview on how to cope with this problem is in Chambers and Skinner (2003). The traditional approach to compositional data as exposed in Aitchison (1986) is model based: the observed compositions are assumed to follow a multivariate logistic normal distribution or a Dirichlet distribution. In this chapter, we take the design-based point of view and thus avoid distribution assumptions on observed compositions.

9.3.1 Weighted arithmetic and geometric means

We concentrate on the estimation of the variance of different types of means. Because of nonresponse, the actual sample size is random and averages are nonlinear statistics of the sample membership indicator. For simplicity, the realized sample is denoted by s instead of $r(s)$ and we drop the * in the inclusion probabilities.

Let t_k be some total amount for unit $k \in U$ (U is a population of size N) and let w_k be the corresponding survey weight. The total amount t_k is distributed among D components $t_k p_k, k = 1, \ldots, N$, where p_k is the corresponding D-part composition.

In the context of surveys, we are interested in means across cases, but in a compositional context, totals have no meaning. So if we want to add cases, we have to go back to the original amounts. For the geometric mean composition on the contrary, the result is the same, whether the amounts are averaged first and then an average composition is computed, or whether the geometric mean of the compositions is computed directly. The D-part composition giving the closed arithmetic mean of amounts and closed geometric mean estimators are, respectively

$$\bar{p}_{am}(S) = \frac{1}{\sum_{i \in U} w_i t_i z_i(S)} \sum_{k \in U} w_k t_k z_k(S) \, p_k, \tag{9.13}$$

$$\bar{p}_{gm}(S) = \frac{1}{\sum_{i \in U} w_i z_i(S)} \odot \prod_{k \in U} [w_k z_k(S)] \odot p_k. \tag{9.14}$$

In Equation (9.14), the product is computed componentwise.

Notice that only the closed geometric mean is compatible with Aitchison's geometry (Pawlowsky-Glahn and Egozcue 2002). On the other hand, when a data set is created, the compositions of interest are frequently obtained from closed sums of amounts. For example, household incomes are sums of the incomes of individual household members. It is necessary to go back to the individual data in order to compute the sampling variability at the household level.

9.3.2 Closed arithmetic mean of amounts

From the design-based viewpoint, it is a ratio of sums. The design-based covariance matrix estimator V_{am} of $\bar{p}_{am}(S)$ in Equation (9.13) is given by Equation (9.11) with:

$$\begin{aligned}
Q'_k &= \frac{\partial \bar{p}_{am}}{\partial z_i} \Big|_{\pi_1, \ldots, \pi_N} = \frac{w_k t_k \, p_k \left(\sum_{i \in U} w_i t_i \pi_i \right) - w_k t_k \left(\sum_{i \in U} w_i t_i \pi_i \, p_i \right)}{\left(\sum_{i \in U} w_i t_i \, \pi_i \right)^2} \\
&= \frac{w_k t_k \left(p_k - \bar{p}_{am}(U) \right)}{\sum_{i \in U} w_i t_i \, \pi_i}, \quad k \in U,
\end{aligned} \tag{9.15}$$

and

$$\widehat{Q'_k} = \frac{w_k t_k \left(p_k - \bar{p}_{am}(S)\right)}{\sum_{i \in S} w_i t_i}, \quad k \in S. \tag{9.16}$$

9.3.3   Centred log-ratio of the geometric mean composition

The transform $\mathrm{clr}(\bar{p}_{gm}(S))$ is obtained from Equation (9.13) with $\mathrm{clr}(p_k)$ replacing p_k and w_k replacing $w_k t_k$. Thus the corresponding covariance matrix V_{clr} is given by Equation (9.11) with

$$\widehat{Q'_k} = \frac{w_k \left[\mathrm{clr}(p_k) - \mathrm{clr}(\bar{p}_{gm}(S))\right]}{\sum_{i \in S} w_i}. \tag{9.17}$$

The covariance matrix of other transforms are obtained along the same lines.

9.3.4   Closed geometric mean composition

Let us write the jth component of $\bar{p}_{gm} = \bar{p}_{gm}(S)$ in Equation (9.14) as:

$$\bar{p}_{gm,j} = C\left\{\exp\left[\mathrm{clr}(\bar{p}_{gm})\right]\right\}_j = \frac{R_j}{\sum_{\ell} R_{\ell}}.$$

$$\frac{\partial R_j}{\partial z_k} = R'_{k,j} = R_j \frac{\partial}{\partial z_k} \mathrm{clr}(\bar{p}_{gm})_j = R_j \widehat{Q'_{k,j}},$$

where $\widehat{Q'_{k,j}}$ is the jth component of $\widehat{Q'_k}$ in Equation (9.17). Then we obtain

$$\frac{\partial}{\partial z_k} \bar{p}_{gm,j} = \bar{p}_{gm,j} \left[\widehat{Q'_{k,j}} - \sum_{\ell} \bar{p}_{gm,\ell} \widehat{Q'_{k,\ell}}\right]$$

and, in vector form,

$$\frac{\partial}{\partial z_k} \bar{p}_{gm}(S) = \left(\mathrm{diag}(\bar{p}_{gm}(S)) - \bar{p}_{gm}(S)\bar{p}_{gm}(S)^{\top}\right) \widehat{Q'_k}. \tag{9.18}$$

The linearized form of the closed geometric mean [Equation (9.18)] is a linear transformation of the linearized clr transform, which implies that, locally, it is compatible with the Aitchison geometry. The estimated linearized covariance matrix of the closed geometric mean composition is thus obtained from the covariance matrix V_{clr} of the clr transforms:

$$V_{gm} \approx \left(\mathrm{diag}(\bar{p}_{gm}(S)) - \bar{p}_{gm}(S)\bar{p}_{gm}(S)^{\top}\right) V_{\mathrm{clr}} \left(\mathrm{diag}(\bar{p}_{gm}(S)) - \bar{p}_{gm}(S)\bar{p}_{gm}(S)^{\top}\right). \tag{9.19}$$

These linearization covariance matrices give good approximations of confidence domains when the sample size is large. In this case, one can rely on asymptotic normal theory for confidence domains and tests (Särndal *et al.* 1992, p. 528). The inference is nonparametric because it does not depend on distributional properties of the measured compositions, but solely on the randomization distribution implied by the sampling design. Two choices are thus proposed for computing linearized design-based confidence domains for $\bar{p}_{gm}$: either stay in the Aitchison's geometry and back-transform the confidence domain obtained from e.g. V_{clr}, or compute an Euclidian approximation based on V_{gm}. Recently, Jarauta-Bragulat and Egozcue (2008, appendix B) have defined compositional derivatives. These are essentially ordinary derivatives in the transformed space (e.g. clr) followed by back-transformation (e.g. clr^{-1}). Confidence domains for $\bar{p}_{gm}$ are then derived from the corresponding domains for $\text{clr}(\bar{p}_{gm})$ by back-transformation. In the present approach, the derivatives are not compositional, but taken with respect to the sample inclusion indicator.

9.3.5   Example: Swiss Earnings Structure Survey (SESS)

In the SESS, $D = 5$ wage components are recorded, among them the gross earnings. Extrapolated totals of the components by economic activity groups are computed and ratios of total wage components to the total gross earnings are published. At the individual level, only the gross earnings is never zero and one seldom observes the presence of all other components together, so that the problem of essential zeros occurs. Yet at the chosen group aggregate levels, zeros do not appear any more and log-ratio transformations can be envisaged. The form of the published results make the additive log-ratio transformation alr the most natural.

The variance computation is described below [see Graf (2005, 2006a,b) for details]. The sampling scheme is a stratified two-stage design, i.e. we have a two-stage design within each stratum, see Examples 9.2.5 and 9.2.6. The design-based covariance matrix is the sum of the within strata covariance matrices. Because in this case, groups are subsets of strata, the sampling scheme is a stratified two-stage design in each group.

Let G designate the economic activity group. We first compute the weighted sum of the cases wage components $c_{ij}, i \in G; j = 1, .., D$ and take the ratio to the last (the weighted sum of gross earnings):

$$Q_j = \sum_i w_i c_{ij} \Big/ \sum_i w_i c_{iD} = \sum_i w_i c_{iD}(c_{ij}/c_{iD}) \Big/ \sum_i w_i c_{iD}$$

$$= \sum_i w_i c_{iD}(p_{ij}/p_{iD}) \Big/ \sum_i w_i c_{iD},$$

where p_{ij} is the part of the jth component in the unit-level composition p_i and $j = 1, \ldots, D - 1$. Collecting the Q_j's into a vector Q_G, we see that $\log(Q_G)$ is the alr transform of a composition $\bar{p}_G = (\bar{p}_{G_1}, \ldots, \bar{p}_{G_D})$, the closed weighted arithmetic mean of amounts as in Equation (9.13). The published ratios are $Q_G = \bar{p}_{G_{-D}}/\bar{p}_{G_D}$, where the index $-D$ means that component D is suppressed. The covariance matrix V_{Q_G} is obtained by linearization along the same lines as above, with:

$$\widehat{Q_{G_k}'} = \frac{w_k c_{kD}(p_{k,-D}/p_{kD} - Q_G)}{\sum_i w_i c_{iD}}.$$

The linearization approximation of the design-based covariance matrix of the alr transform is

$$V_{\text{alr}(\bar{p}_G)} = \text{diag}(Q_G)^{-1} V_{Q_G} \text{diag}(Q_G)^{-1}.$$

In this example, confidence domains were obtained for the alr transform and transformed back to the simplex (Graf 2006b).

Here, group average amounts by components were the basis for the analysis. The data set for compositional data analysis is thus the set of compositions at the group level. The advantage is that essential zeros do not occur any more, but the drawback is that the individual level is lost, except for the assessment of the measurement error. Analysis of compositions at the unit level would require to address the problem of essential zeros (Fry *et al.* 1996, 2000; Martín-Fernández *et al.* 2000).

9.4   Discussion

Another context in which weights are used in connection with compositions is described in Greenacre and Lewi (2009). The authors show how changing the weights of both rows and columns in a clr-transformed compositional data set can improve the visualization of specific effects. They freely choose the weighting of rows and columns in order to downweight large effects, so that not so obvious – and usually more interesting – characteristics can appear on the plot. In their approach, weights can be chosen arbitrarily, because the aim is to emphasize specific characteristics of the data set. Notice that weighting the columns of a clr-transformed data set has no simplicial equivalent.

By contrast, in the present chapter weights on cases are prescribed and correspond to the weights that were set up at the design stage of the survey and modified at the analysis stage in order to take nonresponse and calibration into account. We have shown that a design-based analysis of compositions can be easily obtained from the standard theory of survey sampling. For practical implementation, two R packages for survey sampling and analysis have been developed by Matei and Tillé (2009) and Lumley (2010).

The design-based approach does not make any assumptions on the distribution of compositions. This opens the way to parametrization by general partitions (Aitchison 1986, section 2.7) without the drawback of ad hoc assumptions on multivariate normality (Aitchison 1986, definition 6.7). In household expenditure surveys for instance, a hierarchy of commodities with broad categories are subdivided into more detailed goods. A general partition can follow this organization and may be a more convenient way to convey the information on the surveyed units. The joint probability distribution of transforms of this general partition can be derived from the distribution of the sample inclusion indicator.

References

Aitchison J 1986 *The Statistical Analysis of Compositional Data*. Monographs on Statistics and Applied Probability. Chapman and Hall Ltd (reprinted 2003 with additional material by The Blackburn Press), London (UK). 416 p.

Binder D 1996 Linearization methods for single phase and two-phase samples: a cookbook approach. *Survey Methodology* **22**(1), 17–22.

Chambers RL and Skinner CJ 2003 *Analysis of Survey Data*. John Wiley & Sons, Ltd, Chichester (UK). 376 p.

Demnati A and Rao J 2004 Linearization variance estimators for survey data. *Survey Methodology* **30**(1), 17–26.

Deville JC and Särndal CE 1999 Variance estimation for complex statistics and estimators: linearization and residual techniques. *Survey Methodology* **25**, 193–203.

Egozcue JJ, Pawlowsky-Glahn V, Mateu-Figueras G and Barceló-Vidal C 2003 Isometric logratio transformations for compositional data analysis. *Mathematical Geology* **35**(3), 279–300.

Fry JM, Fry TRL and McLaren KR 1996 Compositional data analysis and zeros in micro data. Centre of Policy Studies (COPS), General Paper No. G-120, Monash University.

Fry JM, Fry TRL and McLaren KR 2000 Compositional data analysis and zeros in micro data. *Applied Economics* **32**(8), 953–959.

Graf M 2005 Assessing the precision of compositional data in a stratified double stage cluster sample: Application to the Swiss earnings structure survey. In *Proceedings of CoDaWork' 05, The 2nd Compositional Data Analysis Workshop* (ed. Mateu-Figueras G and Barceló-Vidal C). http://ima.udg.es/Activitats/CoDaWork05/. University of Girona, Girona (Spain).

Graf M 2006a Swiss Earnings Structure Survey 2002–2004. Compositional data in a stratified two-stage sample: Analysis and precision assessment of wage components. Methodology Report 338-0038. Federal Statistical Office, Neuchâtel, Switzerland.

Graf M 2006b Precision of compositional data in a stratified two-stage cluster sample: Comparison of the Swiss earnings structure survey 2002 and 2004. In *ASA Proceedings of the Joint Statistical Meetings*. American Statistical Association, Alexandria, VA (USA). pp. 3066–3072.

Greenacre M and Lewi P 2009 Distributional equivalence and subcompositional coherence in the analysis of compositional data, contingency tables and ratio-scale measurements. *Journal of Classification* **26**(1), 29–54.

Hron K, Templ M and Filzmoser P 2010 Imputation of missing values for compositional data using classical and robust methods. *Computational Statistics and Data Analysis* **54**(12), 3095–3107.

Jarauta-Bragulat E and Egozcue J 2008 Compositional evolution with mass transfer in closed systems. In *Proceedings of CoDaWork' 08, The 3rd Compositional Data Analysis Workshop* (ed. Daunis-i Estadella J and Martín-Fernández J). http://hdl.handle.net/10256/723. University of Girona, Girona (Spain). CD-ROM.

Lumley T 2010 Survey: Analysis of complex survey samples. R package version 3.21.

Martín-Fernández JA, Barceló-Vidal C and Pawlowsky-Glahn V 2000 Zero replacement in compositional data sets. In *Proceedings of the 7th Conference of the International Federation of Classification Societies (IFCS 2000)* (ed. Kiers H, Rasson J, Groenen P and Shader M). Springer-Verlag, Berlin (Germany). pp. 155–160.

Matei A and Tillé Y 2009 Sampling: Survey sampling. R package version 2.3.

Pawlowsky-Glahn V and Egozcue JJ 2002 BLU estimators and compositional data. *Mathematical Geology* **34**(3), 259–274.

Särndal CE, Swensson B and Wretman J 1989 The weighted residual technique for estimating the variance of the general regression estimator of a finite population total. *Biometrika* **76**, 527–537.

Särndal CE, Swensson B and Wretman J 1992 *Model Assisted Survey Sampling.* Springer-Verlag, New York, NY (USA). 694 p.

Stuart A and Ord JK 1994 *Kendall's Advanced Theory of Statistics, Volume 1, Distribution Theory,* 6th edition. Edward Arnold, London (UK). 439 p.

Tillé Y 2001 *Théorie des sondages: échantillonnage et estimation en populations finies: cours et exercices.* Dunod, Paris (France). 284 p.

10

Notes on the scaled Dirichlet distribution

Gianna Serafina Monti[1], Glòria Mateu-Figueras[2] and Vera Pawlowsky-Glahn[2]
[1]*Department of Statistics, University of Milan-Bicocca, Italy*
[2]*Department of Computer Science and Applied Mathematics, University of Girona, Spain*

10.1 Introduction

Compositional data are by definition parts of some whole. Their sample space can be represented by the simplex, denoted by $\mathcal{S}^D = \{\mathbf{x} = (x_1, \ldots, x_D),\ x_i > 0,\ \sum_{i=1}^{D} x_i = \kappa\}$, where κ is a constant, usually taken to be one. The relevant information for each composition is contained in the ratios between components (Aitchison 1982, 1986). Several operations can be defined in the simplex: an internal operation $\oplus$, called *perturbation* and playing the role of a sum, and an external operation $\odot$, called *powering* and playing the role of a product by real scalars. Also, an inner product $< ., . >_a$ has been defined. With these operations $(\mathcal{S}^D, \oplus, \odot, < ., . >_a)$ has a Euclidean vector space structure of dimension $D - 1$ (Billheimer *et al.* 2001; Pawlowsky-Glahn and Egozcue 2001).

This general structure allows us to represent compositions by coordinates with respect to an orthonormal basis (Egozcue *et al.* 2003) (see also Chapter 3). In particular, given an orthonormal basis $\{\mathbf{e}_1, \ldots, \mathbf{e}_{D-1}\}$ in $\mathcal{S}^D$, a composition $\mathbf{x} \in \mathcal{S}^D$ can be expressed as a perturbation-combination, the counterpart of a linear combination,

$$\mathbf{x} = (\varepsilon_1 \odot \mathbf{e}_1) \oplus (\varepsilon_2 \odot \mathbf{e}_2) \oplus \cdots \oplus (\varepsilon_{D-1} \odot \mathbf{e}_{D-1}) = \bigoplus_{i=1}^{D-1}(\varepsilon_i \odot \mathbf{e}_i).$$

Compositional Data Analysis: Theory and Applications, First Edition. Edited by Vera Pawlowsky-Glahn and Antonella Buccianti.
© 2011 John Wiley & Sons, Ltd. Published 2011 by John Wiley & Sons, Ltd.

The coefficients ε_i are the coordinates of the composition $\mathbf{x} \in \mathcal{S}^D$ with respect to the given orthonormal basis. For a fixed basis, they are uniquely determined, i.e. we can always represent a composition by its coordinates with respect to a fixed orthonormal basis. The vector obtained applying the isometric log-ratio transformation to a composition $\mathbf{x}$, a transformation from $\mathcal{S}^D$ to $\mathbb{R}^{D-1}$ defined by Egozcue *et al.* (2003),

$$\mathrm{ilr}(\mathbf{x}) = (\langle \mathbf{x}, \mathbf{e}_1 \rangle_a, \ldots, \langle \mathbf{x}, \mathbf{e}_{D-1} \rangle_a),$$

is a $D-1$ vector of real coordinates, which are the coefficients with respect to an orthonormal basis.

A measure on the simplex, denoted as λ_a and called the Aitchison measure, which is compatible with the inner vector space structure of the simplex, has been defined by Pawlowsky-Glahn (2003). The same strategy has been used previously to define a Lebesgue type measure on an arbitrary Euclidean space (Eaton 1983). It is absolutely continuous with respect to the Lebesgue measure, denoted by λ (without a subscript) on real space. A density function defined on the simplex and expressed in terms of the Aitchison measure, is the Radon–Nikodym derivative of a probability measure μ with respect to λ_a, denoted by $d\mu/d\lambda_a$. Recall that the Radon–Nikodym theorem is a result in functional analysis which states that, given a measurable space (E, Σ), if a σ-finite measure μ on (E, Σ) is absolutely continuous with respect to a σ-finite measure λ_E on (E, Σ), then there is a measurable function f on E, taking values in $[0, +\infty)$, such that

$$\mu(A) = \int_A f \, d\lambda_E,$$

for any measurable set $A \subset E$ (Athreya and Lahiri 2006). The function f, which is a density function for the measure μ, is called the Radon–Nikodym derivative of μ with respect to λ_E.

The relationship between the two measures, λ_a and λ, is given by the Jacobian:

$$\frac{d\lambda_a}{d\lambda} = \frac{1}{\sqrt{D} x_1 \cdots x_D}. \tag{10.1}$$

In this contribution we analyse the Dirichlet and scaled Dirichlet distributions changing the measure from λ to λ_a. In Section 10.2 we study the genesis of the scaled Dirichlet distribution, starting with a review of the Dirichlet one. In Section 10.3 we analyse this model comparing the classical with the Aitchison representation.

10.2   Genesis of the scaled Dirichlet distribution

The Dirichlet is the best known distribution for modelling categorical data. It plays an important role in compositional data for representing proportions of substances, but also arises in other contexts: from multivariate statistics to distribution theory, such as extreme value distributions. In Bayesian inference the Dirichlet distribution is a natural conjugate prior for the multinomial likelihood (Albert and Gupta 1982; Wong 1998) and is often used as prior for Bayesian life-testing problems with either complete or censored data (Lochner 1975).

This distribution exhibits many convenient mathematical properties including closure under amalgamations of categories and marginalization over a category (Aitchison 1986).

A random composition $\mathbf{X} \in \mathcal{S}^D$ has a Dirichlet distribution with parameter $\boldsymbol{\alpha} = (\alpha_1, \ldots, \alpha_D) \in \mathbb{R}_+^D$, denoted by $\mathbf{X} \sim \mathcal{D}^D(\boldsymbol{\alpha})$, if its density function is of the form

$$f(\mathbf{x}) = \frac{\mathrm{d}\mu}{\mathrm{d}\lambda}(\mathbf{x}) = \frac{\Gamma(\alpha_+)}{\prod_{i=1}^D \Gamma(\alpha_i)} \prod_{i=1}^D x_i^{\alpha_i - 1}, \tag{10.2}$$

where μ is the Dirichlet probability measure, $\alpha_+ = \sum_{i=1}^D \alpha_i$, and Γ denotes the gamma function. The standard form is defined as the distribution of $\mathbf{X} = (X_1, X_2, \ldots, X_D)$, with $X_i = Z_i / \sum_{i=1}^D Z_i$ and Z_i, $(i = 1, 2, \ldots, D)$, D independent, standard gamma-distributed random variables with shape parameters α_i and equally scaled: $Z_i \sim \mathrm{Ga}(\alpha_i, 1)$. When the number of components is $D = 2$ the density (10.2) is known as the beta density (Kotz *et al.* 2000).

Removing the requirement of equal scale parameter of the gamma variables, i.e. $Z_i \sim \mathrm{Ga}(\alpha_i, \beta_i)$, $i = 1, 2, \ldots, D$, a generalization of the Dirichlet distribution, called the scaled Dirichlet distribution, and denoted by $\mathbf{X} \sim \mathcal{SD}^D(\boldsymbol{\alpha}, \boldsymbol{\beta})$, $\boldsymbol{\alpha}, \boldsymbol{\beta} \in \mathbb{R}_+^D$, is obtained.

Proposition 10.2.1 *Let* $\mathbf{X} \sim \mathcal{SD}^D(\boldsymbol{\alpha}, \boldsymbol{\beta})$. *Its density function is*

$$f_s(\mathbf{x}) = \frac{\mathrm{d}\mu_s}{\mathrm{d}\lambda}(\mathbf{x}) = \frac{\Gamma(\alpha_+)}{\prod_{i=1}^D \Gamma(\alpha_i)} \frac{\prod_{i=1}^D \beta_i^{\alpha_i} x_i^{\alpha_i - 1}}{(\sum_{i=1}^D \beta_i x_i)^{\alpha_+}}, \tag{10.3}$$

where μ_s *is the scaled Dirichlet probability measure, and* $\alpha_+ = \sum_{i=1}^D \alpha_i$.

Note that the Dirichlet distribution is just a special case of the scaled Dirichlet one: Equation (10.2) can be obtained from (10.3) when all elements of the vector $\boldsymbol{\beta}$ are equal to a common constant, say β, as such a constant cancels out in Equation (10.3). Note that this is an essential characteristic of compositional data as equivalence classes (Barceló-Vidal *et al.* 2001). An alternative parametrization of (10.3), mathematically equivalent to the standard approach, is helpful to clarify the dimension of the parameter space of the scaled Dirichlet. In fact, it is possible to scale the β_i's to add to one without any change in (10.3), so that $\boldsymbol{\beta}$ becomes an element of $\mathcal{S}^D$. It follows that the number of parameters of the scaled Dirichlet distribution is equal to $2D - 1$.

Equations (10.2) and (10.3) correspond to classical densities, i.e. they are Radon–Nikodym derivatives with respect to the Lebesgue measure in real space (in fact, to the restriction of the Lebesgue measure in real space to the simplex). Using (10.1) it is easy to change the measure, and to express both densities with respect to the measure λ_a (Mateu-Figueras and Pawlowsky-Glahn 2005). The resulting expressions for the Dirichlet and scaled Dirichlet density functions are, respectively,

$$f_a(\mathbf{x}) = \frac{\mathrm{d}\mu}{\mathrm{d}\lambda_a}(\mathbf{x}) = \frac{\Gamma(\alpha_+)\sqrt{D}}{\prod_{i=1}^D \Gamma(\alpha_i)} \prod_{i=1}^D x_i^{\alpha_i}, \tag{10.4}$$

and

$$f_{sa}(\mathbf{x}) = \frac{d\mu_s}{d\lambda_a}(\mathbf{x}) = \frac{\Gamma(\alpha_+)\sqrt{D}\ \prod_{i=1}^{D}(\beta_i x_i)^{\alpha_i}}{\prod_{i=1}^{D}\Gamma(\alpha_i)\ (\sum_{i=1}^{D}\beta_i x_i)^{\alpha_+}}.$$ (10.5)

Proposition 10.2.2   *A scaled Dirichlet distribution can be obtained starting from a perturbed random composition with a Dirichlet density. In fact, let $\mathbf{X} \sim \mathcal{D}^D(\boldsymbol{\alpha})$ be a random composition defined in $\mathcal{S}^D$, and let $\mathbf{p} \in \mathcal{S}^D$ be a composition. Consider the random composition $\widetilde{\mathbf{X}}$ obtained applying a perturbation $\mathbf{p}$ to $\mathbf{X}$,*

$$\widetilde{\mathbf{X}} = \mathbf{p} \oplus \mathbf{X}.$$

The density function of $\widetilde{\mathbf{X}}$ with respect to the Aitchison measure λ_a is then

$$f_a(\widetilde{\mathbf{x}}) = \frac{d\mu_s}{d\lambda_a}(\widetilde{\mathbf{x}}) = \frac{\Gamma(\alpha_+)\sqrt{D}\ \prod_{i=1}^{D}(\frac{\widetilde{x}_i}{p_i})^{\alpha_i}}{\prod_{i=1}^{D}\Gamma(\alpha_i)\ (\sum_{i=1}^{D}\frac{\widetilde{x}_i}{p_i})^{\alpha_+}}.$$ (10.6)

Proof. To prove this result, start from the joint probability density function of $\mathbf{X}$, then make the transformation to new variables:

$$\widetilde{X}_i = \frac{p_i X_i}{\sum_{i-1}^{D} p_i X_i}\quad i = 1, \ldots, D-1,$$

and $\widetilde{X}_D = \sum_{i=1}^{D} p_i X_i$, with Jacobian $J = \dfrac{\partial(x_1, \ldots, x_D)}{\partial(\widetilde{x}_1, \ldots, \widetilde{x}_D)} = \dfrac{\widetilde{x}_D^{D-1}}{\prod_{i=1}^{D} p_i}$. Given the joint probability density function of $\widetilde{\mathbf{X}}$, and integrating out the variable $\widetilde{x}_D$ we obtain the density defined in (10.6).

The density (10.6) coincides with (10.5) setting $\mathbf{p} = \ominus\boldsymbol{\beta}$, where $\ominus$ is, by analogy with standard operations in real space, the inverse operation of perturbation. Thus, the scaled Dirichlet and the Dirichlet distribution define two subclasses of the same parametric family: they do not represent two different classes, as usually considered (Connor and Mosimann 1969; Aitchison 1986; Wong 1998, 2010). Note that this property holds whichever reference measure we use to represent the density.

10.3   Properties of the scaled Dirichlet distribution

In this section, we give some probabilistic properties of the scaled Dirichlet model, starting from a graphical comparison between the classical and the Aitchison representations and providing measures of location and variability.

10.3.1   Graphical comparison

To compare the implications of the two representations, consider the densities in Figure 10.1. When the number of components is $D = 2$ (beta), the densities (10.3) and (10.5) correspond,

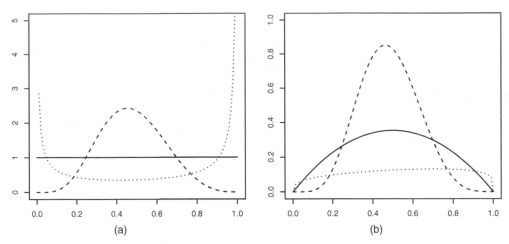

Figure 10.1 Scaled Dirichlet density curves when $D = 2$: (a) with respect to the restriction of the Lebesgue measure λ on the $[0, 1]$ interval; (b) with respect to the Aitchison measure λ_a on S^2. Parameter configurations: $\boldsymbol{\alpha} = (1, 1)$, $\boldsymbol{\beta} = (1, 1)$ (—); $\boldsymbol{\alpha} = (6, 4)$, $\boldsymbol{\beta} = (0.52, 0.3)$ (- - -); $\boldsymbol{\alpha} = (0.3, 0.2)$, $\boldsymbol{\beta} = (0.5, 0.8)$ ($\cdots$).

respectively, to:

$$f_s(\mathbf{x}) = \frac{d\mu_s}{d\lambda}(\mathbf{x}) = \frac{1}{B(\alpha_1, \alpha_2)} \frac{\beta_1^{\alpha_1} x^{\alpha_1-1} \beta_2^{\alpha_2} (1-x)^{\alpha_2-1}}{(\beta_1 x + \beta_2(1-x))^{\alpha_1+\alpha_2}}, \tag{10.7}$$

and

$$f_{sa}(\mathbf{x}) = \frac{d\mu_s}{d\lambda_a}(\mathbf{x}) = \frac{\sqrt{2}}{B(\alpha_1, \alpha_2)} \frac{(\beta_1 x)^{\alpha_1} (\beta_2(1-x))^{\alpha_2}}{(\beta_1 x + \beta_2(1-x))^{\alpha_1+\alpha_2}}. \tag{10.8}$$

Here we recognize a generalization of the beta distribution, which we call, by analogy, the scaled beta distribution.

In Figure 10.1(a) densities (10.7) with respect to the Lebesgue measure λ on the $[0, 1]$ interval are represented, and in Figure 10.1(b) densities (10.8) with respect to the Aitchison measure λ_a on S^2. We observe that, using the density with respect to λ_a, there is always a single mode. This is not the case using the scaled beta distribution with respect to the measure λ, because for $\boldsymbol{\alpha} = (1, 1)$ and $\boldsymbol{\beta} = (1, 1)$ a constant density function is obtained, and for $\boldsymbol{\alpha} = (0.3, 0.2)$ and $\boldsymbol{\beta} = (0.5, 0.8)$ it is a density with vertical asymptotes at 0 and 1 [see Figure 10.1(a)]. A similar behaviour is obtained for $D > 2$.

When $D = 3$, the usual representation for a composition is the ternary diagram. In Figure 10.2(a) the isodensity contour plots of a Dirichlet density with $\boldsymbol{\alpha} = (2, 3, 4)$ (dashed line) and the corresponding perturbed Dirichlet using the perturbation vector $\mathbf{p} = (0.93, 0.05, 0.02)$ (continuous line) in the ternary diagram are represented with respect to the Aitchison measure λ_a on S^3. Figure 10.2(b) represents exactly the same isodensity contour plots in the space of coordinates with respect to an orthonormal basis. Figure 10.2(b) shows that the perturbed Dirichlet density is a translation in the simplex of the original nonperturbed Dirichlet density. This is due to invariance under the internal operation. One interesting

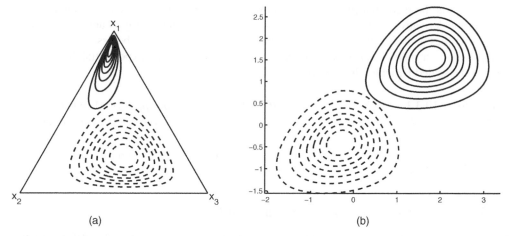

(a)   (b)

Figure 10.2 Scaled Dirichlet density curves when $D = 3$ (a) in the simplex with respect to the Aitchison measure λ_a; (b) in the space of ilr coordinates. Parameter configurations: $\boldsymbol{\alpha} = (2, 3, 4)$, $\boldsymbol{\beta} = (1, 1, 1)$ (- - -); $\boldsymbol{\alpha} = (2, 3, 4)$, $\boldsymbol{\beta} = (1/0.93, 1/0.05, 1/0.02)$ (—).

consequence studied later in this section is that the mode and the expected composition of a scaled Dirichlet can be simply obtained as the perturbed mode and expected composition of a Dirichlet. Note that the invariance under perturbation property only holds when the λ_a measure is used to represent the density.

10.3.2 Membership in the exponential family

Densities of the multidimensional exponential families take the form:

$$f(\mathbf{x}; \boldsymbol{\omega}) = b(\mathbf{x}) \exp \left[\sum_{i=1}^{\ell} \theta_i(\boldsymbol{\omega}) t_i(\mathbf{x}) - a(\boldsymbol{\omega}) \right],$$

where $\boldsymbol{\omega}$ is a parameter, and $\theta(\boldsymbol{\omega})$ and $t(\mathbf{x})$ are vectors of common length, say ℓ.

Many of the commonly used distributions are members of the exponential family. Examples of such distributions are the gamma, beta and Dirichlet distributions. Recall the well known result about the expression of the Dirichlet in terms of an exponential distribution:

$$f(\mathbf{x}) = \exp \left[\sum_{i=1}^{D} (\alpha_i - 1) \log x_i - a(\boldsymbol{\alpha}) \right],$$

where $a(\boldsymbol{\alpha}) = \log \Gamma (\alpha_+) - \sum_{i=1}^{D} \log \Gamma (\alpha_i)$ is a normalizing factor. Differentiating $a(\boldsymbol{\alpha})$ the mean of the sufficient statistic $\log \mathbf{X} = (\log X_1, \ldots, \log X_D)$ is obtained:

$$\mathrm{E}(\log X_i) = \frac{\partial a(\boldsymbol{\alpha})}{\partial \alpha_i} = \psi(\alpha_i) - \psi(\alpha_+), \qquad i = 1, \ldots, D,$$

where $\psi(.)$ is the digamma function (Abramowitz and Stegun 1964). $\psi(.)$ is defined as the logarithmic derivative of the gamma function, which is computable via Taylor approximations.

Sometimes a subfamily is not exponential in its entirety, but the subclasses obtained by fixing a (one- or multidimensional) component of the parameter are exponential. This is the case of the scaled Dirichlet distribution.

Proposition 10.3.1 *Let* $\mathbf{X} \sim \mathcal{SD}^D(\boldsymbol{\alpha}, \boldsymbol{\beta})$. *If we fix the vector* $\boldsymbol{\beta}$ *equal to* $\beta\mathbf{1}$, *where* $\mathbf{1}$ *is a vector of ones and* β *is a scalar, i.e.* $\beta_i = \beta$, $i = 1, \ldots, D$, *the derived model is exponential.*

We observe that the value of β (scalar) is irrelevant. It can be substituted by $(1, 1, \ldots)$ or $\mathcal{C}(1, 1, \ldots)$, or by $\mathcal{C}(\beta, \beta, \ldots)$, where $\mathcal{C}$ is the closure operator (see Chapter 2).

10.3.3   Measures of location and variability

Densities (10.4) and (10.6) can be expressed in terms of the coordinates with respect to an orthonormal basis to compute the expected value of any order and a measure of variability (Mateu-Figueras and Pawlowsky-Glahn 2005). In coordinates, a vector with $(D - 1)$ components is obtained, i.e. $[\mathrm{E}(Y_1), \mathrm{E}(Y_2), \ldots, \mathrm{E}(Y_{D-1})]$. This vector represents the coordinates of the expected composition with respect to the chosen orthonormal basis. To obtain the expected composition, the expected value of a linear combination in the simplex has to be computed:

$$\mathrm{E}_a(\mathbf{x}) = (\mathrm{E}(Y_1) \odot \mathbf{e}_1) \oplus \cdots \oplus (\mathrm{E}(Y_{D-1}) \odot \mathbf{e}_{D-1}). \tag{10.9}$$

In this way an element of the simplex is obtained, such as the mean and the mode of a random composition.

Proposition 10.3.2 *Let* $\mathbf{X} \sim \mathcal{D}^D(\boldsymbol{\alpha})$. *The mode of* $\mathbf{X}$ *with respect to the measures* λ *and* λ_a *is, respectively, equal to*

$$\mathrm{mode}(\mathbf{X}) = \left(\frac{\alpha_1 - 1}{\alpha_+ - D}, \frac{\alpha_2 - 1}{\alpha_+ - D}, \ldots, \frac{\alpha_D - 1}{\alpha_+ - 2} \right)$$

$$\mathrm{mode}_a(\mathbf{X}) = \left(\frac{\alpha_1}{\alpha_+}, \frac{\alpha_2}{\alpha_+}, \ldots, \frac{\alpha_D}{\alpha_+} \right), \tag{10.10}$$

where $\alpha_+ = \sum_{i=1}^{D} \alpha_i$.

Proof. The mode of the Dirichlet distribution can be found by noting that its density is equivalent to the likelihood function from a multinomial distribution. The multinomial distribution gives the probability of each combination of outcomes in n independent trials of a D-outcome process. The probability of each outcome i in any one trial is given by the fixed probabilities p_i where $p_i \geq 0$ and $\sum_{i=1}^{D} p_i = 1$. Let n_i be the number of times that outcome i appeared in n trials. Hence $n = \sum_{i=1}^{D} n_i$. The kernel of the multinomial distribution is $\prod_{i=1}^{D} p_i^{n_i}$. The log-likelihood, ignoring the constant, is $l(\mathbf{p}) = \sum_{i=1}^{D} n_i \log p_i$. When we maximize $l(\mathbf{p})$ we have to be careful since we must enforce the constraint that $\sum_{i=1}^{D} p_i = 1$. Using Lagrange multipliers, the maximum likelihood estimates of each p_i are $\hat{p}_i = n_i / (\sum_{i=1}^{D} n_i)$. In the present case $n_i = \alpha_i - 1$, $(i = 1, \ldots, D)$ and $p_i = x_i$. Therefore the modes of the Dirichlet density occur at

$$x_i = \frac{\alpha_i - 1}{\alpha_+ - D}, \qquad i = 1, \ldots, D.$$

We observe that the expression for mode(**X**) is actually the mode under the conditions that $\alpha_i > 1, (i = 1, \ldots, D)$. We can proceed in the same way to prove the expression for $\text{mode}_a(\mathbf{X})$.

Proposition 10.3.3 *The expected value of* **X** *with respect to the measures* λ *and* λ_a *is, respectively, equal to*

$$E(\mathbf{X}) = \left(\frac{\alpha_1}{\alpha_+}, \frac{\alpha_2}{\alpha_+}, \ldots, \frac{\alpha_D}{\alpha_+} \right)$$

$$E_a(\mathbf{X}) = \mathcal{C} \left(e^{\psi(\alpha_1)}, e^{\psi(\alpha_2)}, \ldots, e^{\psi(\alpha_D)} \right), \tag{10.11}$$

where ψ *is the digamma function seen before, and* $\mathcal{C}$ *is the closure operator.*

Proof. To prove these results it is useful to recall first that, when $\mathbf{X} \sim \mathcal{D}^D(\boldsymbol{\alpha})$, the marginals are beta distributions: $X_i \sim Be(\alpha_i, \alpha_+ - \alpha_i)$. So the expected value of each marginal is equal to $E(X_i) = \frac{\alpha_i}{\alpha_+}$. Secondly, to obtain $E_a(\mathbf{X})$ it is helpful to express the Dirichlet density functions in terms of the coordinates with respect to an orthonormal basis. Then the Dirichlet density on coordinates is obtained, which is a classical density on $\mathbb{R}^2$ with respect to the Lebesgue measure λ. At this point we can calculate the density of each marginal and compute the corresponding expected value in the usual form. Finally we use (10.9) to express again the mean like an element of the simplex.

Note that in Equation (10.10) the mode with respect to the Aitchison measure on the simplex coincides with the expected value with respect to the Lebesgue measure in Equation (10.11).

Proposition 10.3.4 *Consider now* $\widetilde{\mathbf{X}} \sim \mathcal{SD}^D(\boldsymbol{\alpha}, \boldsymbol{\beta})$. *The mode and the expected value of* $\widetilde{\mathbf{X}}$ *with respect to the measure* λ_a *are, respectively, equal to*

$$\text{mode}_a(\widetilde{\mathbf{X}}) = (\ominus \boldsymbol{\beta}) \oplus \text{mode}_a(\mathbf{X})$$

$$E_a(\widetilde{\mathbf{X}}) = (\ominus \boldsymbol{\beta}) \oplus E_a(\mathbf{X}). \tag{10.12}$$

Proof. Given that $\widetilde{\mathbf{X}} \sim \mathcal{SD}^D(\boldsymbol{\alpha}, \boldsymbol{\beta}) = (\ominus \boldsymbol{\beta}) \oplus \mathbf{X}$, where $\mathbf{X} \sim \mathcal{D}^D(\boldsymbol{\alpha})$, the expressions for the mode and the expected value of $\widetilde{\mathbf{X}}$ can be obtained using the linearity of these operators.

In equation (10.12) we can note how $\boldsymbol{\beta}$ operates in a simplicial way. The vector of parameters $\boldsymbol{\beta}$ in fact can be interpreted as a measure of location, instead of as a measure of scale. Moreover, for $\boldsymbol{\alpha}$ a vector of constants $[\boldsymbol{\alpha} = (\alpha, \alpha, \ldots)]$, the mean and the mode with respect to the Aitchison measure coincide and are the neutral element in the simplex. Also, Equation (10.12) shows that through perturbation, any random variable with Dirichlet or scaled Dirichlet distribution, can be either transformed into a random variable centred in the neutral element of the simplex, or into a random variable whose mode is located at the neutral element. These and other properties with respect to the Aitchison geometry and Aitchison measure require further study. There is no closed form for the mode, mode($\widetilde{\mathbf{X}}$), nor the mean, E($\widetilde{\mathbf{X}}$), for a composition $\widetilde{\mathbf{X}} \sim \mathcal{SD}^D(\boldsymbol{\alpha}, \boldsymbol{\beta})$ with respect to the Lebesgue measure λ in real space; in this case one has to compute them using numerical integration.

To compute a measure of variability, like the metric variance, we have to work directly on the coordinates, i.e. on $(D-1)$-part vectors. We cannot proceed as in the previous subsection for the measures of location, because the metric variance is not an element of the simplex; it is only a numerical value that describes the dispersion. In this case, the metric variance has to be computed directly using coordinates. To date, no closed form has been found to compute it directly without choosing an orthonormal basis to represent the composition. For the sake of completeness, we recall the structure of variance and covariance of the Dirichlet distribution with respect to the Lebesgue measure.

Proposition 10.3.5 *Let* $\mathbf{X} \sim \mathcal{D}^D(\boldsymbol{\alpha})$. *The structure of variance and covariance of* $\mathbf{X}$ *are defined by*

$$
\begin{aligned}
\mathrm{Var}(X_i) &= \frac{\alpha_i(\alpha_+ - \alpha_i)}{\alpha_+^2(\alpha_+ + 1)} = \frac{\mathrm{E}(X_i)[1 - \mathrm{E}(X_i)]}{\alpha_+ + 1} \quad i = 1, \ldots, D \\
\mathrm{Cov}(X_i, X_j) &= \frac{-\alpha_i\alpha_j}{\alpha_+^2(\alpha_+ + 1)} \quad i, j = 1, \ldots, D, \ i \neq j.
\end{aligned}
\tag{10.13}
$$

The last expressions show that the covariance structure associated with the Dirichlet model is completely nonpositive. Further we observe that α_+ may be regarded as a precision parameter, as when it increases, the distributions becomes more tightly concentrated around the mean. It follows that the Dirichlet distribution exhibits the maximum form of independence compatible with the unit sum constraint. Among different forms of independence for compositional data, the neutrality concept has a great relevance: it regards the consequences of eliminating a certain number of components on the relative proportions of the remaining ones. Given the vector $\mathbf{X} = (X_1, \ldots, X_D)$, the element i $(i = 1, \ldots, D-1)$ is neutral if X_i and $\left(\frac{X_j}{1 - \sum_{r=1}^{i} X_r}\right)$ are independent for $j > i$ (Connor and Mosimann 1969). A completely neutral vector is one whose elements are all neutral. Note that the ordering of the vector is relevant. Thus neutrality of $\mathbf{X}$ does not imply neutrality of a permutation, $\mathbf{X}' = (X_2, X_1, \ldots, X_D)$. If $\mathbf{X} \sim \mathcal{D}^D(\boldsymbol{\alpha})$, then any permutation of $\mathbf{X}$ is neutral. Connor and Mosimann (1969) started from the concept of neutrality to provide a general form of a random variable with neutrality. Wong (1998) extended their work and showed that the generalized Dirichlet distribution was conjugate to a particular type of sampling experiment.

10.4   Conclusions

In this chapter we have presented a generalization of the Dirichlet distribution, the scaled Dirichlet distribution, and we studied some of its properties comparing the classical approach with an approach based on the Aitchison geometry and measure on the simplex. The most interesting property of the resulting distribution is that, within the Aitchison geometry, a scaled

Dirichlet distribution is obtained perturbing the Dirichlet one. Taking into account the special algebraic-geometric structure of the simplex, the scaled Dirichlet density is just a *translation* of a Dirichlet density (but a translation on the simplex). Consequently the Dirichlet and the scaled Dirichlet belong to the same class of distributions and not to two different classes as we find in the literature. We have provided the density functions and the principal measures of location with respect to both, the Lebesgue measure and the Aitchison measure.

Acknowledgements

This research has been supported by the Spanish Ministry of Science and Innovation (projects CSD2006-00032 and MTM2009-13272) and by the Agència de Gestió d'Ajuts Universitaris i de Recerca of the Generalitat de Catalunya (Ref. 2009SGR424).

References

Abramowitz M and Stegun IA 1964 *Handbook of Mathematical Functions with Formulas, Graphs, and Mathematical Tables*. Dover Publications, New York, NY (USA). 1046 p.

Aitchison J 1982 The statistical analysis of compositional data (with discussion). *Journal of the Royal Statistical Society, Series B (Statistical Methodology)* 44(2), 139–177.

Aitchison J 1986 *The Statistical Analysis of Compositional Data*. Monographs on Statistics and Applied Probability. Chapman and Hall Ltd (reprinted in 2003 with additional material by The Blackburn Press), London (UK). 416 p.

Albert JH and Gupta AK 1982 Mixtures of Dirichlet distributions and estimation in contingency tables. *The Annals of Statistics* 10, 1261–1268.

Athreya KB and Lahiri SN 2006 *Measure Theory and Probability Theory*. Springer, New York, NY (USA). 625 p.

Barceló-Vidal C, Martín-Fernández JA and Pawlowsky-Glahn V 2001 Mathematical foundations of compositional data analysis. In *Proceedings of IAMG'01 – The VII Annual Conference of the International Association for Mathematical Geology* (ed. Ross G). Kansas Geological Survey, Cancun (Mexico). 20 p.

Billheimer D, Guttorp P and Fagan W 2001 Statistical interpretation of species composition. *Journal of the American Statistical Association* 96(456), 1205–1214.

Connor RJ and Mosimann JE 1969 Concepts of independence for proportions with a generalization of the Dirichlet distribution. *Journal of the American Statistical Association* 64(325), 194–206.

Eaton ML 1983 *Multivariate Statistics. A Vector Space Approach*. John Wiley & Sons, Ltd, New York, NY (USA). 512 p.

Egozcue JJ, Pawlowsky-Glahn V, Mateu-Figueras G and Barceló-Vidal C 2003 Isometric logratio transformations for compositional data analysis. *Mathematical Geology* 35(3), 279–300.

Kotz S, Balakrishnan N and Johnson NL 2000 *Continuous Multivariate Distributions. Volume I, Models and Applications*. Wiley Series in Probability and Statistics. Wiley-Interscience, New York, NY (USA). 730 p.

Lochner RH 1975 A Generalized Dirichlet distribution in Bayesian Life Testing. *Journal of the Royal Statistical Society, Series B (Statistical Methodology)* 37, 103–113.

Mateu-Figueras G and Pawlowsky-Glahn V 2005 The Dirichlet distribution with respect to the Aitchison measure on the simplex – a first approach. In *Proceedings of CoDaWork'05, The*

2nd Compositional Data Analysis Workshop (ed. Mateu-Figueras G and Barceló-Vidal C). http://ima.udg.es/Activitats/CoDaWork05/. University of Girona, Girona (Spain). CD-ROM.

Pawlowsky-Glahn V 2003 Statistical modelling on coordinates. In *Proceedings of CoDaWork'03, The 1st Compositional Data Analysis Workshop* (ed. Thió-Henestrosa S and Martín-Fernández JA). http://ima.udg.es/Activitats/CoDaWork03/. University of Girona, Girona (Spain). CD-ROM.

Pawlowsky-Glahn V and Egozcue JJ 2001 Geometric approach to statistical analysis on the simplex. *Stochastic Environmental Research and Risk Assessment (SERRA)* **15**(5), 384–398.

Wong TT 1998 Generalized Dirichlet distribution in Bayesian analysis. *Applied Mathematics and Computation* **97**, 165–181.

Wong TT 2010 Parameter estimation for generalized Dirichlet distributions from the sample estimates of the first and the second moments of random variables. *Computational Statistics and Data Analysis* **54**, 1756–1765.

Part III

THEORY – ALGEBRA AND CALCULUS

11

Elements of simplicial linear algebra and geometry

Juan José Egozcue[1], Carles Barceló-Vidal[2], Josep Antoni Martín-Fernández[2], Eusebi Jarauta-Bragulat[1], José Luis Díaz-Barrero[1] and Glòria Mateu-Figueras[2]

[1] Department of Applied Mathematics III, Technical University of Catalonia, Spain
[2] Department of Computer Science and Applied Mathematics, University of Girona, Spain

11.1 Introduction

Recently, the simplex has been studied as a Euclidean space (Pawlowsky-Glahn and Egozcue 2001; Billheimer *et al.* 2001). These studies were motivated by the use of the simplex as a sample space of compositional data (Aitchison 1986) and related to statistics. The simplex also appears as a parameter space thus reinforcing the interest on its properties in many fields. Understanding the simplex as a Euclidean space permits to organize the operations (perturbation and powering), transformations into real spaces (alr, clr, ilr), and the metrics of the simplex on a systematic and coherent mathematical scheme. Most of these elements of the simplex were introduced by Aitchison in the 1980s addressing statistical problems in compositional data analysis. For instance, perturbation and powering were introduced in Aitchison (1986), with the second operation just a curiosity; simplicial distance was proposed in Aitchison (1983); orthogonal log-contrasts appear incidentally in Aitchison (1986), but they were not referred to any geometric structure. Additional developments of the geometry of the simplex (Egozcue and Pawlowsky-Glahn 2005, 2006; Egozcue *et al.* 2003), then called

Compositional Data Analysis: Theory and Applications, First Edition. Edited by Vera Pawlowsky-Glahn and Antonella Buccianti.
© 2011 John Wiley & Sons, Ltd. Published 2011 by John Wiley & Sons, Ltd.

Aitchison geometry, allowed development of new tools for representation of compositions and their exploratory analysis. However, mathematical and statistical models for compositional data are not completely developed yet (Aitchison and Egozcue 2005). An important tool for this development is so-called calculus, consisting of the initial concepts of limit, convergence, derivatives and integrals involving functions defined on the simplex or with images on the simplex (see Chapters 12 and 13). These elementary concepts relay in a previous geometric setting, i.e. the Euclidean character of the Aitchison geometry of the simplex, and the properties of linear functions. Considering the simplex as a Euclidean space, the study of its geometry and linear algebra on it may seem a trivial task from an abstract point of view. However, the interpretation of this peculiar Euclidean geometry appears a little bit intricate due to, at least, two circumstances: (a) in practice, elements of the simplex are representatives of equivalence classes of elements of n-dimensional real spaces whose positive components are proportional; and (b) the concept of canonical basis in the simplex remains undefined. Both circumstances favour representation of simplicial elements in several different ways, and interpretation often requires the simultaneous use of two or more of these representations.

Section 11.2 exposes succinctly the main elements of the Euclidean geometry of the simplex, called Aitchison geometry. It includes the properties of the coordinate representation with special emphasis on orthonormal basis. Section 11.3 is a study of linear functions, involving the simplex. Whereas linear functions between Euclidean spaces are one part of a general theory when they are expressed using coordinate expressions, simplicial representation requires new notation for matrix products and some properties of the matrix representation of linear functions.

11.2   Elements of simplicial geometry

11.2.1 n-part simplex

Let $\mathcal{S}^n$ be the *simplex* of n parts, or *n-part simplex* for short, defined as

$$\mathcal{S}^n = \left\{ \mathbf{x} = (x_1, x_2, \ldots, x_n) \,\middle|\, x_i > 0,\ i = 1, 2 \ldots, n,\ \sum_{i=1}^{n} x_i = \kappa > 0 \right\}, \qquad (11.1)$$

where $\mathbf{x}$ is a n-tuple in $\mathbb{R}^n$. Elements in $\mathbb{R}^n$ are denoted in boldface and its components in italics, i.e. $\mathbf{x} = (x_1, x_2, \ldots, x_n)$. The same notation is used for elements in $\mathcal{S}^n$ and for those with positive components, i.e. when they are in $\mathbb{R}^n_+$. A simplex element, $\mathbf{x} \in \mathcal{S}^n$, is called *composition*, and its components x_i are called *parts* of $\mathbf{x}$. This notation is extended to other elements of the simplex, e.g. if $\mathbf{z} \in \mathcal{S}^n$, then z_i denote its ith part. When using matrix notation boldface symbols, compositions or real n-tuples, denote column vectors, i.e. $(n, 1)$-matrices, and $(\cdot)^\top$ denotes matrix-transposition. In statistical frameworks compositions have been represented by row vectors, e.g. Egozcue *et al.* (2003). This practice is related to the way that compositional data are represented in data matrices. Here the column-vector standard notation is adopted in order to facilitate the identification with expressions appearing in real analysis. Moreover, real functions like exponential or logarithm are applied to n-tuples componentwise for conciseness, e.g. $\log(\mathbf{x}) = (\log x_1, \log x_2, \ldots, \log x_n)$.

Compositions can be considered representatives of equivalence classes of real vectors with positive components, i.e. in $\mathbb{R}_+^n$ (Barceló-Vidal *et al.* 2001; Aitchison 1992).

Definition 11.2.1 (scale-equivalent vectors)   *Let* x *and* y *be in* $\mathbb{R}_+^n$. *They are scale-equivalent if there is a real constant* $c > 0$ *such that* $\mathbf{x} = c \cdot \mathbf{y}$.

Definition 11.2.2 (closure)   *Let* x *be in* $\mathbb{R}_+^n$. *The canonical representative of its scale-equivalent class is the composition in* $\mathcal{S}^n$

$$\mathcal{C}\mathbf{x} = \left(\frac{\kappa x_1}{\sum_{j=1}^n x_j}, \frac{\kappa x_2}{\sum_{j=1}^n x_j}, \ldots, \frac{\kappa x_n}{\sum_{j=1}^n x_j} \right), \tag{11.2}$$

where κ *is an arbitrarily selected positive constant. The operator* $\mathcal{C}$ *is called the* κ*-closure operator.*

The closure constant κ is arbitrary. Frequent values are 1, 100, 1000, 10^6, etc. Here the default value adopted is $\kappa = 1$ and the simplex is then identified by the so-called *unit-simplex*. In what follows, operations and properties concerning elements of the simplex are equivalently interpreted as operations and results relative to the scale-equivalent classes in $\mathbb{R}_+^n$. As a consequence, closure operator could be suppressed if the canonical representative is not necessary.

11.2.2   Vector space

Aitchison (1986) introduced the operation called *perturbation* between elements of a *n*-part simplex. Connected with perturbation, a *power transformation*, or *powering*, was also suggested as a repeated perturbation. These operations play the role of an addition and a product by real scalars in the simplex.

Definition 11.2.3 (perturbation and powering)   *Let* x, y *be compositions in* $\mathcal{S}^n$, *and* $\alpha \in \mathbb{R}$. *Perturbation and powering, denoted* $\oplus$, $\odot$, *respectively, are defined as*

$$\mathbf{x} \oplus \mathbf{y} = \mathcal{C}(x_1 y_1, x_2 y_2, \ldots, x_n y_n), \quad \alpha \odot \mathbf{x} = \mathcal{C}(x_1^\alpha, x_2^\alpha, \ldots, x_n^\alpha).$$

Note that both perturbation and powering can be performed using scale-equivalent elements of $\mathbb{R}_+^n$; using any representative of these compositions, perturbation and powering yield equal results. With these operations, $\mathcal{S}^n$ is a vector space of dimension $(n - 1)$ (Aitchison *et al.* 2001; Pawlowsky-Glahn and Egozcue 2001). The first assertion is straightforward.

Theorem 11.2.5 (vector space)   *The* n*-part simplex* $\mathcal{S}^n$, *with perturbation* ($\oplus$) *and powering* ($\odot$) *is a vector space on* $\mathbb{R}$. *More explicitly:*

   (a) (associative) $(\mathbf{x} \oplus \mathbf{y}) \oplus \mathbf{z} = \mathbf{x} \oplus (\mathbf{y} \oplus \mathbf{z})$;

   (b) (commutative) $\mathbf{x} \oplus \mathbf{y} = \mathbf{y} \oplus \mathbf{x}$;

   (c) (neutral element) $\mathbf{x} \oplus \mathbf{n} = \mathbf{x} \oplus \mathbf{x}$, *being* $\mathbf{n} = \mathcal{C}(1, 1, \ldots, 1)$;

(d) *(opposite element) the opposite of* $\mathbf{x} = (x_1, x_2, \ldots, x_n)$ *is* $\mathcal{C}(1/x_1, 1/x_2, \ldots, 1/x_n) = (-1) \odot \mathbf{x}$, *and it is denoted* $\ominus\mathbf{x}$;

(e) *(distributive)* $(\alpha \odot \mathbf{x}) \oplus (\alpha \odot \mathbf{y}) = \alpha \odot (\mathbf{x} \oplus \mathbf{y})$;

(f) *(unit)* $1 \odot \mathbf{x} = \mathbf{x}$,

for any $\mathbf{x}$, $\mathbf{y}$, $\mathbf{z}$ *compositions in* $\mathcal{S}^n$ *and any* α *in* $\mathbb{R}$.

To investigate the dimension of the vector space, the additive log-ratio transformation, alr, is used (Aitchison 1982, 1986).

Definition 11.2.5 (alr transformation)   *The transformation* alr $: \mathcal{S}^n \to \mathbb{R}^{n-1}$ *assigns the real* $(n-1)$*-tuple*

$$\mathrm{alr}(\mathbf{x}) = \log\left(\frac{x_1}{x_n}, \frac{x_2}{x_n}, \ldots, \frac{x_{n-1}}{x_n}\right)$$

to the composition $\mathbf{x} \in \mathcal{S}^n$.

Proposition 11.2.6   *The transformation* alr $: \mathcal{S}^n \to \mathbb{R}^{n-1}$ *is one-to-one. If* $\mathbf{x}^* \in \mathbb{R}^{n-1}$, *then the inverse* alr *transformation is*

$$\mathrm{alr}^{-1}(\mathbf{x}^*) = \mathcal{C} \exp\left(x_1^*, x_2^*, \ldots, x_{n-1}^*, 0\right). \tag{11.3}$$

Moreover, the alr *transformation is an isomorphism of vector spaces.*

Proof. To prove that alr is one-to-one is equivalent to checking $\mathrm{alr}^{-1}(\mathrm{alr}(\mathbf{x})) = \mathbf{x}$ for any composition in $\mathcal{S}^n$. This holds from a simple substitution. The isomorphism requires that, for all $\mathbf{x}$, $\mathbf{y}$ in $\mathcal{S}^n$ and real constants α, β,

$$\mathrm{alr}((\alpha \odot \mathbf{x}) \oplus (\beta \odot \mathbf{y})) = \alpha \cdot \mathrm{alr}(\mathbf{x}) + \beta \cdot \mathrm{alr}(\mathbf{y}),$$

which holds from Definitions 11.2.3 and 11.2.5.

Corollary 11.2.7   *As vector space,* $(\mathcal{S}^n, \oplus, \odot)$ *has dimension* $n-1$.

Definition 11.2.8 (perturbation-linear combination)   *The composition in* $\mathcal{S}^n$

$$\mathbf{y} = (\alpha_1 \odot \mathbf{x}_1) \oplus (\alpha_2 \odot \mathbf{x}_2) \oplus \cdots \oplus (\alpha_m \odot \mathbf{x}_m) = \bigoplus_{j=1}^{m} (\alpha_j \odot \mathbf{x}_j),$$

is called perturbation-linear combination of the compositions $\mathbf{x}_1$, $\mathbf{x}_2$, ..., $\mathbf{x}_m$ *in* $\mathcal{S}^n$ *with real coefficients* α_1, α_2, ..., α_m.

It is interesting to express in matrix form the perturbation-linear combination $\mathbf{y} = \bigoplus_{j=1}^{m}(\alpha_j \odot \mathbf{x}_j)$. Define the column-vector $\boldsymbol{\alpha} = (\alpha_1, \alpha_2, \ldots, \alpha_m)^\top$ and $\mathbf{X}$ a (n, m)-matrix whose ith column is the composition $\mathbf{x}_i$, i.e. $\mathbf{X} = (\mathbf{x}_1, \mathbf{x}_2, \ldots, \mathbf{x}_m)$. Then, the perturbation-linear combination $\mathbf{y}$ can be expressed as

$$\mathbf{y}^\top = \boldsymbol{\alpha}^\top \odot \mathbf{X}^\top = \mathcal{C}\left(\prod_{j=1}^m x_{1j}^{\alpha_j}, \prod_{j=1}^m x_{2j}^{\alpha_j}, \ldots, \prod_{j=1}^m x_{nj}^{\alpha_j}\right), \tag{11.4}$$

where x_{ij} is the ith part of $\mathbf{x}_j$. In Equation (11.4), $\odot$ is a matrix product, in which all pluses are translated into $\oplus$ and products by real scalars are translated into $\odot$. It should be noted that $\boldsymbol{\alpha}$ should be placed in front of the matrix $\mathbf{X}$ to be coherent with the use of $\odot$, i.e. real scalars first, compositions afterwards; this causes the transposition of the whole Equation (11.4).

Definition 11.2.9 (perturbation-independent set) *The set of compositions in $\mathcal{S}^n$, $\mathbf{x}_1, \mathbf{x}_2, \ldots, \mathbf{x}_m$, is perturbation-independent if $\mathbf{n} = \bigoplus_{i=1}^m (\alpha_i \odot \mathbf{x}_i)$, implies $\alpha_1 = 0, \alpha_2 = 0, \ldots, \alpha_m = 0$, with $\mathbf{n} = \mathcal{C}(1, 1, \ldots, 1)$, the neutral element in $\mathcal{S}^n$.*

Definitions 11.2.8 and 11.2.9 match exactly standard definitions for vector spaces. Their consequences and proofs are also the standard ones. For instance, a perturbation-independent set in $\mathcal{S}^n$ cannot contain more than $n - 1$ compositions (Corollary 11.2.7); if it contains exactly $n - 1$ compositions, then the set constitutes a basis of $\mathcal{S}^n$. The alr^{-1} transformation provides an interesting example.

Proposition 11.2.10 (alr basis) *Let $\mathbf{y}_1^*, \mathbf{y}_2^*, \ldots, \mathbf{y}_{n-1}^*$ be the $n - 1$ n-tuples of the canonical basis of $\mathbb{R}^{n-1}$. Then*

$$\mathbf{e}_i^{\mathrm{alr}} = \mathrm{alr}^{-1}(\mathbf{y}_i^*) = \mathcal{C}\exp(0, \ldots, 0, 1, 0, \ldots, 0), \quad i = 1, 2, \ldots, n - 1, \tag{11.5}$$

where the component equal to 1 is placed at the ith component, is a basis of $\mathcal{S}^n$. Any composition $\mathbf{x} \in \mathcal{S}^n$ can be written as a perturbation-linear combination of $\mathbf{e}_1^{\mathrm{alr}}, \mathbf{e}_2^{\mathrm{alr}}, \ldots, \mathbf{e}_{n-1}^{\mathrm{alr}}$

$$\mathbf{x} = \bigoplus_{i-1}^{n-1}\left[\log\left(\frac{x_i}{x_n}\right) \odot \mathbf{e}_i^{\mathrm{alr}}\right]. \tag{11.6}$$

Proof. Since alr^{-1} is an isomorphism of $\mathbb{R}^{n-1}$ into $\mathcal{S}^n$ (Proposition 11.2.6) and $\mathbf{y}_1^*, \mathbf{y}_2^*, \ldots, \mathbf{y}_{n-1}^*$ is a basis of $\mathbb{R}^{n-1}$, the set of compositions $\mathbf{e}_i^{\mathrm{alr}} = \mathrm{alr}^{-1}(\mathbf{y}_i^*)$, $i = 1, 2, \ldots, n - 1$, is a basis of $\mathcal{S}^n$. Expression (11.6) holds from Definition 11.2.5. It shows that the coordinates of $\mathbf{x}$ in the basis $\mathbf{e}_i^{\mathrm{alr}}$, $i = 1, 2, \ldots, n - 1$, are the components $\log(x_1/x_n), \log(x_2/x_n), \ldots, \log(x_{n-1}/x_n)$ of alr$(\mathbf{x})$.

An alternative expression to (11.5) is given by

$$\mathbf{e}_i^{\mathrm{alr}} = \mathrm{alr}^{-1}(\mathbf{y}_i^*) = \mathcal{C}\exp\left(\frac{-1}{n}, \ldots, \frac{-1}{n}, \frac{n-1}{n}, \frac{-1}{n}, \ldots, \frac{-1}{n}\right), \tag{11.7}$$

where the component equal to $(n - 1)/n$ is placed at the ith component. The expression (11.7) only differs from (11.5) in the closure constant. Expression (11.7) has been adopted so that the sum of all exponential coefficients add to zero.

The vector space structure permits defining linear manifolds in $\mathcal{S}^n$ and, particularly, straight lines as one-dimensional linear manifolds.

Definition 11.2.11 (linear manifold)   *Consider* $\mathbf{x}_0 \in \mathcal{S}^n$ *and* $\mathbf{x}_i$, $i = 1, 2, \ldots, m$, *be a perturbation-independent set of compositions in* $\mathcal{S}^n$. *The set of compositions* $\mathbf{y}$ *such that*

$$\mathbf{y} = \mathbf{x}_0 \oplus \bigoplus_{i=1}^{m} (\alpha_i \odot \mathbf{x}_i),$$

for any real constants α_i, $i = 1, 2, \ldots, m$, *is called m-dimensional linear manifold whose origin is* $\mathbf{x}_0$. *If* $m = 1$, *then the set is a straight line in* $\mathcal{S}^n$; *for* $m = 2, 3, \ldots, (m < n - 1)$, *they are called plane or hyperplane in* $\mathcal{S}^n$.

Note that a m-dimensional linear manifold in the simplex whose origin is $\mathbf{x}_0 = \mathbf{n}$ is a perturbation-linear subspace of dimension m.

11.2.3   Centred log-ratio representation

There are different ways of representing a composition. In previous subsections compositions have been represented in two ways: using parts of compositions (simplicial representation) and using coordinates with respect to a given basis, Equation (11.6). The centred log-ratio transformation (Aitchison 1986), clr, provides another representation of compositions. The clr representation is a powerful tool in compositional analysis due to its isometric properties. It can be defined in the following way.

Definition 11.2.12 (centred log-ratio)   *For any composition* $\mathbf{x} \in \mathcal{S}^n$, *the real n-tuple* $\mathbf{u}$, *whose components are*

$$u_i = \log(x_i) - \log(g_m(\mathbf{x})), \quad g_m(\mathbf{x}) = \left(\prod_{i=1}^{n} x_i \right)^{1/n}, \tag{11.8}$$

is called the centred log-ratio transformation of $\mathbf{x}$. *It is denoted* $\mathbf{u} = \text{clr}(\mathbf{x})$. *Using matrix notation* $\text{clr}(\mathbf{x}) = \log(\mathbf{x}) - \log(g_m(\mathbf{x})\mathbf{1}_n)$, *where* $\mathbf{1}_n$ *is a n-vector with unit components.*

Proposition 11.2.11   *Let* $\mathbf{x}$ *be a composition in* $\mathcal{S}^n$. *A real n-tuple* $\mathbf{u} = (u_1, u_2, \ldots, u_n)$ *is the centred log-ratio transformation of* $\mathbf{x}$, *i.e.* $\mathbf{u} = \text{clr}(\mathbf{x})$, *if, and only if,*

$$\mathcal{C} \exp(\mathbf{u}) = \mathbf{x} \quad \text{and} \quad \sum_{i=1}^{n} u_i = 0. \tag{11.9}$$

Moreover, $\text{clr}^{-1}(\mathbf{u}) = \mathcal{C} \exp(\mathbf{u}) = \mathbf{x}$.

Proof. Existence of clr(x) is proven substituting the component expression (11.8) into the conditions (11.9). To prove uniqueness, assume that there is $\mathbf{v} \in \mathbb{R}^n$ satisfying conditions (11.9). Therefore, $\mathcal{C} \exp(\mathbf{v}) \ominus \mathcal{C} \exp(\mathbf{u}) = \mathbf{n}$ and $\mathbf{v} - \mathbf{u} = \mathbf{k}$, where $\mathbf{k} \in \mathbb{R}^n$ is a n-tuple with equal components. Since the components of both $\mathbf{v}$ and $\mathbf{u}$ add to zero, $\mathbf{k}$ must be null.

Hereafter, U denotes the $(n - 1)$-dimensional subspace of $\mathbb{R}^n$ defined as $U = \{\mathbf{u} \in \mathbb{R}^n \mid \sum_{i=1}^{n} u_i = 0\}$.

Theorem 11.2.14 *The clr transformation, clr : $\mathcal{S}^n \to U \subset \mathbb{R}^n$, is an isomorphism of $(n - 1)$-dimensional vector spaces.*

Proof. Since clr is one-to-one, the statement is proven if, for any real constants, α, β, and any compositions, $\mathbf{x}$, $\mathbf{y}$,

$$\mathrm{clr}((\alpha \odot \mathbf{x}) \oplus (\beta \odot \mathbf{y})) = \alpha \cdot \mathrm{clr}(\mathbf{x}) + \beta \cdot \mathrm{clr}(\mathbf{y}),$$

being the right-hand side of the equation in U. The right-hand side is in fact in U due to the property (11.9) of clr. The equality is easily checked using (11.8).

11.2.4   Metrics

The simplex $\mathcal{S}^n$ can be structured as a Euclidean space (Billheimer *et al.* 2001; Pawlowsky-Glahn and Egozcue 2001). This means that an inner product compatible with perturbation and powering can be defined (Definition 11.2.3). The inner product is motivated by the definition of orthogonal log-contrasts given by Aitchison (1986). Recall that a log-contrast is any log-linear combination $a_1 \log x_1 + a_2 \log x_2 + \ldots + a_n \log x_n$, such that $\sum_{i=1}^{n} a_i = 0$.

Definition 11.2.15 (Aitchison inner product, norm and distance)   *Let x,y be compositions in $\mathcal{S}^n$ and clr(x), clr(y) their respective clr transformations. The Aitchison inner product is*

$$\langle \mathbf{x}, \mathbf{y} \rangle_a = \sum_{i=1}^{n} \log \frac{x_i}{\mathrm{g_m(x)}} \cdot \log \frac{y_i}{\mathrm{g_m(y)}} = \langle \mathrm{clr}(\mathbf{x}), \mathrm{clr}(\mathbf{y}) \rangle, \tag{11.10}$$

where $\mathrm{g_m(\cdot)}$ is the geometric mean of the argument and $\langle \cdot, \cdot \rangle$ denotes inner product in $\mathbb{R}^n$. Accordingly, the Aitchison norm and distance are

$$\|\mathbf{x}\|_a = \langle \mathbf{x}, \mathbf{x} \rangle_a^{1/2} = \|\mathrm{clr}(\mathbf{x})\|, \quad \mathrm{d}_a(\mathbf{x}, \mathbf{y}) = \|\mathbf{x} \ominus \mathbf{y}\|_a = \mathrm{d}(\mathrm{clr}(\mathbf{x}), \mathrm{clr}(\mathbf{y})), \tag{11.11}$$

where $\| \cdot \|$ and $\mathrm{d}(\cdot, \cdot)$ denote standard norm and distance in $\mathbb{R}^n$.

The subscript a in $\langle \cdot, \cdot \rangle_a$ refers to Aitchison and it is also used in the definitions of the norm and distance in the simplex which are also called the Aitchison norm and distance. The Aitchison distance was introduced (Aitchison 1983, 1986) before the inner product. Here Aitchison norm and distance are defined in agreement with the Aitchison inner product (Definition 11.2.15).

According to (11.10) and (11.11) in Definition 11.2.15, it is clear that the Aitchison inner product, norm and distance in $\mathcal{S}^n$ are defined by the standard inner product, norm and distance of the clr-transformed vectors in $\mathbb{R}^n$. Therefore, the following corollary is immediately deduced from Theorem 11.2.14 and Definition 12.2.15.

Corollary 11.2.16   *The simplex $\mathcal{S}^n$, with perturbation and powering operations, and the metrics defined by the Aitchison inner product is a $(n-1)$-dimensional Euclidean space. Moreover, the clr transformation, clr : $\mathcal{S}^n \rightarrow U \subset \mathbb{R}^n$, is an isometry of $(n-1)$-dimensional Euclidean spaces.*

Proposition 11.2.17   *The Aitchison inner product of $\mathbf{x}$ and $\mathbf{y} \in \mathcal{S}^n$ defined in (11.10) is equal to*

$$\langle \mathbf{x}, \mathbf{y} \rangle_a = \sum_{i=1}^{n} \log x_i \cdot \log y_i - \frac{1}{n} \left(\sum_{j=1}^{n} \log x_j \right) \left(\sum_{k=1}^{n} \log y_k \right) \qquad (11.12)$$

$$= \frac{1}{n} \sum_{i=1}^{n-1} \sum_{j=i}^{n} \log \frac{x_i}{x_j} \cdot \log \frac{y_i}{y_j} \qquad (11.13)$$

$$= \frac{1}{2n} \sum_{i=1}^{n} \sum_{j=1}^{n} \log \frac{x_i}{x_j} \cdot \log \frac{y_i}{y_j} \qquad (11.14)$$

Proof. The sum in (11.14) runs over all pairs i, j but the terms with $i = j$ are null and the log-ratio pair i, j equals the j, i one. Therefore, (11.14) and (11.13) are equal. Log-ratios in (11.14) can be expressed as a difference of logarithms and the product of the two differences can be rearranged as in (11.12). Moreover, the last term in (11.12) is identified as $n \log[g_m(\mathbf{x})] \log[g_m(\mathbf{y})]$; this term is obtained from (11.10) when log-ratios are expanded as differences of logarithms; then (11.10) equals (11.12).

Properties of the Aitchison inner product, norm and distance are those of Euclidean spaces and they hold in the standard way after the definition of the inner product. For completeness, some important properties are listed in the next proposition.

Proposition 11.2.18   *Let $\mathbf{x}, \mathbf{y}, \mathbf{z}$ be compositions in $\mathcal{S}^n$ and α a real constant. Then,*

(a) *(commutative)* $\langle \mathbf{x}, \mathbf{y} \rangle_a = \langle \mathbf{y}, \mathbf{x} \rangle_a$;

(b) *(bilinear)* $\langle \mathbf{x} \oplus \mathbf{z}, \mathbf{y} \rangle_a = \langle \mathbf{x}, \mathbf{y} \rangle_a + \langle \mathbf{z}, \mathbf{y} \rangle_a$; $\langle \alpha \odot \mathbf{x}, \mathbf{y} \rangle_a = \alpha \cdot \langle \mathbf{x}, \mathbf{y} \rangle_a$;

(c) *(positiveness)* $\langle \mathbf{x}, \mathbf{x} \rangle_a \geq 0$. *Equality holds if, and only if $\mathbf{x} = \mathbf{n}$.*

(d) *(Cauchy–Schwartz inequality)*

$$|\langle \mathbf{x}, \mathbf{y} \rangle_a| \leq \|\mathbf{x}\|_a \cdot \|\mathbf{y}\|_a ;$$

(e) *(invariance under perturbation)*

$$d_a(\mathbf{x} \oplus \mathbf{z}, \mathbf{y} \oplus \mathbf{z}) = d_a(\mathbf{x}, \mathbf{y}) ;$$

(f) (scaling)

$$\|\alpha \odot \mathbf{x}\|_a = |\alpha| \cdot \|\mathbf{x}\|_a \;;$$

(g) (Pythagoras) If $\mathbf{x}$, $\mathbf{y}$ *are orthogonal, i.e.* $\langle \mathbf{x}, \mathbf{y} \rangle_a = 0$, *then*

$$\|\mathbf{x} \oplus \mathbf{y}\|_a^2 = \|\mathbf{x}\|_a^2 + \|\mathbf{y}\|_a^2 \;;$$

(h) (triangular inequality)

$$d_a(\mathbf{x}, \mathbf{y}) \le d_a(\mathbf{x}, \mathbf{z}) + d_a(\mathbf{y}, \mathbf{z}).$$

11.2.5   Orthonormal basis and coordinates

Since clr^{-1} transformation is an isometry of U onto $\mathcal{S}^n$, orthonormal bases in $\mathcal{S}^n$ can be obtained from orthonormal bases in U.

Proposition 11.2.19   *Let* $\mathbf{v}_1 = (v_{11}, \ldots, v_{n1})$, $\mathbf{v}_2 = (v_{12}, \ldots, v_{n2})$, $\ldots$, $\mathbf{v}_{n-1} = (v_{1(n-1)}, \ldots,$ $v_{n(n-1)})$ *be* $n - 1$ *n-tuples in* $\mathbb{R}^n$ *constituting an orthonormal basis of the subspace* $U \subset \mathbb{R}^n$, *i.e.* $\sum_{i=1}^{n} v_{ij} = 0$, *for* $j = 1, 2, \ldots, n - 1$, *and* $< \mathbf{v}_i, \mathbf{v}_j > = \delta_{ij}$, *for* $i, j = 1, 2, \ldots, n - 1$, *where* δ_{ij} *is the Kroneker-delta. Then the vectors*

$$\mathbf{e}_i = \mathrm{clr}^{-1}(\mathbf{v}_i) = \mathcal{C}\exp(\mathbf{v}_i), \;\; i = 1, 2, \ldots, n - 1,$$

constitute an orthonormal basis in $\mathcal{S}^n$.

Proof. The vectors $\mathbf{v}_i$, $i = 1, 2, \ldots, n - 1$, are a basis in U and $\mathrm{clr}(\mathbf{e}_i) = \mathbf{v}_i$. After Definition 11.2.15, this is equivalent to orthonormality of compositions $\mathbf{e}_i$, $i = 1, 2, \ldots, n - 1$. Since the dimension of $\mathcal{S}^n$ is $n - 1$, they constitute an orthonormal basis.

Definition 11.2.20 (ilr transformation)   *Let* $\mathbf{e}_i, i = 1, 2, \ldots, n - 1$, *be an orthonormal basis in* $\mathcal{S}^n$. *The coordinate function assigning coordinates with respect to* $\mathbf{e}_i, i = 1, 2, \ldots, n - 1$, *to a composition* $\mathbf{x} \in \mathcal{S}^n$ *is called the isometric log-ratio transformation,* $\mathrm{ilr} : \mathcal{S}^n \to \mathbb{R}^{n-1}$,

$$\mathrm{ilr}(\mathbf{x}) = (\langle \mathbf{x}, \mathbf{e}_1 \rangle_a, \langle \mathbf{x}, \mathbf{e}_2 \rangle_a, \ldots, \langle \mathbf{x}, \mathbf{e}_{n-1} \rangle_a).$$

The isometric log-ratio transformation was first introduced in Egozcue *et al.* (2003). However, in this reference and subsequent ones (Egozcue and Pawlowsky-Glahn 2005; Thió-Henestrosa *et al.* 2008) this name was reserved for some particular bases made of compositions called *balancing element* with coordinates called *balances*. Here, ilr transformation is identified by any coordinate function associated with an orthonormal basis. The following results about ilr transformation in matrix form are also found in Egozcue and Pawlowsky-Glahn (2006) and Tolosana-Delgado *et al.* (2008).

Proposition 11.2.21 (ilr coordinates)   *Let* $\mathbf{e}_i = \mathcal{C}\exp(\mathbf{v}_i)$, $\mathrm{clr}(\mathbf{e}_i) = \mathbf{v}_i, i = 1, 2, \ldots, n - 1$, *be an orthonormal basis in* $\mathcal{S}^n$. *Let* $\mathbf{V}$ *be the* $(n, n - 1)$-*matrix whose columns are* $\mathbf{v}_i = \mathrm{clr}(\mathbf{e}_i)$.

Then, in matrix notation, the vector $\mathbf{x}^*$ *of* $n - 1$ *coordinates of a composition* $\mathbf{x} \in \mathcal{S}^n$ *with respect to* $\mathbf{e}_i, i = 1, 2, \ldots, n - 1$, *is*

$$\mathbf{x}^* = \mathrm{ilr}(\mathbf{x}) = \mathbf{V}^\top \, \mathrm{clr}(\mathbf{x}) = \mathbf{V}^\top \, \log(\mathbf{x}). \tag{11.15}$$

Proof. From Definition 11.2.20 and Equation (11.10), the ith component in $\mathrm{ilr}(\mathbf{x})$ is $x_i^* = \langle \mathrm{clr}(\mathbf{x}), \mathrm{clr}(\mathbf{e}_i) \rangle$, which is the matrix product of the transposed ith column of $\mathbf{V}$ and $\mathrm{clr}(\mathbf{x})$. Components of $\mathrm{clr}(\mathbf{e}_i)$ sum up zero; then, $\mathbf{V}^\top \, \mathrm{clr}(\mathbf{x}) = \mathbf{V}^\top \, \log(\mathbf{x})$; since $\mathrm{clr}(\mathbf{x}) = \log(\mathbf{x}) - \log(g_m(\mathbf{x}), \ldots, g_m(\mathbf{x}))$, the constant vector cancels out when multiplied componentwise by $\mathbf{v}_i = \mathrm{clr}(\mathbf{e}_i)$.

After (11.15), ilr coordinates are a linear combination of logarithms of parts whose coefficients add to zero; therefore, each coordinate is a *log-contrast*. The $(n, n - 1)$-matrix $\mathbf{V}$ is called a *contrast-matrix* associated with the orthonormal basis $\mathbf{e}_i, i = 1, 2, \ldots, n - 1$, because it contains the coefficients of such log-contrasts.

The next proposition concerns the properties of the contrast-matrix $\mathbf{V}$ associated with an orthonormal basis in the simplex.

Proposition 11.2.22   *Let* $\mathbf{v}_i, i = 1, 2, \ldots, n - 1$, *be vectors in the subspace* $U \subset \mathbb{R}^n$ *such that* $\langle \mathbf{v}_i, \mathbf{v}_j \rangle = \delta_{ij}$, *being* δ_{ij} *the Kroneker-delta. Consider the matrix* $\mathbf{V} = (\mathbf{v}_1, \mathbf{v}_2, \ldots, \mathbf{v}_{n-1})$ *whose columns are the vectors* $\mathbf{v}_i$. *Then,*

$$\mathbf{V}\mathbf{V}^\top = \mathbf{I}_n - \frac{1}{n}\mathbf{1}_n\mathbf{1}_n^\top, \quad \mathbf{V}^\top\mathbf{V} = \mathbf{I}_{n-1}, \tag{11.16}$$

where $\mathbf{I}_n$ *denotes the* (n, n)-*identity matrix and* $\mathbf{1}_n$ *a column* n-*vector with unitary components. Moreover,* $\mathbf{V}^\top\mathbf{1}_n = \mathbf{0}_{n-1}$.

Proof. The second equation in (11.16) is equivalent to the orthogonality of $\mathbf{v}_i$ unitary vectors. The singular value decomposition of $\mathbf{V}$ implies that $\mathbf{V}^\top\mathbf{V}$ and $\mathbf{V}\mathbf{V}^\top$ have equal eigenvalues except one null additional eigenvalue in $\mathbf{V}\mathbf{V}^\top$. The equation $\mathbf{V}^\top\mathbf{V} = \mathbf{I}_{n-1}$ implies that non-null eigenvalues are unitary (multiplicity $n - 1$) and the associated eigenvectors are the vectors in the canonical basis of $\mathbb{R}^{n-1}$. Assume that $\mathbf{V}\mathbf{V}^\top = \mathbf{I}_n + \mathbf{M}$, where $\mathbf{M}$ is a constant matrix with entries equal to m. Pre-multiplying by $\mathbf{V}^\top$ gives $\mathbf{V}^\top\mathbf{V}\mathbf{V}^\top = \mathbf{V}^\top + \mathbf{V}^\top\mathbf{M} = \mathbf{V}^\top$, where $\mathbf{V}^\top\mathbf{M}$ is the null matrix because the columns of $\mathbf{V}$ add to zero. Similarly, post-multiplying by $\mathbf{V}$, $\mathbf{V}\mathbf{V}^\top\mathbf{V} = \mathbf{V}$ because $\mathbf{M}\mathbf{V}$ is again null. Therefore, $\mathbf{V}$ is a pseudo-inverse of $\mathbf{V}^\top$. Furthermore, $\mathrm{tr}(\mathbf{V}\mathbf{V}^\top) = \mathrm{tr}(\mathbf{V}^\top\mathbf{V}) = \mathrm{tr}(\mathbf{I}_{n-1}) = n - 1$ and $\mathrm{tr}(\mathbf{I}_n + \mathbf{M}) = n + nm$. Hence $m = -1/n$. Moreover, the equation $\mathbf{V}^\top\mathbf{1}_n = \mathbf{0}_{n-1}$ holds because vectors $\mathbf{v}_i, i = 1, 2, \ldots, n - 1$, are in U.

Proposition 11.2.23 (ilr inverse)   *Let* $\mathbf{V}$ *be a basis-contrast matrix associated with the orthonormal basis* $\mathbf{e}_i = \mathcal{C}\exp(\mathbf{v}_i), i = 1, 2, \ldots, n - 1$, *in* $\mathcal{S}^n$. *The associated* ilr *transformation is one-to-one and the inverse transformation,* $\mathrm{ilr}^{-1} : \mathbb{R}^{n-1} \to \mathcal{S}^n$, *is*

$$\mathbf{x} = \mathrm{ilr}^{-1}(\mathbf{x}^*) = \bigoplus_{i=1}^{n-1}(x_i^* \odot \mathbf{e}_i) = \mathcal{C}\exp(\mathbf{V}\mathbf{x}^*), \tag{11.17}$$

where $\mathbf{x}^*$ *contains the* ilr *coordinates of* $\mathbf{x} \in \mathcal{S}^n$ *with respect to the basis* $\mathbf{e}_1, \mathbf{e}_2, \ldots, \mathbf{e}_{n-1}$.

Proof. Both ilr (11.15) and ilr^{-1} (11.17) are injective. The statement is proven if ilr(ilr^{-1}($\mathbf{x}^*$)) = $\mathbf{x}^*$, for any $\mathbf{x}^* \in \mathbb{R}^{n-1}$, and ilr^{-1}(ilr($\mathbf{x}$)) = $\mathbf{x}$, for any $\mathbf{x} \in \mathcal{S}^n$. Substituting (11.15) into (11.17),

$$\text{ilr}^{-1}(\text{ilr}(\mathbf{x})) = \mathcal{C} \exp(\mathbf{V} \cdot \mathbf{V}^\top \text{clr}(\mathbf{x})) = \mathcal{C} \exp((\mathbf{I}_n - \frac{1}{n}\mathbf{1}_n\mathbf{1}_n^\top)\,\text{clr}(\mathbf{x})) = \mathbf{x},$$

which holds after (11.16). Vice versa, substituting (11.17) into (11.15)

$$\text{ilr}(\text{ilr}^{-1}(\mathbf{x}^*)) = \mathbf{V}^\top \text{clr}(\mathcal{C} \exp(\mathbf{V}\mathbf{x}^*)) = \mathbf{V}^\top \mathbf{V}\mathbf{x}^* = \mathbf{x}^*,$$

due to (11.16).

Non orthonormal basis in $\mathcal{S}^n$ can also be used. An important example are the alr coordinates, which have been used extensively. These coordinates were introduced by Aitchison (1982, 1986) in his seminal log-ratio approach to compositional data analysis. The additive log-ratio transformation (Definition 11.2.5) assigns the coordinates $x_i^* = \log(x_i/x_n)$, $i = 1, 2, \ldots, n-1$, to $\mathbf{x} \in \mathcal{S}^n$. The basis corresponding to these coordinates is $\mathbf{e}_i^{\text{alr}}$, $i = 1, 2, \ldots, n-1$, and (11.6) and (11.7) hold. The alr coordinates x_i^* are computed as inner products with the elements of the dual basis $\mathbf{e}_i^{\text{dalr}}$, $i = 1, 2, \ldots, n-1$, satisfying $\langle \mathbf{e}_i^{\text{dalr}}, \mathbf{e}_j^{\text{alr}} \rangle_a = \delta_{ij}$, being δ_{ij} the Kroneker-delta. Then, $x_i^* = \langle \mathbf{x}, \mathbf{e}_i^{\text{dalr}} \rangle_a$ implies

$$\mathbf{e}_i^{\text{dalr}} = \mathcal{C} \exp(0, 0, \ldots, 0, 1, 0, \ldots, 0, -1),$$

where the 1 is placed at the ith component. The norm $\|\mathbf{e}_i^{\text{dalr}}\|_a = \sqrt{2}$. From the definition of dual basis, the elements $\mathbf{e}_i^{\text{alr}}$ are obtained in Equation (11.7), with norm $\|\mathbf{e}_i^{\text{alr}}\|_a = \sqrt{(n-1)/n}$. The cosine of the angles between different elements of both basis are

$$\frac{\langle \mathbf{e}_i^{\text{alr}}, \mathbf{e}_j^{\text{alr}} \rangle_a}{\|\mathbf{e}_i^{\text{alr}}\|_a \|\mathbf{e}_j^{\text{alr}}\|_a} = -\frac{1}{n-1}, \qquad \frac{\langle \mathbf{e}_i^{\text{dalr}}, \mathbf{e}_j^{\text{dalr}} \rangle_a}{\|\mathbf{e}_i^{\text{dalr}}\|_a \|\mathbf{e}_j^{\text{dalr}}\|_a} = \frac{1}{2}.$$

The angles between elements of the dual basis are 60° (Egozcue and Pawlowsky-Glahn 2005), and for $n = 3$ the angle between elements of the alr basis are 120°; for large n these angles approach 90°.

11.3   Linear functions

The simplex, with perturbation, powering and the Aitchison metrics is a Euclidean space (Corollary 11.2.16). Therefore, linear functions involving real spaces and the simplex, either as domain of definition or final space, are functions between Euclidean spaces. These functions can be characterized using matrix notation. However, some difficulties appear when these linear functions are expressed in terms of compositions and, particularly, when using matrix expressions.

Definition 11.3.1 (linear function in Euclidean spaces)   *Let E_r, E_s be Euclidean spaces on $\mathbb{R}$, with respective dimensions* r *and* s. *A function* $\mathbf{f} : E_r \to E_s$ *is called linear if, for any* α, $\beta \in \mathbb{R}$ *and* $\mathbf{x}, \mathbf{y} \in E_r$,

$$\mathbf{f}(\alpha \cdot \mathbf{x} + \beta \cdot \mathbf{y}) = \alpha \cdot \mathbf{f}(\mathbf{x}) + \beta \cdot \mathbf{f}(\mathbf{y}),$$

where operations $(+, \cdot)$ on the left-hand side correspond to E_r and on the right-hand side correspond to E_s.

A starting point is the following theorem from the linear algebra [e.g. Gantmacher (1959), chapter 3]. Note that, in matrix notation, the elements of $\mathbb{R}^k$ are represented by column vectors.

Theorem 11.3.2   *Let $\mathbf{f} : \mathbb{R}^r \to \mathbb{R}^s$, $r, s \geq 1$, be a linear function. Then, there exists one, and only one, (s, r)-matrix $\mathbf{A}$ with real entries such that, for any $\mathbf{x} \in \mathbb{R}^r$, the expression in coordinates with respect to the canonical bases of both spaces is*

$$\mathbf{f}(\mathbf{x}) = \mathbf{A}\mathbf{x}.$$

11.3.1   Linear functions defined on the simplex

We consider the case of linear functions defined on $\mathcal{S}^n$ with images in $\mathbb{R}^s$.

Theorem 11.3.3   *Let $\mathbf{f} : \mathcal{S}^n \to \mathbb{R}^s$ be a linear function. Then there is a unique real (s, n)-matrix, $\mathbf{A}$, such that*

$$\mathbf{f}(\mathbf{x}) = \mathbf{A} \, \log(\mathbf{x}), \tag{11.18}$$

for any $\mathbf{x} \in \mathcal{S}^n$. Moreover, matrix $\mathbf{A}$ satisfies $\mathbf{A}\mathbf{1}_n = \mathbf{0}_s$.

Proof. Let $\mathbf{e}_i, i = 1, 2, \ldots, n - 1$, be an orthonormal basis in $\mathcal{S}^n$, with basis-contrast matrix $\mathbf{V}$, and $\mathbf{x}^* \in \mathbb{R}^{n-1}$ the ilr coordinates of $\mathbf{x} \in \mathcal{S}^n$ with respect to this basis, i.e. $\mathbf{x} = \sum_{i=1}^{n-1} x_i^* \odot \mathbf{e}_i$. Linearity of $\mathbf{f}$ and Proposition 11.2.21 imply

$$\mathbf{f}(\mathbf{x}) = \mathbf{f}\left(\sum_{i=1}^{n-1} x_i^* \odot \mathbf{e}_i\right) = \sum_{i=1}^{n-1} x_i^* \mathbf{f}(\mathbf{e}_i) = \mathbf{A}^* \mathbf{x}^* = \mathbf{A}^*[\mathbf{V}^\top \, \log(\mathbf{x})] = (\mathbf{A}^* \mathbf{V}^\top) \, \log(\mathbf{x}),$$

where $\mathbf{A}^*$ is a real $(s, n-1)$-matrix containing the s-vectors $\mathbf{f}(\mathbf{e}_i)$ as columns. Taking $\mathbf{A} = \mathbf{A}^* \mathbf{V}^\top$, $\mathbf{A}\mathbf{1}_n = \mathbf{A}^* \mathbf{V}^\top \mathbf{1}_n = \mathbf{0}_s$ (Proposition 11.2.22), thus proving the existence of $\mathbf{A}$. For uniqueness, assume there is another (s, n)-matrix $\mathbf{B}$ such that $\mathbf{f}(\mathbf{x}) = \mathbf{B} \log(\mathbf{x})$, for all $\mathbf{x} \in \mathcal{S}^n$. Then, $(\mathbf{A} - \mathbf{B}) \log(\mathbf{x}) = \mathbf{0}_s$, implying $\mathbf{A} = \mathbf{B}$.

Proposition 11.3.4   *Let $\mathbf{f} : \mathcal{S}^n \to \mathbb{R}^s$ be a function and $\mathbf{A}$ a real (s, n)-matrix, such that $\mathbf{f}(\mathbf{x}) = \mathbf{A} \log(\mathbf{x})$, for any $\mathbf{x} \in \mathcal{S}^n$. If $\mathbf{A}\mathbf{1}_n = \mathbf{0}_s$, then $\mathbf{f}$ is a linear function. Conversely, if $\mathbf{A}\mathbf{1}_n \neq \mathbf{0}_s$, $\mathbf{f}$ is nonlinear.*

Proof. For $\mathbf{A}\mathbf{1}_n = \mathbf{0}_s$, $\mathbf{f}(\alpha \odot \mathbf{x}) = \mathbf{A}[\alpha \log(\mathbf{x}) + \log(c\mathbf{1}_n)]$ and $\mathbf{f}(\mathbf{x}_1 \oplus \mathbf{x}_2) = \mathbf{A}[\log(\mathbf{x}_1) + \log(\mathbf{x}_2) + \log(c\mathbf{1}_n)]$, for some real constant c. The constant terms vanish after matrix multiplication by $\mathbf{A}$, and linearity holds. If condition $\mathbf{A}\mathbf{1}_n = \mathbf{0}_s$ does not hold the constant terms do not vanish and linearity fails.

A linear function $\mathbf{f} : \mathcal{S}^n \rightarrow \mathbb{R}^s$ defined as in (11.18) can also be written as $\mathbf{f}(\mathbf{x}) = \mathbf{A}\, \text{clr}(\mathbf{x})$. In this case, the matrix $\mathbf{A}$ can be substituted by matrices of the form $\mathbf{B} = \mathbf{A} + \mathbf{w}\mathbf{1}_n^\top$, with $\mathbf{w} \in \mathbb{R}^s$, and $\mathbf{f}$ remains invariant. This is due to $\mathbf{1}_n^\top \text{clr}(\mathbf{x}) = 0$.

In practice, one may be interested in the matrix $\mathbf{A}^*$ corresponding to the linear function assigning images in $\mathbb{R}^s$ to the coordinates of the composition. Once $\mathbf{A}$ is given and a basis-contrast matrix $\mathbf{V}$ is selected, $\mathbf{AV} = \mathbf{A}^*\mathbf{V}^\top\mathbf{V} = \mathbf{A}^*$ allows us to compute the $(s, n-1)$-matrix $\mathbf{A}^*$.

11.3.2   Simplicial linear function defined on a real space

We consider the case of linear functions defined on $\mathbb{R}^r$ with images in $\mathcal{S}^n$.

Theorem 11.3.5   *Let $\mathbf{f} : \mathbb{R}^r \rightarrow \mathcal{S}^n$ be a linear function. Then there is a unique real (n, r)-matrix, $\mathbf{A}$, with positive entries, satisfying $\mathbf{1}_n^\top\mathbf{A} = \mathbf{1}_r^\top$, such that*

$$(\mathbf{f}(\mathbf{z}))^\top = \mathbf{z}^\top \odot \mathbf{A}^\top, \tag{11.19}$$

for any $\mathbf{z} \in \mathbb{R}^r$.

Proof. Let $\mathbf{e}_i$, $i - 1, 2, \ldots, n-1$, be an orthonormal basis in $\mathcal{S}^n$, with basis-contrast $(n, n-1)$-matrix $\mathbf{V}$, and $\text{ilr} : \mathcal{S}^n \rightarrow \mathbb{R}^{n-1}$ the corresponding linear isometric log-ratio transformation assigning coordinates associated with $\mathbf{e}_i$, $i = 1, 2, \ldots, n-1$. Then, the composite function $\mathbf{f}^* = \text{ilr} \circ \mathbf{f} : \mathbb{R}^r \rightarrow \mathbb{R}^{n-1}$ is linear and has the expression (Theorem 11.3.2)

$$\mathbf{f}^*(\mathbf{z}) = \mathbf{A}^*\mathbf{z}, \quad \mathbf{z} \in \mathbb{R}^r,$$

where $\mathbf{A}^*$ is a real $(n-1, r)$-matrix. It gives the coordinates of $\mathbf{f}(\mathbf{z})$ with respect to the orthonormal basis $\mathbf{e}_i$, $i = 1, 2, \ldots, n-1$, of $\mathcal{S}^n$. Therefore, according to Proposition 11.2.23,

$$\mathbf{f}(\mathbf{z}) = \mathcal{C}\exp[\mathbf{V}\mathbf{f}^*(\mathbf{z})] = \mathcal{C}\exp(\mathbf{V}\mathbf{A}^*\mathbf{z}), \quad \mathbf{z} \in \mathbb{R}^r. \tag{11.20}$$

Denote $\mathbf{A}^* = (\mathbf{a}_1^*, \mathbf{a}_2^*, \ldots, \mathbf{a}_r^*)$, where $\mathbf{a}_i^*$ is the image of the ith vector of the canonical basis of $\mathbb{R}^r$ by the function $\mathbf{f}^*$, i.e. it is formed by the coordinates of a composition of $\mathcal{S}^n$, $\mathbf{a}_i = \mathcal{C}\exp(\mathbf{V}\mathbf{a}_i^*)$. Consider the (n, r)-matrix, with positive entries, $\mathbf{A} = (\mathbf{a}_1, \mathbf{a}_2, \ldots, \mathbf{a}_r)$ whose columns are compositions in $\mathcal{S}^n$. Therefore, the ith component of $\mathbf{z}$ is powering the composition $\mathbf{a}_i$; and all resulting powered compositions are then perturbed thus forming a perturbation-linear combination of the compositions $\mathbf{a}_i$. This means that (11.20) is equivalently expressed as

$$(\mathbf{f}(\mathbf{z}))^\top = \mathbf{z}^\top \odot \mathbf{A}^\top,$$

as stated in the theorem. Moreover, the closure operator that appears in the definition of matrix $\mathbf{A}$ implies that $\mathbf{1}_n^\top\mathbf{A} = \mathbf{1}_r^\top$ holds. For uniqueness, assume there is another (n, r)-matrix, $\mathbf{B}$, satisfying $\mathbf{1}_n^\top\mathbf{B} = \mathbf{1}_r^\top$, such that $(\mathbf{f}(\mathbf{z}))^\top = \mathbf{z}^\top \odot \mathbf{B}^\top$, for any $\mathbf{z} \in \mathbb{R}^r$. Then, the equality

$$\mathbf{n}^\top = (\mathbf{z}^\top \odot \mathbf{A}^\top) \ominus (\mathbf{z}^\top \odot \mathbf{B}^\top),$$

holds for every $\mathbf{z} \in \mathbb{R}^r$, which implies $\mathbf{A} = \mathbf{B}$.

The condition that the columns of $\mathbf{A}$ must add to one can be removed at the price of uniqueness. Consider a (n, r)-matrix $\mathbf{B}$, with positive entries, such that $\mathbf{1}_n^\top \mathbf{B} = \mathbf{c}^\top \neq \mathbf{1}_r^\top$ has columns that do not add to one but to c_j, $j = 1, 2, \ldots, r$. A (n, r)-matrix in which columns are compositions is readily obtained, $\mathbf{A} = \mathbf{B} \operatorname{diag}(1/c_1, 1/c_2, \ldots, 1/c_r)$. These two matrices, $\mathbf{A}$ and $\mathbf{B}$, operate equivalently in (11.19), i.e. $\mathbf{z}^\top \odot \mathbf{A}^\top = \mathbf{z}^\top \odot \mathbf{B}^\top$.

11.3.3   Simplicial linear function defined on the simplex

In order to express linear functions between simplex $\mathcal{S}^n$ and $\mathcal{S}^m$, a special kind of matrix product in the simplex is introduced.

Definition 11.3.6 (matrix product in the simplex)   *Let $\mathbf{X}$ be a (n, r)-matrix whose r columns are compositions $\mathbf{x}_1, \mathbf{x}_2, \ldots, \mathbf{x}_r$ in $\mathcal{S}^n$, and consider a real (m, n)-matrix $\mathbf{A}$. The matrix-product in the simplex, denoted $\mathbf{Y} = \mathbf{A} \boxdot \mathbf{X}$, is the (m, r)-matrix $\mathbf{Y}$, whose columns $\mathbf{y}_j \in \mathcal{S}^m$ are*

$$\mathbf{y}_j = \mathcal{C}\left(\prod_{i=1}^n x_{ij}^{a_{1i}}, \prod_{i=1}^n x_{ij}^{a_{2i}}, \ldots, \prod_{i=1}^n x_{ij}^{a_{mi}}\right)^\top, \quad j = 1, 2, \ldots, r.$$

Whenever the matrix $\mathbf{X}$ reduces to a single composition, i.e. $r = 1$, the matrix product is

$$\mathbf{y} = \mathbf{A} \boxdot \mathbf{x} = \mathcal{C}\left(\prod_{i=1}^n x_i^{a_{1i}}, \prod_{i=1}^n x_i^{a_{2i}}, \ldots, \prod_{i=1}^n x_i^{a_{mi}}\right)^\top. \tag{11.21}$$

Despite the linear appearance of the matrix product in the simplex, it can correspond to a nonlinear transformation. The reason is that (11.21) is not invariant under scaling, i.e. $\mathbf{A} \boxdot \mathbf{x} \neq \mathbf{A} \boxdot (k\mathbf{x})$, $k > 0$, unless the rows of $\mathbf{A}$ add up to zero. As a consequence, the matrix product in the simplex is only useful whenever $\mathbf{A}\mathbf{1}_n = \mathbf{0}_m$, or equivalently $\sum_{j=1}^n a_{ij} = 0$ for $i = 1, 2, \ldots, m$.

Proposition 11.3.7   *Let $\mathbf{x}, \mathbf{y}$ be compositions in $\mathcal{S}^n$, and $\mathbf{A}$ a real (m, n)-matrix satisfying $\mathbf{A}\mathbf{1}_n = \mathbf{0}_m$. For real constants α and β,*

$$\mathbf{A} \boxdot ((\alpha \odot \mathbf{x}) \oplus (\beta \odot \mathbf{y})) = (\alpha \odot (\mathbf{A} \boxdot \mathbf{x})) \oplus (\beta \odot (\mathbf{A} \boxdot \mathbf{y})).$$

Proof.   Before closure, the ith part of $\mathbf{A} \boxdot ((\alpha \odot \mathbf{x}) \oplus (\beta \odot \mathbf{y}))$ is

$$\prod_{j=1}^n \left(\frac{x_j^\alpha y_j^\beta}{K}\right)^{a_{ij}} = K^{-\sum_{j=1}^n a_{ij}} \prod_{j=1}^n \left(x_j^\alpha y_j^\beta\right)^{a_{ij}},$$

where $K = \sum_{j=1}^m x_j^\alpha y_j^\beta$ is the closure constant for $(\alpha \odot \mathbf{x}) \oplus (\beta \odot \mathbf{y})$. Since $s = \sum_{j=1}^n a_{ij} = 0$, $K^{-s} = 1$ for all $i = 1, 2, \ldots, m$, this constant cancels with closure. The resulting part is readily identified with the perturbation of two powered compositions. Note that, if the condition on the rows of $\mathbf{A}$ does not hold, then, linearity fails.

There are (m, n)-matrices, adding zero row-wise that represent the same linear transformation when used in a matrix product. In addition, the following proposition shows that an equivalent matrix whose rows and columns add to zero always exists.

Proposition 11.3.8   *Let* $\mathbf{A}$ *be a real* (m, n)-*matrix such that* $\mathbf{A}\mathbf{1}_n = \mathbf{0}_m$ *and*

$$\mathbf{A}_0 = \left(\mathbf{I}_m - \frac{1}{m}\mathbf{1}_m\mathbf{1}_m^\top\right)\mathbf{A}. \tag{11.22}$$

Then, for $\mathbf{x} \in \mathcal{S}^n$, $\mathbf{A} \boxdot \mathbf{x} = \mathbf{A}_0 \boxdot \mathbf{x}$. *Moreover,* $\mathbf{A}_0\mathbf{1}_n = \mathbf{0}_m$, $\mathbf{1}_m^\top\mathbf{A}_0 = \mathbf{0}_n$.

Proof. The operation in (11.22) consists of subtracting the mean column value in $\mathbf{A}$ thus implying $\mathbf{1}_m^\top\mathbf{A}_0 = \mathbf{0}_n$. Furthermore, the mean of mean columns is null due to $\mathbf{A}\mathbf{1}_n = \mathbf{0}_m$. When computing $\mathbf{A}_0 \boxdot \mathbf{x}$, before closure, the ith part is

$$\prod_{j=1}^n \left(x_j^{a_{ij}-a_{.j}}\right) = \prod_{j=1}^n x_j^{a_{ij}} \cdot \prod_{k=1}^n x_k^{-a_{.k}},$$

where $a_{.j} = (1/n)\sum_{i=1}^n a_{ij}$. The second factor in this term does not depend on i and it is a common factor in all the parts. Therefore, it cancels after closure.

Proposition 11.3.9   *Let* $\mathbf{A}$ *be a real* (m, n)-*matrix and* $\mathbf{x}$ *a composition in* $\mathcal{S}^n$. *Then,* $\mathbf{A} \boxdot \mathbf{x} = \mathcal{C}\exp(\mathbf{A}\log(\mathbf{x}))$.

Proof. Equation (11.21) is transformed into

$$\mathbf{y} = \mathbf{A} \boxdot \mathbf{x} = \mathcal{C}\exp\left(\sum_{i=1}^n a_{1i}\log x_i, \sum_{i=1}^n a_{2i}\log x_i, \ldots, \sum_{i=1}^n a_{mi}\log x_i\right)^\top = \mathcal{C}\exp[\mathbf{A}\log(\mathbf{x})],$$

independently of the sum of rows and columns in $\mathbf{A}$.

The main result concerning the expression of simplicial linear functions defined on the simplex follows.

Theorem 11.3.10   *Let* $\mathbf{f} : \mathcal{S}^n \to \mathcal{S}^m$ *be a linear function. Then there is a unique real* (m, n)-*matrix,* $\mathbf{A}$, *satisfying* $\mathbf{A}\mathbf{1}_n = \mathbf{0}_m$ *and* $\mathbf{1}_m^\top\mathbf{A} = \mathbf{0}_n^\top$ *such that, for any* $\mathbf{x} \in \mathcal{S}^n$,

$$\mathbf{f}(\mathbf{x}) = \mathbf{A} \boxdot \mathbf{x}. \tag{11.23}$$

Proof. Let $\mathbf{v}_1, \mathbf{v}_2, \ldots, \mathbf{v}_{n-1}$, be an orthonormal basis in $\mathcal{S}^n$, with basis-contrast $(n, n-1)$-matrix $\mathbf{V}$ and let $\mathrm{ilr}_\mathbf{V} : \mathcal{S}^n \to \mathbb{R}^{n-1}$ be the corresponding linear isometric log-ratio transformation assigning the coordinates $\mathbf{x}^*$ to $\mathbf{x} \in \mathcal{S}^n$. Let $\mathbf{u}_1, \mathbf{u}_2, \ldots, \mathbf{u}_{m-1}$, be an orthonormal basis in $\mathcal{S}^m$, with basis-contrast $(m, m-1)$-matrix $\mathbf{U}$ and $\mathrm{ilr}_\mathbf{U} : \mathcal{S}^m \to \mathbb{R}^{m-1}$ the corresponding linear isometric log-ratio transformation assigning coordinates $\mathbf{y}^*$ to $\mathbf{y} \in \mathcal{S}^m$. Then, the composite function $\mathbf{f}^* = \mathrm{ilr}_\mathbf{U} \circ \mathbf{f} \circ \mathrm{ilr}_\mathbf{V}^{-1} : \mathbb{R}^{n-1} \to \mathbb{R}^{m-1}$ is linear and, for any $\mathbf{x}^* \in \mathbb{R}^{n-1}$, it has the expression (Theorem 11.3.2)

$$\mathbf{y}^* = \mathbf{f}^*(\mathbf{x}^*) = \mathbf{A}^*\mathbf{x}^*, \tag{11.24}$$

where $\mathbf{A}^*$ is a real $(m-1, n-1)$-matrix. Taking into account that $\mathbf{y} = \mathcal{C} \exp(\mathbf{U}\mathbf{y}^*)$ (Proposition 11.2.23) and $\mathbf{x}^* = \mathbf{V}^\top \log(\mathbf{x})$ (Proposition 11.2.21) expression (11.24) yields

$$\mathbf{y} = \mathbf{f}(\mathbf{x}) = \mathcal{C} \exp[\mathbf{U}\mathbf{A}^*\mathbf{V}^\top \log(\mathbf{x})], \quad \mathbf{x} \in \mathcal{S}^n.$$

Therefore, according to Proposition 11.3.9, $\mathbf{f}(\mathbf{x}) = \mathbf{A} \boxdot \mathbf{x}$, where $\mathbf{A} = \mathbf{U}\mathbf{A}^*\mathbf{V}^\top$. Moreover, $\mathbf{A}\mathbf{1}_n = \mathbf{0}_m$ because $\mathbf{V}^\top \mathbf{1}_n = \mathbf{0}_{n-1}$ (Proposition 11.2.22). Similarly, $\mathbf{1}_m^\top \mathbf{A} = \mathbf{0}_n^\top$ because $\mathbf{U}^\top \mathbf{1}_m = \mathbf{0}_{m-1}$. Therefore, $\mathbf{1}_m^\top \mathbf{U} = \mathbf{0}_{m-1}^\top$. For uniqueness, assume there are two (m, n)-matrix $\mathbf{A}$ and $\mathbf{B}$, such that their rows and columns add to zero, and $\mathbf{A} \boxdot \mathbf{x} = \mathbf{B} \boxdot \mathbf{x}$, for any $\mathbf{x} \in \mathcal{S}^n$. Then, $(\mathbf{A} \boxdot \mathbf{x}) \ominus (\mathbf{B} \boxdot \mathbf{x}) = \mathbf{n}$. Using the exponential form in Proposition 11.3.9, the uniqueness condition is equivalent to $\mathcal{C} \exp[(\mathbf{A} - \mathbf{B}) \log(\mathbf{x})] = \mathbf{n}$, where the exponent is a n-vector with components adding to zero. The clr is a one-to-one transformation (Theorem 11.2.14). Then, taking clr yields $\mathrm{clr}(\mathbf{n}) = \mathbf{0}_m = (\mathbf{A} - \mathbf{B}) \log(\mathbf{x})$, for any $\mathbf{x} \in \mathcal{S}^n$, thus implying $\mathbf{A} = \mathbf{B}$.

11.4 Conclusions

Concepts of Euclidean spaces have been adapted to the simplex using the perturbation and powering operations and the Aitchison metrics. Most results have been compiled from previous contributions and updated in a formal and systematic way. Particularities of the simplicial formulation of its Euclidean structure are mainly due to: (a) the multiplicity of representations of compositions using simplicial (clr, alr, ilr) representations; and (b) the fact that closure operations introduce conditions on the matrix expressions of coordinates and linear functions involving the simplex.

Acknowledgements

This research has been supported by the Spanish Ministry of Science and Innovation (projects CSD2006-00032 and MTM2009-13272) and by the Agència de Gestió d'Ajuts Universitaris i de Recerca of the Generalitat de Catalunya (Ref. 2009SGR424).

References

Aitchison J 1982 The statistical analysis of compositional data (with discussion). *Journal of the Royal Statistical Society, Series B (Statistical Methodology)* **44**(2), 139–177.

Aitchison J 1983 Principal component analysis of compositional data. *Biometrika* **70**(1), 57–65.

Aitchison J 1986 *The Statistical Analysis of Compositional Data*. Monographs on Statistics and Applied Probability. Chapman and Hall Ltd (reprinted 2003 with additional material by The Blackburn Press), London (UK). 416 p.

Aitchison J 1992 On criteria for measures of compositional difference. *Mathematical Geology* **24**(4), 365–379.

Aitchison J and Egozcue JJ 2005 Compositional data analysis: where are we and where should we be heading?. *Mathematical Geology* **37**(7), 829–850.

Aitchison J, Barceló-Vidal C, Martín-Fernández JA and Pawlowsky-Glahn V 2001 Reply to Letter to the Editor by S. Rehder and U. Zier. *Mathematical Geology* **33**(7), 849–860.

Barceló-Vidal C, Martín-Fernández JA and Pawlowsky-Glahn V 2001 Mathematical foundations of compositional data analysis. In *Proceedings of IAMG'01 – The VII Annual Conference of the International Association for Mathematical Geology* (ed. Ross G). Kansas Geological Survey, Cancun (Mexico). p. 20.

Billheimer D, Guttorp P and Fagan W 2001 Statistical interpretation of species composition. *Journal of the American Statistical Association* **96**(456), 1205–1214.

Egozcue JJ and Pawlowsky-Glahn V 2005 Groups of parts and their balances in compositional data analysis. *Mathematical Geology* **37**(7), 795–828.

Egozcue JJ and Pawlowsky-Glahn V 2006 Simplicial geometry for compositional data. In *Compositional Data Analysis in the Geosciences: From Theory to Practice*. Geological Society, London (UK). pp. 145–159.

Egozcue JJ, Pawlowsky-Glahn V, Mateu-Figueras G and Barceló-Vidal C 2003 Isometric logratio transformations for compositional data analysis. *Mathematical Geology* **35**(3), 279–300.

Gantmacher FR 1959 *The Theory of Matrices*, vol. 2. Chelsea Publishing Co., New York, NY (USA). 276 p.

Pawlowsky-Glahn V and Egozcue JJ 2001 Geometric approach to statistical analysis on the simplex. *Stochastic Environmental Research and Risk Assessment (SERRA)* **15**(5), 384–398.

Thió-Henestrosa S, Egozcue JJ, Pawlowsky-Glahn V, Kovács LO and Kovács G 2008 Balance-dendrogram. a new routine of codapack. *Computer and Geosciences* **34**(12), 1682–1696.

Tolosana-Delgado R, Pawlowsky-Glahn V and Egozcue JJ 2008 Indicator kriging without order relation violations. *Mathematical Geosciences* **40**, 327–347.

12

Calculus of simplex-valued functions

Juan José Egozcue, Eusebi Jarauta-Bragulat and José Luis Díaz-Barrero

Department of Applied Mathematics III, Technical University of Catalonia, Spain

12.1 Introduction

In practice, evolution of compositions in time, space, temperature, etc. occurs very frequently. These situations can be described using functions defined on a real variable with values in the simplex. Changes of a chemical composition, changes of vote intention, evolution of employment in time are examples. Modelling these compositional changes requires the study of these functions and elementary concepts like derivative, describing the rate of change, or integral, to give mean values or summary descriptors. Compositions are suitably represented in the simplex (Aitchison 1986) and its Aitchison geometry (Pawlowsky-Glahn and Egozcue 2001; Egozcue *et al.* 2003) (see also Chapter 11). Accordingly, a proper way of measuring change for compositions is taking log-ratios which is better than taking ordinary differences componentwise. The former option corresponds to the opposite of perturbation (addition in the simplex), and the latter is not a closed operation (a difference can result in a negative part of a composition). Since differences in the simplex appear as log-ratios, the concept of derivative in the simplex should adequately be changed. Similarly, an integral is essentially a sum. Consequently, it should be a perturbation in the simplex. The Aitchison geometry of the simplex is a Euclidean geometry and, therefore, most concepts of the real geometry and calculus of real spaces can be translated into the simplex and vice versa. Most results and

---

Compositional Data Analysis: Theory and Applications, First Edition. Edited by Vera Pawlowsky-Glahn and Antonella Buccianti.
© 2011 John Wiley & Sons, Ltd. Published 2011 by John Wiley & Sons, Ltd.

their proofs are simple translations of standard calculus as developed in many textbooks, e.g. Apostol (1957), Bartle (1976), Estep (2002), Kartachev and Rojdestvenski (1988) and Rudin (1964). Hereafter, we develop some of these concepts, for the so-called simplex-valued (sv) functions. However, there are at least two relevant details that make the development not as simple as initially thought. One of them is the fact that the elements of the simplex are thought of as representatives of equivalence classes (Martín-Fernández *et al.* 2003) of proportional real vectors with positive components. This suggests using the closure operator assigning an element of the simplex to a proportional real vector of positive components. A second disturbing point is the need for simultaneous use of different representations of compositions in order to get an adequate interpretation of the modeled compositional phenomenon. Isometric transformations of the simplex into a real space (e.g. ilr, clr, Aitchison 1986; Egozcue *et al.* 2003; Egozcue and Pawlowsky-Glahn 2006) allow us to perform a complete analysis using standard real calculus according to the so-called *principle of working-in-coordinates* (see Chapter 3), which is based on the theoretical developments by Kolmogorov and Fomin (1957). However, to complete the interpretation, the analyst needs to translate back the results into the simplex, and to have confidence that these results are not dependent on the isometric transformation used. The following development of calculus of sv functions presents definitions and results in simplicial notation thus not depending on the particular representation of coordinates. Moreover, some attention is paid to the commutation of closure with differential and integral operators.

In this section, we introduce some notation and elementary definitions of sv functions. In Section 12.2 elementary concepts of differential calculus (limit, continuity and differentiability) are defined and calculus of derivative is developed. An illustrative example is also presented. In Section 12.3, basic concepts of integral calculus (antiderivative, definite integral) are defined and calculus of antiderivative and definite integral are developed.

Simplex-valued functions are here defined on the real field $\mathbb{R}$, and the ordinary Euclidean distance is considered, $d(a, b) = |a - b|$, for any $a, b \in \mathbb{R}$. The images of sv functions are in the simplex of n parts $\mathcal{S}^n$. The $\mathcal{S}^n$ satisfies:

(1) It is a real finite dimensional vector space of dimension $n - 1$, with operations perturbation $\oplus$ and powering $\odot$.

(2) It is a Euclidean space, with the Aitchison's inner product denoted $\langle \mathbf{x}, \mathbf{y} \rangle_a$, and associate norm $\|\mathbf{x}\|_a = +\sqrt{\langle \mathbf{x}, \mathbf{x} \rangle_a}$.

(3) It is a metric space, with the associated Aitchison distance denoted $d_a(\mathbf{x}, \mathbf{y})$.

An account of the Aitchison geometry of the simplex and the notation hereafter used is found in Chapter 11.

Let I be any interval in $\mathbb{R}$. A positive-component vector-valued function of real variable is defined as a map $\mathbf{f} : I \subseteq \mathbb{R} \to \mathbb{R}^n_+$, $\mathbf{f}(t) = [f_1(t), f_2(t), \ldots, f_n(t)]$, $t \in I$. Each component f_j, $1 \leq j \leq n$ is a positive real-valued function of real variable.

Definition 12.1.1 (Simplex-valued functions) *A sv function of real variable is a map* $\mathbf{f} : I \subseteq \mathbb{R} \to \mathcal{S}^n$ *defined as the composition of a positive-component vector-valued function of real variable and the closure operation $\mathcal{C}$ in the simplex of n parts, $\mathcal{S}^n$. A sv function of real* *variable is expressed as*

$$\mathbf{f}(t) = \mathcal{C}\left(f_1(t), f_2(t), \ldots, f_n(t)\right) = \left[\ldots, \frac{f_i(t)}{\sum_{j=1}^{n} f_j(t)}, \ldots\right], \quad t \in I.$$

The condition to be of real variable is sometimes omitted because it is just assumed. When real functions of real variable (e.g. exp, log) and operators (e.g. $\int$) are applied to a *n*-tuple or a vector-valued function, they are assumed to operate componentwise, e.g. $\exp(a, b, c) = (e^a, e^b, e^c)$. The images of a sv function are elements of the simplex and they are called compositions. The components of sv functions are called parts. Notice that these components add up to the closure constant here assumed to be one.

Example 12.1.2

(1) *Constant functions.* Let $\alpha_1, \alpha_2, \ldots, \alpha_n$ be positive real numbers; the sv function defined by $\mathbf{f}(t) = \mathcal{C}(\alpha_1, \alpha_2, \ldots, \alpha_n)$, is called a constant sv function or simplicial constant for short.

(2) *Polynomials.* Let $\boldsymbol{\beta}_i \in \mathcal{S}^n, 0 \le i \le m$ be compositions of *n* parts. A sv function $\mathbf{P}_m : \mathbb{R} \to \mathcal{S}^n$ defined by $\mathbf{P}_m(t) = \boldsymbol{\beta}_0 \oplus \left(t \odot \boldsymbol{\beta}_1\right) \oplus \left(t^2 \odot \boldsymbol{\beta}_2\right) \oplus \cdots \oplus \left(t^m \odot \boldsymbol{\beta}_m\right)$, is called an *m*-degree polynomial.

(3) *Exponential functions.* Let $a_1, a_2, \ldots, a_n$ be real numbers. The function defined by

$$\mathbf{f}(t) = \mathcal{C}[\exp(a_1 t), \ldots, \exp(a_n t)] = \mathcal{C}\exp(a_1 t, \ldots, a_n t) \in \mathcal{S}^n, \quad t \in \mathbb{R},$$

is called an exponential function.

(4) *Double exponential functions.* Let $\lambda_j, a_j, 1 \le j \le n$, be real numbers. The function $\mathbf{f} : \mathbb{R} \to \mathcal{S}^n$, defined by

$$\mathbf{f}(t) = \mathcal{C}\exp\left[\lambda_1 \exp(a_1 t), \ldots, \lambda_n \exp(a_n t)\right] \in \mathcal{S}^n, \quad t \in \mathbb{R},$$

is called a double exponential function.

(5) Consider the interval $I = [0, 6] \subset \mathbb{R}$ and the function defined by

$$\mathbf{f}(t) = \mathcal{C}\left[\frac{6}{5} + \sin(2t), \frac{10}{1+t}, 1 + \left(\frac{t-1}{2}\right)^2\right]$$

$$= \mathcal{C}\left[\frac{6 + 5\sin(2t)}{5}, \frac{10}{1+t}, \frac{t^2 - 2t + 5}{4}\right], \quad t \in I.$$

Notice that operations involved in the components of this sv function are ordinary operations in $\mathbb{R}$. Figure 12.1 shows different representations of $\mathbf{f}$: parts as functions of t, orbit in the ternary diagram, and in orthonormal coordinates.

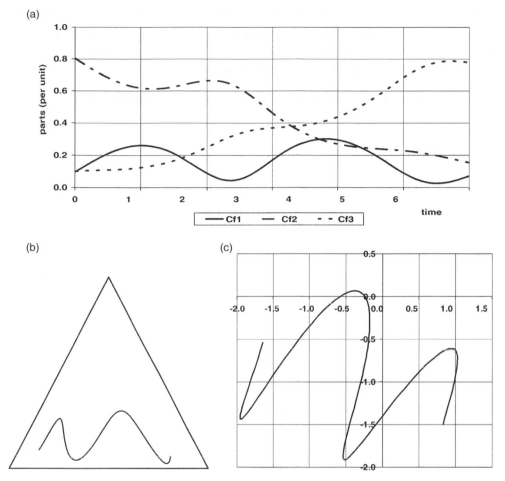

Figure 12.1 Example of sv function. (a) Parts as functions of t; (b) orbit of the function in the ternary diagram; (c) orbit described in orthonormal coordinates.

12.2 Limits, continuity and differentiability

Limit and continuity of sv functions are analogous to standard definitions in calculus. However, simplex operations and Aitchison metrics are here involved and are consequently rewritten.

12.2.1 Limits and continuity

Definition 12.2.1 (Limit of a sv function at a real point)   *Let* $\mathbf{f} : I \subseteq \mathbb{R} \to \mathcal{S}^n$ *be a sv function and let* t_0 *be a limit point of* I. *Function* $\mathbf{f}$ *has limit* $\mathbf{L}_0$ *at* t_0, *written* $\lim\limits_{t \to t_0} \mathbf{f}(t) = \mathbf{L}_0$, *if, and only if, for any* $\varepsilon > 0$, *exists* $\delta > 0$, *such that*

$$0 < |t - t_0| < \delta \;\Rightarrow\; \|\mathbf{f}(t) \ominus \mathbf{L}_0\|_a < \varepsilon. \tag{12.1}$$

One-sided limits, denoted $\lim_{t \to t_0+} \mathbf{f}(t)$ and $\lim_{t \to t_0-} \mathbf{f}(t)$ can be defined as in ordinary calculus.

Definition 12.2.1 is equivalent to $\lim_{t \to t_0} \mathbf{f}(t) = \mathbf{L}_0$ if, and only if, $\lim_{t \to t_0}(\mathbf{f}(t) \ominus \mathbf{L}_0) = \mathbf{n}$, where $\mathbf{n} = \mathcal{C}(1, 1, \ldots, 1)$, is the neutral element of perturbation operation in the simplex, i.e. $\mathbf{x} \oplus \mathbf{n} = \mathbf{x}$. Also, the condition in (12.1) can be written as

$$\|\mathbf{f}(t) \ominus \mathbf{L}_0\|_a = \|\mathrm{clr}(\mathbf{f}(t)) - \mathrm{clr}(\mathbf{L}_0)\| = \|\mathrm{ilr}(\mathbf{f}(t)) - \mathrm{ilr}(\mathbf{L}_0)\| < \varepsilon,$$

where $\| \cdot \|$ is the ordinary Euclidean norm in $\mathbb{R}^n$ and $\mathbb{R}^{n-1}$ (see Chapter 11).

If the interval I is upper unbounded, that is $I = [a, +\infty)$, limit at infinite is also defined: $\lim_{t \to +\infty} \mathbf{f}(t) = \mathbf{L}$ if, and only if, for any $\varepsilon > 0$, exists $M > 0$, such that for any $t > M$ then $\|\mathbf{f}(t) \ominus \mathbf{L}_0\|_a < \varepsilon$. The limit $\lim_{t \to -\infty} \mathbf{f}(t)$ is defined in a similar way.

Proposition 12.2.2 (Properties of limit)

(1) The limit of a function at a point, if it exists, is unique.

(2) Let $\mathbf{f}, \mathbf{g} : I \subseteq \mathbb{R} \to \mathcal{S}^n$ be sv functions and $\lambda \in \mathbb{R}$; if the limits $\lim_{t \to t_0} \mathbf{f}(t)$ and $\lim_{t \to t_0} \mathbf{g}(t)$ exist, then

$$\lim_{t \to t_0}(\mathbf{f} \oplus \mathbf{g})(t) = \left(\lim_{t \to t_0} \mathbf{f}(t) \right) \oplus \left(\lim_{t \to t_0} \mathbf{g}(t) \right); \quad \lim_{t \to t_0}(\lambda \odot \mathbf{f})(t) = \lambda \odot \left(\lim_{t \to t_0} \mathbf{f}(t) \right).$$

$$(12.2)$$

(3) Let $\mathbf{f} : I \subseteq \mathbb{R} \to \mathcal{S}^n$ be a sv function and let $\varphi : I \subseteq \mathbb{R} \to \mathbb{R}$ be a real-valued function; if the limits $\lim_{t \to t_0} \mathbf{f}(t)$ and $\lim_{t \to t_0} \varphi(t)$ exist, then

$$\lim_{t \to t_0}(\varphi \odot \mathbf{f})(t) = \left(\lim_{t \to t_0} \varphi(t) \right) \odot \left(\lim_{t \to t_0} \mathbf{f}(t) \right).$$

$$(12.3)$$

(4) Let $\mathbf{f} : I \subseteq \mathbb{R} \to \mathcal{S}^n$ be a sv function; if $\lim_{t \to t_0} \mathbf{f}(t) = \mathbf{L}_0$, then

$$\lim_{t \to t_0} \mathrm{clr}(\mathbf{f}(t)) = \mathrm{clr}(\mathbf{L}_0); \quad \lim_{t \to t_0} \mathrm{ilr}(\mathbf{f}(t)) = \mathrm{ilr}(\mathbf{L}_0).$$

$$(12.4)$$

Definition 12.2.3 (Continuous functions)   *Let $\mathbf{f} : I \subseteq \mathbb{R} \to \mathcal{S}^n$ be a sv function and let $t_0 \in I$ be a limit point of I; we say that $\mathbf{f}$ is continuous at t_0, if, and only if, $\lim_{t \to t_0} \mathbf{f}(t) = \mathbf{f}(t_0)$.*
The sv function $\mathbf{f}$ is continuous on a set $J \subseteq I$ if, and only if $\mathbf{f}$ is continuous at each point of this subset.

Proposition 12.2.4 (Properties of continuous functions)

(1) (Continuity at a point)

   – *Let $\mathbf{f}, \mathbf{g} : I \subseteq \mathbb{R} \to \mathcal{S}^n$ be sv functions and $\lambda \in \mathbb{R}$; if $\mathbf{f}$ and $\mathbf{g}$ are continuous at $t_0 \in I$, then sv functions $\mathbf{f} \oplus \mathbf{g}$ and $\lambda \odot \mathbf{f}$ are continuous at t_0.*

– Let $\mathbf{f} : I \subseteq \mathbb{R} \to \mathcal{S}^n$ be a sv function and let $\varphi : I \to \mathbb{R}$ be a real-valued function; if $\mathbf{f}$ and φ are continuous at $t_0 \in I$, then the function $\varphi \odot \mathbf{f}$ is continuous at t_0.

– Let $\mathbf{f} : I \subseteq \mathbb{R} \to \mathcal{S}^n$ be a sv function and let $\Phi : J \to I$ be a real-valued function; if Φ is continuous at $t_0 \in J$ and $\mathbf{f}$ is continuous at $\Phi(t_0) \in I$, then the composite function $\mathbf{f} \circ \Phi$ is continuous at t_0.

– Let $\mathbf{f} : I \subseteq \mathbb{R} \to \mathcal{S}^n$ be a sv function; if $\mathbf{f}$ is continuous at $t_0 \in I$, then it is bounded on an interval $(t_0 - \delta, t_0 + \delta) \subseteq I$.

(2) *(Continuity on a closed interval. Weierstrass Theorem.) Let $\mathbf{f} : I \subset \mathbb{R} \to \mathcal{S}^n$ be a sv function; if $\mathbf{f}$ is continuous on a closed interval I, then it is bounded on that interval and its norm attains its maximum and its minimum values at some points in the closed interval I.*

12.2.2 Differentiability

The derivative of a sv function has been previously proposed in Aitchison (1986), Aitchison *et al.* (2002) and Egozcue *et al.* (2008). Here, it is formally defined as follows.

Definition 12.2.5 (Differentiability at a point) *Let $\mathbf{f} : I \subseteq \mathbb{R} \to \mathcal{S}^n$ be a sv function and let $t \in I$ be an interior point of I, namely, $r > 0$ such that $(t - r, t + r) \subset I$; the sv function $\mathbf{f}$ is differentiable at t, if, and only if, the limit*

$$D^{\oplus}\mathbf{f}(t) = \lim_{h \to 0} \left(\frac{1}{h} \odot (\mathbf{f}(t + h) \ominus \mathbf{f}(t)) \right) = \lim_{u \to t} \left(\left(\frac{1}{u - t} \right) \odot (\mathbf{f}(u) \ominus \mathbf{f}(t)) \right), \qquad (12.5)$$

exists. The composition of n parts $D^{\oplus}\mathbf{f}(t) \in \mathcal{S}^n$ is called the simplicial derivative of $\mathbf{f}$ at t.

The right derivative and the left derivative of a sv function at a point, denoted, respectively, $D^{\oplus}_+\mathbf{f}(t)$ and $D^{\oplus}_-\mathbf{f}(t)$ can be also defined as in standard calculus as the one-sided limit in the definition.

Theorem 12.2.6 *Let $\mathbf{f} : I \subseteq \mathbb{R} \to \mathcal{S}^n$ be a function and let $t \in I$ be an interior point of I. If $\mathbf{f}$ is differentiable at $t \in I$, then*

$$D^{\oplus}\mathbf{f}(t) = \mathcal{C} \exp\left(D \, \log(\mathbf{f}(t))\right) = \mathcal{C} \exp\left(\frac{Df_1(t)}{f_1(t)}, \ldots, \frac{Df_n(t)}{f_n(t)} \right). \qquad (12.6)$$

Proof. The incremental ratio in the simplicial derivative can be developed as

$$\frac{1}{h} \odot (\mathbf{f}(t + h) \ominus \mathbf{f}(t)) = \mathcal{C} \left(\ldots, \left(\frac{f_j(t + h)}{f_j(t)} \right)^{1/h}, \ldots \right)$$

$$= \mathcal{C} \exp\left(\ldots, \frac{\log f_j(t + h) - \log f_j(t)}{h}, \ldots \right),$$

and taking the limit,

$$D^{\oplus}\mathbf{f}(t) = \lim_{h \to 0} \left(\frac{1}{h} \odot (\mathbf{f}(t+h) \ominus \mathbf{f}(t)) \right)$$

$$= \mathcal{C} \exp \left(\ldots, \lim_{h \to 0} \frac{\log f_j(t+h) - \log f_j(t)}{h}, \ldots \right)$$

$$= \mathcal{C} \exp \left(\ldots, D \log f_j(t), \ldots \right)$$

$$= \mathcal{C} \exp \left(D \log(\mathbf{f}(t)) \right).$$

Theorem 12.2.7   *Let* $\mathbf{f} : I \subseteq \mathbb{R} \to \mathcal{S}^n$ *be a sv function and let* $t \in I$ *be an interior point of* *I. If* $\mathbf{f}$ *is differentiable at* $t \in I$, *then*

$$D^{\oplus}\mathbf{f}(t) = \mathrm{clr}^{-1} \left(D \ \mathrm{clr}(\mathbf{f}(t)) \right) = \mathrm{ilr}^{-1} \left(D \ \mathrm{ilr}(\mathbf{f}(t)) \right), \tag{12.7}$$

where D *is the ordinary derivative operator for real-valued functions of real variable and it* *is applied componentwise.*

Proof. The first equality follows from the definition of limit combined with the isometric property of clr transformation

$$D^{\oplus}\mathbf{f}(t) = \lim_{h \to 0} \left(\frac{1}{h} \odot (\mathbf{f}(t+h) \ominus \mathbf{f}(t)) \right) = \lim_{h \to 0} \left(\mathrm{clr}^{-1} \left(\frac{1}{h} (\mathrm{clr}(\mathbf{f}(t+h)) - \mathrm{clr}(\mathbf{f}(t))) \right) \right)$$

$$= \mathrm{clr}^{-1} \left(\lim_{h \to 0} \frac{\mathrm{clr}(\mathbf{f}(t+h)) - \mathrm{clr}(\mathbf{f}(t))}{h} \right) = \mathrm{clr}^{-1} \left(D \ \mathrm{clr}(\mathbf{f}(t)) \right).$$

The second equality follows from isometric properties of ilr transformation in a similar way. Isometric properties of clr and ilr are found in Chapter 11.

Proposition 12.2.8 (Properties of differentiability)

(1) Let $\mathbf{f} : I \subseteq \mathbb{R} \to \mathcal{S}^n$ *be a differentiable sv function at* $t_0 \in I$, *an interior point of* I. *Then the sv function is continuous at that point.*

(2) Let $\mathbf{f}, \mathbf{g} : I \subseteq \mathbb{R} \to \mathcal{S}^n$ *be differentiable sv functions at* $t \in I$ *and let* $\lambda \in \mathbb{R}$. *Then:*

$$D^{\oplus}(\mathbf{f} \oplus \mathbf{g})(t) = D^{\oplus}\mathbf{f}(t) \oplus D^{\oplus}\mathbf{g}(t); \quad D^{\oplus}(\lambda \odot \mathbf{f})(t) = \lambda \odot D^{\oplus}\mathbf{f}(t). \tag{12.8}$$

(3) Let $\varphi : I \subseteq \mathbb{R} \to \mathbb{R}$ *a real-valued differentiable function and let* $\mathbf{f} : I \subseteq \mathbb{R} \to \mathcal{S}^n$ *be* *a differentiable sv function at* $t \in I$. *Then,*

$$D^{\oplus}(\varphi \odot \mathbf{f})(t) = (D\varphi(t) \odot \mathbf{f}(t)) \oplus \left(\varphi(t) \odot D^{\oplus}\mathbf{f}(t) \right). \tag{12.9}$$

Proof.

(1) It results from the following computation,

$$\lim_{t \to t_0} (\mathbf{f}(t) \ominus \mathbf{f}(t_0)) = \lim_{t \to t_0} \left((t - t_0) \odot \left(\frac{1}{t - t_0} \odot (\mathbf{f}(t) \ominus \mathbf{f}(t_0)) \right) \right)$$

$$= \left(\lim_{t \to t_0} (t - t_0) \right) \left(\lim_{t \to t_0} \frac{1}{t - t_0} \odot (\mathbf{f}(t) \ominus \mathbf{f}(t_0)) \right)$$

$$= \left(\lim_{t \to t_0} (t - t_0) \right) D^{\oplus} \mathbf{f}(t_0) = \mathbf{n};$$

therefore $\lim_{t \to t_0} \mathbf{f} = \mathbf{f}(t_0)$ and the continuity holds.

(2) The incremental ratio of $D^{\oplus}(\mathbf{f} \oplus \mathbf{g})(t)$ is

$$h^{-1} \odot ((\mathbf{f} \oplus \mathbf{g})(t + h) \ominus (\mathbf{f} \oplus \mathbf{g})(t)) = h^{-1} \odot ((\mathbf{f}(t + h) \ominus \mathbf{f}(t)) \oplus (\mathbf{g}(t + h) \ominus \mathbf{g}(t)))$$

$$= \left(h^{-1} \odot (\mathbf{f}(t + h) \ominus \mathbf{f}(t)) \right) \oplus \left(h^{-1} \odot (\mathbf{g}(t + h) \ominus \mathbf{g}(t)) \right),$$

and after taking limits with $h \to 0$, the first result holds. The second equality is derived likewise.

(3) Applying (12.6) yields

$$D^{\oplus} (\varphi \odot \mathbf{f})(t) = D^{\oplus} \left(\dots, (f_j(t))^{\varphi(t)}, \dots \right) = \mathcal{C} \exp \left(\dots, D \log \left((f_j(t))^{\varphi(t)} \right), \dots \right)$$

$$= \mathcal{C} \exp \left(\dots, D \left(\varphi(t) \log(f_j(t)) \right), \dots \right)$$

$$= \mathcal{C} \exp \left(\dots, D\varphi(t) \log(f_j(t)) + \varphi(t) \frac{Df_j(t)}{f_j(t)}, \dots \right)$$

$$= \mathcal{C} \exp \left(\dots, \log(f_j(t))^{D\varphi(t)}, \dots \right) \oplus \mathcal{C} \exp \left(\dots, \varphi(t) \frac{Df_j(t)}{f_j(t)}, \dots \right)$$

$$= (D\varphi(t) \odot \mathbf{f}(t)) \oplus \left(\varphi(t) \odot D^{\oplus} \mathbf{f}(t) \right).$$

Example 12.2.9

(1) *Derivative of constant functions.* Let $\mathbf{f}(t) = \boldsymbol{\alpha} \in \mathcal{S}^n$ be a constant function; then:

$$D^{\oplus} \mathbf{f}(t) = \mathcal{C} \exp (D \log \boldsymbol{\alpha}) = \mathcal{C} \exp(\mathbf{0}) = \mathbf{n}.$$

(2) *Derivative of polynomial functions.* For a positive integer k, let $\mathbf{p}_k(t) = t^k \odot \boldsymbol{\beta}_k, t \in \mathbb{R}$, $\boldsymbol{\beta}_k \in \mathcal{S}^n$, be a k-degree monomial function. Then,

$$D^{\oplus} \left(t^k \odot \boldsymbol{\beta}_k \right) = \left((Dt^k) \odot \boldsymbol{\beta}_k \right) \oplus \left(t^k \odot D^{\oplus} \boldsymbol{\beta}_k \right)$$

$$= \left(kt^{k-1} \odot \boldsymbol{\beta}_k \right) \oplus \left(t^k \odot D^{\oplus} \boldsymbol{\beta}_k \right) = kt^{k-1} \odot \boldsymbol{\beta}_k.$$

Let $\mathbf{P}_m(t) = \bigoplus_{k=0}^{m} \mathbf{p}_k(t) = \bigoplus_{k=0}^{m}(t^k \odot \boldsymbol{\beta}_k)$ be a m-degree polynomial function. Then,

$$D^{\oplus}\mathbf{P}_m(t) = \bigoplus_{k=1}^{m}(kt^{k-1} \odot \boldsymbol{\beta}_k).$$

(3) *Derivative of exponential functions.* Let

$$\mathbf{f}(t) = \mathcal{C} \exp{(a_1 t, \dots, a_n t)} \in \mathcal{S}^n, \ a_j \in \mathbb{R}, \ j = 1, 2, \dots, n$$

be an exponential function. Then,

$$D^{\oplus}\mathbf{f}(t) = \mathcal{C} \exp{(D \ \log(\exp(a_1 t, \dots, a_n t)))} = \mathcal{C} \exp(a_1, a_2, \dots, a_n).$$

Observe that the derivative of an exponential function at any point $t \in \mathbb{R}$ is constant.

(4) *Derivative of double exponential functions.* Let

$$\mathbf{f}(t) = \mathcal{C} \exp{(\lambda_1 \exp(a_1 t), \dots, \lambda_n \exp(a_n t))} \in \mathcal{S}^n, \ \lambda_j, a_j \in \mathbb{R}, \ j = 1, 2, \dots, n$$

be a double exponential function. Then,

$$D^{\oplus}\mathbf{f}(t) = \mathcal{C} \exp{(D \ \log{(\exp(\lambda_1 \exp(a_1 t), \dots, \lambda_n \exp(a_n t))))}}$$
$$= \mathcal{C} \exp{(D \ (\lambda_1 \exp(a_1 t), \dots, \lambda_n \exp(a_n t)))}$$
$$= \mathcal{C} \exp{(\lambda_1 a_1 \exp(a_1 t), \dots, \lambda_n a_n \exp(a_n t))}.$$

Particularly, if $a_j = a$, $j = 1, \dots, n$, then $D^{\oplus}\mathbf{f}(t) = a \odot \mathbf{f}(t)$, i.e. the double exponential functions with equal a_j's are eigen-functions of the simplicial derivative operator.

(5) The derivative of

$$\mathbf{f}(t) = \mathcal{C}\left(\frac{6 + 5\sin(2t)}{5}, \frac{10}{1+t}, \frac{t^2 - 2t + 5}{4}\right), \ t \in [0, 6] \quad \text{is}$$

$$D^{\oplus}\mathbf{f}(t) = \mathcal{C} \exp\left(\frac{10\cos(2t)}{6 + 5\sin(2t)}, \frac{-1}{1+t}, \frac{2t - 2}{t^2 - 2t + 5}\right), \ t \in (0, 6).$$

Proposition 12.2.10 *Let $\mathbf{f} : [a, b] \subset \mathbb{R} \rightarrow \mathcal{S}^n$ be a sv function differentiable on the open interval (a, b). If $\|D^{\oplus}\mathbf{f}(t)\|_a \leq M$ for every $t \in (a, b)$, $\|D^{\oplus}_{+}\mathbf{f}(a)\|_a \leq M$ and $\|D^{\oplus}_{-}\mathbf{f}(b)\|_a \leq M$, then*

$$\|\mathbf{f}(b) \ominus \mathbf{f}(a)\|_a \leq M(b - a). \tag{12.10}$$

Proof. Condition for simplicial derivative means that for any $\varepsilon > 0$, exists $\delta > 0$ such that if $|h| < \delta$, for any $t \in [a, b)$ we have

$$\left|\frac{1}{h}\right| \cdot \parallel \mathbf{f}(t+h) - \mathbf{f}(t) \parallel_a \leq M,$$

that is

$$\parallel \mathbf{f}(t+h) - \mathbf{f}(t) \parallel_a \leq M \mid h \mid .$$

Taking $t = a, h = b - a$, the statement holds.

Proposition 12.2.11 (Chain rule)   *Let $\varphi : I \subseteq \mathbb{R} \to \mathbb{R}$ be differentiable at $t \in I$ and let $\mathbf{f} : \varphi(I) \subseteq \mathbb{R} \to \mathcal{S}^n$ be a differentiable sv function at $\varphi(t) \in \varphi(I)$. Then $\mathbf{f} \circ \varphi : I \to \mathcal{S}^n$ is a sv function differentiable at t and*

$$D^{\oplus}(\mathbf{f} \circ \varphi)(t) = (D\varphi(t)) \odot D^{\oplus}\mathbf{f}(\varphi(t)). \tag{12.11}$$

Proof. Applying (12.6) yields

$$D^{\oplus}(\mathbf{f} \circ \varphi)(t) = \mathcal{C} \exp(D \, \log(\mathbf{f}(\varphi(t)))) = \mathcal{C} \exp\left(\ldots, D\log(f_j(\varphi(t))), \ldots\right)$$

$$= \mathcal{C} \exp\left(\ldots, \frac{Df_j(\varphi(t))D\varphi(t)}{f_j(\varphi(t))}, \ldots\right)$$

$$= \mathcal{C} \exp\left(\ldots, \log\left(\left(\exp\frac{Df_j(\varphi(t))}{f_j(\varphi(t))}\right)^{D\varphi(t)}\right), \ldots\right)$$

$$= (D\varphi(t)) \odot D^{\oplus}\mathbf{f}(\varphi(t)).$$

Definition 12.2.12 (Differentiability on an open interval. Derivative function)   *Let $\mathbf{f} : I \subseteq \mathbb{R} \to \mathcal{S}^n$ be a sv function and let $J \subseteq I$ be an open interval. The sv function $\mathbf{f}$ is differentiable on the open interval J if, and only if, it is differentiable at each point $t \in J$. The sv function $D^{\oplus}\mathbf{f} : J \subseteq \mathbb{R} \to \mathcal{S}^n$ that assigns the simplicial derivative $D^{\oplus}\mathbf{f}(t)$ to every point $t \in J$ is called the simplicial derivative function of $\mathbf{f}$. The operator $D^{\oplus}$ that assigns the simplicial derivative to any sv function is called the simplicial derivative operator.*

The simplicial derivative operator $D^{\oplus}$ can be applied to a positive-component vector-valued function, because Equation (12.6) is valid since the components of that function are positive. The main result about simplicial derivative is stated and proven in the following theorem.

Theorem 12.2.13   *Let $\mathbf{F} : I \subseteq \mathbb{R} \to \mathbb{R}_+^n$ be a positive-component vector-valued function and let $t \in I$ be an interior point of I. If $\mathbf{F}$ is differentiable at $t \in I$ (ordinary sense) then*

$$D^{\oplus}\mathcal{C}\mathbf{F}(t) = \mathcal{C}D^{\oplus}\mathbf{F}(t) = D^{\oplus}\mathbf{F}(t). \tag{12.12}$$

Proof. On account of (12.6) we have

$$D^{\oplus}\mathbf{F}(t) = \mathcal{C}\exp\left(D\log\mathbf{F}(t)\right) = \mathcal{C}\exp\left(\dots, D\left(\log F_j(t)\right), \dots\right)$$
$$= \mathcal{C}\exp\left(\dots, \frac{DF_j(t)}{F_j(t)}, \dots\right).$$

On the other hand,

$$D^{\oplus}\mathcal{C}\mathbf{F}(t) = D^{\oplus}\left(\dots, \frac{F_j(t)}{\sum_i F_i(t)}, \dots\right) = \mathcal{C}\exp\left(\dots, D\log\frac{F_j(t)}{\sum_i F_i(t)}, \dots\right)$$
$$= \mathcal{C}\exp\left(\dots, D\log F_j(t) - D\log\left(\sum_i F_i(t)\right), \dots\right)$$
$$= \mathcal{C}\left(\dots, \exp\left(D\log F_j(t)\right)\exp\left(-D\log\left(\sum_i F_i(t)\right)\right), \dots\right)$$
$$= \mathcal{C}\exp\left(\dots, \frac{DF_j(t)}{F_j(t)}, \dots\right),$$

because the common factor $\exp\left(-D\log\left(\sum_i F_i(t)\right)\right)$ is included in the closure operation.

Theorem 12.2.13 is very useful in practice because it avoids using the closure operator before derivatives. For instance, when computing the derivative of the function presented in Example 1 (5) (see also Figure 12.1), one can compute the log-derivatives of the components before dividing by an awkward denominator made of the sum of all functional components. This was the method of computing the derivative in Example 12.2.9 (5).

Definition 12.2.14 (Tangent line at a point. Linearization)   *Let* $\mathbf{f} : I \subseteq \mathbb{R} \to \mathcal{S}^n$ *be a sv function, differentiable at* $t_0 \in I$. *The tangent line to the curve* $\mathbf{x} = \mathbf{f}(t)$ *at the point* t_0 *is the sv function*

$$\mathbf{r}(t) = \mathbf{f}(t_0) \oplus \left((t - t_0) \odot D^{\oplus}\mathbf{f}(t_0)\right), \ t \in I. \tag{12.13}$$

In real analysis, a differentiable function can be approximated in a neighbourhood of a point by the tangent function at that point. For sv functions, this fact is expressed by the following property: for $\mathbf{f}$ differentiable at t_0, there exists an open interval $I(t_0) = (t_0 - \delta, t_0 + \delta)$ and a constant $K(t_0)$ such that for all $t \in I(t_0)$

$$\|\mathbf{f}(t) \ominus \mathbf{r}(t)\|_a = \|\mathbf{f}(t) \ominus \left(\mathbf{f}(t_0) \oplus (t - t_0) \odot D^{\oplus}\mathbf{f}(t_0)\right)\|_a \le K(t_0)(t - t_0)^2.$$

The approximation $\mathbf{r}(t)$ is called the linearization of $\mathbf{f}$ at t_0.

Example 12.2.15   Consider the sv function defined in Example 12.2.9 (5),

$$\mathbf{f}(t) = \mathcal{C}\left(\frac{6 + 5\sin(2t)}{5}, \frac{10}{1+t}, \frac{t^2 - 2t + 5}{4}\right), \ t \in [0, 6].$$

The tangent line at point $t_0 = 2$ is

$$\mathbf{r}(t) = \mathbf{f}(2) \oplus \left((t - 2) \odot D^{\oplus}\mathbf{f}(2)\right)$$

$$= \left(\frac{6 + 5\sin(4)}{5}, \frac{10}{3}, \frac{5}{4}\right) \oplus \left((t - 2) \odot \exp\left(\frac{10\cos(4)}{6 + 5\sin(4)}, \frac{-1}{3}, \frac{2}{5}\right)\right)$$

$$= \left(\frac{6 + 5\sin(4)}{5}, \frac{10}{3}, \frac{5}{4}\right) \oplus \left(e^{\frac{10\cos(4)}{6+5\sin(4)}(t-2)}, e^{\frac{-(t-2)}{3}}, e^{\frac{2(t-2)}{5}}\right)$$

Figure 12.2 shows tangent lines at $t_0 = 2$ and at $t_0 = 4.7$ represented by parts. Some non-intuitive features are observed. However, the tangent line represented in coordinates is a straight-line tangent in the ordinary sense to the orbit of the sv function. Furthermore, tangent lines appear curved in this representation, they do not attain negative values and the sum of all their parts add to 1.

12.2.3   Higher order derivatives

Definition 12.2.16 (Higher order derivatives)   *Let $\mathbf{f} : I \subseteq \mathbb{R} \to \mathcal{S}^n$ be a sv function. The function $\mathbf{f}$ is k-differentiable on the open interval $J \subseteq I$ if, for $k \geq 2$, the following*

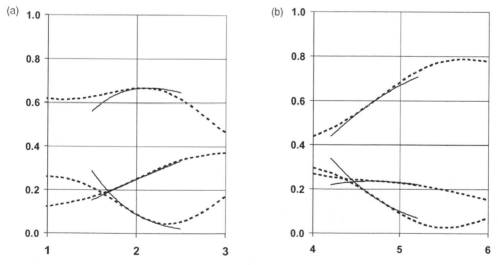

Figure 12.2   Example 12.2.15 sv function (dotted), also plotted in Figure 12.1, and its tangent line (solid) at $t_0 = 2$ (a) and at $t_1 = 4.7$ (b) represented by parts.

limit

$$D^{\oplus k}\mathbf{f}(t) = \lim_{h \to 0} \left(\frac{1}{h} \odot \left(D^{\oplus(k-1)}\mathbf{f}(t+h) \ominus D^{\oplus(k-1)}\mathbf{f}(t) \right) \right),$$

exists. This limit is called the k-order simplicial derivative or the k-simplicial derivative of $\mathbf{f}$ *at t.*

The function $\mathbf{f}$ *is k-differentiable on the open interval* $J \subseteq I$ *if the k-simplicial derivative* $\mathbf{f}$ *exists at any point of J. The sv function defined in J by* $D^{\oplus k}\mathbf{f}(t)$, $t \in J$ *is called the k-simplicial derivative function or the k-order simplicial derivative of* $\mathbf{f}$.

Proposition 12.2.17 (Computing higher order derivatives)   *Let* $\mathbf{f} : I \subseteq \mathbb{R} \to \mathcal{S}^n$ *be a k-differentiable sv function on the open interval* $J \subseteq I$. *The k-order simplicial derivative of* $\mathbf{f}$ *at* $t \in J$ *is*

$$D^{\oplus k}\mathbf{f}(t) = \mathcal{C}\left(\exp\left(D^k \, \log(\mathbf{f}(t)) \right) \right), \quad k = 1, 2, \dots. \tag{12.14}$$

Proof. Reasoning by induction.

Definition 12.2.18 (Taylor polynomial)   *Let* $\mathbf{f} : I \subseteq \mathbb{R} \to \mathcal{S}^n$ *be a m-differentiable sv function at interior point* t_0. *The m-degree Taylor polynomial of* $\mathbf{f}$ *at* t_0 *is*

$$\mathbf{T}_m(\mathbf{f}, t_0)(t) = \mathbf{f}(t_0) \bigoplus_{k=1}^{m} \left(\left(\frac{(t-t_0)^k}{k!} \right) \odot D^{\oplus k}\mathbf{f}(t_0) \right), \quad t \in \mathbb{R}. \tag{12.15}$$

The simplicial difference

$$\mathbf{R}_m(\mathbf{f}, t_0)(t) = \mathbf{f}(t) \ominus \mathbf{T}_m(\mathbf{f}, t_0)(t) \tag{12.16}$$

is called the remainder of the m-degree Taylor polynomial of $\mathbf{f}$ *at* t_0.

A very important property of the remainder is stated in the following theorem.

Theorem 12.2.19 (Taylor's theorem)   *Let* $\mathbf{f} : (a, b) \subset \mathbb{R} \to \mathcal{S}^n$ *be a sv function, m-differentiable on* (a, b) *and let* $\mathbf{T}_m(\mathbf{f}, t_0)(t)$ *the m-degree Taylor polynomial of* $\mathbf{f}$ *at* $t_0 \in (a, b)$. *The residual of the m-degree Taylor polynomial of* $\mathbf{f}$ *at* t_0 *verifies*

$$\lim_{t \to t_0} \left(\frac{1}{|t - t_0|^m} \odot \mathbf{R}_m(\mathbf{f}, t_0)(t) \right) = \mathbf{n}. \tag{12.17}$$

12.3   Integration

12.3.1   Antiderivatives. Indefinite integral

Definition 12.3.1 (Antiderivative of a continuous function)   *Let* $\mathbf{f} : I \subseteq \mathbb{R} \to \mathcal{S}^n$ *be a continuous sv function. A differentiable sv function* $\mathbf{F} : I \subseteq \mathbb{R} \to \mathcal{S}^n$ *is an antiderivative of* $\mathbf{f}$ *on* I, *if, and only if,* $D^{\oplus}\mathbf{F}(t) = \mathbf{f}(t), t \in I$.

As a consequence of this definition, if $\mathbf{F}_1, \mathbf{F}_2$ are antiderivatives of $\mathbf{f}$ on I, then $D^{\oplus}(\mathbf{F}_1(t) \ominus \mathbf{F}_2(t)) = \mathbf{n} \in \mathcal{S}^n$ and any antiderivative of $\mathbf{f}$ has the form $\mathbf{F} \oplus \mathbf{K}$, where $\mathbf{K}$ is a simplicial constant.

Definition 12.3.2 (Indefinite integral of a continuous function)   *Let* $\mathbf{f} : I \subseteq \mathbb{R} \to \mathcal{S}^n$ *be a continuous sv function. The set of antiderivatives*

$$\left\{ \mathbf{F}(t) \oplus \mathbf{K}, \ D^{\oplus}\mathbf{F}(t) = \mathbf{f}(t), \mathbf{K} = \text{constant} \right\},$$

is the indefinite integral of $\mathbf{f}$. *The indefinite integral of a function* $\mathbf{f}$ *is denoted by* $\left(D^{\oplus}\right)^{-1}\mathbf{f}$ *or* $\int^{\oplus}\mathbf{f}$.

Theorem 12.3.3 (Calculus of antiderivatives)   *Let* $\mathbf{f} : I \subseteq \mathbb{R} \to \mathcal{S}^n$ *be a continuous sv function; an antiderivative of* $\mathbf{f}$ *is:*

$$\mathbf{F}(t) = \mathcal{C} \exp\left(D^{-1}\log(\mathbf{f}(t))\right), \ t \in I, \tag{12.18}$$

where D^{-1} *is the antiderivative operator for real-valued functions of real variable, applied componentwise.*

Proof. Applying the definition of antiderivative,

$$D^{\oplus}\mathbf{F}(t) = \mathcal{C} \exp\left(D\left(\log\left(\exp\left(D^{-1}\log(\mathbf{f}(t))\right)\right)\right)\right)$$

$$= \mathcal{C} \exp\left(D\left(D^{-1}\log(\mathbf{f}(t))\right)\right) = \mathcal{C} \exp\left(\log(\mathbf{f}(t))\right) = \mathbf{f}(t)$$

The operator $\left(D^{\oplus}\right)^{-1}$, that assigns an antiderivative to a sv function, $\left(D^{\oplus}\right)^{-1}(\mathbf{f}) = \left(D^{\oplus}\right)^{-1}\mathbf{f}$, is called the simplicial antiderivative operator. Equation (12.18) can be rewritten in a more familiar form using integral notation of antiderivatives

$$\int^{\oplus}\mathbf{f} = \mathcal{C} \exp\left(\int \log(\mathbf{f})\right).$$

This expression was proposed in Aitchison *et al.* (2002) and Egozcue *et al.* (2008) as a definition of simplicial integrals. The pair of transformations (exp, log) can be replaced by $(\text{clr}^{-1}, \text{clr})$ or $(\text{ilr}^{-1}, \text{ilr})$ as shown in Theorems 12.2.6 and 12.2.7 for derivatives. The antiderivative operator $(D^{\oplus})^{-1}$ can be applied to positive vector-valued functions, i.e. before applying the operator C assigning its representative in S^n. The main result for the simplicial antiderivative is stated and proven in the following theorem.

Theorem 12.3.4 *Let $\Phi : I \subseteq \mathbb{R} \to \mathbb{R}^n$ be a continuous positive-component vector-valued function. Then,*

$$\left(D^{\oplus}\right)^{-1} C\Phi(t) = C \left(D^{\oplus}\right)^{-1} \Phi(t) = \left(D^{\oplus}\right)^{-1} \Phi(t). \tag{12.19}$$

Proof. We denote $\Phi(t) = \left(\dots, \Phi_j(t), \dots\right)$, $D^{-1} \log \left(\Phi_j(t)\right) = \Psi_j(t)$, $j = 1, 2, \dots, n$ and $\Psi(t) = \left(\dots, \Psi_j(t), \dots\right)$. From (12.18) we get

$$\left(D^{\oplus}\right)^{-1} \Phi(t) = C \exp \left(D^{-1} \log \Phi(t)\right) = C \exp \left(\dots, D^{-1}\left(\log \Phi_j(t)\right), \dots\right)$$
$$= C \exp \left(\dots, \Psi_j(t), \dots\right) = C \exp \left(\Psi(t)\right).$$

On the other hand,

$$\left(D^{\oplus}\right)^{-1} C\Phi(t) = \left(D^{\oplus}\right)^{-1} \left(\dots, \frac{\Phi_j(t)}{\sum \Phi_j(t)}, \dots\right) = C \exp \left(\dots, D^{-1} \log \left(\frac{\Phi_j(t)}{\sum \Phi_j(t)}\right), \dots\right)$$
$$= C \exp \left(\dots, D^{-1} \log \left(\Phi_j(t)\right) - D^{-1} \log \left(\sum \Phi_j(t)\right), \dots\right)$$
$$= C \left(\dots, \exp \left(\Psi_j(t)\right) \exp \left(-D^{-1} \log \left(\sum \Phi_j(t)\right)\right), \dots\right)$$
$$= C \exp \left(\dots, \Psi_j(t), \dots\right) = C \exp \left(\Psi(t)\right),$$

because the common factor $\exp \left(-D^{-1} \log \left(\sum \Phi_j(t)\right)\right)$ is included in the closure operation.

12.3.2 Integration of continuous sv functions

Definition 12.3.5 (Integral on a closed interval) *Let $f : I = [a, b] \subset \mathbb{R} \to S^n$ be a continuous sv function. The Newton's integral of f on the closed interval I is*

$$\int_I^{\oplus} (dt \odot f(t)) = \int_{[a,b]}^{\oplus} (dt \odot f(t)) = F(b) \ominus F(a), \tag{12.20}$$

where F is any antiderivative of f.

Proposition 12.3.6 (Basic properties of the integral) *Let f, g be sv functions continuous on $I = [a, b]$ and let $\lambda, \mu \in \mathbb{R}$ be real numbers. The integral (12.20) has the following properties:*

(1) The integral of **f** *does not depend on the chosen antiderivative, i.e., if* $\mathbf{F}_1, \mathbf{F}_2$ *are antiderivatives of* **f** *on I, then*

$$\int_{[a,b]}^{\oplus} (dt \odot \mathbf{f}(t)) = \mathbf{F}_1(b) \ominus \mathbf{F}_1(a) = \mathbf{F}_2(b) \ominus \mathbf{F}_2(a).$$

(2) Linearity.

$$\int_{[a,b]}^{\oplus} (dt \odot ((\lambda \odot \mathbf{f})(t) \oplus (\mu \odot \mathbf{g})(t)))$$

$$= \left(\lambda \odot \int_{[a,b]}^{\oplus} (dt \odot \mathbf{f}(t))\right) \oplus \left(\mu \odot \int_{[a,b]}^{\oplus} (dt \odot \mathbf{g}(t))\right).$$

(3) Any integrable sv function on I is also integrable on any closed subinterval $J \subseteq I$.

(4) For any c such that $a < c < b$

$$\int_{[a,b]}^{\oplus} (dt \odot \mathbf{f}(t)) = \left(\int_{[a,c]}^{\oplus} (dt \odot \mathbf{f}(t))\right) \oplus \left(\int_{[c,b]}^{\oplus} (dt \odot \mathbf{f}(t))\right).$$

(5) Norm inequality.

$$\left\|\int_{[a,b]}^{\oplus} (dt \odot \mathbf{f}(t))\right\|_a \le \int_{[a,b]} \|\mathbf{f}(t)\|_a \, dt.$$

(5) Normalization property. Let **p** *be a simplicial constant in* $\mathcal{S}^n$, *and let* $[a, b]$ *be a closed interval of* $\mathbb{R}$. *Then,*

$$\int_{[a,b]}^{\oplus} (dt \odot \mathbf{p}) = (b - a) \odot \mathbf{p}.$$

After Proposition 12.3.6 (6), the integral of a piecewise constant sv function, **f**, on $[a, b]$ is readily written

$$\int_{[a,b]}^{\oplus} (dt \odot \mathbf{f}(t)) = \sum_{i=1}^{k} ((a_i - a_{i-1}) \odot \mathbf{f}(t_i)), \qquad (12.21)$$

where $a_0, a_1, \ldots, a_k$, $a = a_0$, $b = a_k$ are the discontinuity points of **f** in $[a, b]$ and t_i is an interior point in $[a_{i-1}, a_i]$. When a continuous sv function is approximated by a step function, their respective integrals approach each other. Therefore, the right-hand side of (12.21) is the form of Riemann sums for the simplicial definite integral.

Proposition 12.3.7 (Mean value theorem)   *If **f** is a sv function continuous on $[a, b]$, then there exists a composition $\mathbf{m} \in \mathcal{S}^n$, called the simplicial mean value of **f** on $[a, b]$, such that*

$$\mathbf{m} = \left(\frac{1}{b-a}\right) \odot \left(\int_{[a,b]}^{\oplus} (dt \odot \mathbf{f}(t))\right). \tag{12.22}$$

Proof. If **F** is an antiderivative of **f** then

$$\int_{[a,b]}^{\oplus} (dt \odot \mathbf{f}(t)) = \mathbf{F}(b) \ominus \mathbf{F}(a).$$

Setting $\mathbf{F}(b) \ominus \mathbf{F}(a) = (b-a) \odot \mathbf{m}$ the statement holds.

Theorem 12.3.8 (Fundamental theorem of calculus)   *Let $\mathbf{f} : [a, b] \subset \mathbb{R} \to \mathcal{S}^n$ be a sv function continuous on $[a, b]$. The function $\mathbf{F}(t) = \int_{[a,t]}^{\oplus} (ds \odot \mathbf{f}(s)), a < t \leq b$ with $\mathbf{F}(a) = \mathbf{n}$ is differentiable on (a, b) and $D^{\oplus}\mathbf{F}(t) = \mathbf{f}(t), t \in (a, b)$.*

Proof. Since **f** is continuous on $[a, b]$, for any $t_0 \in [a, b]$ and given $\varepsilon > 0$, there exists $\delta > 0$ such that, for $t \in [a, b]$ with $t > t_0$ and $|t - t_0| < \delta$, we have $\|\mathbf{f}(t) \ominus \mathbf{f}(t_0)\|_a < \varepsilon$. Then,

$$\left\|\left(\left(\frac{1}{t - t_0}\right) \odot (\mathbf{F}(t) \ominus \mathbf{F}(t_0))\right) \ominus \mathbf{f}(t_0)\right\|_a$$

$$= \left(\frac{1}{t - t_0}\right) \left\|\int_{[t_0,t]}^{\oplus} (ds \odot \mathbf{f}(s)) \ominus \int_{[t_0,t]}^{\oplus} (ds \odot \mathbf{f}(t_0))\right\|_a$$

$$\leq \left(\frac{1}{t - t_0}\right) \int_{t_0}^{t} \|\mathbf{f}(s) \ominus \mathbf{f}(t_0)\|_a \, ds < \varepsilon.$$

12.4   Conclusions

The calculus of sv functions of a real variable is analogous to the standard calculus of vector-valued real functions due to the isometry between the real Euclidean space and the simplex. However, the operations in the simplex (perturbation and powering) determine how changes of a sv function are measured, thus leading to new concepts of derivative and integral. These concepts become relevant in practice since many phenomena are described as simplicial functions. Moreover, the expressions of standard theorems in simplicial notation may become unfamiliar, and this development can be useful for a scientist involved in compositional analysis.

A key point is that the simplicial derivative and integral can be applied to vector-valued functions with positive components and then these operators commute with the closure which assigns a representative in the simplex (Theorems 12.2.13 and 12.3.4). This is true even if the closure constant depends on the argument of the function.

Finally, the expression of simplicial derivatives and integrals and its elementary properties in simplicial notation proves that these results are invariant under changes of representation of compositions.

Acknowledgements

This research has been supported by the Spanish Ministry of Science and Innovation (projects CSD2006-00032 and MTM2009-13272) and by the Agència de Gestió d'Ajuts Universitaris i de Recerca of the Generalitat de Catalunya (Ref. 2009SGR424).

References

Aitchison J 1986 *The Statistical Analysis of Compositional Data*. Monographs on Statistics and Applied Probability. Chapman and Hall Ltd (reprinted 2003 with additional material by The Blackburn Press), London (UK). 416 p.

Aitchison J, Barceló-Vidal C, Egozcue JJ and Pawlowsky-Glahn V 2002 A concise guide for the algebraic-geometric structure of the simplex, the sample space for compositional data analysis. In *Proceedings of IAMG'02 – The VIII Annual Conference of the International Association for Mathematical Geology* (ed. Bayer U, Burger H and Skala W), vol. I and II. Selbstverlag der Alfred-Wegener-Stiftung, Berlin (Germany). pp. 387–392.

Apostol TM 1957 *Mathematical Analysis*. Addison-Wesley, Reading, MA (USA). 553 p.

Bartle RG 1976 *The Elements of Real Analysis*, 2nd edition. John Wiley & Sons, Ltd, New York, NY (USA). 496 p.

Egozcue JJ and Pawlowsky-Glahn V 2006 Simplicial geometry for compositional data *Compositional Data Analysis in the Geosciences: From Theory to Practice*. Geological Society, London (UK) pp. 145–159.

Egozcue JJ, Pawlowsky-Glahn V and Díaz Barrero JL 2008 Otros espacios euclídeos. *La Gaceta de la Real Sociedad Matemática Española* 11(2), 263–267.

Egozcue JJ, Pawlowsky-Glahn V, Mateu-Figueras G and Barceló-Vidal C 2003 Isometric logratio transformations for compositional data analysis. *Mathematical Geology* 35(3), 279–300.

Estep D 2002 *Practical Analysis in One Variable*. Springer-Verlag, New York, NY (USA). 621 p.

Kartachev A and Rojdestvenski B 1988 *Analyse Mathématique*. (French trans. from Russian). Editions Mir Moscou. 472 p.

Kolmogorov AN and Fomin SV 1957 *Elements of the Theory of Functions and Functional Analysis, Vol. I and II*. Graylock Press (reprinted 1999 by Dover Publications), Rochester, NY (USA). 127 p.

Martín-Fernández JA, Barceló-Vidal C and Pawlowsky-Glahn V 2003 Dealing with zeros and missing values in compositional data sets using nonparametric imputation. *Mathematical Geology* 35(3), 253–278.

Pawlowsky-Glahn V and Egozcue JJ 2001 Geometric approach to statistical analysis on the simplex. *Stochastic Environmental Research and Risk Assessment (SERRA)* 15(5), 384–398.

Rudin W 1964 *Principles of Mathematical Analysis*. McGraw-Hill, New York (USA). 370 p.

13

Compositional differential calculus on the simplex

Carles Barceló-Vidal, Josep Antoni Martín-Fernández and Glòria Mateu-Figueras

Department of Computer Science and Applied Mathematics, University of Girona, Spain

13.1 Introduction

In some studies, the analyst assumes that the response variable y depends on the proportions $x_1, \ldots, x_n$ of n ingredients or components present in a specific mixture and not on its total amount. These proportions are often expressed by volume, weight, mole fraction, etc. In mathematical terms, the response variable y is a real or vector-valued function φ whose domain is a subset of the simplex space. In many practical situations, the expression $y = \varphi(x_1, \ldots, x_n)$ is unknown and the emphasis is on fitting the simplest model to the experimental data. In other scenarios, the function φ can be deduced from physical laws or from other sources. In such cases, the standard interpretation of function φ from differential calculus is not directly applicable because of the constraint $x_1 + \cdots + x_n = 1$ of its components. For example, the partial derivative $(\partial \varphi / \partial x_i)(\mathbf{x})$ of φ at $\mathbf{x}$ with respect to x_i cannot be interpreted as the variation of the function φ with respect to the change in x_i when the other components are held constant because any change in x_i requires changing at least one of the other components.

Basic mathematical concepts, first introduced in Barceló-Vidal and Martín-Fernández (2002), are here extended. In the following sections, we develop the main topics of compositional differential calculus of vector-valued functions without delving into mathematical aspects related to the problems of existence and uniqueness. For succinctness, most proofs of

Compositional Data Analysis: Theory and Applications, First Edition. Edited by Vera Pawlowsky-Glahn and Antonella Buccianti.
© 2011 John Wiley & Sons, Ltd. Published 2011 by John Wiley & Sons, Ltd.

propositions have been omitted, but for some of them brief suggestions have been included. We explain how the introduction of the new concept of *compositional derivative* of a function defined on the simplex, together with the use of simplicial operators, allows us to handle the differential calculus of these functions by analogy to standard differential calculus. The topics of differential calculus introduced here are an adaptation of classical concepts that can be found in any standard textbook on multivariate differential calculus, e.g. Marsden and Tromba (1996). For more details about the terminology and notation associated with the metric vector space structure of the simplex $\mathcal{S}^n$, see Chapter 11.

13.2   Vector-valued functions on the simplex

After Aitchison (1986), it is well known that one of the most important principles in compositional data analysis – hereafter, CODA – is the scale-invariant principle. This principle is due to the assumption that in CODA, the total amount of the composition is irrelevant to the analysis. In this section, we show that vector-valued functions on $\mathcal{S}^n$ are uniquely related to a special type of vector-valued function on $\mathbb{R}^n_+$.

13.2.1   Scale-invariant vector-valued functions on $\mathbb{R}^n_+$

Let f be a vector-valued function whose domain is a subset U of $\mathbb{R}^n_+$ and with range contained in $\mathbb{R}^m$. That is, f assigns to each $\mathbf{w}$ in U a value $f(\mathbf{w}) = (f_1(\mathbf{w}), \ldots, f_m(\mathbf{w}))^\top$, an m-tuple in $\mathbb{R}^m$.

Definition 13.2.1   *We say that* f *is* scale-invariant – *or* homogeneous of degree 0 – *if* $f(k\mathbf{w}) = f(\mathbf{w})$, *for every* $k \in \mathbb{R}^+$, *and* $\mathbf{w} \in U$ *such that* $k\mathbf{w} \in U$.

It is clear that the function components $f_1, \ldots, f_m$ of a scale-invariant vector-valued function f are also real-valued scale-invariant functions. Conversely, if $f_1, \ldots, f_m$ are real-valued scale-invariant functions, then the vector-valued function f also has this property.

Definition 13.2.2   *Let* U *be a subset of* $\mathbb{R}^n_+$. *We define the* scale expansion *of* U, *denoted by* U_+, *as the set*

$$U_+ = \{k\mathbf{w} : \ k > 0, \mathbf{w} \in U\}.$$

A subset $U \subset \mathbb{R}^n_+$ *is* scale-closed *if* $U_+ = U$. *In particular,* U_+ *is a scale-closed set.*

Any scale-invariant vector-valued function f with domain $U \subset \mathbb{R}^n_+$ can be extended to the scale expansion U_+ of U. It suffices to define $f(k\mathbf{w}) = f(\mathbf{w})$, for $k > 0$ and $\mathbf{w} \in U$. This extended function with domain U_+ is also scale-invariant. Therefore, throughout this chapter, we implicitly suppose that the domain of any scale-invariant function is always a scale-closed subset of $\mathbb{R}^n_+$.

13.2.2   Vector-valued functions on $\mathcal{S}^n$

Proposition 13.2.1   *Let $f : U \longrightarrow \mathbb{R}^m$ be a vector-valued function defined on a subset $U \subset \mathbb{R}^n_+$. Let $\underline{U} = \{\mathcal{C}\mathbf{w} : \mathbf{w} \in U\} \subset \mathcal{S}^n$ the compositional closure of U, where $\mathcal{C}$ denotes the closure operator. If f is scale-invariant, then it induces a vector-valued function $\underline{f} : \underline{U} \longrightarrow \mathbb{R}^m$. It suffices to define*

$$\underline{f}(\mathbf{x}) = f(\mathbf{w}) \quad (\mathbf{x} \in \underline{U}),$$

where $\mathbf{w}$ is any element in U such that $\mathcal{C}\mathbf{w} = \mathbf{x}$.

Proposition 13.2.2   *Let $\varphi : U \longrightarrow \mathbb{R}^m$ be a vector-valued function defined on $U \subset \mathcal{S}^n$ with values in $\mathbb{R}^m$. Let U_+ be the scale expansion of U. Then φ induces a scale-invariant function f defined on $U_+ \subset \mathbb{R}^n_+$, such that $\underline{f} = \varphi$. It suffices to define*

$$f(\mathbf{w}) = \varphi(\mathcal{C}\mathbf{w}) \quad (\mathbf{w} \in U_+).$$

In consequence, we can suppose that any vector-valued function defined on a subset of $\mathcal{S}^n$ is derived from a scale-invariant vector-valued function defined on a scale-closed set in $\mathbb{R}^n_+$, and vice versa. Hereafter, we symbolise any vector-valued function on $\mathcal{S}^n$ by $\underline{f}$, underlining the scale-invariant f on $\mathbb{R}^n_+$ associated with it.

13.3 $\mathcal{C}$-derivatives on the simplex

In this section, we define the principal topics of compositional differential calculus that are necessary to introduce the concept of the $\mathcal{C}$-derivative of a vector-valued function with domain on the simplex. We begin by recalling some properties satisfied by partial derivatives of scale-invariant functions. These properties are then used to show those of $\mathcal{C}$-derivatives. We end the section with the interpretation of the $\mathcal{C}$-gradient of a $\mathcal{C}$-differentiable real-valued function on the simplex and the characterisation of its critical points.

13.3.1   Derivative of a scale-invariant vector-valued function on $\mathbb{R}^n_+$

It is well known that if a vector-valued function $f = (f_1, \ldots, f_m)^\top$ with domain an open subset U in $\mathbb{R}^n_+$ is differentiable at $\mathbf{w} \in U$, then its derivative is the $m \times n$ matrix

$$\mathbf{D}f(\mathbf{w}) = \begin{bmatrix} \dfrac{\partial f_1}{\partial w_1}(\mathbf{w}) & \cdots & \dfrac{\partial f_1}{\partial w_n}(\mathbf{w}) \\ \vdots & \ddots & \vdots \\ \dfrac{\partial f_m}{\partial w_1}(\mathbf{w}) & \cdots & \dfrac{\partial f_m}{\partial w_n}(\mathbf{w}) \end{bmatrix}.$$

This matrix, known as the *Jacobian matrix* of the partial derivatives of f at $\mathbf{w}$, is associated with the linear function L from $\mathbb{R}^n$ to $\mathbb{R}^m$ defined as $\mathbf{h} \rightarrow L(\mathbf{h}) = \mathbf{D}f(\mathbf{w})\mathbf{h}$. The linear function

L is the only one that satisfies the condition

$$\lim_{\mathbf{h} \to 0} \frac{\| f(\mathbf{w} + \mathbf{h}) - f(\mathbf{w}) - L(\mathbf{h}) \|}{\| \mathbf{h} \|} = 0.$$

In particular, the derivative of a real-valued function f becomes the $1 \times n$ vector

$$\mathbf{D} f(\mathbf{w}) = \left[\frac{\partial f}{\partial w_1}(\mathbf{w}), \dots, \frac{\partial f}{\partial w_n}(\mathbf{w}) \right],$$

called the *gradient* of f at $\mathbf{w}$.

Because a scale-invariant function is a homogeneous function of degree zero, we can apply to this family of functions the results of the well-known *Euler's theorem*. Therefore, if f is scale-invariant and differentiable at $\mathbf{w} \in U$, it holds that

$$\sum_{j=1}^{n} w_j \frac{\partial f_i}{\partial w_j}(\mathbf{w}) = 0 \quad (i = 1, \dots, m). \tag{13.1}$$

Thus, only $m \times (n - 1)$ of the $m \times n$ partial derivatives of the Jacobian matrix of f at $\mathbf{w}$ are really independent, as the remaining m partial derivatives can be obtained from Equation (13.1). Furthermore, f is also differentiable at $k\mathbf{w}$, for each $k > 0$, and

$$\frac{\partial f_i}{\partial w_j}(k\mathbf{w}) = \frac{1}{k} \frac{\partial f_i}{\partial w_j}(\mathbf{w}) \quad (i = 1, \dots, m)\,(j = 1, \dots, n). \tag{13.2}$$

The following result derives from the application of Euler's theorem.

Proposition 13.3.1 *Let $f = (f_1, \dots, f_m)^\top$ be a differentiable vector-valued function with domain an open subset U in $\mathbb{R}_+^n$. If f is scale-invariant, then the functions*

$$\mathbf{w} \to w_k \frac{\partial f_i}{\partial w_j}(\mathbf{w}) \quad (i = 1, \dots, m)\,(j, k = 1, \dots, n) \tag{13.3}$$

are also scale-invariant. Moreover, if f is twice differentiable, then it holds that

$$\sum_{k=1}^{n} w_k \frac{\partial^2 f_i}{\partial w_k \partial w_j}(\mathbf{w}) = -\frac{\partial f_i}{\partial w_j}(\mathbf{w}) \quad (i = 1, \dots, m)\,(j = 1, \dots, n). \tag{13.4}$$

Proof. From (13.2) follows the scale-invariance of functions (13.3). It suffices to apply (13.1) to functions (13.3) to deduce (13.4). □

As in the case of derivatives of first order, it is unnecessary to calculate all of the second-order partial derivatives. Only $\frac{1}{2} \times m \times (n - 1) \times n$ are really independent, as the remaining second-order partial derivatives can be obtained from Equation (13.4) and from the equality of mixed partial derivatives.

13.3.2   Directional C-derivatives

Definition 13.3.2   *Let f be a real-valued function defined on a C-open subset U of S^n. Let $\mathbf{u}$ be a C-unitary vector of S^n. The directional C-derivative of $\underline{f}$ at $\mathbf{x} \in U$ along the direction $\mathbf{u}$, denoted by $\mathbf{D}_{C,\mathbf{u}} \underline{f}(\mathbf{x})$, is defined as*

$$\mathbf{D}_{C,\mathbf{u}} \underline{f}(\mathbf{x}) = \left(\frac{d\underline{f}(\mathbf{x} \oplus (t \odot \mathbf{u}))}{dt} \right)_{t=0} = \lim_{t \to 0} \frac{\underline{f}(\mathbf{x} \oplus (t \odot \mathbf{u})) - \underline{f}(\mathbf{x})}{t}, \tag{13.5}$$

if this limit exists.

Proposition 13.3.3   *Let U be a C-open subset of S^n, and let $\underline{f}$ be a real-valued function on U. If f is differentiable at $\mathbf{x} \in \mathbb{R}^n_+$, then all directional C-derivatives of $\underline{f}$ exist. Moreover, if $\mathbf{u}$ is a C-unitary vector of S^n, the directional C-derivative of $\underline{f}$ at $\mathbf{x}$ along the direction $\mathbf{u}$ is given by*

$$\mathbf{D}_{C,\mathbf{u}} \underline{f}(\mathbf{x}) = \sum_{j=1}^{n} x_j \frac{\partial f}{\partial w_j}(\mathbf{x}) \log u_j. \tag{13.6}$$

Proof.   From Definition 13.3.2, $\mathbf{D}_{C,\mathbf{u}} \underline{f}(\mathbf{x})$ is equal to

$$\lim_{t \to 0} \frac{1}{t} \left(f \left(\frac{x_1 u_1^t}{s(t)}, \ldots, \frac{x_n u_n^t}{s(t)} \right) - \underline{f}(\mathbf{x}) \right), \tag{13.7}$$

where $s(t) = x_1 u_1^t + \cdots + x_n u_n^t$. That is, it is the derivative at $t = 0$ of the function

$$g : t \to f \left(\frac{x_1 u_1^t}{s(t)}, \ldots, \frac{x_n u_n^t}{s(t)} \right).$$

The differentiability of f at $\mathbf{x}$ implies the differentiability of g at $t = 0$. Then it suffices to apply the chain rule to compute the derivative of g at $t = 0$ on account of (13.1).

For $j = 1, \ldots, n$, let

$$\mathbf{e}_j = C(\underbrace{1, \ldots, 1}_{j-1}, \exp(1), \underbrace{1, \ldots, 1}_{n-j})^\top \tag{13.8}$$

be the vector of S^n 'pointing' towards the jth component. Vector $\mathbf{e}_j$ is not C-unitary because

$$\|\mathbf{e}_j\|_a = \left(\frac{n-1}{n} \right)^{1/2} \quad (j = 1, \ldots, n).$$

For $j = 1, \ldots, n$, let

$$\mathbf{u}_j = \left(\frac{n}{n-1} \right)^{1/2} \odot C\mathbf{e}_j \tag{13.9}$$

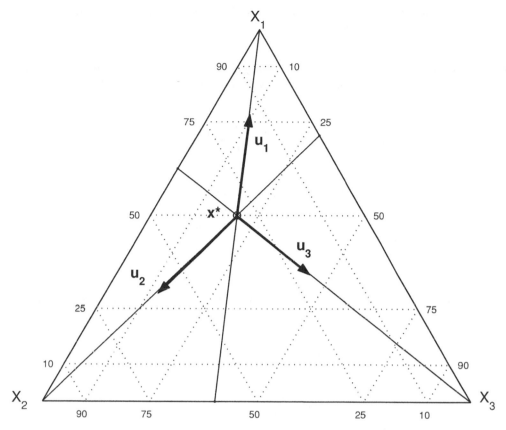

Figure 13.1 The C-unitary composition vectors $\mathbf{u}_1$, $\mathbf{u}_2$ and $\mathbf{u}_3$ in $\mathcal{S}^3$ show the directions associated with the directional C-derivatives $\mathbf{D}_{C,\mathbf{u}_1}\underline{f}(\mathbf{x}^*)$, $\mathbf{D}_{C,\mathbf{u}_2}\underline{f}(\mathbf{x}^*)$ and $\mathbf{D}_{C,\mathbf{u}_3}\underline{f}(\mathbf{x}^*)$ at point $\mathbf{x}^*$ of a real-valued function $\underline{f}$.

be the C-unitary vector of $\mathcal{S}^n$ associated with vector $\mathbf{e}_j$ defined in (13.8). The $\{\mathbf{x} \oplus (t \odot \mathbf{u}_j):$ $t \in \mathbb{R}\}$ is a C-linear path in $\mathcal{S}^n$ through the point $\mathbf{x}$ in the direction given by the vector $\mathbf{u}_j$. According to the definition of the vector $\mathbf{u}_j$, this C-linear path is obtained by applying a multiplicative change to the jth component of $\mathbf{x}$ and holding constant the ratios x_i/x_k of the remaining components (Figure 13.1).

Proposition 13.3.4 *Let U be a C-open subset of $\mathcal{S}^n$, and let $\underline{f}$ be a real-valued function at $\mathbf{x} \in U$. If f is differentiable at $\mathbf{x} \in \mathbb{R}_+^n$ the directional C-derivatives $\mathbf{D}_{C,\mathbf{u}_j}\underline{f}(\mathbf{x})$ $(j = 1, \ldots, n)$ exist and are equal to*

$$\mathbf{D}_{C,\mathbf{u}_j}\underline{f}(\mathbf{x}) = \left(\frac{n}{n-1}\right)^{1/2} x_j \frac{\partial f}{\partial w_j}(\mathbf{x}) \quad (j = 1, \ldots, n). \tag{13.10}$$

Proof. It suffices to apply Proposition 13.3.3 to the direction given by the vector $\mathbf{u}_j$ defined in (13.9).

The part C-derivatives of function $\underline{f}$ can also be computed directly from the ordinary partial derivatives of the function $\underline{f}$.

Proposition 13.3.5   *Let U be a C-open subset of S^n, and let $\underline{f}$ be a real-valued function at $\mathbf{x} \in U$. If $\underline{f}$ is differentiable at $\mathbf{x} \in \mathbb{R}^n_+$, any part C-derivative of $\underline{f}$ exists and is equal to*

$$\frac{\partial_C \underline{f}}{\partial x_j}(\mathbf{x}) = x_j \left(\frac{\partial \underline{f}}{\partial x_j}(\mathbf{x}) - \sum_{i=1}^n x_i \frac{\partial \underline{f}}{\partial x_i}(\mathbf{x}) \right) \quad (j = 1, \ldots, n). \tag{13.11}$$

Proof. The scale-invariant function f associated with $\underline{f}$ is the composition of the closure function $C : \mathbb{R}^n_+ \to S^n$ and the function $\underline{f}$. By applying the chain rule to the composite function $\underline{f} \circ C$, we obtain the partial derivatives of f at $\mathbf{w}$ in terms of ordinary partial derivatives of $\underline{f}$ at $\mathbf{x} = C\mathbf{w}$. That is,

$$\frac{\partial f}{\partial w_j}(\mathbf{w}) = \left(\sum_{i=1}^n w_i \right)^{-2} \left(\left(\sum_{i=1}^n w_i \right) \frac{\partial \underline{f}}{\partial x_j}(C\mathbf{w}) - \sum_{i=1}^n w_i \frac{\partial \underline{f}}{\partial x_i}(C\mathbf{w}) \right) \quad (j = 1, \ldots, n).$$

In particular, the jth partial derivative of f at $\mathbf{x}$ is equal to

$$\frac{\partial f}{\partial w_j}(\mathbf{x}) = \frac{\partial \underline{f}}{\partial x_j}(\mathbf{x}) - \sum_{i=1}^n x_i \frac{\partial \underline{f}}{\partial x_i}(\mathbf{x}) \quad (j = 1, \ldots, n). \tag{13.12}$$

To obtain (13.11), it suffices to substitute this value of $(\partial f / \partial w_j)(\mathbf{x})$ into (13.10). $\blacksquare$

13.3.3 C-derivative

Definition 13.3.6   *Let $\underline{f} = (\underline{f}_1, \ldots, \underline{f}_m)^\top$ be a vector-valued function defined on a C-open subset $U \subset S^n$. The function $\underline{f}$ is C-differentiable at $\mathbf{x} \in U$ if there exists a $m \times p$ matrix $\mathbf{A} = (a_{ij})$, satisfying $\mathbf{A}\mathbf{1}_n = \mathbf{0}_m$, such that*

$$\lim_{\mathbf{u} \xrightarrow{C} \mathbf{n}} \frac{\| \underline{f}(\mathbf{x} \oplus \mathbf{u}) - \underline{f}(\mathbf{x}) - \mathbf{A} \log \mathbf{u} \|}{\| \mathbf{u} \|_a} = 0,$$

where $\mathbf{n} = C(1, \ldots, 1)^\top$ is the neutral element of $(S^n, \oplus)$. The matrix $\mathbf{A}$ is the C-derivative of $\underline{f}$ at $\mathbf{x}$ and will be denoted by $\mathbf{D}_C \underline{f}(\mathbf{x})$. The (i, j)-entry of $\mathbf{D}_C \underline{f}(\mathbf{x})$ will be symbolised by $\frac{\partial_C \underline{f}_i}{\partial x_j}$, the jth part C-derivative of $\underline{f}_i$ at $\mathbf{x}$, and the matrix $\mathbf{D}_C \underline{f}(\mathbf{x})$ is the Jacobian matrix of part C-derivatives of $\underline{f}$ at $\mathbf{x}$.

Proposition 13.3.7   *If f is differentiable at* **x** *then* $\underline{f}$ *is C-differentiable at* **x** *and the Jacobian matrix* $\mathbf{D}_C \underline{f}(\mathbf{x})$ *is equal to*

$$
\mathbf{D}_C \underline{f}(\mathbf{x}) = \begin{bmatrix} x_1 \dfrac{\partial f_1}{\partial w_1}(\mathbf{x}) & \cdots & x_n \dfrac{\partial f_1}{\partial w_n}(\mathbf{x}) \\ \vdots & \ddots & \vdots \\ x_1 \dfrac{\partial f_m}{\partial w_1}(\mathbf{x}) & \cdots & x_n \dfrac{\partial f_m}{\partial w_n}(\mathbf{x}) \end{bmatrix}.
$$

In consequence, the Jacobian matrix $\mathbf{D}_C \underline{f}(\mathbf{x})$ is singular, and therefore only $n-1$ of the n part C-derivatives of $\underline{f_i}$ $(i = 1, \ldots, m)$ are independent. This matrix is associated with the linear function L_C from the vector space $(\mathcal{S}^n, \oplus, \odot)$ to $\mathbb{R}^m$ defined as $\mathbf{u} \to L_C(\mathbf{u}) = \mathbf{D}_C \underline{f}(\mathbf{x}) \log \mathbf{u}$. As in standard differential calculus, L_C is the only C-linear function that satisfies the condition

$$
\lim_{\substack{\mathbf{u} \xrightarrow{C} \mathbf{n}}} \frac{\| \underline{f}(\mathbf{x} \oplus \mathbf{u}) - \underline{f}(\mathbf{x}) - L_C(\mathbf{u}) \|}{\|\mathbf{u}\|_a} = 0.
$$

In the Proposition 13.3.8, we show the expression of the Jacobian matrix when the compositions are referred to a C-orthonormal basis in $\mathcal{S}^n$.

Proposition 13.3.8   *Let* $\underline{f} = (\underline{f_1}, \ldots, \underline{f_m})^\top$ *be a vector-valued function defined on a C-open subset* $U \subset \mathcal{S}^n$. *Let* **V** *be a* $n \times (n-1)$ *basis-contrast matrix associated with a C-orthonormal basis in* $\mathcal{S}^n$, *and let* $\mathrm{ilr}_\mathbf{V}$ *be the C-isometric log-ratio transformation associated with this basis. If the function* $\underline{f}$ *is C-differentiable at* $\mathbf{x} \in U$, *then the composite function* $\underline{f} \circ \mathrm{ilr}_\mathbf{V}^{-1}$ *from* $\mathbb{R}^{n-1}$ *to* $\mathbb{R}^m$ *is differentiable, and its Jacobian matrix is equal to* $\mathbf{D}_C \underline{f}(\mathbf{x})\mathbf{V}$.

Proof.   This result follows from the properties of linear functions detailed in Chapter 11.

When all of the part C-derivatives of $\underline{f}$ are defined at all points of a C-open subset $U \subset \mathcal{S}^n$, then each part C-derivative $\partial_C f_i / \partial x_j$ can be interpreted as a function $\mathbf{x} \to (\partial_C f_i / \partial x_j)(\mathbf{x})$, from U with values in $\mathbb{R}^m$. As in standard differential calculus, the following result holds.

Proposition 13.3.9   *Let* $\underline{f} = (\underline{f_1}, \ldots, \underline{f_m})^\top$ *be a vector-valued function defined on a C-open subset* $U \subset \mathcal{S}^n$. *Assume that all part C-derivatives* $\partial_C f_i / \partial x_j$ $(i = 1, \ldots, m)$ $(j = 1, \ldots, n)$ *exist and are C-continuous at any* $\mathbf{x} \in U$. *Then the function* $\underline{f}$ *is C-differentiable at any* $\mathbf{x} \in U$.

From Definition 13.3.6 immediately follows the *first-order Taylor approximation* of a real-valued function $\underline{f}$ in a neighbourhood of any point in its domain.

Proposition 13.3.10 (Taylor's approximation)   *Let* U *be a C-open subset of* $\mathcal{S}^n$, *and let* $\underline{f}$ *be a C-differentiable real-valued function at* $\mathbf{x} \in U$. *Then it holds that*

$$
\underline{f}(\mathbf{x} \oplus \mathbf{u}) = \underline{f}(\mathbf{x}) + \sum_{j=1}^{n} \log u_j \left(\frac{\partial_C \underline{f}}{\partial x_j}(\mathbf{x}) \right) + R_1(\mathbf{x}; \mathbf{u}),
$$

where $R_1(\mathbf{x}; \mathbf{u})/\|\mathbf{u}\|_a \to 0$ in $\mathbb{R}$, as $\mathbf{u} \overset{\mathcal{C}}{\to} \mathbf{n}$ in $\mathcal{S}^n$, where $\mathbf{n} = \mathcal{C}(1, \ldots, 1)^\top$ is the barycentre of the simplex.

13.3.4 $\mathcal{C}$-gradient

Definition 13.3.11   *Let U be a $\mathcal{C}$-open subset of $\mathcal{S}^n$ and $\underline{f}$ a $\mathcal{C}$-differentiable real-valued function on U. Then the row-vector of part $\mathcal{C}$-derivatives of $\underline{f}$ at $\mathbf{x}$*

$$\mathbf{D}_{\mathcal{C}}\underline{f}(\mathbf{x}) = \left[\frac{\partial_{\mathcal{C}}\underline{f}}{\partial x_1}(\mathbf{x}), \ldots, \frac{\partial_{\mathcal{C}}\underline{f}}{\partial x_n}(\mathbf{x}) \right]$$

is called the $\mathcal{C}$-gradient of $\underline{f}$ at $\mathbf{x}$.

Proposition 13.3.12   *Let U be a $\mathcal{C}$-open subset of $\mathcal{S}^n$ and $\underline{f}$ a $\mathcal{C}$-differentiable real-valued function on U. Then all directional $\mathcal{C}$-derivatives of $\underline{f}$ exist. Moreover, if $\mathbf{u} = (u_1, \ldots, u_n)^\top$ is a $\mathcal{C}$-unitary vector of $\mathcal{S}^n$, the directional $\mathcal{C}$-derivative of $\underline{f}$ at $\mathbf{x}$ along the direction $\mathbf{u}$ is given by*

$$\mathbf{D}_{\mathcal{C}, \mathbf{u}}\underline{f}(\mathbf{x}) = \left(\sum_{j=1}^{n} \frac{\partial_{\mathcal{C}}\underline{f}}{\partial x_j}(\mathbf{x}) \log u_j \right). \tag{13.13}$$

In particular, it holds that

$$\frac{\partial_{\mathcal{C}}\underline{f}}{\partial x_j}(\mathbf{x}) = \left(\frac{n-1}{n} \right)^{1/2} \mathbf{D}_{\mathcal{C}, \mathbf{u}_j}\underline{f}(\mathbf{x}) \quad (j = 1, \ldots, n),$$

where $\mathbf{u}_j$ $(j = 1, \ldots, n)$ are the $\mathcal{C}$-unitary vectors of $\mathcal{S}^n$ defined in (13.9).

Proposition 13.3.13   *Let U be a $\mathcal{C}$-open subset of $\mathcal{S}^n$, and let $\underline{f}$ be a $\mathcal{C}$-differentiable real-valued function on U. If the $\mathcal{C}$-gradient of $\underline{f}$ at $\mathbf{x}$ is different from $\overline{\mathbf{0}}$, then the composition vector $\mathcal{C}\{\exp((\mathbf{D}_{\mathcal{C}}\underline{f}(\mathbf{x}))^\top)\}$ points in the direction of the greatest rate of increase of the function $\underline{f}$ in a $\mathcal{C}$-neighbourhood of $\mathbf{x}$, i.e., the maximum directional $\mathcal{C}$-derivative of $\underline{f}$ at $\mathbf{x}$ is attained along the direction given by the $\mathcal{C}$-unitary vector parallel to $\mathcal{C}\{\exp((\mathbf{D}_{\mathcal{C}}\underline{f}(\mathbf{x}))^\top)\}$.*

Proof.   The proof follows by applying the method of Lagrange multipliers to find the maximum of the function $F(u_1, \ldots, u_n) = \sum_{j=1}^{n} x_j \frac{\partial f}{\partial w_j}(\mathbf{x}) \log u_j$, defined on $\mathbb{R}_+^n$, constrained by $\|\mathbf{u}\|_a^2 = 1$.

13.3.5   Critical points of a $\mathcal{C}$-differentiable real-valued function on $\mathcal{S}^n$

The topology induced in $\mathcal{S}^n$ by the Euclidean metric structure $(\mathcal{S}^n, \oplus, \odot)$ is equivalent to the topology induced by the ordinary Euclidean metric. Therefore, in practice, the distinction between the compositional and ordinary neighbourhood of a composition $\mathbf{x} \in \mathcal{S}^n$ is unnecessary. In consequence, the local extrema of a real-valued function defined on $\mathcal{S}^n$ will be independent of the topology under consideration.

Definition 13.3.14 *Let U be a $\mathcal{C}$-open subset of $\mathcal{S}^n$ and $\underline{f}$ a $\mathcal{C}$-differentiable real-valued function on U. The composition $\mathbf{x} \in U$ is a C-critical point of $\underline{f}$ if the $\mathcal{C}$-gradient at $\mathbf{x}$ is $\mathbf{0}$. That is, $\partial_{\mathcal{C}} \underline{f} / \partial x_j(\mathbf{x}) = 0$, for $j = 1, \ldots, n$.*

Note that, according to Proposition 13.3.4, if $\mathbf{x}$ is a $\mathcal{C}$-critical point of $\underline{f}$, then $k\mathbf{x}$ is also a critical point of function f, for any $k > 0$.

Proposition 13.3.15 *Let U be a $\mathcal{C}$-open subset of $\mathcal{S}^n$, and $\underline{f}$ a $\mathcal{C}$-differentiable real-valued function at $\mathbf{x} \in U$. If $\mathbf{x}$ is a local extremum of $\underline{f}$, then $\mathbf{D}_{\mathcal{C}} \underline{f}(\mathbf{x}) = \mathbf{0}$. That is, $\mathbf{x}$ is a $\mathcal{C}$-critical point of $\underline{f}$.*

Proof. Let $\mathbf{x}$ be a local maximum of $\underline{f}$. For any $\mathbf{u} \in \mathcal{S}^n$, the real function $t \to g(t) = \underline{f}(\mathbf{x} \oplus (t \odot \mathbf{u}))$ is differentiable and has a local maximum at $t = 0$. Therefore, it holds that $\dot{g}(0) = 0$. Then any directional $\mathcal{C}$-derivative of $\underline{f}$ at $\mathbf{x}$ must be 0. Therefore, $\mathbf{D}_{\mathcal{C},\mathbf{u}} \underline{f}(\mathbf{x}) = 0$, for any $\mathbf{u} \in \mathcal{S}^n$. It follows that $\mathbf{D}_{\mathcal{C}} \underline{f}(\mathbf{x}) = \mathbf{0}$. The reasoning is analogous in the case that $\underline{f}$ reaches a local minimum at $\mathbf{x}$.

13.4 Example: experiments with mixtures

In the context of experiments with mixtures (Cornell, 1990), it is necessary to analyse the behaviour of a measured response η depending only on the proportions $x_1, \ldots, x_n$ of the ingredients present in the mixture and not on the total amount of ingredients. In much experimental work involving multicomponent mixtures, the emphasis is on studying their physical characteristics, such as the shape or the highest point of the measured response surface. In these scenarios, one assumes that there exists some functional relationship between η and $x_1, \ldots, x_n$ given by

$$\eta = \varphi(x_1, \ldots, x_n). \tag{13.14}$$

Often, polynomial functions are used for dealing with $\varphi(x_1, \ldots, x_n)$. The reason is that one can expand $\varphi(x_1, \ldots, x_n)$ using Taylor series, and thus a polynomial can be used as an approximation. Most of the polynomials used are of degree one or two. In turn, according to Proposition 13.3.10, the function $\varphi(x_1, \ldots, x_n)$ can also be approximated by first-degree polynomials in $\log x_1, \ldots, \log x_n$. We devote this section to the study of such type of functions.

13.4.1 Polynomial of degree one

Suppose the response η is given by a polynomial of degree one

$$\eta = \varphi(x_1, \ldots, x_n) = \beta_0 + \sum_{i=1}^{n} \beta_i x_i. \tag{13.15}$$

In that case, the response surface is depicted by a hyperplane over the simplex $\mathcal{S}^n$. Note that the coefficients $\beta_1, \ldots, \beta_n$ simply just the standard partial derivatives of φ with respect to components $x_1, \ldots, x_n$. Nevertheless, the amount $\beta_i \times \delta$ cannot be interpreted as usual, as the change of the function φ when the component x_i changes (additively) by an amount δ and the other components remain constant. Only in some particular cases can the standard partial derivatives be interpreted. For example, it is possible to interpret the difference $\beta_i - \beta_j$ ($i \neq j$). Thus, $(\beta_i - \beta_j) \times \delta$ is the change in the response φ when x_i increases (additively) by an amount δ, and simultaneously x_j decreases (additively) by the same amount, while the other components remain constant.

The jth part C-derivative of function (13.15) at point $\mathbf{x}$ becomes equal to

$$\frac{\partial_C \varphi}{\partial x_j}(\mathbf{x}) = \left(\beta_j - \sum_{i=1}^{n} \beta_i x_i \right) x_j, \tag{13.16}$$

for $j = 1, \ldots, n$. Note that the value of $(\partial_C \varphi / \partial x_j)(\mathbf{x})$ is not constant on the simplex because it depends on the point $\mathbf{x}$ where it is evaluated. In particular, at the barycentre $\mathbf{n}$ of the simplex, the C-gradient of the function (13.15) becomes equal to

$$\mathbf{D}_C \varphi(\mathbf{n}) = \frac{1}{n} [\beta_1 - \overline{\beta}, \ldots, \beta_n - \overline{\beta}], \tag{13.17}$$

where $\overline{\beta} = \frac{1}{n} \sum_{i=1}^{n} \beta_i$. When one moves from the barycentre $\mathbf{n}$ by increasing (in a multiplicative form) only the component x_i and preserving the ratios between the other components, the response changes (additively) by an amount proportional to $\beta_i - \overline{\beta}$. This change will be positive, negative or zero depending on whether β_i is greater, less than or equal to $\overline{\beta}$. Finally, starting at point $\mathbf{n}$, we must move along the direction given by $C \left(\exp(\beta_1), \ldots, \exp(\beta_n) \right)^\top$ to get the fastest (additive) change in the response.

13.4.2   Polynomial of degree two

First, we assume that the response η is given by

$$\eta = \varphi(x_1, \ldots, x_n) = \sum_{i=1}^{n} \beta_{ii} x_i^2. \tag{13.18}$$

The jth part C-derivative of this polynomial at point $\mathbf{x}$ is equal to

$$\frac{\partial_C \varphi}{\partial x_j}(\mathbf{x}) = 2 \left(\beta_{jj} x_j - \sum_{i=1}^{n} \beta_{ii} x_i^2 \right) x_j, \quad (j = 1, \ldots, n). \tag{13.19}$$

In particular, the C-gradient of the function (13.18) at the barycentre $\mathbf{n}$ becomes equal to

$$\mathbf{D}_C \varphi(\mathbf{n}) = \frac{2}{n^2} [\beta_{11} - \overline{\overline{\beta}}, \ldots, \beta_{nn} - \overline{\overline{\beta}}], \tag{13.20}$$

where $\overline{\overline{\beta}} = \frac{1}{n} \sum_{i=1}^{n} \beta_{ii}$.

Secondly, we assume that the response η is a monomial

$$\eta = \varphi(x_1, \ldots, x_n) = \beta_{ij} x_i x_j, \tag{13.21}$$

where $i < j$. Then the ith and jth part C-derivatives of this polynomial at point $\mathbf{x}$ are, respectively, equal to

$$\frac{\partial_C \varphi}{\partial x_i}(\mathbf{x}) = \beta_{ij}(1 - 2x_i) x_i x_j,$$

$$\frac{\partial_C \varphi}{\partial x_j}(\mathbf{x}) = \beta_{ij}(1 - 2x_j) x_i x_j,$$

and the kth ($k \neq i, j$) part C-derivative is equal to

$$\frac{\partial_C \varphi}{\partial x_k}(\mathbf{x}) = -2\beta_{ij} x_i x_j x_k. \tag{13.22}$$

Finally, because any polynomial of degree two is the sum of polynomials of types (13.15), (13.18) and (13.21), its C-derivative can be obtained by adding the derivatives in (13.16), (13.19), (13.22) and (13.22) of these elementary polynomials.

Note that although functions (13.15), (13.18) and (13.21) do not explicitly depend on x_k, their kth part C-derivatives are not necessarily equal to zero.

13.4.3 Polynomial of degree one in logarithms

Suppose the response η is given by a first-degree polynomial in $\log x_1, \ldots, \log x_n$. That is,

$$\eta = \varphi(x_1, \ldots, x_n) = \gamma_0 + \sum_{i=1}^{n} \gamma_i \log x_i. \tag{13.23}$$

The jth part C-derivative of this function at point $\mathbf{x}$ is equal to

$$\frac{\partial_C \varphi}{\partial x_j}(\mathbf{x}) = \gamma_j - x_j \sum_{i=1}^{n} \gamma_i, \quad (j = 1, \ldots, n). \tag{13.24}$$

In particular, the C-gradient of the function (13.23) at the barycentre $\mathbf{n}$ becomes equal to

$$\mathbf{D}_C \varphi(\mathbf{n}) = [\gamma_1 - \overline{\gamma}, \ldots, \gamma_n - \overline{\gamma}], \tag{13.25}$$

where $\overline{\gamma} = \frac{1}{n} \sum_{i=1}^{n} \gamma_i$.

Note that if $\sum_{i=1}^{n} \gamma_i = 0$, then the C-gradient of the function (13.23) is constant in the simplex $\mathcal{S}^n$ because

$$\frac{\partial_C \varphi}{\partial x_j}(\mathbf{x}) = \gamma_j, \quad (j = 1, \ldots, n),$$

for any $\mathbf{x} \in \mathcal{S}^n$. This is logical because (13.23) represents a $\mathcal{C}$-affine variety in $\mathcal{S}^n$, when $\sum_{i=1}^{n} \gamma_i = 0$.

13.4.4   A numerical example

The case presented here is based on an example from Cornell (1990): three constituents – polyethylene (x_1), polystyrene (x_2) and polypropylene (x_3) – were blended together and the resulting fibre material was spun to form yarn for draperies. The response of interest is the thread elongation η of the yarn measured in kilograms of force applied. From the data collected in an experimental design, it has been concluded that the response answer can be fitted fairly well by a second-degree polynomial function $\varphi(x_1, x_2, x_3)$. The equation is

$$\eta = \varphi(x_1, x_2, x_3) = 11.7 x_1 + 9.4 x_2 + 16.4 x_3 + 19.0 x_1 x_2 + 11.4 x_1 x_3 - 9.6 x_2 x_3.$$

We assume that the current operating condition is the blend consisting of $x_1 = 0.5$, $x_2 = 0.3$ and $x_3 = 0.2$. We are interested in exploring the thread elongation of the yarn in a neighbourhood of this specific blend $\mathbf{x}^* = (0.5, 0.3, 0.2)^\top$ (Figure 13.2).

The value of function φ at point $\mathbf{x}^*$ is 15.364, and its $\mathcal{C}$-gradient is equal to [0.390, $-0.467, 0.076$]. Because $(\partial_{\mathcal{C}} \varphi / \partial x_1)(\mathbf{x}^*) = 0.390 > 0$, when x_1 increases from the initial value 0.500 and the values of x_2 and x_3 decrease, but the ratio of x_2 and x_3 (0.3/0.2) remains constant, the response η will increase. For example, if we move from $\mathbf{x}^*$ to $\mathbf{x}^{(1)} = (0.550, 0.270, 0.180)^\top$,

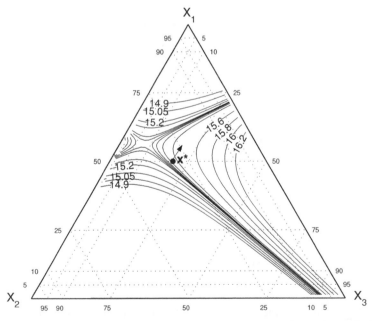

Figure 13.2   Contour lines of the function $\eta = \varphi(x_1, x_2, x_3)$ from Cornell (1990). The plotted vector points at the direction of the composition vector $\mathcal{C}\{\exp((\mathbf{D}_{\mathcal{C}} \varphi(\mathbf{x}^*))^\top)\} = (0.375, 0.285, 0.339)^\top$, where $\mathbf{D}_{\mathcal{C}} \varphi(\mathbf{x}^*) = [0.390, -0.467, 0.076]$ is the $\mathcal{C}$-gradient of φ at $\mathbf{x}^* = (0.5, 0.3, 0.2)^\top$.

increasing by 10% the component x_1, function φ grows from 15.364 to 15.409. Similarly, because $(\partial_C\varphi/\partial x_2)(\mathbf{x}^*) = -0.467 < 0$, when x_2 increases from the initial value 0.3 and the values of x_1 and x_3 decrease, but the ratio $(0.5/0.2)$ of x_1 and x_3 remains constant, the response η will decrease. Indeed, if we increase by 10% the component x_2 moving from $\mathbf{x}^*$ to $\mathbf{x}^{(2)} = (0.479, 0.330, 0.191)^\top$, the function φ decreases from 15.364 to 15.279. The interpretation of $(\partial_C\varphi/\partial x_3)(\mathbf{x}^*)$ would be done in the same way. Finally, to get the fastest increase of the function φ, we have to move from $\mathbf{x}^*$ along the direction given by the composition vector $\mathcal{C}(\exp(0.390), \exp(-0.467), \exp(0.076))^\top = (0.375, 0.285, 0.339)^\top$. For example, if we move in this direction from $\mathbf{x}^*$ to $\mathbf{x}^{(3)} = (0.550, 0.251, 0.199)^\top$, increasing by 10% the component x_1, function φ increases to 15.449.

13.5   Discussion

The constraint $x_1 + \cdots + x_n = 1$ on the components of a function $\eta = \varphi(x_1, \ldots, x_n)$ defined on the simplex $\mathcal{S}^n$ prevents the application of standard differential calculus generally applied to functions without restrictions. Although it is formally possible to calculate the jth partial derivative $\partial\varphi/\partial x_j$ at a point $\mathbf{x}$, it cannot be interpreted as usual. Nevertheless, it would be possible to calculate the derivative along some special directions. For example, we can interpret the difference $(\partial\varphi/\partial x_i)(\mathbf{x}) - (\partial\varphi/\partial x_j)(\mathbf{x})$ of two partial derivatives $(i \neq j)$ because it gives the rate of change of the function φ at $\mathbf{x}$ when x_i increases additively and simultaneously x_j decreases by the same amount as x_i grows, while the other components remain constant. Box and Cox (1971) highlighted the difficulty of interpreting the individual parameters in ordinary polynomial response functions for experiments with mixtures. They proposed to measure the effect of an individual component on the response by the change in value of the response to a change (additive) in the proportion of that component, while holding constant the relative proportions of the other components. That is, they proposed to move (additively) in the direction of the $\mathcal{C}$-unitary vector (13.9) used to define the jth part $\mathcal{C}$-derivative of a real-valued function on the simplex. Cornell (1990) includes and extends that proposal. However, while Box and Cox (1971) analyses the effect of the response to additive changes on a component, we analyse this effect by multiplicative changes, according to the compositional approach to the problem. This difference is crucial because it allows to consistently define and interpret the concept of the $\mathcal{C}$-derivative of a vector-valued function on the simplex.

Usually, in a standard approach, the computation of the local extremum values of a differentiable function $\eta = \varphi(x_1, \ldots, x_n)$ defined on the simplex $\mathcal{S}^n$ is done using the method of Lagrange multipliers. It leads to the resolution of the system of equations

$$\frac{\partial\varphi}{\partial x_1}(\mathbf{x}) = \frac{\partial\varphi}{\partial x_2}(\mathbf{x}) = \cdots = \frac{\partial\varphi}{\partial x_n}(\mathbf{x}), \quad x_1 + \cdots + x_n = 1. \tag{13.26}$$

However, according to Proposition 13, the local extremum values of $\eta = \varphi(x_1, \ldots, x_n)$ are $\mathcal{C}$-critical points of φ. Therefore, from a compositional perspective, any $\mathbf{x} \in \mathcal{S}^n$ where the function φ reaches a local extremum value should be a solution of the equation $\mathbf{D}_\mathcal{C}\varphi(\mathbf{x}) = \mathbf{0}$, as in standard differential calculus. That is, according to Equation (13.11), it must be a solution of the system of equations

$$x_j \left(\frac{\partial \varphi}{\partial x_j}(\mathbf{x}) - \sum_{i=1}^{n} x_i \frac{\partial \varphi}{\partial x_i}(\mathbf{x}) \right) = 0 \quad (j = 1, \ldots, n), \quad x_1 + \cdots + x_n = 1.$$

It is easy to prove that this system of equations is equivalent to (13.26) because our sample space is the open simplex $\mathcal{S}^n$ and, therefore, $x_j > 0$ $(j = 1, \ldots, n)$.

Acknowledgements

This research has been supported by the Spanish Ministry of Science and Innovation (projects CSD2006-00032 and MTM2009-13272) and by the Agència de Gestió d'Ajuts Universitaris i de Recerca of the Generalitat de Catalunya (Ref. 2009SGR424).

References

Aitchison J 1986 *The Statistical Analysis of Compositional Data*. Monographs on Statistics and Applied Probability. Chapman and Hall Ltd (reprinted 2003 with additional material by The Blackburn Press), London (UK). 416 p.

Barceló-Vidal C and Martín-Fernández JA 2002 Differential calculus on the simplex. In *Terra Nostra* (ed. Bayer U, Burger H and Skala W), vol. I and II. Selbstverlag der Alfred-Wegener-Stiftung, Berlin (Germany). pp. 393–399.

Box GEP and Cox DR 1971 A note on polynomial response functions for mixtures. *Biometrika* **58**, 155–159.

Cornell JA 1990 *Experiments with Mixtures: Designs, Models, and the Analysis of Mixture Data*, 2nd edition. John Wiley & Sons , Ltd, New York, NY (USA). 632 p.

Marsden JE and Tromba AJ 1996 *Vector Calculus*, 4th edition. W.H. Freeman and Company, New York, NY (USA). 627 p.

Part IV

APPLICATIONS

14

Proportions, percentages, ppm: do the molecular biosciences treat compositional data right?

David Lovell[1], Warren Müller[1], Jen Taylor[2], Alec Zwart[1] and Chris Helliwell[2]

[1]*CSIRO Mathematics, Informatics and Statistics, Canberra, Australia*
[2]*CSIRO Plant Industry, Canberra, Australia*

14.1 Introduction

Compositional data analysis has its roots in the geosciences where geologists faced a challenge of how to analyse and interpret measurements of the mineral content of rocks – samples would be described in terms of percentages of different components, or in parts per million (or billion) for trace elements.

It can take a long time for deep methodological knowledge developed in one domain to be applied elsewhere. This chapter is motivated by the concern that there is a lot of compositional data in 'the omics' (genomics, transcriptomics, etc.) but little awareness of the pitfalls of ignoring compositional constraints, or even that compositional constraints are at play. We are concerned that molecular biology not be led astray by findings that are more to do with artifacts of the measurement and analysis processes than the biological system being measured. Readers unfamiliar with modern molecular bioscience may wish to consult Lesk (2007).

So how widespread is compositional data in the molecular biosciences? Examples include:

- *Fixed size/volume samples of different components*, e.g. 1 g of tissue (containing different kinds of cells); 1 μg of total RNA (containing different species of RNAs);

Compositional Data Analysis: Theory and Applications, First Edition. Edited by Vera Pawlowsky-Glahn and Antonella Buccianti.
© 2011 John Wiley & Sons, Ltd. Published 2011 by John Wiley & Sons, Ltd.

1 μg of metagenomic DNA (containing DNA from different genomes); 1 ml of blood (containing different metabolites).

- *Constrained counts*, such as counts of different bases or codons in a fixed-length DNA sequence.

- *Proportions*, say, of different k-mers in genomes, gene ontology (GO) terms in samples, or different reads in next-generation sequencing runs.

Compositional data are commonplace in molecular biology; currently, evidence of principled approaches to analysing this kind of data is scarce.

The biosciences can benefit greatly from ground gained in the geosciences, but biology has some important differences to geology as far as compositions are concerned. Molecular biology frequently produces compositions with tens- if not hundreds-of-*thousands* of components, whereas geosciences data usually has much lower dimension (tens to hundreds). That said, Kliebenstein (2009) writes in relation to transcriptomics that 'the majority of nucleic acids in most RNA samples is contained in fewer than 40 transcripts, and even fewer transcripts represent the bulk of a sample.' (Kliebenstein goes on to note problems due to the compositional nature of this data: '. . . a polymorphism that affects the abundance of one of these high-accumulating transcripts has the potential to cause significant differences in all other less-abundant transcripts because of the bulk standardization of RNA amount.')

A more fundamental difference between compositions in the bio- and geosciences is that biologists are generally interested in *living* organisms, whose cells *produce* DNA, RNA, proteins and metabolites. Often, their interest centres on the productivity of these cells. Aitchison (2008) writes that '[when] we say that a problem is compositional we are recognizing that the sizes of our specimens are irrelevant'; in the biosciences, the size (sometimes referred to as the *absolute abundance*) of specimens is often *highly* relevant because it pertains to the productivity of cells. However, many methods of sample preparation or measurement remove information about size, leaving only relative information behind. We begin with a thought experiment to highlight this.

14.2   The Omics Imp and two bioscience experiment paradigms

Small enough to fit into a cell, yet somehow able to wield pencil and paper, The Omics Imp is a molecular accountant par excellence. Without disrupting biological processes, the Imp can tally the molecules it observes. Figure 14.1(a) shows it counting messenger RNA (mRNA) as it emerges from the nucleus of a cell. The Imp can help experimentalists by counting the different types of mRNA molecules that it sees in a specified time interval. Clearly, these counts are non-negative and constrained only by productivity of the nucleus in that time interval. [Aitchison (1986) referred to such a vector of non-negative values as a *basis*, however this usage is not common today. We call such vectors *unclosed*, and their components *raw*.]

The Imp can also work in other styles of experiment. Figure 14.1(b) shows it counting mRNA collected in a bucket *after* emerging from the nucleus of a cell. The Imp can help experimentalists by counting the different types of mRNA molecules it sees in this *full* bucket. Once again, these counts are non-negative but, unlike the scenario in Figure 14.1(a), they are constrained by the (arbitrary) size of the bucket, they carry no information about the *absolute*

(a)   (b)

Figure 14.1   (a) The Omics Imp tallying the different kinds of messenger RNA molecules emerging from the nucleus of a cell. (b) The Omics Imp tallying the different kinds of messenger RNA molecules in a full bucket collected from the nucleus of a cell. (Illustration of mRNA courtesy of the National Human Genome Research Institute.)

rate of mRNA production, and they are not independent of each other – if the amount of one kind of mRNA in the full bucket increases, the amounts of one or more other kinds of mRNA must decrease. The *sum-constrained* vector of counts produced in this experiment is a *composition*, and this constraint has a profound impact on both the information carried by the counts, and their subsequent interpretation.

Without The Omics Imp, most molecular bioscience measurement processes follow the *bucket-survey* paradigm shown in Figure 14.1(b), often using a series of buckets (e.g. extracting a fixed mass of total RNA), with filters (e.g. RNA size fractionation by gel electrophoresis) and multipliers [e.g. polymerase chain reaction (PCR) amplification] between them. However, it is easy to lose sight of the sum-constraints being placed on experimental material, particularly as there is no strong tradition of concern for these issues in molecular biology and bioinformatics.

The current raft of nucleotide-counting sequencing technologies (also known as *next-generation*, *short-read*, or *deep* sequencing) also give the impression that a biologist can count, or at least estimate the count of the different types of DNA or RNA sequences produced by a sample of cells. But some thought about the sample preparation and DNA/RNA extraction process should make it clear that there are some different buckets constraining the numbers of molecules under measurement including: (1) starting with a fixed weight or volume tissue sample (ignoring cellularity); (2) extracting a fixed weight or volume of DNA/RNA; and (3) concluding with a finite (if very large) number of sequence fragment reads.

The terms *under-* and *over-expression* are often used in gene expression analysis to refer to mRNAs that are less/more expressed in comparison with some reference situation. These mRNAs are also described as being *down-/up-regulated* by processes that control their level of expression. Figure 14.2 emphasises the perils of confusing these terms with under- and over-*production*. The bucket-survey suggests that, in comparison with Treatment A, mRNA z is over-expressed in Treatment B, even though it is being produced at exactly the same rate in both situations. To make comprehensive statements about gene expression, we have to know the total amount of mRNA being produced (or, as Aitchison terms it, the *size* of the specimen) as well as the relative abundances of different mRNA species.

Note that, in Figure 14.2, mRNA z could be thought of as an ideal *control* or *reference transcript* – an mRNA derived from a *housekeeping gene* (i.e. a gene ubiquitously expressed in all tissue/cell types) that is also expressed at a constant rate. Control transcripts have been

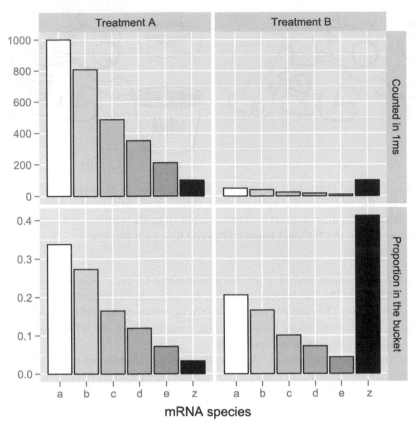

Figure 14.2   (Fictitious) data obtained by the Omics Imp on different mRNA species (a–z) from a cell nucleus under two different treatments (A and B). The top panel shows counts of the mRNAs observed over 1 ms. (Note that the Imp counted exactly the same number of mRNA z in both treatments.) The bottom panel shows the findings of the corresponding bucket-survey, expressed as the proportion of each mRNA species collected in the Imp's (full) bucket.

considered the 'gold standard' means by which to normalise (i.e. render comparable) mRNA levels across different samples (de Jonge *et al.* 2007). However, *actual* control transcript expression levels still vary across experimental conditions and, as we shall discuss further below, simply dividing through by one (or a combination) of these amounts has drawbacks as an analysis strategy.

Relative abundance data can also make statistically independent components appear correlated. In the fictitious data of Figure 14.2 the absolute amount of mRNA z remained constant in both Treatments A and B while mRNAs a–e changed dramatically. Figure 14.3 shows the opposite scenario: the absolute abundances of mRNAs a–e remain constant across Treatments C–F while mRNA z changes dramatically. A naïve interpretation of the mRNA proportions in the bucket would describe mRNAs a–e as positively correlated with each other and negatively correlated with mRNA z across the four treatments – the proportions of mRNAs a–e increase together while the proportion of mRNA z decreases. All this, despite the fact that the absolute number of copies of mRNAs a–z are statistically independent in the four treatments. This is another manifestation of the sum constraint imposed by looking at the contents of a

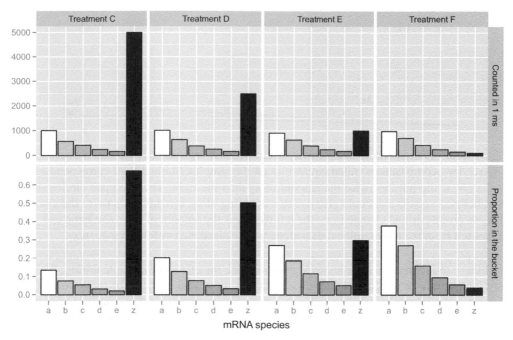

Figure 14.3 (Fictitious) data obtained by the Omics Imp on different mRNA species (a–z) from a cell nucleus under four different treatments (C–F). Top and bottom panels show the same kind of measurements as in Figure 14.2. Note that the Omics Imp counted approximately the same number of mRNAs a–e in all treatments.

full bucket. We repeat: if the amount of one kind of mRNA in the full bucket increases, the amounts of one or more other kinds of mRNA must decrease.

14.3 The impact of compositional constraints in the omics

Biology is complex, and biological systems have many sources of variability. Probability and statistics provide a principled framework for dealing with variation and uncertainty in the biosciences but, to paraphrase Aitchison (1986), standard statistical procedures usually lead to misinterpretation and doubtful inferences when applied to compositions.

In this section, we look at how various statistics used in the molecular biosciences are affected by *closure*, the constraint that all components sum to a constant. We will assume the reader has a grasp of compositional data analysis concepts and terminology set out earlier in this book and, for simplicity, deal only with compositions of positive components that sum to 1, i.e. that are in the simplex.

Where possible, we will relate the statistics and statistical methods to prevailing practices of analysing bioscience data, particularly the use of log-transformation which, as we shall see, is no panacea for the analysis of purely relative information.

14.3.1 Univariate impact of compositional constraints

Even though many molecular bioscience measurements are highly multivariate, *univariate* statistical methods – such as the *t*-test – are workhorses in the omics, albeit with additional

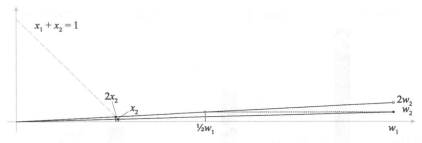

Figure 14.4 Did a twofold change in x_2 occur because w_2 doubled? Or was it because w_1 halved? Two compositions (filled circles) lie on the simplex (dashed line) corresponding to $x_1 + x_2 = 1$: the left has twice the amount of x_2 than the right. The two open circles show two unclosed vectors that could give rise to a doubling of x_2: the right uses twice the amount of w_2 to give a twofold change in x_2; the left uses *half* the amount of w_1 to the same end. We cannot tell from the compositional data alone what caused the twofold change in x_2.

machinery to incorporate prior knowledge or mitigate the effects of multiple testing. We can gain insight into the impact of sum-constraints on univariate statistics of these very high-dimensional mixtures by working with a two-part composition $\mathbf{x} = (x_1, x_2)$ derived from $\mathbf{w} = (w_1, w_2)$ through its closure: $\mathbf{x} = \mathcal{C}(\mathbf{w})$. This can be regarded as reducing a high-dimensional composition to two dimensions by making x_1 be the amalgamation of all components *except* x_2.

As Witten and Tibshirani (2007) point out, despite the wealth of available methods, biologists show a fondness for fold-change and the t-statistic for univariate analysis – presumably because of their simplicity and interpretability. However, as we can see in Figure 14.4 and (in more detail) in Lovell *et al.* (2010), twofold change in a component of a composition (i.e. x_2) does not, on its own, tell us what has happened in the raw components. We can say that $x_2 \mapsto kx_2$ implies $w_2 \mapsto kw_2$ provided (1) x_2 is a relatively small component, (2) the rest of the raw components stay the same and (3) k is 'near' 1. This means that we are reasonably safe to do univariate statistics on components of very large (i.e. high dimensional) compositions that are not changing dramatically. [Filzmoser *et al.* (2009) describe strategies for univariate data analysis when this is not the case, using the isometric log-ratio transformation (Egozcue *et al.* 2003) to represent compositional data in Euclidean space where standard statistical methods can be applied.]

In practice, this is exactly the situation with 'spike-in' experiments, where a known concentration of a readily identifiable molecule (or cocktail of molecules) is added to samples of a mixture containing unknown concentrations of molecules, usually to assess the sensitivity of different microarray technologies (McCall and Irizarry 2008) or infer sample mRNA concentrations (Bissels *et al.* 2009).

We believe that samples in many gene expression experiments will be much more variable than in carefully controlled spike-in evaluations. Are the amounts of mRNA produced by the cells under study similar enough in size and composition not to confound univariate statistical analysis? We do not know. But we suspect that the mRNA products of cells from different tissues (e.g. brain/liver, cancer/noncancer) or tissues at different stages of life (e.g. dormant/germinating) are quite different in both size and composition, calling into question the whole paradigm of testing for 'significant differential expression' using only measures of relative abundance.

14.3.2 Impact of compositional constraints on multivariate distance metrics

Molecular bioscience is replete with multivariate data, including microarray data in which each sample is represented by a point in a space of as many dimensions as there are spots on the array. Multivariate distance metrics underpin clustering methods (e.g. hierarchical clustering) by telling us how 'close' multivariate points are to each other.

14.3.2.1 Properties of three distance metrics with compositional data

Here we explore how different multivariate distance metrics are affected by compositional constraints using two compositions $\mathbf{x}$, $\mathbf{y}$ whose positive components are denoted x_i, y_i, respectively. Their perturbation difference is $\mathbf{z} = \mathbf{x} \ominus \mathbf{y}$, that is

$$\mathbf{z} = \mathcal{C}(x_1/y_1, x_2/y_2, \ldots, x_D/y_D).$$

We will consider Aitchison's distance between $\mathbf{x}$ and $\mathbf{y}$, Euclidean distance between $\mathbf{x}$ and $\mathbf{y}$, and Euclidean distance between $\log(\mathbf{x})$ and $\log(\mathbf{y})$, working with squared distances for clarity.

The Aitchison distance between $\mathbf{x}$ and $\mathbf{y}$ can be written in terms of $\mathbf{z}$:

$$d_a^2(\mathbf{x}, \mathbf{y}) = \|\text{clr}(\mathbf{x}) - \text{clr}(\mathbf{y})\| = \|\text{clr}(\mathbf{z})\|$$

$$= \sum_i \left(\log \frac{x_i}{\text{g}_m(\mathbf{x})} - \log \frac{y_i}{\text{g}_m(\mathbf{y})} \right)^2 = \sum_i \left(\log \frac{z_i}{\text{g}_m(\mathbf{z})} \right)^2.$$

Being a function of $\mathbf{z}$ alone, Aitchison's distance tells us only about *relative* differences (i.e. *ratios*) of corresponding components.

The Euclidean distance between $\mathbf{x}$ and $\mathbf{y}$ is

$$d_e^2(\mathbf{x}, \mathbf{y}) = \sum_i (x_i - y_i)^2 = \sum_i \left(\frac{x_i}{y_i} - 1 \right)^2 y_i^2.$$

Euclidean distance depends not only on ratios of corresponding components, but also the value of one of the compositions. Furthermore, since the largest Euclidean distance is attained when $\mathbf{x}$ and $\mathbf{y}$ are at two different vertices of the simplex [e.g. $\mathbf{x} = (1, 0, 0, \ldots)$, $\mathbf{y} = (0, 1, 0, \ldots)$], d_e is bounded by $\sqrt{2}$.

Imagine very high-dimensional compositions, e.g. RNA-seq counts of thousands of different mRNA species. By necessity, each component will be a very small proportion of the total number of mRNA sequence reads. Consequently, the Euclidean distance between any two such compositions will be very small also, even though they may have components that are many-fold different in relative abundance. Aitchison's distance, with its focus on the ratio of corresponding components, will emphasise these differences in relative abundance much more effectively.

A common approach to analysing count data is to adopt a log-linear Poisson model (McCullagh and Nelder 1989). It is also common in the biosciences to analyse and present strictly positive data using a logarithmic transformation, without necessarily referring to an underlying probabilistic model. 'Logged data' is commonplace in microarray analysis and now, RNA-seq data analysis (Robinson and Smyth 2007; Marioni *et al.* 2008). With this in mind, we look at the Euclidean distance between log(**x**) and log(**y**):

$$
d_e^2(\log(\mathbf{x}), \log(\mathbf{y})) = \sum_i (\log(x_i) - \log(y_i))^2
$$

$$
= \sum_i \left(\log \frac{x_i}{g_m(\mathbf{x})} - \log \frac{y_i}{g_m(\mathbf{y})} + \log \frac{g_m(\mathbf{x})}{g_m(\mathbf{y})} \right)^2
$$

$$
= \sum_i \left(\log \frac{z_i}{g_m(\mathbf{z})} \right)^2 + 2 \log \frac{g_m(\mathbf{x})}{g_m(\mathbf{y})} \sum_i \left(\log \frac{z_i}{g_m(\mathbf{z})} \right) + D \log^2 \left(\frac{g_m(\mathbf{x})}{g_m(\mathbf{y})} \right)
$$

$$
= d_a^2(\mathbf{x}, \mathbf{y}) + D \log^2 \left(\frac{g_m(\mathbf{x})}{g_m(\mathbf{y})} \right) \quad \geq d_a^2(\mathbf{x}, \mathbf{y}), \tag{14.1}
$$

since the terms $\log(z_i/g_m(\mathbf{z}))$ are the components of clr(**z**) and sum to zero. So the Euclidean distance between logged compositions is, like Aitchison's distance, a function of **z** but with an explicit dependence on D, the number of components in the composition, and the geometric means of **x** and **y**.

Microarray data are conventionally dealt with on a $\log_2$ scale. So even though compositional constraints may be at play in these data (because they are derived from fixed weights of total RNA), distance-based analyses (e.g. hierarchically clustered heatmaps) are likely to be quite similar to what we would get with a purely compositional approach – more of which below.

14.3.2.2   Three distance metrics on some two-part compositions

To provide additional insight into the three distance metrics we have considered, we apply them to the simplest possible compositions: those with only two parts. We look at distances between the composition $\mathbf{w} = \mathcal{C}(1024, w)$ and two other (fixed) compositions $\mathbf{x} = \mathcal{C}(1024, 1)$ and $\mathbf{y} = \mathcal{C}(1024, 1024)$. Figure 14.5 shows the distances $d_a(\mathbf{w}, \cdot)$, $d_e(\mathbf{w}, \cdot)$, and $d_e(\log(\mathbf{w}), \log(\cdot))$ to **x** and **y**.

As we are working with only two components, we can go a little further in understanding when $d_e(\log(\mathbf{w}), \log \mathbf{x}) = d_a(\mathbf{w}, \mathbf{x})$, *i.e.*, when $g_m(\mathbf{x}) = g_m(\mathbf{w})$ [Equation (14.1)]. Algebra shows that this occurs for $\mathbf{x} = \mathcal{C}(1024, 1)$ and $\mathbf{x} = \mathcal{C}(1024, 1024^2)$ as seen in Figure 14.5(a).

The three main points to observe are: that Euclidean distance does not reflect relative changes in components very well; Aitchison's distance clearly depends only on the relative abundance of corresponding components; and the Euclidean distance between log-components is bounded below by Aitchison's distance.

14.3.2.3   Three distance metrics on some high-dimensional compositions

While we cannot think of a way to systematically visualise the behaviour of different distance measures on higher dimensional compositions, we can look at a particular data set that has already been used to exemplify differences between distance metrics.

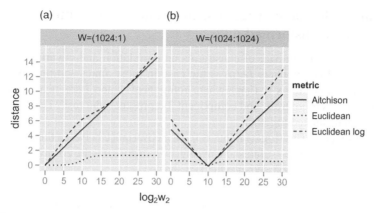

Figure 14.5   Plots of three different distance metrics between composition $\mathbf{w} = \mathcal{C}(1024, w)$ and $\mathbf{x} = \mathcal{C}(1024, 1)$ (a), and $\mathbf{y} = (1024, 1024)$ (b), as w varies. Note that $d_e(\mathbf{w}, \mathbf{x})$ approaches $\sqrt{2}$, the length of the edge of a D-part simplex, and that $d_e(\mathbf{w}, \mathbf{y})$, the Euclidean distance from $\mathbf{w}$ to $\mathbf{y}$, the barycentre of the simplex, approaches $\sqrt{(D-1)/D} = 1/\sqrt{2}$.

Vêncio *et al.* (2007) proposed Aitchison's simplicial distance as an alternative to Euclidean distance in clustering digital gene expression data, and demonstrated that Aitchison's distance clustered simulated RNA-seq data in essentially the same way as a Euclidean clustering of data from the corresponding microarray experiment. They also showed that Aitchison's distance clustered simulated RNA-seq data more interpretably than clustering of that data based on Euclidean distance.

Lovell *et al.* (2010) replicate the results obtained by Vêncio *et al.*, and show that the Euclidean distance between the logged RNA counts gives essentially identical results to those using Aitchison's distance. The geometric means of the 36 vectors of RNA counts are within 1% of each other so, by Equation (14.1), d_a should be very close to d_e on log-transformed data.

As mentioned above, we expect distance-based analysis of microarray fluorescence data on a $\log_2$ scale to be quite similar to a purely compositional approach. Vêncio *et al.* (2007) advocate Aitchison's distance as a metric for RNA-seq and other forms of enumeration-based gene expression data on the grounds that they are compositional data. As we shall see in the next section, we think issues to do with correlation and covariance provide stronger reasons for using compositional methods.

14.4   Impact of compositional constraints on correlation and covariance

Compositional constraints are notorious for their impacts on the covariance and correlation structures of data (Aitchison 1986, Section 3.3). A major motivation for the investigations described here is to understand better how compositional constraints in omics data might affect our estimates of covariance when working with log-transformed data, as is common practice in the omics.

14.4.1   Compositional constraints, covariance, correlation and log-transformed data

To help us understand the impact of constraints on correlation and covariance where there are large numbers of components D (as is the case in most omics settings), we will work with a three-part composition $\mathbf{x} = \mathcal{C}(\mathbf{w})$, where components 1 and 2 are measurements whose pairwise correlation is of interest, and component 3 represents 'the rest', i.e. the other $D - 2$ measurements amalgamated together.

We consider first how $\text{Cov}(\log(x_1), \log(x_2))$ – the covariance between the two components of interest in the composition – relates to $\text{Cov}(\log(w_1), \log(w_2))$ – the covariance between the two *raw* components of interest. We use the fact that

$$\text{Cov}(A + C, B + C) = \text{Cov}(A, B) + \text{Cov}(A, C) + \text{Cov}(B, C) + \text{Var}(C)$$

to write:

$$\begin{aligned} \text{Cov}(\log(x_1), \log(x_2)) &= \text{Cov}\left(\log(w_1/t), \log(w_2/t)\right) \\ &= \text{Cov}(\log(w_1), \log(w_2)) - \text{Cov}(\log(w_1), \log(t)) \\ &\quad - \text{Cov}(\log(w_2), \log(t)) + \text{Var}(\log(t)). \end{aligned} \tag{14.2}$$

This highlights the effect of variation in the *size* t of the D-part unclosed vector (Aitchison 1986, section 9.2). Equation (14.2) shows explicitly that $\text{Cov}(\log(x_1), \log(x_2)) = \text{Cov}(\log(w_1), \log(w_2))$ when t is constant, in other words, *when* $\mathbf{w}$ *is already a composition* (but one constrained to sum to t rather than 1).

Unfortunately, the *correlation* between the (log) components of interest cannot be expressed as neatly as in Equation 14.2. However, we note that $\text{Corr}(\log(x_1), \log(x_2)) = \text{Corr}(\log(w_1), \log(w_2))$ when t is constant.

Symbolic understanding of the relationship between $\log(\mathbf{x})$ and $\log(\mathbf{w})$ is useful, but we also felt a need to visualise possible relationships between raw components and compositions to explore the extent to which $\text{Cov}(\log(x_1), \log(x_2))$ could mislead us about what is going on in the underlying unclosed vector.

14.4.2   A simulation approach to understanding the impact of closure

Simulation has been used previously to explore compositional data (Skala 1977; Brehm *et al.* 1998) but, as far as we know, not to investigate the properties of *log-transformed* compositions. Simulation gives us complete control over the statistical properties of the data at the expense of losing connection to real experimental data. Unfortunately, at this time, we know of no one who has actual experimental data on both raw and compositional measurements in a molecular biology setting.

When it comes to simulating data from a three-part unclosed vector, we are confronted by having to assume a distribution for $\mathbf{w}$. Clearly, w_1, w_2, w_3 must all be positive. For simplicity, we decided to ensure w_1, w_2 could have some straightforward statistical dependence while remaining statistically independent of w_3. We see this as 'one step along' from the scenario of completely independent parts.

We think that a trivariate log-normal distribution for $\mathbf{w}$ is the simplest but most general way to create a three-part vector that can be used to explore the impact of closure on the pairwise relationship between w_1, w_2. In this scenario $\log(\mathbf{w}) \sim \mathcal{N}_3(\boldsymbol{\mu}, \boldsymbol{\Sigma})$, where $\boldsymbol{\mu}^{\mathrm{T}} = (\mu_1 \ \mu_2 \ \mu_3)$ and

$$\boldsymbol{\Sigma} = \begin{pmatrix} \sigma_1^2 & \rho_{12}\sigma_1\sigma_2 & 0 \\ \rho_{12}\sigma_1\sigma_2 & \sigma_2^2 & 0 \\ 0 & 0 & \sigma_3^2 \end{pmatrix}.$$

To understand the impact that parameter changes had in different parts of this 7-dimensional space ($\mu_{1,2,3}$, $\sigma_{1,2,3}$, ρ_{12}) we developed interactive simulation software (Lovell *et al.* 2010). We have used this software to find two extreme (but plausible) situations that characterise how the analysis of log-transformed compositional data could lead to incorrect inferences about the relationship between the components of interest, x_1 and x_2:

$w_1, w_2 \gg w_3$. When the unclosed vector is dominated by the components of interest, $\log(x_1)$ and $\log(x_2)$ tend to move towards their upper limits, i.e. the boundary defined by $\log(x_2) = \log(1 - x_1)$ for $x_1 \in (0, 1)$. This imposes a negative bias on the $\mathrm{Corr}(\log(x_1), \log(x_2))$ in comparison with $\mathrm{Corr}(\log(w_1), \log(w_2))$. This can be explored by sweeping ρ_{12} through its range, with $\log(\mu_{1,2,3}) = (4, 4, 0)$ and $\log(\sigma_{1,2,3}) = (0, 0, 0)$ [see Figure 14.6(a)]. This situation is easy to detect in compositional data because

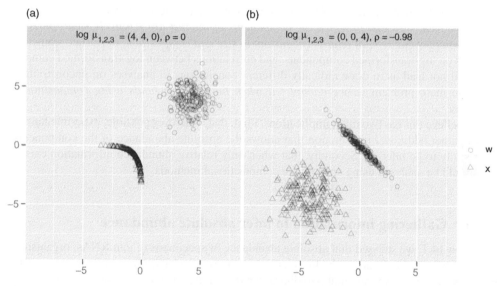

Figure 14.6 Two plots adapted from interactive simulation software [see Lovell *et al.* (2010)] showing 200 samples from a trivariate log-normal distribution ($\mathbf{w}$), and the corresponding composition ($\mathbf{x}$). Only components 1 and 2 are shown, and axes are drawn on a log scale. (a) Samples where $\log(\mu_{1,2,3}) = (4, 4, 0)$ and $\log(\sigma_{1,2,3}) = (0, 0, 0)$. While the sample correlation between $\log(w_1)$ and $\log(w_2)$ is zero, the correlation observed between $\log(x_1)$ and $\log(x_2)$ is around -0.75. (b) Samples where $\log(\mu_{1,2,3}) = (0, 0, 4)$ and $\log(\sigma_{1,2,3}) = (0, 0, 0)$. While the sample correlation between $\log(w_1)$ and $\log(w_2)$ is -0.98, the correlation observed between $\log(x_1)$ and $\log(x_2)$ is around 0. Reprinted with permission of David Lovell.

$x_1 + x_2$ will be close to 1. However, nothing can be done to infer the relationship between the raw variables w_1 and w_2 using the compositional data alone.

$w_3 \gg w_1, w_2$. When the unclosed vector is not dominated by the components of interest, the degree of correspondence between $\text{Corr}(\log(x_1), \log(x_2))$ and $\text{Corr}(\log(w_1), \log(w_2))$ depends on the variance of $\log(w_3)$. As σ_3 increases, $\text{Corr}(\log(x_1), \log(x_2))$ tends to be positively biased in comparison with $\text{Corr}(\log(w_1), \log(w_2))$. This can be explored by sweeping $\log(\sigma_3)$ through its range, with $\log(\boldsymbol{\mu}) = (0, 0, 4)$, $\log(\sigma_{1,2}) = (0, 0)$ and $\rho_{12} = -0.98$ [see Figure 14.6(b)]. While it is again easy to detect this situation in compositional data (this time $x_1 + x_2$ will be close to 0), there is nothing in that data to tell us about the variance of $\log(w_3)$.

In summary, the correlation between log-transformed components of interest in the composition will be approximately the same as that of their raw counterparts only when $w_3 \gg w_1, w_2$ and $\text{Var}(\log(w_3))$ is small – in other words, when w_1 and w_2 are small parts of a relatively constant total. *We are unable to tell when that is the case using the compositional data alone.*

14.5   Implications

We have explored the potential for sum-constrained data to lead analyses of omics data astray. We have seen that *provided the components of interest are relatively small parts of mixture samples that remain relatively constant in size and composition*, univariate statistics, distances on log-transformed components, and correlations between log-transformed components will not lead us to draw radically different conclusions to analyses on unconstrained data. The main problem is: *we cannot tell when that proviso holds using compositional data alone.*

We believe this has two main implications. First, that, wherever possible, experimentalists should gather additional information that allows the absolute abundance of the components under study to be inferred. Secondly, that when only relative abundance information exists, data should be analysed using appropriate compositional methods.

14.5.1   Gathering information to infer absolute abundance

In Section 14.1, we stressed that absolute abundance of specimens (e.g. mRNAs, organisms, etc.) is often very important in the biosciences. We introduced The Omics Imp as a means to show how different experimental paradigms can determine whether absolute abundance can be inferred, and how relative abundance alone does not tell us about how many copies of an mRNA are being produced.

Kanno *et al.* (2006) and Miura *et al.* (2008) describe methods to measure mRNA absolute abundance, so that a cell's transcriptome could be described in terms of the counts of each different kind of transcript present in that cell. To the best of our knowledge, application of methods to measure mRNA absolute abundance is not yet commonplace, but we hope that this chapter will serve as an argument for these, and other absolute abundance techniques to be employed more often in the pursuit of understanding biological systems.

14.5.2   Analysing compositional omics data appropriately

There are circumstances where omics data are truly relative (e.g. metabolite concentrations within the bloodstream), or when interest genuinely centres on comparing relative amounts (e.g. the nucleotide or codon composition of samples of genomic DNA). There are also many circumstances where measurements have been made in ways that ensure that data carry only relative information (e.g. RNA-seq or microarray data obtained from fixed volumes of total RNA). In their seminal paper on RNA-seq, Mortazavi *et al.* (2008) explicitly render their data compositional by working in terms of reads per kilobase of exon per million mapped sequence reads (RPKM). [By working with fixed weight aliquots of mRNA and using a sequencing platform that has limits (albeit very large ones) on the number of sequences that can be read, the data were already constrained to be compositional.]

In these situations, we think much more needs to be done to apply compositional data analysis methods instead of analysis techniques that assume data are unconstrained. We have shown that simply log-transforming the compositional data is not a panacea – we need to be sure that the components of interest are relatively small parts of mixture samples that remain relatively constant in size and composition, and this cannot be determined using compositional data alone.

Aitchison (1986) has pioneered methods for compositional data analysis, founded upon *log-ratios* of components. Among other things, this approach provides a sound alternative to the analysis of covariance in compositions (cf. Section 14.4) by using the *compositional variation array* (Aitchison 1986, p. 69), *centred log-ratio covariance matrix* (Aitchison 1986, p. 78), or other decomposition of total variance.

It is important to note that, while founded on ratios, the log-ratio analyses described by Aitchison and others are not about 'dividing through' by some normalising measurement and treating the resulting values as though they were absolute quantities (the philosophy behind control transcript normalisation mentioned in Section 14.2). If we accept that we are working with data constrained to sum to a constant, then we must accept that we have lost all information about absolute abundance and retain only relative information. [The principle of Vandesompele *et al.* (2002) that 'the expression ratio of two proper control genes should be identical in all samples regardless of the experimental condition or cell type' is in accord with this, and their approach comes tantalisingly close to full log-ratio analysis.]

We conjecture that bringing log-ratio analysis into play with omics data would mean, for example:

- Working with (log) ratios of fluorescence intensities *between* spots within a microarray. This would be an explicitly multivariate treatment of the data rather than, say, the conventional approach of multiple univariate analyses that seek to test for significant differential expression. (One of the beliefs that has to be abandoned in working with compositional data is the idea that a single component means anything in isolation – it is only meaningful *relative* to other components.) We wonder also whether adopting this approach would obviate or simplify the process of microarray normalisation that seeks to render arrays comparable within and across experiments.

- Working with (log) ratios of mRNA counts within RNA-seq runs.

- Adopting Aitchison's distance as a metric for compositional comparison. Given the relationship between Aitchison's distance and Euclidean distance with log-transformed data [Equation (14.1)], and the fact that omics data is often log-transformed before hierarchical clustering or other distance-based methods are applied, this may not lead to dramatically different results across the board. However, in areas that use Euclidean distance on (untransformed) compositional data, we expect the application of Aitchison's distance to provide more meaningful insights.

We can see that omics data poses challenges to compositional data analysis methods. Datasets often contain zero measurements – either because a component was not present, or because it was present but not sampled, or because some measurement error occurred. The problem of zeros becomes more pernicious the less that samples have in common, e.g. metagenomic samples drawn from very different environments. Of course, this is not so much a defect of compositional data analysis methods as a sharp reminder that comparing samples with different attributes is an ill-posed problem. Furthermore, we are optimistic that recent work by Scealy and Welsh (2010) will provide a practical alternative to log-ratio analyses that can also tolerate zero values to a greater extent.

A second challenge posed by omics data to compositional (indeed *any*) analysis methods is the paucity of independent samples (n) in comparison with the abundance of measurements (p). Modern bioscience data is as notoriously high-dimensional as modern bioscience data collection is underfunded, and $p \gg n$ datasets are commonplace. The industrialisation of biology has us, at present, in a situation where it is feasible to make millions of measurements on a few individuals, but not vice versa.

We acknowledge these challenges, both methodological and financial. However, our primary aim is to ensure that bioscientists are not lead astray by artifacts of the measurement process, and we hope through this, and subsequent publications that awareness will be raised about the need to handle compositional data from the molecular biosciences appropriately.

Acknowledgements

We gratefully acknowledge colleagues who provided us with feedback on our ideas, in particular, Rob Knight (University of Colorado, Boulder) and Ian Saunders (CSIRO). Our sincere thanks go to reviewers Juan José Egozcue (especially for bringing our notation and terminology up to date) and Dean Billheimer (particularly for highlighting the use of housekeeping genes in normalisation).

References

Aitchison J 1986 *The Statistical Analysis of Compositional Data*. Monographs on Statistics and Applied Probability. Chapman and Hall Ltd (reprinted 2003 with additional material by The Blackburn Press), London (UK). 416 p.

Aitchison J 2008 The single principle of compositional data analysis, continuing fallacies, confusions and misunderstandings and some suggested remedies. In *Proceedings of CoDaWork'08, The 3rd Compositional Data Analysis Workshop* (ed. Daunis-i Estadella J and Martín-Fernández J). http://hdl.handle.net/10256/723. University of Girona, Girona (Spain).

Bissels U, Wild S, Tomiuk S, Holste A, Hafner M, Tuschl T and Bosio A 2009 Absolute quantification of microRNAs by using a universal reference. *RNA* **15**(12), 2375–2384.

Brehm J, Gates S and Gomez B 1998 A Monte Carlo comparison of methods for compositional data analysis. *1998 Annual Meeting of the Society for Political Methodology.* http://polmeth.wustl.edu/retrieve.php?id=295.

de Jonge HJM, Fehrmann RSN, de Bont ESJM, Hofstra RMW, Gerbens F, Kamps WA, de Vries EGE, van der Zee AGJ, te Meerman GJ and ter Elst A 2007 Evidence based selection of housekeeping genes. *PLoS ONE* **2**(9), e898.

Egozcue JJ, Pawlowsky-Glahn V, Mateu-Figueras G and Barceló-Vidal C 2003 Isometric logratio transformations for compositional data analysis. *Mathematical Geology* **35**(3), 279–300.

Filzmoser P, Hron K and Reimann C 2009 Univariate statistical analysis of environmental (compositional) data: Problems and possibilities. *Science of The Total Environment* **407**(23), 6100–6108.

Kanno J, Aisaki K, Igarashi K, Nakatsu N, Ono A, Kodama Y and Nagao T 2006 *Per cell* normalization method for mRNA measurement by quantitative PCR and microarrays. *BMC Genomics* **7**, 64.

Kliebenstein D 2009 Quantitative genomics: Analyzing intraspecific variation using global gene expression polymorphisms or eqtls. *Annual Review of Plant Biology* **60**(1), 93–114.

Lesk A 2007 *Introduction to Genomics*, 1st edition. Oxford University Press, New York, NY (USA).

Lovell D, Müller W, Taylor J, Zwart A and Helliwell C 2010 Caution! Compositions! Can constraints on omics data lead analyses astray? Technical Report EP10994, CSIRO. http://www.csiro.au/David.Lovell.

Marioni JC, Mason CE, Mane SM, Stephens M and Gilad Y 2008 RNA-seq: an assessment of technical reproducibility and comparison with gene expression arrays. *Genome Research* **18**, 1509–1517.

McCall MN and Irizarry RA 2008 Consolidated strategy for the analysis of microarray spike-in data. *Nucleic Acids Research* **36**(17), e108

McCullagh P and Nelder JA 1989 *Generalized Linear Models*, 2nd edition. Chapman and Hall, London (UK).

Miura F, Kawaguchi N, Yoshida M, Uematsu C, Kito K, Sakaki Y and Ito T 2008 Absolute quantification of the budding yeast transcriptome by means of competitive PCR between genomic and complementary DNAs. *BMC Genomics* **9**, 574.

Mortazavi A, Williams BA, McCue K, Schaeffer L and Wold B 2008 Mapping and quantifying mammalian transcriptomes by RNA-Seq. *Nature Methods* **5**(7), 621–628.

Robinson MD and Smyth GK 2007 Moderated statistical tests for assessing differences in tag abundance. *Bioinformatics* **23**(21), 2881–2887.

Scealy JL and Welsh AH 2010 Regression for compositional data by using distributions defined on the hypersphere. http://onlinelibrary.wiley.com/doi/10.1111/j.1467-9868.2010.00766.x/pdf.

Skala W 1977 A mathematical model to investigate distortions of correlation coefficients in closed arrays. *Mathematical Geology* **9**(5), 519–528.

Vandesompele J, Preter KD, Pattyn F, Poppe B, Roy NV, Paepe AD and Speleman F 2002 Accurate normalization of real-time quantitative RT-PCR data by geometric averaging of multiple internal control genes. *Genome Biology* **3**(7), research0034.1–0034.11.

Vêncio R, Varuzza L, de B Pereira C, Brentani H and Shmulevich I 2007 Simcluster: clustering enumeration gene expression data on the simplex space. *BMC Bioinformatics* **8**(1), 246

Witten DM and Tibshirani R 2007 A comparison of fold-change and the t-statistic for microarray data analysis. Technical Report, Stanford University. http://www-stat.stanford.edu/tibs/ftp/FCTComparison.pdf.

15

Hardy–Weinberg equilibrium: a nonparametric compositional approach

Jan Graffelman[1] and Juan José Egozcue[2]

[1]*Department of Statistics and Operations Research, Technical University of Catalonia, Spain*
[2]*Department of Applied Mathematics III, Technical University of Catalonia, Spain*

15.1 Introduction

More than one hundred years ago, Hardy (1908) and Weinberg (1908) independently formulated the principle that the genotypes AA, AB and BB at a bi-allelic locus are expected to occur in the proportions p^2, $2pq$ and q^2, respectively, where p is the allele frequency of A, and $q = 1 - p$ the allele frequency of B. This principle, known as Hardy–Weinberg equilibrium (HWE), will hold whenever there are no disturbing forces such as selection, nonrandom mating, mutation, migration, etc. Let f_{AA}, f_{AB}, and f_{BB}, represent the relative population frequencies of the genotypes AA, AB and BB respectively. By squaring the heterozygote frequency, an alternative formulation of the Hardy–Weinberg law is obtained as

$$f_{AB}^2 = (2pq)^2 = 4p^2q^2 = 4 f_{AA} \cdot f_{BB}. \tag{15.1}$$

This equation shows that the Hardy–Weinberg law relates the three elements of a three-way composition. The Hardy–Weinberg law is a cornerstone of modern genetics (Crow 1988). A

Compositional Data Analysis: Theory and Applications, First Edition. Edited by Vera Pawlowsky-Glahn and Antonella Buccianti.
© 2011 John Wiley & Sons, Ltd. Published 2011 by John Wiley & Sons, Ltd.

recent interesting historical account of the Hardy–Weinberg law is given by Edwards (2008). The issue of testing genetic markers for HWE has attracted the attention of both geneticists and statisticians. Several statistical tests are available for testing genetic markers for HWE, such as Pearson's χ^2 test, the likelihood ratio test, the exact test, and modifications of these (Weir 1996, chapter 3). Comparisons of these test procedures have been described in the literature (Li 1976; Elston and Forthofer 1977; Emigh 1980; Haber 1981; Hernández and Weir 1989). A bayesian procedure to test for HWE has been described by Lindley (1988). Recent bayesian work on HWE can be found in Ayres and Balding (1998), Shoemaker *et al.* (1998) and Wakefield (2010).

Modern genetic studies often use large numbers of single nucleotide polymorphisms (SNPs). These markers are typically bi-allelic, and SNPs are routinely screened for deviations from HWE, prior to their use in marker-disease association studies. SNPs found to deviate significantly from HWE often show problems in the laboratory assay (Wigginton *et al.* 2005; Teo *et al.* 2007). Often such SNPs are discarded in the posterior analysis. This is crucial in association studies, as HW disequilibrium may also result from disease association (Wittke-Thompson *et al.* 2005).

In this contribution we briefly describe the structure of SNP databases (Section 15.2) and summarize the classical tests for HWE (Section 15.3). In Section 15.4 we describe a compositional characterization of HWE. An important issue is which equilibrium genotype composition is the nearest to the observed sample. As a consequence, we propose a nonparametric compositional approach to test for HWE which can compete with the classical tests. An example with empirical data is given in Section 15.5.

15.2   Genetic data sets

Most SNPs are bi-allelic markers. Because there are four different bases (A, T, G and C), there exist six possible bi-allelic polymorphisms. Thus, each SNP is characterized as an A/T or A/G, etc. polymorphism. The most simple situation concerns the study of a single marker, possibly sampled repeatedly over different groups. Modern genotyping technology generates tremendous amounts of SNP data. Currently, over 3 million SNPs have been described for the human genome (The International HapMap Consortium 2007). Table 15.1 shows the nature of the data obtained in a modern genotyping study.

The data in Table 15.1 are seen to be of multivariate categorical nature. Often individuals come in groups (e.g. different biological populations, cases and controls, males and females, etc.). SNPs can also be grouped according to their particular position (e.g. on which chromosome, within which gene, etc.). SNPs are typically tested for HWE in a one-by-one fashion, disregarding their possible correlation (linkage disequilibrium).

Table 15.1   SNP data. The genotypes of some individuals for four different SNPs.

Individual	SNP_1	SNP_2	SNP_3	SNP_4
I_1	AA	GT	CA	TT
I_2	AT	GG	CC	TT
I_3	AT	TT	CC	TT

15.3   Classical tests for HWE

We briefly summarize the most important statistical tests for HWE. These are the chi-square test and the exact test. The Pearson χ^2 statistic for a test for HWE is given by:

$$X_c^2 = \sum_{i=1}^{3} \frac{(|n_i - e_i| - c)^2}{e_i},$$ (15.2)

where n_i represents one of the three genotype counts (n_{AA}, n_{AB} and n_{BB}), and e_i the respective expected value under HWE. This is a test for the goodness-of-fit of a multinomial distribution. Parameter c represents the continuity correction. With $c = 0$ we obtain the ordinary χ^2 statistic, and with $c = \frac{1}{2}$ the corrected χ^2 statistic is obtained. The p-value of the test is obtained by comparing the χ^2 statistic with a χ^2 distribution with one degree of freedom.

The exact test for HWE (Levene 1949; Haldane 1954; Weir 1996) uses the conditional distribution of the number of heterozygotes N_{AB}, given the allele count N_A, and is given by

$$P(N_{AB} = n_{AB}|N_A = n_A) = \frac{n!\, n_A!\, n_B!\, 2^{n_{AB}}}{(2n)!\, n_{AB}!\, \left(\frac{1}{2}(n_A - n_{AB})\right)!\, \left(\frac{1}{2}(n_B - n_{AB})\right)!},$$ (15.3)

where n_A and n_B are the sample counts of allele A and B, respectively ($n_A = 2n_{AA} + n_{AB}$, $n_B = 2n_{BB} + n_{AB}$). The p-value of the exact test is computed as the sum of the probabilities of all possible samples with the same allele count that are as likely or less likely than the observed sample. Fast recursive procedures exist to compute the probabilities according to Equation (15.3) for different numbers of heterozygotes (Elston and Forthofer 1977; Wigginton *et al.* 2005).

15.4   A compositional approach

The Italian statistician Bruno de Finetti (1926) recognized that the genotype frequencies for a bi-allelic locus can be represented in a ternary plot, and the latter plot is therefore in genetics also known as a *de Finetti* diagram (Cannings and Edwards 1968). The Hardy–Weinberg law [Equation (15.1)] defines a parabola inside the ternary plot. A genetic marker that is in exact HWE is represented by a point on the parabola. The ternary plot is a useful tool for exploring a set of samples that has been typed for a particular marker. It constitutes a joint plot of the three genotype frequencies and the allele frequencies. Some examples of ternary plots are given later. Compositional data sets are data sets where the observations form part of a whole, and are therefore subject to a constant-sum constraint. For genetic markers, the obvious constraints are unit-sum constraints concerning the relative (population) allele frequencies and the relative (population) genotype frequencies:

$$p + q = 1 \quad \text{and} \quad f_{AA} + f_{AB} + f_{BB} = 1.$$

In genetics, the population genotype frequencies are typically estimated by sample genotype frequencies, $\hat{f}_{AA}$, $\hat{f}_{AB}$ and $\hat{f}_{BB}$, and the allele frequency p is estimated from the sample genotype frequencies by $\hat{p} = \hat{f}_{AA} + 0.5 \hat{f}_{AB}$. Here, we will use Jeffreys' (1946) estimators of the genotype frequencies given by

$$\hat{f}_{AA} = \frac{n_{AA} + \frac{1}{2}}{n + \frac{3}{2}}, \quad \hat{f}_{AB} = \frac{n_{AB} + \frac{1}{2}}{n + \frac{3}{2}}, \quad \hat{f}_{BB} = \frac{n_{BB} + \frac{1}{2}}{n + \frac{3}{2}}. \tag{15.4}$$

The use of Jeffreys' estimators for the genotype frequencies in Equation (15.4) has the advantage that the resulting genotypic compositions contain no zeros. Thus, log-ratios can always be computed, and the *zero problem*, well-known in compositional data analysis, is effectively avoided. An alternative solution to the zero problem is to replace the zeros by a small amount, while retaining the ratios of the remaining frequencies constant. This alternative is described in detail by Martín-Fernández and Thió-Henestrosa (2006) and by Fry *et al.* (2000). By taking log-ratios [e.g. of the form $\ln(\hat{f}_{AA}/\hat{f}_{AB})$, $\ln(\hat{f}_{BB}/\hat{f}_{AB})$] we obtain transformed variables that are free of the unit-sum constraint. Standard statistical analysis can then be applied to the transformed data. Let **x** be the 3-part composition containing the genotype counts: $\mathbf{x} = (f_{AA}, f_{AB}, f_{BB})$. By applying the additive (alr) or centred (clr) log-ratio transformation (Aitchison 2003), or the isometric log-ratio (ilr) transformation (Egozcue *et al.* 2003) (see also Chapter 2), we obtain

$$\mathrm{alr}(\mathbf{x}) = \begin{cases} \left(\ln \dfrac{f_{AA}}{f_{AB}}, \ln \dfrac{f_{BB}}{f_{AB}} \right), \\[2mm] \left(\ln \dfrac{f_{AB}}{f_{AA}}, \ln \dfrac{f_{BB}}{f_{AA}} \right), \\[2mm] \left(\ln \dfrac{f_{AA}}{f_{BB}}, \ln \dfrac{f_{AB}}{f_{BB}} \right), \end{cases} \tag{15.5}$$

$$\mathrm{clr}(\mathbf{x}) = \left(\ln \dfrac{f_{AA}}{g_{\mathrm{m}}(\mathbf{x})}, \ln \dfrac{f_{AB}}{g_{\mathrm{m}}(\mathbf{x})}, \ln \dfrac{f_{BB}}{g_{\mathrm{m}}(\mathbf{x})} \right), \tag{15.6}$$

$$\mathrm{ilr}(\mathbf{x}) = \begin{cases} \left(\dfrac{1}{\sqrt{2}} \ln \dfrac{f_{AA}}{f_{BB}}, \dfrac{1}{\sqrt{6}} \ln \dfrac{f_{AA}f_{BB}}{f_{AB}^2} \right), \\[2mm] \left(\dfrac{1}{\sqrt{2}} \ln \dfrac{f_{AB}}{f_{BB}}, \dfrac{1}{\sqrt{6}} \ln \dfrac{f_{AB}f_{BB}}{f_{AA}^2} \right), \\[2mm] \left(\dfrac{1}{\sqrt{2}} \ln \dfrac{f_{AA}}{f_{AB}}, \dfrac{1}{\sqrt{6}} \ln \dfrac{f_{AA}f_{AB}}{f_{BB}^2} \right), \end{cases} \tag{15.7}$$

where $g_{\mathrm{m}}(\cdot)$ denotes the geometric mean of its arguments. There are three alr and three ilr transformations. In the remainder of this chapter, we will use the first alr and the first ilr transformation given in Equations (15.5) and (15.7), respectively. The choice between the three transformations is immaterial for the final analysis. Under the condition of exact HWE, the log-ratio coordinates are either constant or linearly related. The clr coordinates under HWE are given by $(\frac{1}{3} \ln 2 + \ln \frac{p}{1-p}, \frac{2}{3} \ln 2, -\frac{1}{3} \ln 2 - \ln \frac{p}{1-p})$. For these coordinates HWE implies that the second dimension is constant, and that the first and the third are

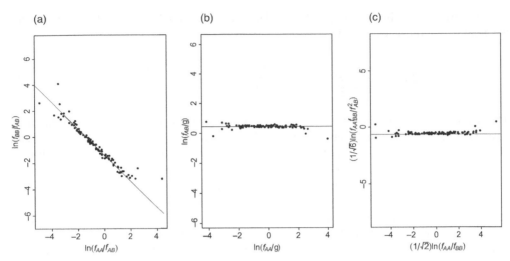

Figure 15.1 Log-ratio coordinates under HWE. Simulated data, 100 SNPs, 1000 individuals, uniform allele frequency. (a) alr coordinates; (b) clr coordinates; (c) ilr coordinates.

colinear. Under HWE the alr and ilr coordinates are $(-\ln 2 + \ln \frac{p}{1-p}, -\ln 2 - \ln \frac{p}{1-p})$ and $(\sqrt{2}\ln \frac{p}{1-p}, -\sqrt{2/3}\ln 2)$, respectively. All relationships are illustrated in Figure 15.1, with a simulated data set. One hundred SNPs were simulated by drawing samples of size 1000 from a multinomial distribution under the assumption of HWE, using a uniformly distributed allele frequency. Figure 15.1 shows that the Hardy–Weinberg parabola in the ternary plot corresponds to a straight line in log-ratio coordinates. The samples are seen to cluster around the HWE line, with increased dispersion for extreme allele frequencies. Under HWE, alr coordinates are linearly related, whereas for clr and ilr the second coordinate is constant.

A natural compositional measure for HW disequilibrium is the Euclidean distance between the log-ratio coordinates of a sample and its projection onto the HWE line (Mackenzie *et al.* 2008). The Euclidean distance in ilr coordinates corresponds to the Aitchison distance between the genotypic sample and a point on the Hardy–Weinberg parabola in the ternary plot. A distance of zero implies exact HWE, and large distances imply disequilibrium. We note that the projection of a sample point onto the HWE line in ilr coordinates corresponds, in general, to an *oblique* projection of the sample point onto the HWE parabola in the ternary plot. This is illustrated in Figure 15.2. Points B and C will coincide if the allele frequency is 0.5. Point B can be interpreted as the genotypic composition that would be reached after one generation of random mating, given the current allele frequency. Point C is the genotypic composition in HWE that requires the smallest change in genotype frequencies with respect to the current genotype frequencies using Aitchison geometry of the simplex (Pawlowsky-Glahn and Egozcue 2001) (see also Chapter 11).

In order to decide whether the sample deviates significantly from HWE, we need the distribution of this distance, or the distribution of the second ilr coordinate. Mackenzie *et al.* (2008) used the Aitchison distance as a criterion for deviation from HWE. The log-ratio representation in Figure 15.2 suggests that the second ilr coordinate can be used as a measure of the deviation from HWE. We explore this possibility in some more detail. Note that this

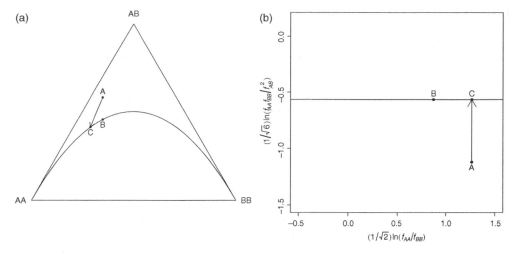

Figure 15.2 (a) Ternary plot of compositions A, B and C. (b) ilr plot of the same data. A, a genotypic composition; B, the expected composition under HWE for a χ^2 test; C, the expected composition under HWE for a compositional test. Reprinted with permission from Wiley UK Ltd.

compositional measure of deviation implies that the allele frequency of the corresponding *expected composition under equilibrium* (C in Figure 15.2) is different from the observed sample allele frequency. This is in contrast to the classical χ^2 test where the allele frequency of the composition expected under HWE (B in Figure 15.2) has the same allele frequency as the observed sample. Point C is an orthogonal projection (Aitchison geometry), and its ilr coordinates are readily computed: the first one is equal to $\sqrt{2}\ln[p/(1-p)]$, and the second coordinate is equal to $-\sqrt{2/3}\ln 2$. It can be shown that the allele frequency of point C, $\tilde{p}$, can be computed from the sample genotype frequencies (point A) as

$$\tilde{p} = \frac{\sqrt{\hat{f}_{AA}}}{\sqrt{\hat{f}_{AA}} + \sqrt{\hat{f}_{BB}}},$$

whereas the allele frequency corresponding to points A and B is the usual one $\hat{p}_A = \hat{f}_{AA} + 0.5\hat{f}_{AB}$. The statistical distribution of the second ilr coordinate is not known, and we therefore use a nonparametric approach. By bootstrapping the observed genetic composition, we generate a bootstrap distribution of the second ilr coordinate. To correct for bias, the bootstrap distribution is centred onto its expected value under the null hypothesis, $-\sqrt{(2/3)}\ln 2$. This is usually referred to as the bootstrap shift method which applies a translation of the bootstrap distribution onto the null hypothesis distribution. See Noreen (1989) and Efron and Tibshirani (1993) for more details on this translation and on bootstrap methods in general. A bootstrap p-value can be computed by referring the second ilr coordinate of the observed sample to this bootstrap distribution. HWE is rejected if the bootstrap p-value is smaller than some prespecified value (e.g. 0.05). In order to gain some insight in the performance of the

214   HARDY–WEINBERG EQUILIBRIUM

Table 15.2 Monte Carlo simulation results (10 000 simulations) for a test for HWE based on the second ilr coordinate. p, allele frequency; α, Monte Carlo rejection rate; Bias, difference between the mean of the bootstrap samples and the null value $-\sqrt{(2/3)}\ln(2)$.

	$n = 50$		$n = 100$		$n = 500$	
p	α	Bias	α	Bias	α	Bias
0.1	0.0640	0.2025	0.1673	0.0286	0.0568	−0.04490
0.2	0.1172	−0.0363	0.0522	−0.0471	0.0411	−0.00756
0.3	0.0411	−0.0398	0.0351	−0.0168	0.0433	−0.00444
0.4	0.0329	−0.0256	0.0434	−0.0109	0.0477	−0.00201
0.5	0.0402	−0.0165	0.0440	−0.0108	0.0502	−0.00190
0.6	0.0311	−0.0233	0.0438	−0.0120	0.0495	−0.00256
0.7	0.0415	−0.0416	0.0367	−0.0156	0.0476	−0.00410
0.8	0.1238	−0.0449	0.0493	−0.0440	0.0438	−0.00812
0.9	0.0619	0.2000	0.1612	0.0323	0.0543	−0.04233

bootstrap test, we performed a Monte Carlo study. We generated genotypic compositions under the HWE assumption by drawing samples from a multinomial distribution, using three different sample sizes (multinomial distributions with 50, 100 or 500 trials) for a given allele frequency (0.1, 0.2, ..., 0.9). We used 10 000 Monte Carlo simulations, and estimated the significance level (α) as well as the bias from the simulations. The results are shown in Table 15.2.

Table 15.2 shows that for large samples the bias [difference between the mean of the bootstrap distribution and the null value $-\sqrt{(2/3)}\ln 2$] is small, and that the rejection rate of the bootstrap test is close to the theoretical level of 5%. For smaller samples (50, 100) and extreme allele frequencies (≤ 0.2 or ≥ 0.8) the rejection rate of the bootstrap test is above the nominal level. We note that under these conditions there will be more zero counts in the data set.

15.5   Example

We apply the bootstrap test of the previous section to an empirical data set, and compare the result with those obtained by the classical tests. We consider bloodgroup data for the MN locus as an example. The bloodgroup of an English sample of 1000 donors was determined (Cleghorn 1960; Hedrick 2005), giving the result (298,489,213) for MM, MN and NN individuals, respectively. The classical tests do not reject HWE. The χ^2 test for HWE, with continuity correction, gives a test statistic of $\chi^2 = 0.1790$ and the corresponding p-value is 0.6723. The exact test gives a p-value of 0.6723. This is the p-value for a two-sided test, the dost p-value (Graffelman 2010). In Figure 15.3(a) we illustrate our nonparametric compositional test by depicting 1000 bootstrap samples in a ternary plot. Bootstrap samples for which a χ^2 test is significant are dark, whereas nonsignificant bootstrap samples are not. The curves around the HWE parabola delimit the acceptance region for the χ^2 test (Graffelman and Morales-Camarena 2008). Figure 15.3(b) shows the bootstrap samples in ilr coordinates. The original observed sample is indicated by a triangle and has a second ilr coordinate of −0.5407. The bootstrap p-value is 0.63, and this is in close agreement with both classical tests.

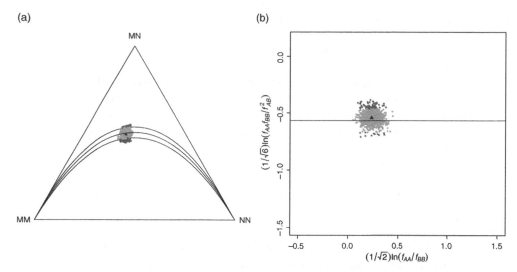

Figure 15.3 Ternary plot (a) and scatterplot of ilr coordinates (b) of 1000 bootstrap samples of the MN blood group data. The triangle represents the original sample. Bootstrap samples with a significant χ^2 test are dark.

15.6 Conclusion and discussion

We have highlighted that statistical inference for HWE is subject to an important decision: what genetic composition is chosen as *the* reference equilibrium composition? The classical tests condition on allele frequency, and obtain the reference composition by vertical projection onto the HWE parabola inside the ternary plot. Compositional considerations based on Aitchison's geometry imply a different equilibrium composition, and this can affect the inference being made. A problem of the compositional approach is that a reference distribution for the log-ratio coordinates is lacking, though this can be resolved by bootstrapping. The results of the Monte Carlo study show that the estimation of the equilibrium value for the log-ratio coordinates is biased for extreme allele frequencies. However, the bootstrap estimator can be corrected for this bias. The empirical example shows the bootstrap procedure to agree with the classical tests, but this is for allele frequencies around 0.5. For extreme allele frequencies, a large amount of zeros will appear in the data and consequently bias will become more manifest.

Acknowledgements

This research has been supported by the Spanish Ministry of Science and Innovation (projects SEC2003-04476, CSD2006-00032 and MTM2009-13272) and by the Agència de Gestió d'Ajuts Universitaris i de Recerca of the Generalitat de Catalunya (Ref. 2009SGR424).

References

Aitchison J 2003 *The Statistical Analysis of Compositional Data*. The Blackburn Press, Caldwell, NJ (USA). 435 p.

Ayres KL and Balding DJ 1998 Measuring departures from Hardy–Weinberg: a Markov-chain-Monte Carlo method for estimating the inbreeding coefficient. *Heredity* **80**, 769–777.

Cannings C and Edwards AWF 1968 Natural selection and the de Finetti diagram. *Annals of Human Genetics* **31**, 421–428.

Cleghorn 1960 MNSs gene frequencies in english blood donors. *Nature* **187**(4738), 701.

Crow JF 1988 Eighty years ago: the beginnings of population genetics. *Genetics* **119**, 473–476.

de Finetti B 1926 Considerazioni matematiche sull'ereditarietà mendeliana. *Metron* **6**(3), 3–41.

Edwards AWF 2008 G. H. Hardy (1908) and Hardy–Weinberg Equilibrium. *Genetics* **179**, 1143–1150.

Efron B and Tibshirani RJ 1993 *An Introduction to the Bootstrap*. Chapman & Hall, New York, NY (USA). pp. 124–133, 224–227.

Egozcue JJ, Pawlowsky-Glahn V, Mateu-Figueras G and Barceló-Vidal C 2003 Isometric logratio transformations for compositional data analysis. *Mathematical Geology* **35**(3), 279–300.

Elston RC and Forthofer R 1977 Testing for Hardy–Weinberg equilibrium in small samples. *Biometrics* **33**(3), 536–542.

Emigh TH 1980 A comparison of tests for Hardy–Weinberg equilibrium. *Biometrics* **36**, 627–642.

Fry JM, Fry TRL and McLaren KR 2000 Compositional data analysis and zeros in micro data. *Applied Economics* **32**(8), 953–959.

Graffelman J 2010 The number of markers in the hapmap project: some notes on chi-square and exact tests for Hardy–Weinberg equilibrium. *The American Journal of Human Genetics* **86**, 813–818.

Graffelman J and Morales-Camarena J 2008 Graphical tests for Hardy–Weinberg equilibrium based on the ternary plot. *Human Heredity* **65**(2), 77–84. DOI: 10.1159/000108939.

Haber M 1981 Exact significance levels of goodness-of-fit tests for the Hardy–Weinberg equilibrium. *Human Heredity* **31**, 161–166.

Haldane JBS 1954 An exact test for randomness of mating. *Journal of Genetics* **52**, 631–635.

Hardy GH 1908 Mendelian proportions in a mixed population. *Science* **28**, 49–50.

Hedrick PW 2005 *Genetics of Populations*, 3rd edition. Jones and Bartlett, Sudbury, MA (USA). 69 p.

Hernández JL and Weir BS 1989 A disequilibrium coefficient approach to Hardy–Weinberg testing. *Biometrics* **45**(1), 53–70.

Jeffreys H 1946 An invariant form for the prior probability in estimation problems. *Royal Society of London Proceedings Series A* **186**, 453–461.

Levene H 1949 On a matching problem arising in genetics. *Annals of Mathematical Statistics* **20**, 91–94.

Li CC 1976 *First Course in Population Genetics*. Boxwood Press, Pacific Grove, CA (USA).

Lindley DV 1988 Statistical inference concerning hardy-weinberg equilibrium. In *Bayesian Statistics 3* (ed. Bernardo JM, DeGroot MH, Lindley DV and Smith AFM), Oxford University Press, New York, NY (USA), pp. 307–326.

Mackenzie JR, Egozcue JJ, Heilbronner R, Hielscher R, Müller A and Schaeben H 2008 Quantifying rock fabricks - a test of independence of the spatial distribution of crystals. In *Proceedings of CoDaWork'08, The 3rd Compositional Data Analysis Workshop* (ed. Daunis-i Estadella J and Martín-Fernández J). http://hdl.handle.net/10256/723. University of Girona, Girona (Spain). CD-ROM.

Martín-Fernández JA and Thió-Henestrosa S 2006 Rounded zeros: some practical aspects for compositional data. In *Compositional Data Analysis in the Geosciences: From Theory to Practice* (ed. Buccianti A, Mateu-Figueras G and Pawlowsky-Glahn V). Geological Society, London (UK). pp. 191–201.

Noreen EW 1989 *Computer Intensive Methods for Testing Hypotheses. An Introduction*. John Wiley & Sons, Ltd, New York, NY (USA). pp. 66–69.

Pawlowsky-Glahn V and Egozcue JJ 2001 Geometric approach to statistical analysis on the simplex. *Stochastic Environmental Research and Risk Assessment (SERRA)* **15**(5), 384–398.

Shoemaker J, Painter I and Weir BS 1998 A bayesian characterization of Hardy–Weinberg disequilibrium. *Genetics* **149**, 2079–2088.

Teo YY, Fry AE, Clark TG, Tai ES and Seielstad M 2007 On the usage of HWE for identifying genotyping errors. *Annals of Human Genetics* **71**(5), 701–703.

The International HapMap Consortium 2007 A second generation human haplotype map of over 3.1 million SNPs. *Nature* **449**, 851–861.

Wakefield J 2010 Bayesian methods for examining Hardy–Weinberg equilibrium. *Biometrics* **66**, 257–265.

Weinberg W 1908 Über den Nachweis der Vererbung beim Menschen. *Jahreshefte des Vereins für Vaterländische. Naturkunde in Württemberg, Stuttgart* **64**, 369–382.

Weir BS 1996 *Genetic Data Analysis II*. Sinauer Associates, Sunderland, MA (USA). pp. 92–102.

Wigginton JE, Cutler DJ and Abecasis GR 2005 A note on exact tests of Hardy–Weinberg equilibrium. *American Journal of Human Genetics* **76**, 887–893.

Wittke-Thompson JK, Pluzhnikov A and Cox NJ 2005 Rational inferences about departures from Hardy–Weinberg equilibrium. *American Journal of Human Genetics* **76**, 967–986.

16

Compositional analysis in behavioural and evolutionary ecology

Michele Edoardo Raffaele Pierotti[1] and Josep Antoni Martín-Fernández[2]
[1]Department of Biology, East Carolina University, USA
[2]Department of Computer Science and Applied Mathematics, University of Girona, Spain

16.1   Introduction

Despite the ubiquity of the compositional problem in the natural sciences, the application of compositional data analysis – hereafter CODA – in ecology and evolutionary biology is still patchy and largely dependent on the interest and statistical literacy of the individual researcher rather than a well established methodology when the research question is dealing with proportions or parts.

The class of gradient analysis methods includes correspondence analysis and related methods which are frequently applied in those studies where the vector of proportions derives from counts. Most of these methods are supported by several software packages with perhaps the more ecologically oriented being CANOCO (ter Braak and Šmilauer 2002), and the R-packages 'VEGAN' (Oksanen *et al.* 2010) and 'ade4' (Dray and Dufour 2007). A basic and relevant property shared by these ordination methods is that they preserve chi-square distances between count data. For example, ter Braak and Schaffers (2004) noted that 'In such a plot, distances among sites and among species represent chi-square distances and

Compositional Data Analysis: Theory and Applications, First Edition. Edited by Vera Pawlowsky-Glahn and Antonella Buccianti.
© 2011 John Wiley & Sons, Ltd. Published 2011 by John Wiley & Sons, Ltd.

points of different items allow, for all practical purposes, a biplot interpretation'. In the same paper the authors recognize that chi-square distances are the Euclidean distances if the data are appropriately transformed through a linear function. In fact, Legendre and Gallagher (2001) showed that chi-square distance is a member of the class of weighted Euclidean distances between vectors of proportions. In this sense, chi-square distance inherits not only the positive properties from Euclidean distance, but also the negative ones, most notably the subcompositional incoherence. It is well known that any appropriate distance between two vectors of proportions – compositions – should satisfy particular requirements (Aitchison et al. 2000) which include the perturbation invariance and the subcompositional dominance. In ecology, Jackson (1997) recognized this type of difficulties when classical tools (i.e. those based on Euclidean distances) are applied to vectors of proportions. It is therefore surprising that, in the same paper, the author recommends applying Correspondence Analysis in order to circumvent those difficulties. Nowadays, there are available CODA strategies (e.g. clr-transformation and working on coordinates) that simultaneously allow the application of classical tools and the verification of data requirements. Similar considerations could be made for other methods based on Euclidean or chi-square distances such as the least square algorithm (Mooijaart et al. 1999) and the reduced-rank vector generalized linear models (Yee and Hastie 2003). In addition, the latter proposed a new regression technique in the context of a multinomial logit model (MLM) for count data. This encourages the application of CODA tools because a detailed examination of equation (11) in Yee and Hastie (2003) allows the interpretation of the MLM as a classical linear regression model applied to the alr-transformed data. The search for this type of relationship between classical and CODA methods has motivated a number of studies. Among the most interesting are by Greenacre (2009) (see also Chapter 8). In our opinion, the most important difference between other techniques and the log-ratio methodology is that in CODA the vector of proportions represents realizations of a random vector for which the sample space is the simplex.

Our intent here is not comprehensiveness: recent work on gene expression analysis (e.g. Liebscher 2008), life history tables (Oeppen 2008) and Chapters 14 and 15 are just a few examples of the wide range of topics in ecology and genetics where ad hoc compositional techniques are being successfully developed. The purpose of this chapter is to present areas of research in behavioural and evolutionary ecology where CODA would represent a useful analytical tool with many promising applications. We will also provide examples where (early) compositional methods have long been used but for which the integration with more recent CODA developments might allow more flexibility (e.g. working on coordinates or in handling zero data), wider range of applications and to incorporate more complex and realistic statistical models. For information about software packages that support new advances in CODA see Chapters 24–26.

16.2   CODA in population genetics

Genes or particular genetic sequences (in general, loci) come in various 'versions' or alleles. By measuring the frequency of certain alleles of a locus in different populations and by looking at many loci, it is possible to infer whether migration between populations is taking place and whether individuals of the studied populations interbreed (in doing so, exchanging genes), whether selection is acting on those loci, and to estimate population sizes and other processes such as random drift. It follows that the analysis of allele frequencies in population genetics is (or should be) an inherently compositional statistical problem.

The issue of zeros in the analysis of population genetic data is of considerable relevance. Some authors (Gasparini and Di Gaetano 2003) state that zeros might be both 'false zeros' (or 'count zeros': alleles undetected because of their very low frequency in a population) and 'structural zeros', the latter condition when the allele is truly absent from the population (of course, here not including zeros stemming from missing data). However, it is debatable whether such dichotomy needs to be for zeros in population genetic data sets: population genetic theory fully assimilates low undetected frequency with absence of the allele and only rarely the absence of an allele is considered qualitatively different from very low frequency. An exception, however, might be represented, under certain conditions, by the 'private alleles' approach in which genetic distance between populations is inferred from the number of alleles that are unique to a population.

We believe that zeros in population genetic data sets can be modelled as 'count' zeros with no loss of information. However, the more compelling issue remains the possibly large number of zeros inevitably associated with the use of high dimensionality of the compositions characterizing these data sets. It is well established that the power to detect weak genetic structuring and microevolutionary processes, as well as its reliability, are more strongly affected by the number of loci used rather than by the allelic diversity per locus (Bernatchez and Duchesne 2000). In the recent past, the relatively high costs in terms of time and funding associated with developing microsatellite markers and their relatively high per marker information content, led to the use of 6–20 markers in an average population structure study (with notable exceptions such as in human population genetic studies and more recently mapping studies where hundreds of microsatellites are detected and developed by high throughput sequencing). More recently, markers with extremely low per marker information content but widespread distribution across the genome, high resolution and relative ease of development have forced a reconsideration of many statistical tools routinely used in the analysis of marker data.

Last generation markers such as single nucleotide polymorphisms (SNPs) are typically used in sets of hundreds to hundreds of thousands of loci (Nelis *et al.* 2009). Therefore, from a compositional point of view, microsatellite data sets are very different from SNPs data sets. This difference is not only in terms of the number of compositions (number of loci) involved simultaneously as observed above, but the number of possible states (i.e. the size of the composition): in population genetics, the frequency of each of a number of alleles at each locus is the quantity of interest and the number of alleles in informative microsatellite loci is generally between 4–5 and 50; on the other hand, with SNPs, the possible different variants (alleles) per locus can only be 4 (i.e. the number of existing nucleotides in nature, representing the letters of the genetic alphabet) and 2–3 is the norm.

We believe CODA might provide significant contributions to the analysis of such inherently compositional data sets but the likely presence of a large number of zero values, the high dimensionality of the data sets and the interest in considering individual-level patterns of variation (multilocus genotypes) by comparing compositions of compositions, might require novel theoretical developments in compositional research.

Given the level of sophistication of the statistical tools commonly employed in population genetics and the widespread use of multivariate analysis, it is surprising that almost no attention has yet been paid to the consequences of not recognizing the compositional nature of allele frequencies at a locus so that, for example, principal component analysis (PCA) is commonly performed on the raw frequencies. Even a recent review on multivariate analysis of genetic markers (Jombart *et al.* 2009), while recognizing the existence of the compositional problem and its general dismissal in the population genetic literature, suggests that the results of a PCA of allele frequencies could be 'improved' by applying transformations developed by Aitchison

(Aitchison 1983) rather than pointing out that such ordination on the raw frequencies would be simply meaningless (Aitchison 1983, 1986). Beside the direct application of multivariate techniques to the sets of allele frequencies, more commonly such statistical tools are used after derivation of so-called genetic distances between populations at each locus. A commonly used estimator, the Cavalli-Sforza and Edwards chord distance (Cavalli-Sforza and Edwards 1967) considers populations as points in a D-dimensional Euclidean space each specified by specific sets of D allele frequencies. For populations $\mathbf{X}$ and $\mathbf{Y}$, with r loci and D alleles at each locus, let x_{ik} and y_{ik} be the frequencies of the ith allele at the kth locus in populations $\mathbf{X}$ and $\mathbf{Y}$, respectively. The Cavalli-Sforza distance for the kth locus is

$$D_{CH}(\mathbf{X}, \mathbf{Y}) = \frac{2}{\pi} \sqrt{2 \left(1 - \sum_i \sqrt{x_{ik} y_{ik}} \right)},$$

represented by the angle between these points. Also, frequently used are the following genetic distance estimators:

$$D_R(\mathbf{X}, \mathbf{Y}) = \sqrt{\frac{\sum_l (x_{ik} - y_{ik})^2}{2}} \qquad \text{(Rogers 1972)},$$

$$D_N(\mathbf{X}, \mathbf{Y}) - -\ln \left(\frac{\sum_k \sum_i x_{ik} y_{ik}}{\sqrt{\sum_k \sum_i x_{ik}^2 \sum_k \sum_i y_{ik}^2}} \right) \qquad \text{(Nei 1972)},$$

$$\theta_w(\mathbf{X}, \mathbf{Y}) = \sqrt{\frac{\sum_k \sum_i (x_{ik} - y_{ik})^2}{2 \sum_k \left(1 - \sum_i x_{ik} y_{ik} \right)}} \qquad \text{(Reynolds et al. 1983)},$$

$$D_S(\mathbf{X}, \mathbf{Y}) = \sum_{i \neq j} |i - j| \left(x_{ik} y_{jk} - \frac{x_{ik} x_{jk} + y_{ik} y_{jk}}{2} \right) \qquad \text{(Shriver et al. 1995)},$$

$$(\delta\mu)^2(\mathbf{X}, \mathbf{Y}) = \frac{\sum_k \left(\sum_i i x_{ik} - \sum_i i y_{ik} \right)^2}{r} \qquad \text{(Goldstein et al. 1995)}.$$

However, generating a matrix of genetic distances per locus among populations and applying multivariate methods such as PCA or DFA to these distances does not represent a solution to the problem of the compositional nature of the data (here allele frequencies). All these distances are inconsistent with the geometric properties of the space characterizing vectors of proportions, ratios, frequencies and, in particular, scale invariance and subcompositional dominance (see Chapter 2 for a description of these principles). Unfortunately, the use of the above estimators in conjunction with multivariate analysis represents a well-established approach in population genetic studies with consequences on the robustness of the biological inferences made on these incorrect analyses that are hard to quantify. The solution to this issue is relatively straightforward. Two classes of measures handle appropriately compositional vectors such as allele frequencies. The first, Aitchison's distance, follows from the geometry of the simplex, the second, the C-KL, derives from measures of dissimilarity (Palarea-Albaladejo et al. 2011). The adequacy of different dissimilarities in the simplex, together with the behaviour of the common log-ratio transformations, are also discussed in Palarea-Albaladejo *et al.* (2011). It is therefore clear that the introduction of CODA techniques in population genetics is overdue and is likely to become the standard approach, once ad-hoc user-friendly software for the multivariate analysis of genetic markers becomes available. Actually, the state of the art at the moment mirrors the state of the geological sciences in the early 1980s when the pioneering work of John Aitchison was seen with scepticism and the results of even large expensive research programmes could be compromised by a failure in recognizing the compositional nature of the data (Aitchison 1986). Landscape genetics and the multivariate analysis of spatially referenced genetic data, represent a critical avenue for future research in which CODA might give significant contributions. Some signs of growing awareness are nevertheless emerging and it is particularly notable that this is happening in theoretical population genetics. Broquet *et al.* (2009) present a novel method to infer recent migration rates with multilocus genotype data. The authors' approach consists in calculating pairwise migration rates between a set of populations by sampling at two time points that span the putative migration events and deriving estimates of migration from changes in population structure that have occurred in this time span. Each individual in each population is sampled at a number of independent marker loci. Broquet *et al.* (2009) set to estimate the matrix of migration rates, i.e. the proportion of multilocus individuals sampled in a population X_2 at a time t_2 that were present in population X_1 at the time t_1 before dispersal occurred, for all pairs of populations. Recognizing the compositional nature of both the multilocus frequency data and the migration rates, the authors model them with Aitchison's lognormal distributions. They find this model more appropriate and flexible than the Dirichlet distribution, commonly used for genetic frequencies, for example allowing estimates of the covariances between migration rates when populations are sampled over more than two time points. This approach seems very promising particularly because it requires very few demographic and population genetic assumptions and because the distribution of multilocus frequencies are described by more realistic (compositional) models.

16.3   CODA in habitat choice

In the context of wildlife management, population ecology and animal behaviour, habitat or resource selection studies are particularly important since they can potentially provide important indications on the life history, physiology and ethological traits of the focal species. They also allow the identification of the relevant environmental parameters that affect the local distribution of individuals of the species of interest. Habitat or resource choice is studied at the individual level by comparing the composition of utilized habitats (resource) with that of the available habitats (resource units). Data are frequently gathered with GIS databases and radio-tracking of individual animals. Aebischer *et al.*'s (1993) influential paper established log-ratio analysis (CODA) in studies of habitat choice as the most appropriate technique. At the same time, they critically discussed the inadequacy of very popular techniques such as χ^2 and Friedman's tests for data that are evidently compositional. In fact, for a given home range of an individual animal, D habitats will be available in proportions described by the composition $x_{a1}, x_{a2}, \ldots, x_{aD}$ and the individual's habitat use by the composition $x_{u1}, x_{u2}, \ldots, x_{uD}$. Considering the corresponding sets of log-ratios $(y_{a1}, y_{a2}, \ldots, y_{aD-1})$ and $(y_{u1}, y_{u2}, \ldots, y_{uD-1})$, Aebischer *et al.* (1993) proposed a set of pairwise differences $d_i = y_{ui} - y_{ai}$, generated for each habitat and focal individual. A population level assessment of habitat choice is obtained by testing whether the vector of mean d_i values is significantly different from a zero vector. More in general, a MANOVA/MANCOVA approach can be used allowing the study of within-population variation in habitat use deriving from individual variation and effects such as morph type, age, sex, environmental conditions or daily or seasonal differences, etc. Wilks' lambda or, when the distribution is not multivariate normal, randomization tests, were proposed for pairwise comparisons between habitat types. While, thanks to this seminal paper, the compositional approach to habitat choice analysis is widely applied, the ecologists' community has at least partly failed to follow the rapid evolution of compositional techniques in the last decade. This fact has allowed the propagation in some literature of the impression that CODA is of limited applicability in habitat choice studies given the frequent occurrence of cases when an individual animal does not use an available habitat, resulting in zero values in the utilized habitat composition, or when a certain individual cannot access a particular habitat (i.e. one habitat is not available for that individual), resulting in zero values on both utilized and available compositions. Aebischer *et al.* (1993) suggested merging habitat types or excluding some, rather over-penalizing strategies for dealing with missing data, or alternatively, substitution with a small positive value 'less than the smallest recorded nonzero proportion'. In the case of missing data resulting from unavailability of a habitat (i.e. zero value in both utilized and available habitat compositions), replacement with the mean of all non missing values for that log-ratio was suggested. Today, more efficient approaches are available (see Chapter 4). The issue of missing data and zeros is undoubtedly critical in CODA but solutions do exist. Most of the recent improvements are inspired by Little and Rubin's (2002) monograph recommending specific treatments depending on the nature and amount of missing values. This is consistent with more recent CODA methods (Martín-Fernández *et al.* 2003) when the number of missing values and zero values in the data set are not very large (<10%). When the amount is large, other methods should be preferred (Palarea-Albaladejo and Martín-Fernández 2008; Hron *et al.* 2010). Emblematic of this hiatus between the habitat choice literature and the developments in zero replacement in compositional data is Bingham *et al.* (2007). The authors point the attention of the research community on the fact that compositional analysis leads to misclassified resource selection when unused habitats are present. By progressively substituting zero values with smaller values, Bingham *et al.* (2007) observed that the MANOVA method is sensitive. This difficulty is shared by all the multivariate techniques and it is well known since Aitchison (1986) who recommended a sensitivity analysis for the inputed small values. In fact, more recent developments in the treatment of zeros and missing values (Martín-Fernández *et al.* 2003) share a focus on addressing this difficulty with the minimum distortion of the covariance and distance structures of the data set.

More in general, the time seems ripe for the application of more recent compositional tools to the analysis of habitat choice introduced by Aebischer *et al.*'s (1993) alr study, by taking advantage of perturbation and the other operations in the simplex and by using coordinates on an orthonormal basis. For example, Billheimer *et al.* (2001), and also, independently, Pawlowsky-Glahn and Egozcue (2001), introduced the structure of Euclidean space to the simplex sample space in the context of an analysis of the abundance of species in communities with different predator compositions. Nowadays, this structure is supporting the adaptation of multivariate methods to CODA (see Chapter 3).

The genetic origin of stocks and the ecological correlates of population genetic structure are of extreme importance for the management and conservation of vulnerable or endangered species. A commonly used approach is mixed stock analysis, in which the relative contribution of various source populations to the focal mixed population is estimated (Bowen and Karl 2007; Baker 2008). Okuyama and Bolker (2005) set to study the ecological underpinnings of the genetic contributions of different source populations (rookeries) to sea turtle stocks. Typically, sea turtles aggregate for most of their life cycle in feeding mixed populations with contributions from a large number of source rookeries distributed on a large geographical scale (Bowen *et al*. 1996), a considerable challenge for population geneticists and conservation biologists. A possible solution, spearheaded by the work of Okuyama and Bolker (2005), is to include ecological information in Bayesian hierarchical models of mixed stock analysis. The authors develop a covariate model in which the prior for the distribution of rookery contributions is Aitchison's logistic normal (rather than the Dirichlet's characterizing previous models) allowing the consideration of important ecological factors such as rookery size and the effect of ocean currents in the model. As pointed out by the authors, the compositional approach offers great potential for developments of such essential tools in conservation genetics, for example by extending models to the spatial dimension, i.e. including the geographical distances between rookeries and applying Billheimer *et al*.'s (2001) conditional autoregressive models of the compositional rookery contributions.

16.4   Multiple choice and individual variation in preferences

As pointed out by Aebischer *et al*. (1993), in habitat preference studies the unit of replication is the individual animal. Recent developments in compositional analysis of individual variation in mating preferences, detailed below, hold promise for the statistical treatment of habitat preference data. Pierotti *et al*. (2009) examined within-population variation in male mating preferences for females belonging to three different colour morphs in a Lake Victoria cichlid fish species. This scenario, continuing the parallel with habitat preference data sets, is analogous to the study of habitat preferences (three habitats available) by individuals of a deer population. Pierotti *et al*. (2009) tested each individual male multiple times in simultaneous three-way choice trials with different combinations of females belonging to each morph and recorded the number of courtship displays towards each female during each trial. Therefore, a single male's mating preference in every trial is represented by the set of proportions of displays to each female morph, i.e. a composition. Dealing with preferences, one is not interested in the absolute number of displays: in fact, for example, in two different days an individual might be more or less willing to invest resources into courtship (e.g. as opposed to foraging, escaping predators, etc.) but this has little to do with its preferences. In other words, what matters is the individual's ranking of possible options [female morphs, in Pierotti *et al*. (2009)], rather than the intensity with which the choice is made or willingness to choose (Widemo and Sæther 1999). Each individual male was represented by its compositional geometric mean of preferences and confidence regions under log-ratio normality assumptions were calculated. A MANOVA was used to explore the differences between the geometric means of individual male preferences and hierarchical clustering adopting log-ratio Mahalanobis' distance was employed to identify the presence of preference classes in the population.

By subjecting each individual male to multiple trials with different triads of females, it was possible to generate a measurement of individual variation in compositional preferences by developing compositional multivariate forms of consistency and repeatability. In particular, using Lessells and Boag (1987) and Boake (1989) univariate repeatabilities, Pierotti *et al*.

(2009) replaced the univariate variances with the corresponding trace of the covariance matrix. For the univariate case, repeatability can be calculated as

$$R_u = \frac{s_A^2}{s_A^2 + s^2},$$

where s_A^2 is the *among-groups* variance component and s^2 is the *within-group* variance component. These variance components are calculated (Lessells and Boag 1987) from an analysis of variance as: $s_A^2 = (MS_B - MS_W)/n_0$ and $s^2 = MS_W$, where MS_B is the mean squares between-groups and MS_W the mean squares within-groups. The parameter n_0 is related to the sample size per group by:

$$n_0 = [1/(G-1)] \left[\sum_1^G n_i - \left(\frac{\sum_1^G n_i^2}{\sum_1^G n_i} \right) \right],$$

where n_i is the size sample of group i and G is the number of groups. For balanced experiments n_0 is equal to N, the total sample size. The expression for R_u suggests that the repeatability approximates unity when the variability among-groups (i.e. among-individuals) is larger than the variability deriving from the error part, that is, the individual males. As s_A^2 decreases (i.e. similarity between individual preferences increases) or as s^2 increases (i.e. individual preferences are weaker) then repeatability measure will yield values close to zero.

Repeatability R_u can be expressed in terms of means squares as

$$R_u = \frac{MS_B - MS_W}{MS_B + (n_0 - 1)MS_W}.$$

For the multivariate case in CODA, the mean squares are defined in terms of the trace of the matrices of variability of the ilr-transformed data set (Egozcue *et al.* 2003). We consider $MS_B = \text{Trace}(B)/(G-1)$ and $MS_W = \text{Trace}(W)/(N-G)$, where B and W are the matrices of variability 'between' and 'within'; N is the sample size and G the number of groups. Therefore a multivariate version of measure of repeatability R_m can be defined as

$$R_m = \frac{\frac{\text{Trace}(B)}{G-1} - \frac{\text{Trace}(W)}{N-G}}{\frac{\text{Trace}(B)}{G-1} + (n_0 - 1)\frac{\text{Trace}(W)}{N-G}}.$$

In addition to repeatability, a multivariate version of consistency of individual preferences was introduced (Pierotti *et al.* 2009) as the multivariate coefficient of variation (CV*; Hallgrímsson and Hall 2005):

$$\text{CV}^* = \frac{\sqrt{\text{totvar}(\mathbf{X})}}{\|\text{mean}(\mathbf{X})\|}.$$

where small values of CV* indicate high consistency in preferences. In the CV* expression, the term $\|\text{mean}(\mathbf{X})\|$ represents the norm of the centre of the data set $\mathbf{X}$, that is, the distance between mean($\mathbf{X}$) vector and the origin of coordinates. The dispersion of the trials results (total variance, totvar) can be measured as the trace of the variance-covariance matrix. If the data are proportions (i.e. compositional data), the data set is ilr-transformed and the totvar is the trace of its corresponding variance-covariance matrix. The centre of the sample set (set of trials of an individual) is the geometric mean and the distance is Aitchison's (Aitchison *et al.* 2000). In other words, for compositional data, $\|\text{mean}(\mathbf{X})\|$ is equivalent to calculating the distance between the geometric mean of a data set $\mathbf{X}$ and the centre of the simplex or no-preference point (i.e. $[1/D, 1/D, \ldots, 1/D]$), using Aitchison distance. This measure CV^* generalizes to the case where the no-preference point is any distribution $\mathbf{X}_0$ (e.g. vector of proportions of resources) using the expression

$$CV^* = \frac{\sqrt{\text{totvar}(\mathbf{X})}}{\|\text{mean}(\mathbf{X}) - \mathbf{X}_0\|}.$$

The derivation of this expression is fully equivalent to first centring the data and then calculating the CV^*:

$$CV^* = \frac{\sqrt{\text{totvar}(\mathbf{X} - \mathbf{X}_0)}}{\|\text{mean}(\mathbf{X} - \mathbf{X}_0)\|}$$

In CODA the centring consists of applying the perturbation operation by the inverse of the composition $\mathbf{X}_0$. When one progressively varies the value of $\|\text{mean}(\mathbf{X} - \mathbf{X}_0)\|$ with simultaneously the totvar($\mathbf{X}$) remaining constant, contour levels of consistency can be obtained. Figure 16.1 shows these contours when totvar $= 1$ and $\mathbf{X}_0 = [1/3, 1/3, 1/3]$. Note that the consistencies of two individuals $\mathbf{X}_1$ and $\mathbf{X}_2$ with the same totvar amount (0.4) are different because their distances from the centre of the distribution $\mathbf{X}_0$ are different. On the other hand, two individuals $\mathbf{X}_2$ and $\mathbf{X}_3$ that are at the same contour level have different consistencies because their variability is different – 0.4 and 0.2, respectively – the individual $\mathbf{X}_3$ with less totvar's value being more consistent (Figure 16.1).

Finally, in order to relate the concepts of preference and consistency, the authors analysed the correlation between the coefficients of variation for each individual and the corresponding distances between each individual and the centre of the compositional space, i.e. the no-preference point. In general, high values of correlation indicate that individuals with stronger preferences also exhibit more consistent preferences.

Since limited number of trials and strong male preferences for a single female morph led frequently to no courtship displays towards one or both 'other' females, these zeros were modelled as follows. It is observed that zeros are present because in some particular trial the male does not court one of the three morphs. But in another trial the same male does court it suggesting that in those trials with a zero courtship the cause could be the limit on trial time (15 min). Therefore absence of displays represent 'count zeros' (see Chapter 4). These null values were modelled with a mixed Bayesian multiplicative approach to minimizing distortions in the covariance structure of the data by preserving the true total number of displays per trial and the ratios between the variables.

The growing interest in individual variation in behavioural traits has also led various authors to investigate so-called behavioural syndromes, suites of correlated behaviours reflecting consistency between individuals across situations. In the context of preferences (association, mating, attack, diet, habitat use, etc.), the implementation of compositional

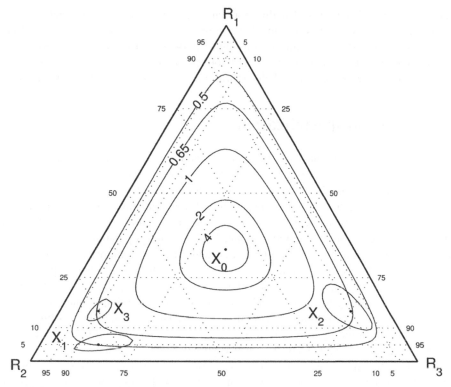

Figure 16.1 Consistency levels, when totvar $= 1$, from population composition $X_0 =$ [1/3, 1/3, 1/3]. Three individual compositions X_1, X_2, and X_3 with the corresponding contour of variability.

techniques provides new tools to fully characterize the behavioural types (personalities) within a certain behavioural syndrome. For example, behavioural syndromes might affect the allocation of time to different activities in the presence or absence of predators or across other ecological axes of variation. Under different predation risk, the individual might prefer to redirect a certain amount of energy/time to different activities. The identification of behavioural types within a population would then require a compositional analysis of individual activity preferences. A measure of individual consistency in the way the individual changes its time budget under predation can be derived by repeated measures under alternative treatments. A cluster analysis of the individual compositional strategies (Pierotti *et al.* 2009) provides a robust tool for the identification of behavioural types within the population syndrome and for the study of within- and between-individual variance associated with single behavioural types. Ultimately, by applying this individual-based approach, it should be possible to assess whether such preference types extend over a range of domains (Sih *et al.* 2004) or they are limited to a specific (mating, fighting, foraging, etc.) context. For example, in species polymorphic for traits with dual function of ornaments and armaments, there might be a correlation between compositional mating preference, association preference and aggression preference bias towards different morphs at the individual level. Once individual preferences are assessed for each behaviour and for each individual, regression analysis on the compositional preferences can identify patterns of within-population covariation in multiple behaviours. In general, when behavioural syndromes involve multiple contexts, the strength and variation of preference types and how each contributes to the composition of preference syndromes in a population might inform us on the presence of pleiotropy and/or trade-offs between behaviours as well as on the nature of co-evolved behavioural trait complexes (e.g. alternative tactics).

16.5   Ecological specialization

A similar paradigm shift emphasizing individual variation is emerging in niche theory and foraging theory, spawning a series of recent exciting studies on specialization (Bolnick *et al.* 2003; Svanbäck and Persson 2004). An animal's diet is composed of various items in different proportions, maximizing the rate of energy gain and the acquisition of important nutrients in the appropriate quantities (Stephens and Krebs 1986). When the distribution of resource types is discrete, the problem is compositional. Each species/population/individual will be characterized by a composition of food items that will reflect its requirements and the availability of the single items in its habitat. This will define the foraging niche. Populations (or individuals) representing foraging specialists have narrower foraging niches than the species' (population's) niche (Roughgarden 1972; Bolnick *et al.* 2003). An important distinction between fundamental and realised niche was introduced by Hutchinson (1957): a fundamental niche is the n-dimensional hypervolume whose axes are the set of conditions and resources within which the population can sustain positive growth rate. The realized niche is the multidimensional set that the population actually occupies and it is frequently smaller than the fundamental niche because the population has not yet adapted or exploited a certain resource or because access to the resource has been prevented by a competing species. The deviation of an individual's realized niche from its fundamental niche could be evaluated as the overlap of the individual's compositional distributions of resource use in both scenarios. This approach has the potential to reveal and quantify within-population divergence in the fundamental niche as well as within-population divergence in the realised niche from a single or a set of shared fundamental niches (Bolnick *et al.* 2002, 2003; Sargeant 2007). Intraspecific competition deriving from diet overlap could be measured for a certain individual as the proportion of the distribution of its diet composition that overlaps with the diet distribution of other individuals in the simplex space. For example, Bolnick *et al.* (2002) proposed an adaptation of Schoener's (1968) Proportional Similarity (PS) index, a measure of interspecific diet overlap, to individual specialization within a population. Let X be the matrix of proportion diet data, with elements p_{ij} describing the proportion of the jth resource category in individual i's diet. Let q_j be the proportion of the jth resource category in the population's niche. The PS index adapted to individual i and the population is

$$\mathrm{PS}_i = 1 - 0.5 \sum_j |p_{ij} - q_j| = 1 - 0.5 d_M(\mathbf{p}_i, \mathbf{q}),$$

where $d_M(\cdot, \cdot)$ is the well-known Manhattan or City-Block distance, and $\mathbf{p}_i$ and $\mathbf{q}$ are the corresponding row vectors of proportions. Note that the row vectors are in fact compositions and the PS index is a simple scaled version – between 0 and 1 – of the Manhattan distance between the individual and population compositions. Unfortunately, we know now that this distance is not fully compatible with the essential operations in the simplex: perturbation and subcomposition, a limitation that, of course, affects both Bolnick *et al.*'s (2002) within-population and Schoener's (1968) between-species indices.

When resources become less abundant, a population is expected to diversify its diet (Stephens and Krebs 1986). This could be achieved by all individuals expanding their range of items similarly (i.e. with high consistency between individuals in diet composition) but it could also derive from individual diversification, as shown by Tinker *et al.* (2008) in sea otters. In this work, the authors showed that such context-dependent individual variation is purely behavioural. It would be interesting to examine the within-individual consistency in such diet shifts in response to replicated changes in resource abundance. Are always the same individuals shifting to a certain behavioural diet specialization? Compositional data expressions for individual consistency in diet might then help assess whether an individual carries a pre-existing tendency towards one particular specialization.

If a reliable assessment of diet specialization is to be made, not only the breadth of food types available to the individual forager is important but their relative availability too. Individuals foraging on a restricted number of food types might be interpreted as specialists only to find afterwards that their narrow diet was simply reflecting local unevenness of food type availability. This is best represented by a compositional set of relative abundances for each food item. Specialization is then assessed by matching the compositional set of relative item preferences of an individual with the compositional set of relative abundances. We can then consider a perfect match between item preference composition and item abundance to underscore a condition of no preference (absolute generalist).

The level of specialization of a certain individual could then be measured as its compositional (i.e. Aitchison's) distance from the absolute generalist (diet item composition completely matching item availability) in compositional space (i.e. in the simplex). It is also important to note that a satisfactory measure of specialization should also account for individual differences in preference consistency (Schindler *et al.* 1997).

16.6   Time budgets: more on specialization

Early in the development of compositional statistics, Aitchison identified the analysis of time budgets as a fertile ground for application (Aitchison 1983). Here, we wish only to draw the attention to the fact that, in the analysis of population variation in time budgets, one could develop CODA strategies analogous to those applied to the study of habitat choice data in wildlife management, individual mating preferences and foraging strategies. Two recent studies, by Dornhaus (2008) and Knudsen *et al.* (2006), highlight how the use of this compositional approach might provide a lot more insight to the biological phenomena hidden in the data.

Dornhaus (2008) analyses the relationship between individual specialization and efficiency in a series of ant colonies by measuring how often individual ants engaged in each of four different tasks, the time lag between two consecutive tasks and the individual performance at each task. Performance in a task was assessed by recording the amount of time required by the individual ant in order to complete the task. Dornhaus (2008) also defined a measure of individual specialization on a certain task as the proportion of trips allocated to that task over the total number of trips. For each task, individuals are ranked by specialization and tested for a correlation between degree of specialization for a task and performance on the same task. The problem with such data sets and the definition of specialization above is that the degree of specialization of an individual ant for one of the tasks is not independent from its degree of specialization on the other tasks but all specializations on different tasks of each individual ant are subject to the unit sum constraint (they are a set of percentages derived from the proportions of trips allocated to a task over the total number of trips, for that individual). Directly comparing a single proportion (or percentage) from individual 1 on task A with a second proportion from individual 2 on the same task A, would be incorrect: compositional transformations on the whole set of proportions across individuals allow to solve this problem and to make use of powerful multivariate statistics on the transformed data. If one assumes that an individual can devote only a finite amount of time to foraging, different items of diet represent compositional strategic allocation of foraging effort.

Moreover, the existence of possible trade-offs between tasks and, at the other opposite, possible single adaptations favouring specializations to multiple tasks (e.g. resulting from similar handling requirements, energy demands or acquired skills), suggests that a compositional treatment of the data might allow more inferential power and, given the authors' experimental design, represents the most robust approach.

A potentially interesting extension of their analysis could be to employ a multivariate linear regression model between a compositional vector (specialization) and performance as the dependent variable. A MANOVA could then be applied to test for effects such as colony size. The compositional approach would make full use of the data (instead of ranks) and allow a statistically sound comparison between individuals and colonies.

Knudsen *et al.* (2006) studied ecological speciation in Arctic charr in a Norwegian postglacial lake. There, a profundal and a littoral morph have diverged in behaviour, foraging ecology, and time and habitat for reproduction. While the profundal morph is restricted to deep waters, some small and young individuals of the littoral morph move in the profundal zone during the warm season. The authors set to compare the diet of the littoral migrants with the resident profundal morph in order to explore potential within-habitat divergence in feeding ecology. They also compared the diet of the littoral morph migrants from the polymorphic lake with charr populations from monomorphic lakes lacking a profundal morph, across months: if the profundal morph had originated by niche invasion, deep benthic prey should be absent from the diet of the monomorphic populations likely reflecting the ancestral littoral diet of Arctic charr.

Knudsen *et al.* (2006) present their results as bar graphs suggesting a clear pattern: the diet of the ancestral littoral morph in the polymorphic and in monomorphic populations is similar and does not include benthic prey, supporting an invasion scenario for the divergence of Arctic charr ecomorphs. In addition, the distribution of prey type in the bar graphs over a number of years suggests stability of diet composition on a relatively long time scale. No statistical analysis is performed on this rich set of diet data.

This interesting paper provides a good example where the application of compositional analysis to the same data could provide additional insight. By making full use of diet data from each sampled individual, one could calculate geometric mean and confidence limits around the mean for each ecological morph and therefore fully characterize the foraging niche of each morph. It would be then interesting to contrast the amount of variation in diet preferences associated with profundal and littoral morphs as well as between littoral morph in the polymorphic Lake Fjellfrøsvatn and the three other monomorphic lakes studied by Knudsen *et al.* (2006). The application of a metric such as the Mahalanobis distance to the ilr-transformed data of the two groups would allow quantitative estimates of divergence between morphs in diet space. Extending this for a time series (monthly diet data set) could provide insights in the magnitude of diet shift experienced by the two co-occurring morphs over the year. Measures of compositional repeatability as developed in Pierotti *et al.* (2009) could then be used to analyse the stability of morph diet composition over several years.

16.7   Conclusions

In the course of this brief and necessarily incomplete survey, we have at various places stressed the need for ecologists and evolutionary biologists to become aware of the compositional problem in their data and follow CODA developments relevant to their field of investigation. However, here we wish to rebalance the burden of responsibility with an example and a warning.

Recently, a paper on the analysis of fish diet data presents a 'novel' approach by employing what the authors called '%PCA', derived from Aitchison's compositional analysis (de Crespin de Billy *et al.* 2000). Despite the authors' claims, little of Aitchison's insight is left in their work and, worryingly, the very issue that triggered one of Aitchison's most influential papers (Aitchison 1983), namely the distortions associated with the use of PCA with proportion data, is turned on its head: the authors suggest the use of a 'centred' PCA on the proportion data for the analysis of diet compositions with appealing graphical representations of the results. Unfortunately, their analysis making use of means, variances, distances and angles all referred to raw proportions is hardly an application of Aitchison's compositional analysis. Therefore it is of serious concern that this 'compositional' approach has been embraced by various ecologists. Since the publication of de Crespin de Billy *et al.* (2000), a notable number of studies, all in good ecological journals, apply their '%PCA method'. It is not our intention here to criticize a specific work but to advocate reciprocal attention of the two communities of scientists, ecologists and statisticians, to their work. While the first should take advantage of the possibilities CODA provides and ensure a correct application of compositional techniques in their studies, CODA developers need to 'branch out' and exert some scrutiny over the (inappropriate) use of compositional analysis. There is no doubt that editors and reviewers of ecological journals bear a large part of this responsibility. This book and the primary literature referred to in it is good evidence that this branching out is already taking place and is a testament to John Aitchison's insights into the potential applications of CODA to the natural sciences.

Acknowledgements

We wish to thank Hitoshi Araki for helpful comments on an earlier draft. This research has been supported by the Spanish Ministry of Science and Innovation (projects CSD2006-00032 and MTM2009-13272) and by the Agència de Gestió d'Ajuts Universitaris i de Recerca of the Generalitat de Catalunya (Ref. 2009SGR424).

References

Aebischer NJ, Robertson PA and Kenward RE 1993 Compositional analysis of habitat use from animal radio-tracking data. *Ecology* **74**, 1313–1325.

Aitchison J 1983 Principal component analysis of compositional data. *Biometrika* **70**(1), 57–65.

Aitchison J 1986 *The Statistical Analysis of Compositional Data*. Monographs on Statistics and Applied Probability. Chapman and Hall Ltd (reprinted 2003 with additional material by The Blackburn Press), London (UK). 416 p.

Aitchison J, Barceló-Vidal C, Martín-Fernández JA and Pawlowsky-Glahn V 2000 Logratio analysis and compositional distance. *Mathematical Geology* **32**(3), 271–275.

Baker CS 2008 A truer measure of the market: the molecular ecology of fisheries and wildlife trade. *Molecular Ecology* **17**, 3985–3998.

Bernatchez L and Duchesne P 2000 Individual-based genotype analysis in studies of parentage and population assignment: how many loci, how many alleles? *Canadian Journal of Fisheries and Aquatic Sciences* **57**, 1205–1214.

Billheimer D, Guttorp P and Fagan W 2001 Statistical interpretation of species composition. *Journal of the American Statistical Association* **96**(456), 1205–1214.

Bingham RL, Brennan LA and Ballard BM 2007 Misclassified resource selection: compositional analysis and unused habitat. *Journal of Wildlife Management* **71**, 1369–1374.

Boake C 1989 Repeatability: its role in evolutionary studies of mating behavior. *Evolutionary Ecology* **3**, 173–182.

Bolnick D, Svanbäck R, Fordyce J, Yang LH, Davis JM, Hulsey C and Forister M 2003 The ecology of individuals: incidence and implications of individual specialization. *American Naturalist* **161**, 1–28.

Bolnick DI, Yang LH, Fordyce JA, Davis JM and Svanbäck R 2002 Measuring individual-level resource specialization. *Ecology* **83**, 2936–2941.

Bowen B and Karl S 2007 Population genetics and phylogeography of sea turtles. *Molecular Ecology* **16**, 4886–4907.

Bowen BW, Bass AL, Garcia-Rodriguez A, Diez CE, van Dam R, Bolten AB, Bjorndal KA, Miyamoto MM and Ferl RJ 1996 Origin of hawksbill turtles in a caribbean feeding area as indicated by genetic markers. *Ecological Applications* **6**, 566–572.

Broquet T, Yearsley J, Hirzel AH, Goudet J and Perrin N 2009 Inferring recent migration rates from individual genotypes. *Molecular Ecology* **18**, 1048–1060.

Cavalli-Sforza L and Edwards A 1967 Phylogenetic analysis: models and estimation procedures. *American Journal of Human Genetics* **19**, 233–257.

de Crespin de Billy V, Dolédec S and Chessel D 2000 Biplot presentation of diet composition data: an alternative for fish stomach contents analysis. *Journal of Fish Biology* **56**, 961–973.

Dornhaus A 2008 Specialization does not predict individual efficiency in an ant. *PLoS Biology* **6**, 285.

Dray S and Dufour A 2007 The ade4 package: implementing the duality diagram for ecologists. *Journal of Statistical Software* **22**(4), 1–20.

Egozcue JJ, Pawlowsky-Glahn V, Mateu-Figueras G and Barceló-Vidal C 2003 Isometric logratio transformations for compositional data analysis. *Mathematical Geology* **35**(3), 279–300.

Gasparini M and Di Gaetano C 2003 On the use of principal components in contemporary population genetics: a case study. In *Proceedings of CoDaWork'03, The 1st Compositional Data Analysis Workshop* (ed. Thió-Henestrosa S and Martín-Fernández JA). http://ima.udg.es/Activitats/CoDaWork03/. University of Girona, Girona (Spain). CD-ROM.

Goldstein D, Linares A, Cavalli-Sforza L and Feldman M 1995 Genetic absolute dating based on microsatellites and the origin of modern humans. *Proceedings of the National Academy of Sciences of the United States of America* **92**, 6723–6727.

Greenacre M 2009 Distributional equivalence and subcompositional coherence in the analysis of compositional data, contingency tables and ratio-scale measurements. *Journal of Classification* **26**(1), 29–54.

Hallgrímsson B and Hall BK 2005 *Variation: A Central Concept in Biology*. Elsevier Academic Press, Burlington, MA (USA). 568 p.

Hron K, Templ M and Filzmoser P 2010 Imputation of missing values for compositional data using classical and robust methods. *Computational Statistics and Data Analysis* **54**(12), 3095–3107.

Hutchinson GE 1957 Concluding remarks. *Cold Spring Harbor Symposia on Quantitative Biology* **22**, 415–427.

Jackson DA 1997 Compositional data in community ecology: the paradigm or peril of proportions? *Ecology* **78**, 929–940.

Jombart T, Pontier D and Dufour AB 2009 Genetic markers in the playground of multivariate analysis. *Heredity* **102**, 330–341.

Knudsen R, Klemetsen A, Amundsen P and Hermansen B 2006 Incipient speciation through niche expansion: an example from the arctic charr in a subarctic lake. *Proceedings of the Royal Society, Series B* **22**, 2291–2298.

Legendre P and Gallagher ED 2001 Ecologically meaningful transformations for ordination of species data. *Oecologia* **129**, 271–280.

Lessells CM and Boag P 1987 Unrepeatable repeatabilities: a common mistake. *Auk* **104**, 116–121.

Liebscher V 2008 Compositions in life science data. In *Proceedings of CoDaWork'08, The 3rd Compositional Data Analysis Workshop* (ed. Daunis-i-Estadella J and Martín-Fernández J). University of Girona, Girona (Spain). CD-ROM.

Little RJA and Rubin DB 2002 *Statistical Analysis with Missing Data*, 2nd edition. John Wiley & Sons, Ltd, New York, NY (USA). 381 p.

Martín-Fernández JA, Barceló-Vidal C and Pawlowsky-Glahn V 2003 Dealing with zeros and missing values in compositional data sets using nonparametric imputation. *Mathematical Geology* **35**(3), 253–278.

Mooijaart A, van der Heijden P and van der Ark A 1999 A least squares algorithm for a mixture model for compositional data. *Computational Statistics & Data Analysis* **30**, 359–379.

Nei M 1972 Genetic distance between populations. *American Naturalist* **106**, 283–292.

Nelis M, Esko T, Mägi R, Zimprich F and Zimprich A 2009 Genetic structure of europeans: A view from the north-east. *PLoS ONE* **4**(5), 5472.

Oeppen J 2008 Coherent forecasting of multiple-decrement life tables: a test using japanese cause of death data. In *Proceedings of CoDaWork'08, The 3rd Compositional Data Analysis Workshop* (ed. Daunis-i-Estadella J and Martín-Fernández J). University of Girona, Girona (Spain). CD-ROM.

Oksanen J, Guillaume-Blanchet F, Kindt R, Legendre P, O'Hara RB, Simpson GL, Solymos P, Stevens MHH and Wagner H 2010 *vegan: Community Ecology Package*. R package version 1.17-4.

Okuyama T and Bolker BM 2005 Combining genetic and ecological data to estimate sea turtle origins. *Ecological Applications* **15**, 315–325.

Palarea-Albaladejo J and Martín-Fernández J 2008 A modified EM alr-algorithm for replacing rounded zeros in compositional data sets. *Computers and Geosciences* **34**(8), 902–917–1861.

Palarea-Albaladejo J, Martín-Fernández J and Soto J 2011 Dealing with distances and transformations for fuzzy c-means clustering of compositional data. *Journal of Classification* (in press).

Pawlowsky-Glahn V and Egozcue JJ 2001 Geometric approach to statistical analysis on the simplex. *Stochastic Environmental Research and Risk Assessment (SERRA)* **15**(5), 384–398.

Pierotti MER, Martín-Fernández JA and Seehausen O 2009 Mapping individual variation in male mating preference space: multiple choice in a colour polymorphic cichlid fish. *Evolution* **63**(9), 2372–2388.

Reynolds J, Weir B and Cockerham C 1983 Estimation of the coancestry coefficient: Basis for a short-term genetic distance. *Genetics* **105**, 767–779.

Rogers J 1972 Measures of genetic similarity and genetic distances. *Studies in Genetics, University of Texas Publication* **7213**, 145–153.

Roughgarden J 1972 Evolution of niche width. *American Naturalist* **106**, 683–718.

Sargeant B 2007 Individual foraging specialization: niche width versus niche overlap. *Oikos* **116**, 1431–1437.

Schindler DE, Hodgson JR and Kitchell JF 1997 Density-dependent changes in individual foraging specialization of largemouth bass. *Oecologia* **110**, 592–600.

Schoener TW 1968 The anolis lizards of Bimini: resource partitioning in a complex fauna. *Ecology* **49**, 704–726.

Shriver M, Jin L, Boerwinkle E, Deka R, Ferrel R and Chakraborty R 1995 A novel measure of genetic distance for highly polymorphic tandem repeat loci. *Molecular Biology and Evolution* **12**(5), 914–920.

Sih A, Bell A and Johnson JC 2004 Behavioral syndromes: an ecological and evolutionary overview. *Trends in Ecology and Evolution* **19**, 372–378.

Stephens DW and Krebs JR 1986 *Foraging Theory*. Princeton University Press, Princeton, NJ (USA). 247 p.

Svanbäck R and Persson L 2004 Individual diet specialization, niche width and population dynamics: implications for trophic polymorphisms. *Journal of Animal Ecology* **73**, 973–982.

ter Braak CJF and Schaffers AP 2004 Co-correspondence analysis: A new ordination method to relate two community compositions. *Ecology* **85**(3), 834–846.

ter Braak CJF and Šmilauer P 2002 CANOCO Reference Manual and CanoDraw for Windows User's Guide: Software for Canonical Community Ordination (version 4.5). Microcomputer Power, Ithaca, NY (USA).

Tinker MT, Bentall G and Estes JA 2008 Food limitation leads to behavioral diversification and dietary specialization in sea otters. *Proceedings of the National Academy of Sciences of the United States of America* **105**, 560–565.

Widemo F and Sæther SA 1999 Beauty is in the eye of the beholder: causes and consequences of variation in mating preferences. *Trends in Ecology and Evolution* **14**, 26–31.

Yee T and Hastie T 2003 Reduced-rank vector generalized linear models. *Statistical Modelling* **3**, 15–41.

17

Flying in compositional morphospaces: evolution of limb proportions in flying vertebrates

Luis Azevedo Rodrigues[1], Josep Daunis-i-Estadella[2], Glòria Mateu-Figueras[2] and Santiago Thió-Henestrosa[2]
[1]Secondary School Gil Eanes, Lagos, Portugal
[2]Department of Computer Science and Applied Mathematics, University of Girona, Spain

17.1   Introduction

In this chapter, we will use compositional data analysis (CODA) to document the geometric variation of limb proportions in ternary morphospaces and in linear bivariate spaces.

This chapter will reanalyse the data of Dyke *et al.* (2006) and McGowan and Dyke (2007) using CODA (Aitchison 1986), specifically designed to deal with the statistical properties of proportions. CODA is appropriate for studying the evolution of flight mechanics because the functional properties of wings and hindlimbs can be expressed as the proportion of one limb segment to another. The limb element lengths of the specimens used by Dyke *et al.* (2006) and McGowan and Dyke (2007) have been used to infer biomechanic similarities and differences among three flying vertebrate groups, namely birds (Aves), pterosaurs and bats. McGowan and Dyke (2007) proposed there was competitive exclusion between extinct and living flying vertebrates. Dyke *et al.* (2006) attempted to determine whether the extinct pterosaur flew in a 'bird-like' mode (only forelimb involved) or in a 'bat-like' mode (both fore- and hindlimbs). Dyke *et al.* (2006) used an individual of *Sordes pilosus* – one of the few pterosaur specimens with a preserved flight membrane – as a model, in order to contrast each pterosaur flight paradigm.

17.2   Flying vertebrates – general anatomical and functional characteristics

In contrast to the other flying taxa analysed herein, Aves have a flying module – the forelimb – independent from the hindlimb and tail. Unlike bats and pterosaurs, the wings are not membranous, but are composed of feathers. Bird adult forelimb morphology is characterised by three ossified digits, and digit III is the longest (Figure 17.1).

Bats comprise about one-quarter of the present mammalian diversity, with more than a thousand species (Mickleburgh *et al.* 2002). The forelimb zeugopodium of bats is dominated by the radius and the ulna is vestigial. Chiroptera wings have a membrane supported primarily by the II–V forelimb digits as well as by the hindlimb.

The monophyletic Pterosauria clade is divided into two groups: Pterodactyloidea and the paraphyletic Rhamphorhynchoidea. Originally small, pterodactyloids developed morphological innovations in the forelimb as well as a reduction/loss of the tail which permitted better functional performance than that of rhamphorhynchoids. The Rhamphorhynchoidea pterosaurs were broadly characterised by their long tails, which enabled dynamic stability and a considerable degree of maneuverability (Wellnhofer 1991; Witmer *et al.* 2003). In Rhamphorhynchoid digit V was longer than digit I; some authors have argued that pedal digit V controlled the uropatagium, and was therefore functionally implicated in pterosaur flight (Unwin 1988; Bakhurina and Unwin 1992). Broadly there are two functional paradigms of pterosaur flight: the first posits that the wing membrane incorporates the hindlimb with the forelimb (Wellnhofer 1991; Unwin and Bakhurina 1994; Unwin 1999; Unwin 2006), and the second asserts that the hindlimb does not contribute to flight, due to the absence of wing membrane attachment of the forelimb to the hindlimb (Padian 1983). Pterosauria's primary morphological feature in the forelimb is the extensive development of digit IV, with the corresponding metacarpal generally longer in Pterodactyloidea and shorter in Rhamphorhynchoidea (Gatesy and Middleton 2007). This extensively developed digit supported the wing membrane that permitted active flight in pterosaurs.

17.3   Materials

The data analysed in this work were selected from previously published sets of measurements (Dyke *et al.* 2006; McGowan and Dyke 2007). The total data set is composed of 955 total specimens: 603 Aves non-passerines, 97 Aves passerines, 217 Chiroptera (184 Microchiroptera and 33 Megachiroptera), 13 Rhamphorhynchoidea, 11 Pterodactyloidea and 14 Theropoda [see Dyke *et al.* (2006) supplementary material].

Since birds and nonavian dinosaurs are subsets from within the same larger clade, specimens from Theropoda were included, in order to contrast patterns of morphospace occupation and to include a phylogenetic control. Theropoda specimens were selected due to the completeness of the limb elements required for this analysis and data were compiled from several databases (Rodrigues 2009, appendix II). Preliminary results indicated that the Chiroptera

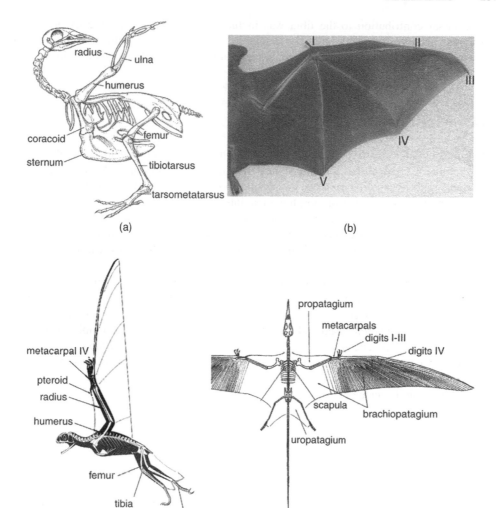

Figure 17.1 (a) General morphology of an adult bird. Adapted from Martin (2006). (b) Forelimb morphology of an adult bat, *Carollia perspicillata*. Adapted from Weatherbee *et al.* (2006). Copyright (2006) National Academy of Sciences, USA. (c) Pterosaur *Jeholopterus ningchengensis* general appendicular morphology. Adapted from Mike Hanson (unpublished). (d) Pterosaur *Rhamphorhynchus muensteri* limbs and wing membrane morphology. Adapted from Wellnhofer (1991).

sample should be analysed in greater detail; therefore, in some analyses the Chiroptera data set was divided into two subsamples, each corresponding to a suborder: Megachiroptera and Microchiroptera. For the taxonomical setting of the bat specimens the following works were adopted: Burkitt (1995); and Schutt and Simmons (1998); Giannini and Simmons (2005).

The limb elements analyzed for each specimen are: for the forelimb, humerus, radius or ulna and metacarpal IV (pterosaurs) or metacarpal III (theropds and bats) or carpumetacarpus (birds); for the hindlimb, femur, tibia and metatarsal III, for all groups.

The tarsal contribution to the tibia was included for all taxa, with the exception of Theropoda. In pterosaurs and bats, whose feet are not fused, the length of metatarsal III was considered the equivalent to the avian tarsometatarsus and used in analysis (Gatesy and Middleton 1997).

17.4   Methods

CODA considers the relative magnitude and variations between component, rather than their absolute value. CODA allows to: (1) evaluate and quantify positioning between specimens/ groups and limb occupation patterns within morphospace; (2) quantify the morphological disparity; and (3) infer aspects of morphological integration.

Two log-ratio transformations were used: the centred log-ratio transformation (clr) and the isometric log-ratio transformation (ilr). Although its interpretation is not straightforward for nonspecialists, a specific kind of ilr transformation, known as balances, was used in these analyses.

Projected samples were summarised in a dendrogram-type graph indicating: (a) grouping parts methods; (b) the explanatory contributions of subcompositions generated in the partitioning process; (c) the decomposition of the variance; and (d) the center and quantiles of each balance. The equations used and the fundamentals of data analyses employed will be briefly introduced (Egozcue et al. 2003; Egozcue and Pawlowsky-Glahn 2005a,b, 2006).

Principal Component Analysis (PCA) and corresponding biplots were used to analyze our data following the interpretation rules of Aitchison and Greenacre (2002).

The Aitchison distance defined as

$$d_a^2(\mathbf{x}, \mathbf{x}^*) = \frac{1}{D} \sum_{i<j} \left(\ln \frac{x_i}{x_j} - \ln \frac{x_i^*}{x_j^*} \right)^2,$$

was used and interpreted as a disparity index.

Disparity can also be defined as the degree of morphological differentiation between taxa within groups (Foote 1999; Eble 2000; Ciampaglio et al. 2001). Morphological disparity and morphospace occupations are similar concepts, and each is widely used in macroevolutionary studies for different purposes (Foote 1991, 1993, 1994, 1999; Wills et al. 1994). The most common of them being to confront those values with the diversity within lineages. Two aspects of morphological disparity and morphospace patterning must be taken into account in any analysis: variance and range. The variance captures the average dissimilarity between forms in morphospace while the range reflects the amount of morphospace occupied (Foote 1991).

CODA allows comparison between specimens in the morphospace quantified as the total variance (sum of univariate variance) in the distinct computed proportions. Therefore, in this work (and others) (Van Valen 1974; Smith and Bunje 1999; Eble 2000) the morphological disparity will be quantified as the total variance (sum of univariate variances) in the distinct computed morphospace proportions. Further, the term 'disparity' is used here with the same meaning as 'variance'.

We performed two types of statistical tests: two-sample t-test comparisons of the intragroup Aitchison distances and MANOVA tests of the ilr variables. We interpreted the Aitchison distance as a limb proportions disparity index, which revealed distinct disparities

Table 17.1 Geometric center, by percentage, for fore- and hindlimb elements (fore/hind). Non-pass., non-passerines; Pass., passerines; Thero., Theropoda; Chirop., Chiropera; Rham., Rhamphorhynchoidea; Ptero., Pterodactyloidea; H, humerus; R/U, radius/ulna; MC, metacarpal III; F, femur; T, tibia; MT, metatarsal III.

	Non-pass.	Pass.	Thero.	Chirop.	Rham.	Ptero.
Stylopodium (H-F)	39/26	35/26	51/38	18/44	10/34	14/34
Zeugopodium (R/U-T)	39/46	42/44	32/40	30/47	15/46	18/50
Autopodium (MC-MT)	22/28	23/30	17/22	52/9	75/20	68/16

within the proportions morphospaces. The t-tests allowed us to compare patterns of disparity between the different groups, that is, the morphospace occupation patterns. ilr was used in the MANOVA tests, instead of clr, since the clr covariance matrix is, among other peculiarities, singular. The ilr MANOVA tests demonstrated the existence of differences between the bone proportions.

All of the specific CODA analyses as log-ratio transformations, balances dendrograms, biplots and some plots were performed using the freeware package CoDaPack (Thió-Henestrosa *et al.* 2008).

17.5   Aitchison distance disparity metrics

Geometric centroids for each distinct taxa were calculated both for the fore- and hindlimbs (Table 17.1). Intragroup Aitchison distances were calculated based on each specimen and its group centroid.

The intragroup Aitchison distances for both limbs means, standard deviation and maximum values were calculated and analyzed (Table 17.2).

17.5.1   Intragroup Aitchison distance

The passerines represented the most tightly clustered group in terms of forelimb proportions. This group was followed by Chiroptera, Pterodactyloidea and the non-passerines. The most

Table 17.2 Intragroup Aitchison distance (fore/hind) mean, standard deviation (SD) and maximum (Max.).

	Mean	SD	Max.
Non-passerines ($n = 603$)	0.148/0.263	0.102/0.177	0.861/0.913
Passerines ($n = 97$)	0.110/0.149	0.066/0.086	0.315/0.431
Theropoda ($n = 14$)	0.167/0.147	0.057/0.102	0.275/0.355
Chiroptera ($n = 217$)	0.117/0.178	0.085/0.095	0.817/0.513
Rhamphorhynchoidea ($n = 13$)	0.248/0.199	0.107/0.109	0.420/0.393
Pterodactyloidea ($n = 11$)	0.123/0.200	0.082/0.141	0.308/0.503

Table 17.3 Intergroups Aitchison distance for fore- and hindlimb elements (fore/hind). Non-pass., non-passerines; Rham., Rhamphorhynchoidea; Ptero., Pterodactyloidea.

	Non-pass.	Passerines	Chiroptera	Rham.	Ptero.
Passerines	0.140/0.099				
Chiroptera	1.216/1.198	1.122/1.273			
Rham.	1.956/0.459	1.879/0.527	0.781/0.746		
Ptero.	1.674/0.640	1.601/0.721	0.534/0.563	0.286/0.219	
Theropoda	0.412/0.503	0.550/0.534	1.576/0.833	2.275/0.224	1.988/0.408

disparate is Rhamphorhynchoidea, closely followed by theropod dinosaurs. These distinct Aitchison distances indicate that both bird groups and bats represent a more compact distribution in the forelimb morphospace, while pterosaur and theropod individuals are more spread out. Rhamphorhynchoidea presents an intragroup Aitchison distance nearly twice that of Pterodactyloidea. This discrepancy in forelimb disparity[1] could have resulted from distinct levels of phylogenic groupings, since Rhamphorhynchoidea is not considered to be a true clade. Thus, comparing Rhamphorhynchoidea and Pterodactyloidea may represent a comparison within two levels of classification. Although we analyzed for the forelimb Aitchison distance as a single group, the Chiroptera sample integrates dozens of distinct species and exhibits lower Aitchison distances than other groups with higher taxonomical diversity – non-passerines. Thus, bats exhibit less forelimb morphological disparity than non-passerines, but higher morphological disparity than passerines.

In analyzing hindlimb morphology, we found that theropods and passerine birds show the lowest values of Aitchison distances. Non-passerine birds showed the highest values of Aitchison distances followed by Pterodactyloidea and Rhamphorhynchoidea. Both pterosaur groups show nearly identical hindlimb Aitchison distance, indicating that both groups of extinct fliers showed similar disparity indices. Bats revealed a hindlimb dissimilarity index higher than passerine birds and theropods, each of which presented equivalent Aitchison distances.

17.5.2 Intergroup Aitchison distance

In order to reduce the limitations of 'visual analysis' and the absence of an adequate numeric quantification of the constructed morphospace, the intergroup Aitchison distances (distances between group centroids) was computed to evaluate the morphological disparity between groups (Table 17.3).

The clear difference between pterodactyloids and rhamphorhynchoids indicated by Dyke *et al.* (2006) could not be confirmed by the intercentroid group Aitchison distances. Forelimb intercentroid Aitchison distances were smaller (half of the Aitchison distance) among the bird groups than among the pterosaurs. Comparing Aitchison distances between pterosaurs and birds showed that Pterodactyloidea was morphologically more similar to the extant fliers than to Rhamphorhynchoidea. Pterodactyloidea filled a more restricted area of the morphospace than did Rhamphorhynchoidea, which was more disperse and presented extreme

---

[1] A correction for phylogenetic autocorrelation should be performed for confirmation.

relative values particularly in metacarpal length. There was a large amount of dispersion and specimen overlap among the bird groups, and a small group of nine non-passerine specimens, all belonging to the families Apodidae and Trochilidae, was substantially separated from the rest of the bird species and was classified as Aitchison distance extreme values (Figure 17.2). Theropods occupied an area close to both bird groups and, despite there dispersion, were closer to non-passerines than to passerines. Although closely related to both birds and theropods among the clade Archosauria, pterosaurs occupied an extreme region of morphospace and were closer to bats than to archosaurians. The Chiroptera cluster revealed a distinct trend in its morphospace dispersion [Figure 17.2(c)]. This variation trend was identified roughly as a variation in relative metacarpal length. Some specimens fell out of the cluster, including the most primitive bat – *Icaronycteris index*. Bats revealed a trend in variation similar to that of pterosaurs and bat metacarpal variation ranged within the upper limit of more than 60% to the lower limit of less than 40% of *Taphozous flaviventris*. For most of the bat specimens, variation mainly ranged from 50–60% in metacarpal to 25–35% in radius/ulna, with an almost constant humerus relative length of 15–20%. The microchiroptera cluster was less spread out than the Megachiroptera cluster.

The hindlimb morphospace was perceptibly different than that of the forelimb, with most specimens occupying two major areas [Figure 17.2(d)]. Despite some continuity in those two areas, one was occupied primarily by archosaurian specimens (theropods, birds and pterosaurs) and the other was filled by mammals (bats). The limit region was mainly occupied by pterosaurs, with theropods occupying a specific region of the hindlimb morphospace. Despite some overlap, the two groups of bats occupied distinct areas of morphospace, with Megachiroptera individuals distributed in a broader area [Figure 17.2(e,f)]. Thus, Microchiroptera inhabited a more compact region of morphospace, spanning the relative lengths of the femur from 34 to 57%, the tibia from 36 to 53%, and the metatarsal from 5 to 14%. Megachiroptera relative length limits ranged from 37 to 45% of the femur, 47 to 57% of the tibia and 7 to 15% of the metatarsal. Aves morphospace area varied primarily along the femur axis, even though there was an observable variation along the other two axes. Passerine morphospace was more compact than that of non-passerines. The lowest intergroup Aitchison distance was observed between passerine and non-passerine birds, reflecting the close association in hindlimb element proportions (Table 17.3). This relationship of hindlimb elements' ratio between the two groups was slightly less than the forelimb ratio, which could indicate that the observed differences in bone proportions were primarily due to differences in the forelimb. Both pterosaur groups occupied contiguous and overlapping morphospace regions but, nonetheless, Rhamphorhynchoidea exhibited lower percentages of tibia and higher percentages of metatarsal, implying that, for pterosaurs, the relative length of the femur was roughly constant. Theropods exhibited Aitchison distances closer to pterosaurs than to birds despite being more closely related to birds. This close relationship between the hindlimb morphospace could have resulted from functional constraints experienced by flying vertebrates (birds and pterosaurs).

Concerning the observations on variation patterns in combined limbs, although both pterosaur groups are the closest to bats in the fore- and hindlimb morphospaces, Chiroptera showed Aitchison distances more similar to Pterodactyloidea than to Rhamphorhynchoidea pterosaurs. Despite the differences between Theropoda and Pterosauria in forelimb intercentroid, the Aitchison distances for the hindlimb are considerably reduced. This may have been due to large functional differences in the hindlimb proportions of the two groups. Conversely, the proportions of the forelimb were more related in pterosaurs and theropods.

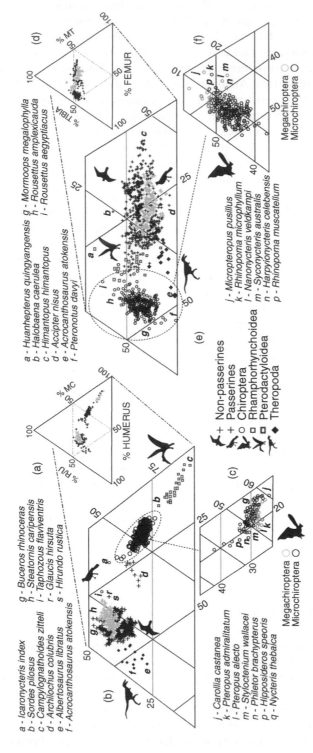

a - Icaronycteris index
b - Sordes pilosus
c - Campylognathoides zitteli
d - Archilochus colubris
e - Albertosaurus libratus
f - Acrocanthosaurus atokensis

g - Buceros rhinoceras
h - Steatornis caripensis
i - Taphozous flaviventris
r - Glaucis hirsuta
s - Hirundo rustica

a - Huanhepterus quingyangensis
b - Halobaena caerulea
c - Himantopus himantopus
d - Accipter nisus
e - Acrocanthosaurus atokensis
f - Pteronotus davyi

g - Mormoops megalophylla
h - Rousettus amplexicauda
i - Rousettus aegyptiacus

j - Carollia castanea
k - Pteropus admiralitatum
l - Pteropus alecto
m - Styloctenium wallacei
n - Philetor brachypterus
p - Hipposideros speoris
q - Nycteris thebaica

j - Micropteropus pusillus
k - Rhinopoma microphyllum
l - Nanonycteris veldkampi
m - Syconycteris australis
n - Harpyionycteris celebensis
p - Rhinopoma muscatellum

+ Non-passerines
+ Passerines
○ Chiroptera
□ Rhamphorhynchoidea
■ Pterodactyloidea
◆ Theropoda

Megachiroptera ○
Microchiroptera ○

Figure 17.2 (a) Empirical morphospace of forelimb parts of different flying vertebrates. (b) All forelimb morphospace occupation for all specimens. (c) Chiroptera groups' forelimb morphospace occupation. Specimens in the morphospace outskirts are identified. (d) Empirical morphospace of hindlimb elements of different flying vertebrates. (e) All hindlimb morphospace occupation for all specimens. (f) Chiroptera groups' hindlimb morphospace occupation. Specimens in the morphospace outskirts are identified.

17.6   Statistical tests

In order to compare intragroup forelimb Aitchison distance means two-sample t comparisons were performed. These tests confirmed that there are significant differences between the Aitchison distance means of the two pterosaur groups ($t = 3.157$, $P = 0.005$). The same test confirmed no significant differences between the hindlimb Aitchison distance means ($t = -0.012$, $P = 0.990$). Therefore, Rhamphorhynchoidea and Pterodactyloidea revealed different disparity indices in forelimb proportions. Rhamphorhynchoidea occupied a larger morphospace area than Pterodactyloidea. The two-sample t comparisons between the two bat groups indicated significant differences in the forelimb morphospace disparity patterns ($t = -4.310$, $P = 0.000$). The two-sample t comparisons of the two bat group hindlimb morphospace disparities indicated no significant differences ($t = -0.770$, $P = 0.448$). Thus, the two Chiroptera groups revealed distinct disparities between forelimb morphospace and identical disparities in hindlimb morphospace.

To examine the limb proportions among the six groups we used the isometric log-ratio transformation that allowed us to apply the standard techniques as MANOVA. The scatterplot of ilr coordinates (Figure 17.3) suggested differences in limb proportions. In terms of the forelimb, MANOVA indicated highly significant differences between the six groups means – Wilks' lambda $= 0.035$, $F[10,1896] = 822.365$, $P < 0.001$. Comparing the forelimb means of non-passerines and passerines, there were still significant differences between group means – Wilks' lambda $= 0.819$, $F[2,697] = 77.196$, $P < 0.001$. Moreover, the significant differences were reflected in each of the three bones that were compared. MANOVA indicated no significant differences between the two groups of pterosaurs means – Wilks' lambda $= 0.875$, $F[2,21] = 1.496$, $P = 0.247$. MANOVA indicated significant differences between the two groups of bats means – Wilks' lambda $= 0.813$, $F[2,214] = 24.452$, $P < 0.001$.

In terms of the hindlimb, MANOVA analysis of the ilr coordinates indicated highly significant differences in hindlimb element proportions between the six groups means – Wilks' lambda $= 0.147$, $F[10,1896] = 305.032$, $P < 0.001$. Comparing the hindlimb means of non-passerines and passerines there were still significant differences between group means – Wilks' lambda $= 0.885$, $F[2,697] = 45.172$, $P < 0.001$. MANOVA indicated significant differences among the two groups of pterosaurs hindlimb group means – Wilks' lambda $= 0.633$, $F[2,21] = 6.096$, $P = 0.008$. The MANOVA analysis of the ilr coordinates indicated highly significant differences in hindlimb element proportions between the two groups of Chiroptera means – Wilks' lambda $= 0.724$, $F[2,214] = 40.719$, $P < 0.001$.

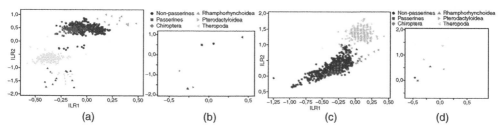

Figure 17.3   ilr coordinates plot of: (a) forelimb proportions of all specimens; (b) forelimb group mean proportions; (c) hindlimb proportions of all specimens; (d) hindlimb group mean proportions.

For both, the fore- and hindlimb, MANOVA analysis of the ilr coordinates indicated highly significant differences in both of the limb element proportions between the six groups means – Wilks' lambda = 0.010, $F[25, 3512] = 340.688$, $P < 0.001$. Comparing the proportions of the six bones within the two groups of birds the MANOVA of the ilr coordinates indicated highly significant differences – Wilks' lambda = 0.689, $F[5,694] = 62.630$, $P < 0.001$. Comparing the proportions of the six bones within the two pterosaur groups we found that the MANOVA analysis of the ilr coordinates showed highly significant differences – Wilks' lambda = 0.528, $F[5,18] = 3.206$, $P < 0.030$.

17.7   Biplots

The joint study of the six limb parts (fore and hind proportions) started with the clr biplot, where patterns among parts and the variability of clr parts were described. This study of bone variability will be discussed in detail in Section 17.8.

17.7.1   Chiroptera

The two main axes were very similar in importance (38% and 30%, respectively), and therefore the variables associated with each axis explain an equivalent variability (Figure 17.4). PC1 is mainly influenced by metatarsal and, to a lesser degree, by femur. Other bones contributed to this axis to a much lesser degree. Metatarsal had the longest ray which exposes its large influence in the total variability among individuals being followed, in importance, by the metacarpal and tibia. PC2 was mostly influenced by the metacarpal and tibia, although, as with PC1, other bones explained the variability of the second axis. Most of the total variability among bat individuals was due to hindlimb bone proportions. Forelimb log-centred variables were associated in the same quadrant and are related to PC2. The two groups of bats exhibited a considerable number of specimens spread along both axes but one can roughly state that Megachiroptera was less disperse along PC1 than PC2, with the former coupled chiefly with metatarsal.

The relative importance of the femur on the total variability was larger for Microchiroptera than for the combined sample [Figure 17.4(a,b)]. This may have been due to the strong influence of the femur within Megachiroptera. In the Megachiroptera data set, PC1 was primarily influenced by metatarsal followed by the log-centred variables of the femur and tibia, which are practically collinear and with their vertices very close implying that the femur

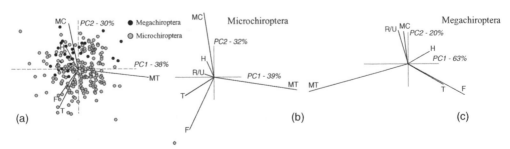

(a)   (b)   (c)

Figure 17.4 Biplots of the clr-transformed space of the first two principal components (PC1 vs PC2) of: (a) Chiroptera, six limb parts with all specimens; (b) Microchiroptera subsample, six limb parts; (c) Megachiroptera subsample, six limb parts. H, humerus; R/U, radius/ulna; MC, metacarpal III; F, femur; T, tibia; MT, metatarsal III.

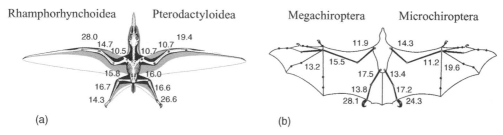

Figure 17.5   Bone proportion variability expressed as percentages of clr variance for: (a) Pterodactyloidea and Rhamphorhynchoidea. Generic pterosaur silhouette adapted from John Conway's illustration of *Nemicolopterus crypticus* (unpublished); (b) Megachiroptera and Microchiroptera. Generic bat silhouette adapted from Habersetzer and Storch (1987). Bones analyzed: the humerus, the radius and the metacarpal III, for the forelimb; the femur, the tibia and the metatarsal III, for the hindlimb.

and tibia parts have an almost constant proportion (0.811) [Figure 17.4(c)]. Similarly, we observed nearly constant proportion of femur relative to tibia (i.e. log-ratios variance close to zero) for both pterosaur groups (0.947). The importance of the metacarpal on the total variability of Megachiroptera was roughly equivalent to the radius and considerably smaller than that of Microchiroptera. This implies that the largest forelimb digit presented a more conservative pattern in Megachiroptera than in Microchiroptera.

Both bat groups showed less variation in forelimb proportions than in hindlimb proportions [Figure 17.5(b)]. Microchiroptera revealed greater variability in forelimb proportions than did Megachiroptera, and the variability increased distally in the former group. In both groups the most variable bone was the metatarsal.

17.7.2   Pterosauria

Rhamphorhynchoids revealed a similar pattern of variation as Pterodactyloidea although reverse PCs (Figure 17.6). PC1 was primarily influenced by the metacarpal and, controlling as well PC2, femur and tibia. The metatarsal influenced both PC1 and PC2 and its degree of influence on total variability is equivalent to the femur and tibia. In rhamphorhynchoids, PC2 was mainly influenced by the radius/ulna, opposite to what is observed on pterodactyloids.

An approximately constant ratio of femur to tibia was observed for groups of pterosaurs [Figure 17.6(a,b)]. Although some common patterns were observed, these biplots showed different relationships between the limb parts of the two groups of pterosaurs. In both groups the autopodial elements were the most important factor in the total variability although the Rhamphorhynchoidea's metatarsal exhibited less influence than it did in Pterodactyloidea. The Pterodactyloids' main axis of variability was primarily influenced by the metatarsal and the radius/ulna and, sequentially with reduced influence by the metacarpal, tibia, femur and humerus, which were controlling PC2.

Regarding the explained variability for the first two axes both groups were roughly equivalent, although revealing different percentages for the first two individual axes. Both groups of pterosaurs exhibited an approximately constant ratio between femur and tibia (0.75 for Rhamphorhynchoidea and 0.66 for Pterodactyloidea). Comparing pterosaur and bat groups [Figure 17.5(a,b)], the variability of bone parts proportions was quite distinct. Through different approaches trends and patterns have been identified that can be generally systematised

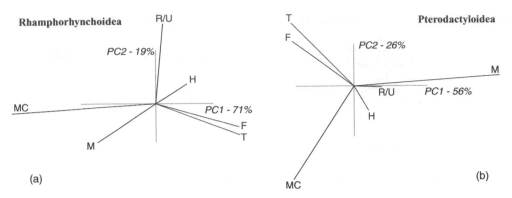

Figure 17.6 Biplot of the clr-transformed space for the first two principal components (PC1 vs PC2) of: (a) Rhamphorhynchoidea subsample, six limb parts; (b) Pterodactyloidea subsample, six limb parts. H, humerus; R/U, radius/ulna; MC, metacarpal III; F, femur; T, tibia; M, metatarsal III.

as follows: almost half of the total variability in bone proportions originates in the autopodial bones; bats' forelimb combined proportions were more conservative than the hindlimb combined proportions; Megachiroptera revealed higher variability than Microchiroptera, mainly in metatarsal III and femur; Microchiroptera showed higher variability in forelimb proportions than Megachiroptera, due mainly to metacarpal III variability.

17.8   Balances

We studied the balance of our complete data set. The balances dendrogram and the table of the variance decomposition are shown if Figure 17.7 and Table 17.4, respectively. The sequential binary partition is detailed in the first column of Table 17.4 and illustrates that greatest balance

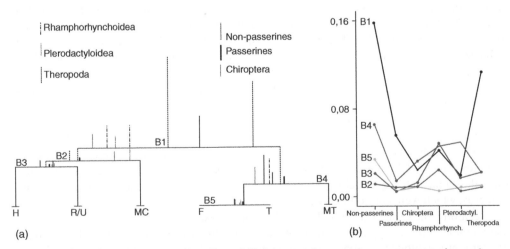

Figure 17.7   (a) Balances dendrogram of flying vertebrates: Aves non-passerines; Aves passerines; Chiroptera; Rhamphorhynchoidea; Pterodactyloidea; and Theropoda. F, femur; H, humerus; MC, metacarpal III/IV; MT, metatarsal III; R/U, radius/ulna; T, tibia; (b) Variance for each balance and the complete sample.

Table 17.4 Variance decomposition for each group and respective balances. Non-pass., non-passerines; Pass., Passerines; Megachi., Megachiroptera; Microchi., Microchiroptera; Rham., Rhamphorynchoidea; Ptero., Pterodactyloidea; Thero., Theropoda. F, femur; H, humerus; MC, metacarpal III/IV; MT, metatarsal III; R/U, radius/ulna; T, tibia.

	Non-pass.		Pass.		Megachi.		Microchi.		Rham.		Ptero.		Thero.		var total	var %
	var	%	var	%	var	%	var	%	var	%	var	%	var	%	(by balance)	(by balance)
B1 (fore vs hind)	0.158	54.3	0.056	64.4	0.011	19.3	0.022	26.8	0.042	25.5	0.019	19.4	0.113	64.6	0.421	44.1
B2 (H and R/U vs MC)	0.021	7.2	0.007	8.0	0.002	3.5	0.018	22.0	0.062	37.6	0.018	18.4	0.020	11.4	0.148	15.5
B3 (H vs R/U)	0.011	3.8	0.005	5.7	0.006	10.5	0.004	4.9	0.010	6.1	0.003	3.1	0.011	6.3	0.05	5.2
B4 (F and T vs MT)	0.066	22.7	0.014	16.1	0.034	59.6	0.031	37.8	0.046	27.9	0.050	51.0	0.021	12.0	0.262	27.4
B5 (F vs T)	0.035	12.0	0.005	5.7	0.004	7.0	0.007	8.5	0.005	3.0	0.008	8.2	0.010	5.7	0.074	7.7
var total (by groups)	0.291		0.087		0.057		0.082		0.165		0.098		0.175			
var % (by groups)	30.5		9.1		6.0		8.6		17.3		10.3		18.3			

in terms of variance is B1, followed by B4. The least important balance are the homologous B5 and B3. Note that balance B3 corresponds to the proportional brachial index and is the least variable balance. This index is informative for power flight requirements, and therefore, this fact could be the justification for the least variability since it represents a very strong selective pressure factor.

The balance of the forelimb versus hindlimb (B1) was the most important variability factor in both Aves groups, as well as in Theropoda [Figure 17.7(b)]. B1 constitutes the second most important balance for both groups of bats and for Pterodactyloidea. The relative variability of B1 in bats and pterosaurs was not as significant as the relative variability of B2 or B4. Thus, the major contribution for the total variability among bat and pterosaur individuals is primarily derived from the balance between the hindlimb parts and the balance between the humerus and the radius/ulna. B4 revealed consistently higher variability than B2 for all groups, except Rhamphorhynchoidea, in which B2 showed greater variability than either B4 or B1. This was due to the greater variability of the metacarpal. B3 and B5 presented opposing relative importance within the two groups of pterosaurs and in the two groups of bats. Rhamphorhynchoidea individuals showed higher relative variability in B3, while Pterodactyloidea presented higher variability in the equivalent ratios of the hindlimbs. This alternation among the ratios of stylopodium and zeugopodium balances could be similarly verified in bats, since Megachiroptera presented greater variability within B3 – associated with the brachial index – while Microchiroptera show a similar trend in B5. In both groups, the relative intervals between B3 and B5 were equivalent. In birds and theropods the deviations involving B3 and B5 were distinct from those of bats and pterosaurs. The contribution of B3 to the total variability was considerably higher in non-passerines than in passerines and theropods, each of which showed similar percentages of ratio variability. B1 represented more than half of the total variability in birds. The remaining balances followed the hierarchical tendencies of the complete sample, with the exception of the ratio between the femur and the tibia in non-passerines, which was the third most important, exhibiting more than twice the percentage of variance of the equivalent balance in passerines. Finally, the study by groups revealed that the pterosaurs and bats variability was originated mainly by B2 and B4. Non-passerines were the most variant sampled group while the bat groups were the two least variant taxa. The highlimb proportion variability was due mainly to the balance between limbs, indicating that non-passerines were functionally very dissimilar in forelimb versus hindlimb. Non-passerine individuals exhibited diverse locomotion abilities that allowed them to exploit different ecological niches, and this could be the source of the variability.

The greatest source of variability among bats was detected in balance B4 revealing that the stylopodium and zeugopodium of the hindlimb were similar in proportion, despite the aforementioned clr variability of the femur in Megachiroptera. Megachiroptera was the bat group that sums the biggest percentage of variability in the hindlimb and consequently had the least variation among groups in the forelimb. Microchiroptera exhibited a total hindlimb variability comparable with that of non-passerines. Comparing B3 among bats, we observed that Megachiroptera showed higher variability in this log-ratio than did Microchiroptera. The main source of variability in Pterosaurs arose from the three balances B1, B2 and B4. More than two-thirds of the total variability between Rhamphorhynchoidea individuals was originated by B2, followed by B4. Thus, more than half of the total variability in rhamphorhynchoids arose when the autopodial bones were considered. More than half of Pterodactyloidea variability came from B4, which could be attributed to the metatarsal proportion, since B5 variability is very low.

Table 17.5 Equations for each group between B3 ilr–forelimb length (log-transformed) and B3 ilr–hindlimb length (log-transformed); r, Pearson's correlation coefficient. Coefficients significant at $P < 0.01$ and $P < 0.05$ are indicated. B3, balance B3; H, humerus; R, radius.

y	x (size)	Group with significant size	$r;p$	Equation
B3 (H/R)	Forelimb	Megachiroptera	$0.646; < 0.01$	$y = -0.881 + 0.243 * x$
		Non-passerines	$0.123; < 0.01$	$y = -0.118 + 0.051 * x$
		Passerines	$0.225; < 0.05$	$y = -0.315 + 0.131 * x$
	Hindlimb	Megachiroptera	$0.650; < 0.01$	$y = -0.682 + 0.217 * x$
		Microchiroptera	$-0.198; < 0.01$	$y = -0.203 - 0.107 * x$
		Rhamphorhynchoidea	$0.559; < 0.05$	$y = -0.853 + 0.291 * x$

17.9 Size effect

Various non-autopodial elements of the forelimb were reduced in size through the evolutionary history of birds. Some authors propose an inverse correlation between humerus length and aerial maneuverability, positing that birds with a longer humerus (i.e. auks, loons, cuckoos, grebes, and albatrosses) are poor maneuvering fliers (Middleton and Gatesy 2000; Gatesy and Middleton 2007).

One particularly informative ratio is the brachial index: the ratio of the humerus to radius length (Howell 1944). This ratio can be used to infer power requirements in birds (Rayner and Dyke 2002), such that bird wings with low brachial indices have low moments of inertia, which should reduce power requirements. In order to test the influence of size on distinct balances, we performed several regression analyses (linear regression model Type I) on the ilr variables (B3, corresponding to the brachial index) corresponding to the size of the total forelimb or the hindlimb. The total length of each limb was previously log-transformed. Balance B3 is directly related to the brachial index, since it is the ratio of the humerus to the radius proportions.

The significant correlation between B3 and size (Table 17.5) indicated that Megachiroptera with larger forelimbs showed higher brachial indices with consequently more powerful flight requirements. Similarly, Megachiroptera showed the most variation in balance B3, indicating that there are distinct flight performances among bats. Megachiroptera is the only group in which we found a significant correlation between forelimb size and balance B3. This group of bats also revealed positive and significant correlations between hindlimb size and ilrcoefficients from balance B3. In contrast to forelimb, the size of the hindlimb is significantly correlated with balance B3 in several groups. It is positively correlated in both Aves groups (low correlation) and in Rhamphorhynchoidea, and is negatively correlated (low correlation) in Microchiroptera.

17.10 Final remarks

Despite the fact that it is not well known among palaeontologists or biologists, CODA should be regarded as a standard form of analysis for data sets in which the values are expressed as proportions or percentages and for which there is a desire to summarise the structure of such data in a linear space.

17.10.1   All groups

The Aitchison distances of hindlimbs are considerably larger than the Aitchison distances of forelimbs for all groups, except theropods and rhamphorhynchoids pterosaurs. Hindlimb morphological disparity is generally greater than forelimb morphological disparity. With the exception of theropods, the primary locomotor module in the analyzed taxa is the forelimb; nonetheless, the forelimb is more stable in proportions and respective Aitchison distances, than the secondary module, the hindlimb. This may be due to greater selective pressure on the primary locomotor function, contributing to a more conservative proportion pattern and correspondingly lower variability in morphospace occupation. The balance B1 reveals high variance in both bird groups, implying this that both limbs show low levels of morphological integration. In contrast, the bats and pterosaurs groups showed lower B1 variance, indicating higher levels of morphological integration between the fore- and hindlimbs.

17.10.2   Aves

The bone proportions Aitchison distances MANOVA confirmed that there are significant differences in limb parts proportions, for both fore- and hindlimbs between the two groups of birds. Each bird group reveals different Aitchison distances for the hindlimbs, indicating a difference in morphospace occupation. Our disparity assessment quantifies the functional discrepancies described in a previous study (Middleton and Gatesy 2000). The authors distinguished more maneuverable fliers (passerines) from less maneuverable fliers (non-passerines). Despite being not directly linked to flight,[2] the morphological similarity of pterosaur and bird hindlimbs could suggest that bird hindlimbs are more conditioned by their function than by the phylogeny. Diverse groups of birds reveal ecological adaptations primarily resulting from selective pressure on hindlimb morphology [e.g. species whose habitat affiliation is mainly the ground, tree or swimmer, as noted by Zeffer *et al.* (2003)]. The majority of species that were identified as hindlimb proportional outliers were classified as belonging to habitats associated with intensive use of the hindlimbs.

17.10.3   Pterosauria

The *t*-tests performed on the intragroup Aitchison distances confirmed that the two groups of pterosaurs each show different patterns of morphospace occupation. The distinct disparity indices in the pterosaur groups could derive from different functional performances between the two groups: Pterodactyloidea forelimb morphology could have reached a functional evolutionary peak at which morphological disparity would have been stabilised. The MANOVA performed on the bone proportions confirmed that there were no significant differences in forelimb parts proportions between the two pterosaurs groups, but demonstrated a significant difference in hindlimb parts. Considering the similarity of the forelimb morphospace occupation for pterosaur groups we conclude that pterosaur groups occupy different forelimb

---

[2] This is more evident in birds, since there is evidence of membrane attachment in pterosaurs hindlimbs, indicating that there is an effective contribution by the hindlimb to pterosaur flight.

morphospaces despite the fact that they each possess similar bone part proportions. In the hindlimb, pterosaurs occupy similar morphospace areas although they reveal distinct bone parts proportions. The difference in variability between the pterosaur groups' autopodium may be due to distinct areas of wing membrane attachment. Assuming the paradigm of hindlimb attachment of pterosaurs flight membrane, the difference in autopodial variability between the pterosaur groups may be due to differing modes of membrane attachment. Pterodactyloids are thought to have had no hindlimb membrane connection and their autopodium could therefore vary more than that of the rhamphorhynchoids, which were likely to have had some hindlimb influence on the membrane attachment. We have identified consistent differences in both fore- and hindlimbs proportion morphospace patterning and distances of group centroids between pterosaurs and bats. These differences have never been previously quantified. Additionally, Pterosaurs and Megachiroptera bats both exhibited a nearly constant proportion between the log-centred variables femur and tibia. In both pterosaur groups, the major contributions to the total variability between individuals are derived from proportions of the three bones of each limb, and on a small scale from the log-ratio between the two limbs. The difference between pterodactyloids and rhamphorhynchoids (Dyke *et al.* 2006) could not be confirmed by comparing the intercentroid group Aitchison distances since the Aitchison distances of forelimb and hindlimb are considerably smaller between the two groups of birds than between the pterosaur groups.

17.10.4   Chiroptera

We identified a trend in variability within the pterosaur sample and we showed that the variability increased distally in the proportions of both limbs. The exception of this trend was Rhamphorhynchoidea: the metatarsal III showed lower variability than the femur or tibia. In Rhamphorhynchoidea, about half of the variability of the metatarsal III of Pterodactyloidea was observable. Middleton and Gatesy (2000) and Gatesy and Middleton (2007), analyzed taxa morphospaces similar to the ones in the present work, and concluded that Chiroptera are a less disparate group in forelimb proportions than either Aves or Pterosauria. These previous studies used a disparity index with weaknesses described by Rodrigues (2009) and were primarily focused on the application of non-CODA techniques in discriminating and testing hypotheses in compositional data morphospaces. The Aitchison distance disparity index employed by the present study partially contradicts the conclusions of previous studies, as we found the forelimb to be the less disparate group. Using a CODA methodology we found the lowest Aitchison distance for passerine birds (Aitchison distance = 0.110), followed by bats (Aitchison distance = 0.117) and Pterodactyloids (Aitchison distance = 0.123). The MANOVA performed on the bone proportions confirmed that the two groups of bats are distinct both in the fore- and hindlimbs parts proportions, despite the fact that they show identical morphospace occupation pattern for the hindlimb. The bat's chief locomotor module is the forelimb through active flight, this function constituting its main and almost exclusive type of locomotion, although certain exceptions include the common vampire bat (*Desmodus rotundus*) and the New Zealand short-tailed bat (*Mystacina tuberculata*), which have evolved the ability to move well on the ground, using a method differing from that of birds (Riskin *et al.* 2006). Variability within bats limbs should not be as high as in birds since the bat hindlimb does not contribute as actively to the locomotion function as do bird hindlimbs, although there is some influence of the bat hindlimb on flight stability. This discrepancy can be observed in Figure 17.7 and Table 17.4.

Acknowledgements

We thank Angela Delgado Buscalioni (Universidad Autónoma de Madrid, Spain) for the endless scientific discussions on disparity, morphological integration and morphospaces, which made this chapter possible, for L.A. Rodrigues' thesis supervision and all the support for this chapter; Vera Pawlowsky-Glahn (Universitat de Girona, Spain), for reading and commenting on an earlier draft of this chapter and for being the one responsible for entering flying animals into compositional morphospaces; Norman MacLeod (Natural History Museum, UK) for reading and commenting on L.A. Rodrigues' thesis chapter on which this chapter is based; P. David Polly (University of Indiana, USA) for critically reviewing the manuscript and contributing several improvements and for suggesting modifications in the title; Janice L. Pappas (University of Michigan, USA) for reviewing the manuscript; G. J. Dyke (University College Dublin, Ireland), R. L. Nudds and J. M. V. Rayner (University of Leeds, UK) for providing the data sample; and T. R. Holtz (University of Maryland, USA) and K.M. Middleton (California State University, USA) for theropods measurements. This research has been supported by the Spanish Ministry of Science and Innovation (projects CSD2006-00032 and MTM2009-13272) and by the Agència de Gestió d'Ajuts Universitaris i de Recerca of the Generalitat de Catalunya (Ref. 2009SGR424).

References

Aitchison J 1986 *The Statistical Analysis of Compositional Data*. Monographs on Statistics and Applied Probability. Chapman and Hall Ltd (reprinted 2003 with additional material by The Blackburn Press), London (UK). 416 p.

Aitchison J and Greenacre M 2002 Biplots for compositional data. *Applied Statistics* **51**(4), 375–392.

Bakhurina N and Unwin D 1992 *Sordes pilosus* and the function of the fifth toe in pterosaurs. *Journal of Vertebrate Paleontology* **12**(3), 18A

Burkitt JH 1995 *Mammals: A World Listing of Living and Extinct Species*, 2nd edition. Tennessee Department of Agriculture, Nashville, TN (USA).

Ciampaglio C, Kemp M and McShea D 2001 Detecting changes in morphospace occupation patterns in the fossil record: characterizations and analysis of measures of disparity. *Paleobiology* **27**, 695–715.

Dyke G, Nudds R and Rayner J 2006 Limb disparity and wing shape in pterosaurs. *Journal of Evolutionary Biology* **19**, 1339–1342.

Eble G 2000 Contrasting evolutionary flexibility in sister groups: disparity and diversity in mesozoic atelostomate echinoids. *Paleobiology* **26**, 56–79.

Egozcue JJ and Pawlowsky-Glahn V 2005a Coda-dendrogram: a new exploratory tool. In *Proceedings of CoDaWork'05, The 2nd Compositional Data Analysis Workshop* (ed. Mateu-Figueras G and Barceló-Vidal C). http://ima.udg.es/Activitats/CoDaWork05/. University of Girona, Girona (Spain).

Egozcue JJ and Pawlowsky-Glahn V 2005b Groups of parts and their balances in compositional data analysis. *Mathematical Geology* **37**(7), 795–828.

Egozcue JJ and Pawlowsky-Glahn V 2006 Simplicial geometry for compositional data. In *Compositional Data Analysis in the Geosciences: From Theory to Practice*. Geological Society, London (UK). pp. 145–159.

Egozcue JJ, Pawlowsky-Glahn V, Mateu-Figueras G, Barceló-Vidal C 2003 Isometric logratio transformations for compositional data analysis. *Mathematical Geology* **35**(3), 279–300.

Foote M 1991 Morphologic patterns of diversification: examples from trilobites. *Palaeontology* **34**, 461–485.

Foote M 1993 Contributions of individual taxa to overall morphological disparity. *Palaeontology* **19**, 403–419.

Foote M 1994 Morphological disparity in ordovician-devonian crinoids and the early saturation of morphological space. *Palaeontology* **20**, 320–344.

Foote M 1999 Morphological diversity in the evolutionary radiation of paleozoic and post-paleozoic crinoids. *Paleobiology Memoirs, Supplement to Paleobiology* **25**(2), 1–115.

Gatesy S and Middleton K 1997 Bipedalism, flight, and the evolution of theropod locomotor diversity. *Journal of Vertebrate Paleontology* **17**, 308–329.

Gatesy S and Middleton K 2007 *Skeletal Adaptations for Flight* (ed. Hall BK). University of Chicago Press, Chicago, IL (USA). pp. 269–283.

Giannini NP and Simmons NB 2005 Conflict and congruence in a combined DNA-morphology analysis of megachiropteran bat relationships (mammalia: Chiroptera: Pteropodidae). *Cladistics* **21**, 411–437.

Habersetzer J and Storch G 1987 Klassifikation und funktionelle Flügelmorphologie paläogener Fledermäuse (Mammalia, Chiroptera). *Courier Forschungsinstitut Senckenberg* **91**, 117–150.

Howell AB 1944 *Speed in Animals*. University of Chicago Press, Chicago, IL (USA). 270 p.

Martin AJ 2006 *Introduction to the study of Dinosaurs*, 2nd edition. Blackwell Publishing Ltd, Malden, MA (USA). 576 p.

McGowan AJ and Dyke GJ 2007 A morphospace-based test for competitive exclusion among flying vertebrates: did birds, bats and pterosaurs get in each other's space? *Journal of Evolutionary Biology* **20**, 1230–1236.

Mickleburgh SP, Hutson AM, Racey PA 2002 A review of the global conservation status of bats. *Oryx* **36**(01), 18–34.

Middleton K, Gatesy SM 2000 Theropod forelimb design and evolution. *Zoological Journal of the Linnean Society* **128**(2), 149–187.

Padian K 1983 A functional analysis of flying and walking in pterosaurs. *Paleobiology* **9**, 218–239.

Rayner J and Dyke G 2002 Evolution and origin of diversity in the modern avian wing. In *Vertebrate Biomechanics and Evolution* (ed. Bels V, Gasc JP and Casinos A). Bios Scientific Publishing Ltd, Oxford (UK). pp. 297–317.

Riskin D, Parsons S, Schutt WJ, Carter G and Hermanson J 2006 Terrestrial locomotion of the new zealand short-tailed bat *Mystacina tuberculata* and the common vampire bat *Desmodus rotundus*. *The Journal of Experimental Biology* **209**, 1725–1736.

Rodrigues L 2009 Sauropodomorpha (Dinosauria, Saurischia) appendicular skeleton disparity: a theoretical morphology and Compositional Data Analysis study. PhD thesis, University of Madrid.

Schutt WJ and Simmons N 1998 Morphology and homology of the chiropteran calcar, with comments on the phylogenetic relationships of archaeopteropus. *Journal of Mammalian Evolution* **5**(1), 1–32.

Smith L and Bunje P 1999 Morphologic diversity of inarticulate brachiopods through the phanerozoic. *Paleobiology* **25**, 396–408.

Thió-Henestrosa S, Egozcue JJ, Pawlowsky-Glahn V, Kovács LO and Kovács G 2008 Balancedendrogram a new routine of codapack. *Computer and Geosciences* **34**(12), 1682–1696.

Unwin D 2006 *The Pterosaurs From Deep Time*. Pi Press, New York, (NY) USA.

Unwin D and Bakhurina N 1994 *Sordes pilosus*and the nature of the pterosaur flight apparatus. *Nature* **371**, 62–64.

Unwin DM 1988 New remains of the pterosaur dimorphodon (pterosauria: Rhamphorhynchoidea) and the terrestrial locomotion of early pterosaurs. *Modern Geology* **13**, 57–68.

Unwin DM 1999 Pterosaurs: back to the traditional model? *Trends in Ecology and Evolution* **14**(7), 263–268.

Van Valen L 1974 Multivariate structural statistics in natural history. *Journal of Theoretical Biology* **45**, 235–247.

Weatherbee S, Behringer R, Rasweiler J and Niswander L 2006 Interdigital webbing retention in bat wings illustrates genetic changes underlying amniote limb diversification. *Proceedings of the National Academy of Science of the United States of America* **103**, 15103–15107.

Wellnhofer P 1991 *The Illustrated Encyclopedia of Pterosaurs*. Salamander Books, London (UK). 192 p.

Wills M, Briggs D and Fortey R 1994 Disparity as an evolutionary index: A comparison of cambrian and recent arthropods. *Paleobiology* **20**, 93–130.

Witmer L, Chatterjee S, Franzosa J and Rowe T 2003 Neuroanatomy of flying reptiles and implications for flight, posture and behavior. *Nature* **425**, 950–953.

Zeffer A, Johansson L and Marmebro Å 2003 Functional correlation between habitat use and leg in birds (*Aves*). *Biological Journal of the Linnean Society* **79**, 461–484.

18

Natural laws governing the distribution of the elements in geochemistry: the role of the log-ratio approach

Antonella Buccianti
Department of Earth Sciences, University of Florence, Italy

18.1 Introduction

The problem of the peculiar character of chemical laws and theories is a central topic in chemistry and geochemistry. The law discovered by Mendeleev in 1869 is a typical example of a law of nature. Trying to establish a general law concerning the properties of chemical elements, Mendeleev took measurable quantities, like the atomic weights of element and the molecular formulas of compounds (i.e. the atomic composition of the compounds), as a starting point. In this sense a law of nature is a generalization of our best knowledge and belief about the way the world works. Every element has a place determined by the group (designated by a Roman numeral) and the line (Arabic numeral) where it is situated in the periodic system. They indicate the value of the atomic weight, the properties and form of its higher oxide and hydrogen compounds, as well as those of its other compounds, in a word – the main quantitative and qualitative characteristics of the element. The periodic law has facilitated a rational explanation of the chemical properties of the elements and of their observed distribution in nature. This point was central in geochemistry from the early twentieth century when V. M. Goldschmidt indicated that the main role of this discipline

Compositional Data Analysis: Theory and Applications, First Edition. Edited by Vera Pawlowsky-Glahn and Antonella Buccianti.
© 2011 John Wiley & Sons, Ltd. Published 2011 by John Wiley & Sons, Ltd.

was the determination of the distribution of the elements in the different reservoirs of the Earth and the explanation of the partition phenomena. In this framework chemical reactions, represented by algebraic equations, are able to express the balance of mass and charge of the reactants and products and describe the natural tendency to progress toward equilibrium. In order to reach this condition, the reactants are consumed and products are formed such that the amounts of both change with time. At the equilibrium state the amounts of reactants and products per unit weight or unit volume become constant. If a chemical reaction is disturbed by changes in the physical and chemical conditions a new state of equilibrium is achieved, by counteracting the disturbance. This property of the chemical reactions was expressed by Henry Le Châtelier in 1888.

In the general case of a reaction at equilibrium, represented for example by:

$$H_2 + Cl_2 \leftrightarrow 2HCl, \tag{18.1}$$

the *law of mass action*, first formulated by Guldberg and Waage in 1863, takes the form:

$$\frac{(HCl)^2}{(H_2)(Cl_2)} = K, \tag{18.2}$$

where the components in parentheses are in appropriate units of concentration. The law of mass action was initially stated only as a generalization of experimental results and did not have the force of a scientific law. However, later on it was derived from the first and second laws of thermodynamics for reactions among ideal gases at equilibrium. Consequently the Gibbs free energy was recognized as the function of the state of a system which, by combining enthalpy and entropy, defines the progress of a reaction. Equation (18.2) has an equivalent logarithmic condition:

$$\ln \frac{(H_2)}{(HCl)} + \ln \frac{(Cl_2)}{(HCl)} = -\ln(K). \tag{18.3}$$

This encourages the view that a sensible way to identify patterns in compositional data sets is to search for log-ratios. Following Aitchison's (1986, 1999) ideas, the role of log-ratio analysis is to provide an appropriate tool for identifying natural patterns by sound statistical procedures. Whenever there is a change in a composition by an independent process able to produce random variation, log-ratioed data tend to become normally distributed. For geological materials it is easy to assume that independent compositional changes have occurred from their origin to the time of investigation. Under this hypothesis the most important inferential methods can be applied, since probabilistic models are available.

This chapter seeks to develop a comprehensive and understandable framework for applying simple statistical techniques to several geochemical matrices using log-ratios. The aim is to demonstrate that the approach may be the key to really understanding what is happening when chemical elements move among natural materials.

18.2   Geochemical processes and log-ratio approach

Most geological processes involve chemical reactions for a wide range of natural phenomena governing the cycles of the elements. Their distribution in different geological materials is the result of a long history during which many dilution/dispersion stages occurred. Processes such as chemical and mechanical weathering are at the base of these changes generating dilution/dispersion in the environment. When the frequency distribution of an element or chemical species is observed, what we are analysing is the repeated action of these processes in time and space. In this context the common presence of right-skewed distributions of concentrations was interpreted as a natural law explaining the distribution of elements (Ahrens 1954a,b; Miesch 1977). The role of asymmetric distributions is also discussed under the theory of successive random dilutions (Ott 1995) usually applied to an element released at initially high concentrations into a carrier medium, that undergoes dilution in successive independent stages. From a general point of view, dilution is the process by which one constituent is mixed with others, causing the components of the compositions to spread out over a larger volume than before, with a consequent reduction in concentration. The process always accompanies the diffusion, because molecules tend to spread out, occupying an ever-increasing volume. However, dilution can also occur in processes that do not primarily involve diffusion such as the simple mechanical mixing of an aqueous solution with another. In general, if n dilutions repeatedly occur, the resulting concentration c_n will be the product of n individual dilution factors, α_i, and the original concentration c_0:

$$c_n = \alpha_1 \alpha_2 \cdots c_0 = c_0 \Pi_{i=1}^{n} \alpha_i, \tag{18.4}$$

or in logarithmic form:

$$\ln c_n = \ln c_0 + \ln \alpha_1 + \cdots + \ln \alpha_n = \ln c_0 + \sum_{i=1}^{n} \ln \alpha_i. \tag{18.5}$$

The product occurs because the ending mixture becomes the beginning mixture for the next dilution step, thus generating a proportionate process. If α_i are random variables and $0 < \alpha_i \leq 1$ for all i, then logarithms of the dilution factors will be random variables ranging from slightly greater than minus infinity to 0. The central limit theorem implies that the process of adding independent random variables gives rise asymptotically to a normal distribution. In the limit, as n approaches infinity, $\ln c_n$ will equal a normally distributed random variable plus a constant. Because the logarithm of the concentration will asymptotically be normally distributed, the concentration itself, in the limit, will be logarithmic-normally distributed. In summary, a series of n independent successive random dilutions of an initial concentration creates a distribution which is approximately lognormal. This mechanism, developed under the theory of successive random dilutions, may be viewed as a special case of Kapteyn's law of proportional effect (Aitchison and Brown 1957). The theory applies to any process in which the value of the variate in each state is a random proportion of the value of the variate in the previous state.

However, by taking only the logarithm of concentrations, the proportional nature of the acting processes may be not fully captured. The log-ratio approach allow us to overcome this difficulty putting at the core of the analysis the relative magnitudes and variations of the components of a composition rather than their absolute values. An important step in the analysis of compositional data was consequently made by Aitchison (1982) introducing the additive log-ratio (alr) and centred log-ratio (clr) transformations. In both cases the resulting log-ratios are real variables that can be analysed using standard statistical techniques. However it is important to note that alr coordinates cannot be mapped onto orthogonal axes, because they are actually at $60°$ and that clr coordinates lie on a plane in D-dimensional real space leading to singularity of covariance matrices. A representation of a composition by its coordinates with respect to a given orthonormal basis is a transformation from the simplex into a real space; this has been called isometric log-ratio (ilr) transformation (Egozcue *et al.* 2003) and does not show the disadvantages of the previous ones. They are coordinates in an orthogonal system and thus any classical statistical technique can be used (see Chapter 3). Since some times they can be difficult to interpret from a geochemical point of view, their use is not so for accepted. With the aim to encourage the analysis of compositional data by using this approach some application examples concerning different geological materials will be now presented and discussed. Simple alternative diagrams to investigate geochemical processes, on which statistical evaluations can be performed, are consequently proposed.

18.3   Log-ratio approach and water chemistry

When water samples are collected from a defined area or their composition is monitored in time, each observation can be considered as the result of an independent successive random dilution process if time and distance are sufficiently large. In this context, the log-ratio approach can be used to describe the proportionality law underlying the process as described by chemical reactions. The water chemical compositions are often analysed by considering binary diagrams with common denominator ratios (molar ratio diagrams). An application example can be found in Gaillardet *et al.* (1999) for the most important rivers of the world. Data are drawn from the GEMS/WATER Global Register of River Inputs (Meybeck and Ragu 1997), with the exception of the Mekong and Indus rivers. Each river selected in this study was characterized by a number of sodium-normalized molar ratios. Elemental ratios were preferred instead of concentrations as intensive parameters able to permit comparison between rivers draining areas of high run-off (Amazon, Democratic Republic of Congo, Orinoco) and rivers draining arid areas (Limpopo, Murray). Moreover since most sodium-normalized ratios fluctuate over three orders of magnitude a logarithmic scale was preferred to represent data on binary diagrams. This approach in the treatment of compositional data is a first approximation since the axes of alr-transformed data are not perpendicular. If statistical considerations following Euclidean geometry have to be performed, the three variables involved in the common denominator ratio diagrams have to be transformed into two ilr coordinates. New diagrams to visualize the results can be consequently proposed.

As an example, the relationship between Ca^{2+}, HCO_3^- and SO_4^{2-} is analysed for the dissolved phase of the Arno river water (Tuscany, central Italy). Data from Nisi *et al.* (2008) have been transformed into ilr coordinates, thus obtaining Figure 18.1. It represents an alternative to molar ratio plots and is a space where classical statistical evaluation, based on Euclidean geometry, can be performed. Moreover, the investigation of this subcomposition can allow the analysis of the behaviour of two important anions with respect to a cation that shows the common ion effect (participation in more than one chemical equilibrium involving the two previous anions).

The hydrographic catchment covers a surface of $8228 \ km^2$ with an average elevation of 353 m. The river is 242 km long, springs out from Northern Apennines at an elevation of 1650 m and flows into the Ligurian Sea, 10 km west from Pisa and 110 km from Florence. The annual rainfall pattern is typical of the Mediterranean area, with low regime in summer and two peaks of precipitation in winter (December and February). Mean annual rainfall values range from 600 mm, mainly in the low lands, up to 3000 mm on the Apennine ridge. The out-cropping rocks are predominantly sedimentary folded and faulted Mesozoic and Tertiary units resulting from the formation of the Apennine chain. The subsequent extensional tectonic phase has produced a NW–SE oriented horst and graben system, made of Cretaceous to Paleogene allochtonous units, belonging to the Ligurian, Sub-Ligurian and Tuscan domains, being overthrusted mostly in the Early Miocene (Nisi *et al.* 2008).

In Figure 18.1 a complex behaviour is highlighted so that nonlinear functions are required for modelling. In the part of Figure 18.1 marked with the letter A, the ilr coordinates on the x-axis increase until the zero value ($HCO_3^- \cong SO_4^{2-}$), and a similar behaviour is shown by the values on the y-axis up to about 0.6–0.7. On this axis the relationship between the variables moves from $HCO_3^-/Ca^{2+} \cong 2Ca^{2+}/SO_4^{2-}$ to $HCO_3^-/Ca^{2+} \cong 5.5Ca^{2+}/SO_4^{2-}$. From a geochemical point of view, the contribution of sources of SO_4^{2-}, given by gyps-anhydrite rich evaporites, presence of thermal waters, mixing with sea water and pollution, tends to decrease and water composition tends to be dominated by Ca^{2+} and HCO_3^-, both of which are predominantly derived from the dissolution of carbonates bearing sedimentary rocks. In the area of Figure 18.1 marked with the letters B and C, an inverse relationship is shown and the increase in the ilr coordinates on the x-axis is associated with a decrease on the y-axis. The relationship between the variables on the y-axis moves from $HCO_3^-/Ca^{2+} \cong 5.5Ca^{2+}/SO_4^{2-}$ to $HCO_3^-/Ca^{2+} \cong Ca^{2+}/SO_4^{2-}$, until $HCO_3/Ca^{2+} \cong 0.2Ca^{2+}/SO_4^{2-}$, thus characterizing waters with low SO_4^{2-} content or with low total dissolved solid content (springs). The maximum of the quadratic or cubic function corresponds to samples collected at Pisa, near to the mouth of the river, where the origin of sulfate is related to the influence of mixing processes with sea water. On the left of the maximum all the samples affected by the presence of SO_4^{2-} due mainly to rocks or pollution can be found. The source of the Arno river is located in the B quadrant of Figure 18.1.

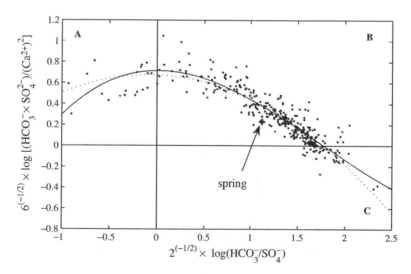

Figure 18.1   ilr transformation applied to SO_4^{2-}, HCO_3^- and Ca^{2+} for the dissolved phase of the water pertaining to the Arno river basin. Points have been modelled by using quadratic (dashed line) and cubic (continuous line) functions.

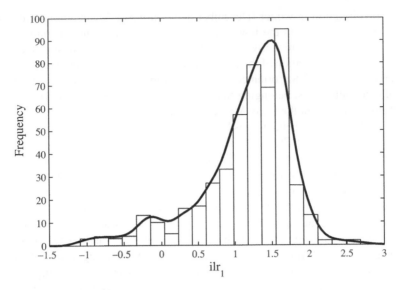

Figure 18.2 Frequency distribution of the ilr$_1$ coordinate and probability density estimated using a kernel smoothing method.

The analysis of the frequency distribution of the ilr coordinate values gives further information about geochemical processes. In Figure 18.2 the histogram of the *x*-axis coordinate, ilr$_1$, together with the probability density estimated using a kernel smoothing method, is reported. In Figure 18.3 the same analysis has been performed for the *y*-axis coordinate, ilr$_2$. The ilr$_1$ coordinate cannot be described with a normal model (Kolmogorov–Smirnov

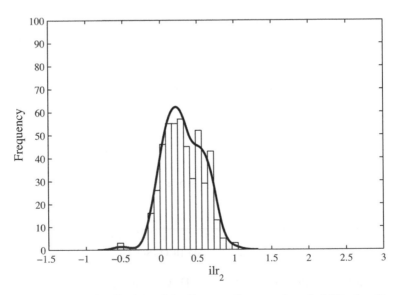

Figure 18.3 Frequency distribution of the ilr$_2$ coordinate and probability density estimated using a kernel smoothing method.

test, $p < 0.05$) due to its high skewness and kurtosis. The behaviour of the two anions is, consequently, partially disjointed and the presence of homogeneous groups of data characterized by lower HCO_3^-/SO_4^{2-} ratios is pointed out as well as the role of the different sources of SO_4^{2-}. On the other hand, the ilr_2 coordinate can be described by a normal model, indicating that the process of competition of two anions for the same cation appears to be more homogeneous.

18.4   Log-ratio approach and volcanic gas chemistry

The composition of gases from volcanic discharges is a function of deep processes, such as vapour–melt separation during the generation and rise of the magma and shallow processes, active within the volcanic apparatus. Generally the variability in the composition of volcanic gases is largely due to shallow processes such as re-equilibration in response to cooling and dilution by meteoric water, and interaction with fluids of associated hydrothermal systems. The evolution of eruptive processes is primarily conditioned by the solute states of the various volatile species in the magma. The effects of vapour exsolution on the relative proportions of a magmatic volatile i transferred to the vapour phase v, or remaining in the melt m, are usually discussed on the basis of their solubility S_i, for a given temperature, or the Henry's law constant h_i. Henry's law, $a_i = h_i X_i$ describes the solution behaviour in which the activities a_i (concentration of a component adjusted for any effects of nonideality) of dissolved species become directly proportional to their concentrations X_i at high dilution; h_i is a proportionality constant. Other commonly used variables to describe vapour–melt distribution processes are the dimensionless mass/volume Ostwald-type distribution coefficient $Q_i = c_{i,v}/c_{i,m}$ as used by Zhang and Zindler (1989) or the mass/mass distribution coefficient $D_i = c_{i,v}/c_{i,m}$ as defined by Kilinc and Burnham (1972), both related to the Henry's law constant through the molecular weight of the species i, density of the melt, total pressure of the gases, and mole fraction X_i of the gas i in the vapour phase. Since the composition of volcanic gases is expressed by using concentrations and the relative presence of chemical species is attributable to a partition phenomenon, the use of log-ratios appears to represent an adequate tool.

In the investigation of the chemical composition of volcanic gas discharges the ilr transformation can be used to investigate the relative changes in some important components as H_2O, CO_2 and SO_2 (Giggenbach 1996). Increased concentrations of CO_2 and SO_2 are often considered gas precursors of eruptions because H_2O may derive from magmatic or meteoric sources. However, an increase in SO_2 is not always observed prior to volcanic events due to the presence of groundwater or surface water that scrubs magmatic gases during its path toward the surface. Geochemical modelling (Symonds *et al.* 2001) suggests that CO_2 is the main species to monitor if scrubbing phenomena affect volcanic discharges since concentrations in SO_2 will be significantly influenced. The log-ratio:

$$ilr_1 = \frac{1}{\sqrt{2}} \ln \frac{CO_2}{H_2O} \tag{18.6}$$

allows the analysis of the variability in the vapour–melt ratio and therefore the degree of degassing of magma batches feeding the vents. When the ratio is high an enrichment in CO_2

can be assumed while its decreasing can be attributable to excess H_2O due to shallow meteoric sources or addition of groundwater. The log-ratio:

$$ilr_2 = \frac{1}{\sqrt{6}} \ln \frac{CO_2 \times H_2O}{(SO_2)^2} \tag{18.7}$$

is able to balance the previous behaviour with respect to the relative presence of a magmatic component such as SO_2. A binary diagram of the ilr coordinates such as that reported in Figure 18.4 represents a real space in which data patterns pertaining to the Vulcano system (Sicily, southern Italy) can be investigated. Different subplots are related to fumarolic

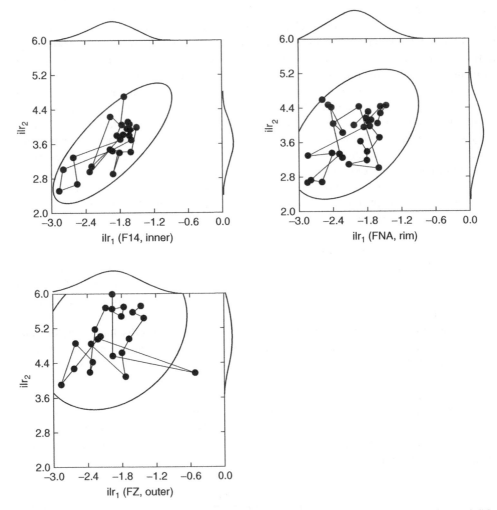

Figure 18.4 Application of ilr transformation for relative changes in H_2O, CO_2 and SO_2 per F14, FNA and FZ fumarolic discharges of Vulcano island (Sicily, southern Italy). Points are connected in time order and on the axes the marginal normal probability distributions are reported. Ellipse represents the confidence region of 95%.

discharges discriminated by their spatial position. In particular, F14 is located on the inner part of the crater rim, FNA on the rim itself and FZ on its outer part. The last eruption of Vulcano dates back to 1888–1890 with the main event being located in the crater named La Fossa. Since then, the main fumarolic field, located on the northern inner and outer flanks of the La Fossa crater, covering an area of about 9000 m^2, has exhibited particularly high variability with regard to total gas emission rates and location and composition of the gas discharges (Buccianti *et al.* 2006).

As we can see, data variability is different depending on the spatial position of the fumarolic discharge. Samples pertaining to F14 (inner) and FNA (rim) fumaroles follow the bivariate normal distribution (Mardia test, $p \gg 0.05$) and are characterized by a significant anisotropy (ratio between major and minor axis of the 95% confidence ellipse). The anisotropy decreases from the inner part of the crater towards the rim. On the other hand, samples drawn from the FZ fumarole located on the outer part of the crater rim do not follow the bivariate normal distribution ($p \ll 0.05$), the dispersion is high (low anisotropy) and an anomalous observation (July 2002) occurred. From this simple diagram, corroborated by statistical evaluations, it is possible to point out that the spatial position of the fumaroles is important to plan monitoring programmes of volcanic activity. Systems that are characterized by high variability in time are not so efficient in detecting important changes. Moreover since points are connected in time order from 2000 to 2010 (mostly more than two samples per year), a cyclical behaviour of the fumarolic discharges is evident. Each system after a given time interval tends to return back to a previous state and this path is affected by low or high variability depending on space.

18.5 Log-ratio approach and subducting sediment composition

Trace elements are fundamental tracers of geochemical processes mainly because they are dilute and their behaviour depends primarily on the trace element–matrix interaction (for example Rb and host feldspar, Sr and host calcite, and so on) and very little on interaction among traces (Rb-Rb, Sr-Sr). Consequently, the distribution of trace elements among natural phases largely obeys Henry's law. The modelling of trace elements in various geological environments (magmas, hydrothermal fluids, seawater, and so on) relies on three different aspects: (1) the total mass of each element distributed among several subsystems such as phases (minerals, melts, fluids) or reservoirs constituting a closed system is preserved; (2) the equilibrium distribution of a trace element i between two phases α and β is usually handled through the law of diluted solutions $C_\beta^i / C_\alpha^i = K_{\beta/\alpha}^i(T, P)$, where $K_{\beta/\alpha}^i(T, P)$ is the temperature-pressure dependent Berthelot–Nerst distribution or partition coefficient; (3) compatible elements are easily hosted by the structure of the crystallizing minerals ($K_{sol/liq}^i > 1$) while incompatible elements are rejected into the liquid ($K_{sol/liq}^i \ll 1$). In all these hypothesis partitioning processes dominate the data variability and the log-ratio approach may be useful to describe statistically how the natural systems work.

Subducted sediments represent an important tool to investigate arc magmatism and the crust–mantle recycling. Understanding involved mass and chemical fluxes is, in fact, useful to constrain models of continental growth. The bulk composition of sediment columns approaching trenches (Plank and Langmuir 1998) has been analysed to propose an alternative

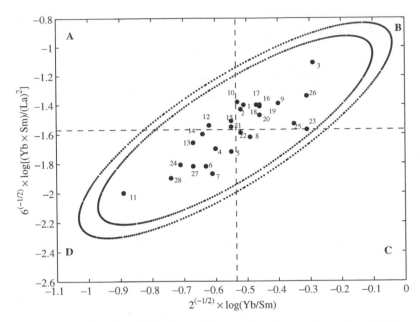

Figure 18.5 REE pattern for subducting sediments after ilr transformation. 1, Tonga; 2, Kermadec; 3, Vanuatu; 4, E. Sunda; 5, Java; 6, Sumatra; 7, Andaman; 8, Makran; 9, Philippines; 10, Ryuku; 11, Nankai; 12, Marianas (801); 13, Marianas (800); 14, Marianas; 15, Izu-Bonin; 16, Japan; 17, Kurile; 18, Kamchatka; 19, Aleutians; 20, Alaska; 21, Cascades; 22, Mexico; 23, C. America; 24, Colombia; 25, Peru; 26, S. Sandwich; 27, N. Antilles; 28, S. Antilles.

diagram to that used to evaluate the effect of different sediment components and processes on rare earth element (REE) patterns. The usual way to describe this process is to consider a binary diagram where La/Sm values are analysed versus Yb/Sm. These ratios reflect the fractionation processes between light (La/Sm) and heavy (Yb/Sm) rare earths providing information about their potential sources. Generally, REE patterns are largely due to detrital sources (volcanic and continental), but abundances may also be affected by biological productivity and preservation of carbonate and opal as well as sedimentation rate. In order to verify the presence of some pattern in the data to be modelled by classical statistics, the ilr transformation has been used. Results are reported in Figure 18.5.

In Figure 18.5, data are characterized by a high positive correlation and most of them are located in the quadrants labelled with letters B and D. The Yb/Sm ratio on the x-axis moves from 0.41 to 0.74, describing the continuous enrichment in heavy REEs towards the right. Simultaneously, the Yb/La ratio moves from 0.13La/Sm to 0.3La/Sm on the y-axis. Since the joint distribution follows the bivariate normal model ($p \gg 0.01$), confidence regions for 0.90 and 0.95 probability can be reported. No value falls out of the 0.95 probability region and cases 15 (Izu-Bon), 21 (Cascades) and 22 (Mexico) are representative of the barycentre of the joint distribution; their subduction rates ranges from 35 to 52 mm year^{-1} (Plank and Langmuir 1998). A representative value for the subduction rate considering all the data is about 60 mm year^{-1} (Ryuku, Aleutine), if the value of Tonga (170 mm year^{-1}) is not considered in calculus due to its distance from the others. Data for Izu-Bon, Cascades and Mexico are below this average value, indicating that the barycentre for the relationship among La, Sm and Yb does not correspond with that of the subduction rate. On the whole, subduction rates are positively related to the ilr coordinates ($r = 0.60$, $p_{uncorr} = 0.001$) and, consequently, to an enrichment in heavy REEs associated with increment in volcanic arc contribution with respect to continental detritus. Thus, ilr coordinates are able to indicate something about the proximity of detrital sources, volcanic and continental, on a statistical basis. No significant relationship is present between ilr coordinates and other parameters as thickness (m) of sediment columns subducting at trenches, length of the trench (km) while a weak inverse relationship characterizes sediment density (g cm^{-3}) (Plank and Langmuir 1998). A final consideration concerns the frequency distribution of the ilr coordinates, which is bivariate normal for the joint distribution and normal for the marginals ($p \gg 0.01$). The result implies for ratios of concentrations the presence of the fingerprint of the theory of successive random dilutions. If the chemical composition of the subducting sediment is considered on a global scale, low values for Yb/Sm, Yb/La and Sm/La are preferred indicating that processes of enrichment of light REEs (presence of continental detritus) are dominant.

18.6   Conclusions

Simple methods to investigate the geochemical behaviour of elements or chemical species in subcompositions have been proposed. Their use is motivated not only by the fact that they represent a correct geometrical framework where statistical analysis can be performed, but also because the use of log-ratios appears to capture in an exhaustive way the proportionality that characterizes the apportionment of the elements in different geological materials. In this way the natural physical–chemical laws that are at the base of the changes in concentrations in time and space can be tentatively described by taking into account the uncertainty. The discussed examples have demonstrated that classical statistical methods, when applied in the correct sample space, are able to put the basis for modelling oriented on a probabilistic basis. This point of view allows us to take into account that our approach is limited. Investigation in fact concerns a sample drawn from a population and following the log-ratio approach the uncertainty affecting the evolution of natural systems can be managed and quantified.

Acknowledgements

This research has been financially supported by Italian MIUR *(Ministero dell'Istruzione, dell'Università e della Ricerca Scientifica e Tecnologica)*, PRIN 2007, through project 2007M4K94A-002 (A.B.). Orlando Vaselli and Raimon Tolosana-Delgado are thanked for suggestions that have highly improved the manuscript.

References

Ahrens L 1954a The lognormal distribution of the elements (1). *Geochimica et Cosmochimica acta* **5**, 49–73.

Ahrens L 1954b The lognormal distribution of the elements (2). *Geochimica et Cosmochimica Acta* **6**, 121–131.

Aitchison J 1982 The statistical analysis of compositional data (with discussion). *Journal of the Royal Statistical Society, Series B (Statistical Methodology)* **44**(2), 139–177.

Aitchison J 1986 *The Statistical Analysis of Compositional Data*. Monographs on Statistics and Applied Probability. Chapman and Hall Ltd (reprinted 2003 with additional material by The Blackburn Press), London (UK). 416 p.

Aitchison J 1999 Logratios and natural laws in compositional data analysis. *Mathematical Geology* **131**(5), 563–580.

Aitchison J and Brown JAC 1957 *The Lognormal Distribution*. Cambridge University Press, Cambridge (UK). 176 p.

Buccianti A, Tassi F and Vaselli O 2006 Compositional changes in a fumarolic field, Vulcano Island, Italy: a statistical case study. *Geological Society, London, Special Publications* **264**, 67–77.

Egozcue JJ, Pawlowsky-Glahn V, Mateu-Figueras G and Barceló-Vidal C 2003 Isometric logratio transformations for compositional data analysis. *Mathematical Geology* **35**(3), 279–300.

Gaillardet J, Dupré B, Louvat P and Allègre C 1999 Global silicate weathering and CO_2 consumption rates deduced from the chemistry of large rivers. *Chemical Geology* **159**, 3–30.

Giggenbach WF 1996 Chemical composition of volcanic gases. In *Monitoring and Mitigation of Volcano Hazards* (ed. Scarpa R and Tilling R). Springer-Verlag, Berlin (Germany). pp. 221–256.

Kilinc I and Burnham C 1972 Partitioning of chloride between a silicate melt and coexisting aqueous phase from 2 to 8 kilobars. *Economic Geology* **67**, 231–235.

Meybeck M and Ragu A 1997 River discharges to the oceans: an assessment of suspended solids, major ions and nutrients. *United Nations Environment Programme* **298**.

Miesch AT 1977 Log transformations in geochemistry. *Mathematical Geology* **9**(2), 191–194.

Nisi B, Buccianti A, Vaselli O, Perini G, Tassi F, Minissale A and Montegrossi G 2008 Hydrogeochemistry and strontium isotopes in the arno river basin (Tuscany, Italy): Constraints on natural controls by statistical modeling. *Journal of Hydrology* **360**, 166–183.

Ott WR 1995 *Environmental Statistics and Data Analysis*. Lewis Publishers, Boca Raton, FL (USA). 313 p.

Plank T and Langmuir C 1998 The chemical composition of subducting sediment and its consequences for the crust and mantle. *Chemical Geology* **145**, 325–394.

Symonds RB, Gerlach TM and Reed MH 2001 Magmatic gas scrubbing: implications for volcano monitoring. *Journal of Volcanology and Geothermal Research* **108**, 303–341.

Zhang Y and Zindler A 1989 Noble gas constraints on the evolution of earth's atmosphere. *Journal of Geophysical Research* **94**, 13710–13737.

19

Compositional data analysis in planetology: the surfaces of Mars and Mercury

Helmut Lammer[1], Peter Wurz[2], Josep Antoni Martín-Fernández[3] and Herbert Iwo Maria Lichtenegger[1]

[1]*Space Research Institute, Austrian Academy of Sciences, Austria*
[2]*Institute of Physics, University of Bern, Switzerland*
[3]*Department of Computer Science and Applied Mathematics, University of Girona, Spain*

19.1   Introduction

Planetary surfaces are generally altered by various chemical and physical processes over the planet's history. Relevant processes which modify the surface of a planet could be global tectonics, chemical weathering, dust storms, liquids, winds, volcanism, and space weathering due to solar radiation and particle surface exposure. Figure 19.1 illustrates these various processes and their related efficiency for a planet which has experienced significant atmospheric and climate changes during its history (Mars) and an airless body like Mercury whose surface was exposed to solar radiation and the particle environment over billions of years.

19.1.1   Mars

Due to the lack of plate tectonics since more than 4 Gyrago, present day Mars is considered a one-plate planet (Banerdt *et al.* 1992). Therefore, the Martian surface represents a long-term

Compositional Data Analysis: Theory and Applications, First Edition. Edited by Vera Pawlowsky-Glahn and Antonella Buccianti.
© 2011 John Wiley & Sons, Ltd. Published 2011 by John Wiley & Sons, Ltd.

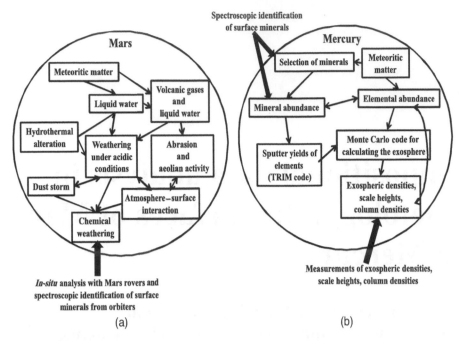

Figure 19.1 Illustration of various weathering processes which have affected the surfaces and their composition on Mars (a) and Mercury (b). While on Mars volcanic activity in conjunction with liquid water was most likely the main chemical weathering process, in the case of an airless body such as Mercury mainly solar radiation and particle exposure type space weathering permanently affects and modifies the unprotected uppermost surface layers.

archive of exogenic processes and contains a record of the evolution of surface conditions on ancient Mars. Recent discoveries by the Pathfinder rover, the Mars Exploration rovers (MER-A, MER-B) and the Mars Express and Mars Reconnaissance orbiters show clear evidence for the involvement of liquid water in chemical weathering on the early Martian surface. These findings are based mainly on the discoveries of coarse crystalline weathering phases such as hematite spherules (Christensen *et al.* 2000, 2004) and the hydrated minerals and sediments in the oldest sections of the Martian surface areas (Bibring *et al.* 2005).

The above-mentioned rovers and orbiters also discovered widespread, massive layered sulfate-rich sediments but no massive carbonate deposits. The high sulfur concentrations in Martian soils (Toulmin *et al.* 1977; Brückner *et al.* 2003; Foley *et al.* 2003; Gellert *et al.* 2004; Rieder *et al.* 2004) further indicate the chemical interaction between surface minerals and acidic volcanic exhalations (Banin *et al.* 1997). Bullock and Moore (2007) studied the formation of the sulfate-rich Meridiani sediments via coupled atmospheric and geochemical calculations and concluded from their results that large scale volcanism from the build-up of Tharsis during the late Noachian would have injected large amounts of SO_2 into the atmosphere. In such a case an efficient photochemical conversion of SO_2 to H_2SO_4 would have caused widespread sulfuric acid/water clouds, similar to those seen on present day Venus. Precipitation from these clouds and acidification of surface H_2O would have sustained a thick, warm CO_2 atmosphere via carbonate inhibition. Such an atmosphere could have been subjected to loss to space via impact erosion, sputtering, hot atom escape and other nonthermal atmospheric loss processes. Once atmospheric SO_2 gas production dropped and waters became more alkaline, the remaining CO_2 in the atmosphere collapsed to form poorly consolidated carbonate patinas on rock surfaces and in open fractures.

While the chemical composition of the Martian soil appears to be uniformly distributed over the whole planet, the chemical data on the rocks show a considerable scatter (Brückner et al. 2003; Rieder et al. 2004). Morris et al. (2000) described the uniformity of soil compositions as a homogenization of weathering products in the course of aeolian activity. In the case of Mars chemical weathering seems to be the major process for modifying the planetary surface.

19.1.2   Mercury

As one can see from Figure 19.1 in the case of a body without a significant atmosphere, like Mercury, the surface is affected by different processes that are mainly related to the radiation and plasma environment of the Sun and to micrometeorites, which are delivered to Mercury's surface (Milillo et al. 2005; Killen et al. 2007). When a planetary surface is freshly exposed to ion bombardment, sputtering of the different surface elements will lead to an enrichment of those elements with low sputtering yields in the top-most atomic layers of the surface. For that reason, exospheric densities resulting from particle sputtering, will give us quantitative compositional information about all elements of the bulk surface, including the refractory elements when a steady-state situation is established. The sputter yields of released minerals can be calculated by using the TRIM.SP software (Ziegler 2004). After the minerals are released with a velocity distribution the exosphere density can be calculated by a Monte Carlo code.

In such a case it can be assumed that the composition of Mercury's thin collisionless atmosphere, the exosphere, is related to the composition of the planetary crustal materials. If so, then inferences regarding the bulk chemistry of the planet can be made from a study of atoms and molecules in the exosphere after they are released from the mineral surface by a variety of release processes. One difficult challenge is the identification of the main source of some elements like H, He, Na or K. Generally it is believed that H and He come primarily from the solar wind, while Na and K originate from volatilised materials partitioned between Mercury's crust and impacts from meteorites (Hodges 1980). Besides the above-mentioned elements corresponding to spectroscopic observations (Warell 2003; Warell and Blewett 2004) and experiments with soil analogues (Burbine et al. 2002), other elements such as O, Na, Mg, Al, Si, P, S, K, Ca, Ti, Cr, Fe, Ni, Zn and OH should also be related to Mercury's surface soils (Wurz et al. 2010, and references therein).

The main processes which release various elements into the exosphere are thermal vaporization, solar photon-stimulated desorption (PSD), impact vaporization, and solar wind ion sputtering. Each of these processes has its own temporal and spatial dependence, which is also related to the planet's orbital location and rotation. Due to these orbital parameters, Mercury's surface experiences large fluctuations in temperature and differences of insolation with longitude. Because it is intended to measure these exospheric particles in situ with particle detectors on board of ESA's BepiColombo Mercury Planetary Orbiter (MPO) spacecraft for reconstructing the surface composition, we will focus on sputtered atoms in the following sections because sputtered atoms are expected to dominate the exosphere at altitudes around

1000 km (Wurz and Lammer 2003), with the exception of thermally released H_2 and He, which are sufficiently light and abundant to be observed also at higher altitudes.

19.1.3   Analysis of surface composition

To show how different weathering processes affect different planetary surface environments we discuss the latest stage of compositional data analysis methods for Mars and Mercury. Compositional data analysis has been employed for the interpretation of chemical data returned from Mars' surface (Clark 1993; Morris *et al.* 2000) and recently for Mercury's surface minerals (Sprague *et al.* 2009). In these cases the applied method was based on simple correlation methods, which do not exploit the full potential of the available data. In addition, the closed nature of compositional data, i.e. the assumption that component concentrations have to sum up to 100% in an analysis, bears important implications for the statistical analysis of compositional data, which do not seem to have been sufficiently appreciated until now.

To investigate the default of the classical additive analysis method our research group recently applied (Kolb *et al.* 2006; Wurz *et al.* 2010) a more realistic multiplicative method (Aitchison 1986) based on the Euclidean space geometry of the simplex (see Chapter 11). Our recent results presented in detail in Kolb *et al.* (2006) for Mars and Wurz *et al.* (2010) for Mercury are summarized in this chapter, highlighting the relevance of composition data analysis in planetology.

19.2   Compositional analysis of Mars' surface

The most exciting results in the recent studies of Martian surface mineralogy were obtained by the analysis of surface materials and rocks by the US-Mars Exploration Rovers (MER-A and MER-B), the Mars Odyssey Gamma Ray Spectrometer (GRS) measurements, and the spectroscopic observations European Mars Express mission. There is the clear indication for the presence of liquid water that acted as an agent of chemical weathering on ancient Mars. This exobiological highly relevant discoveries are based on several lines of evidence, including the occurrence of coarse crystalline weathering phases such as hematite spherules (Christensen *et al.* 2000, 2004), the hydrated minerals and sediments discovered in the oldest sections of the Martian surface (Bibring *et al.* 2005), and the extensive regions of enhanced near-surface hydrogen abundance that GRS has discovered, which are estimated to contain up to 5–10% water by mass (Feldman *et al.* 2002).

In the sense of Aitchison (1986) the compositional data, which are returned from the planetary surface, must be considered as subcompositions. The correlation coefficients between element concentrations can be essentially influenced by the normalization to a constant sum, where the closure operation disturbs correlations in the covariance structure of a given set of data (Aitchison 1997; Kolb *et al.* 2006). For that reason classical multivariate techniques, such as conventional correspondence analysis or biplot techniques can be misleading when applied to subcompositions. In such cases the visualization of compositional data with subcompositions may be deceptive (Aitchison 1997).

Kolb *et al.* (2006) applied a centred log ratio (clr) transformation prior to statistical analysis that translates compositional data from a constrained Simplex space to a reality representing

Euclidean space. clr transformation of a compositional vector $\mathbf{C}$ with components c_i can be defined after Aitchison (1986) as

$$\mathrm{clr}(\mathbf{C}) = \left[\log\left(\frac{c_1}{g_m(\mathbf{C})} \right), \ldots, \log\left(\frac{c_D}{g_m(\mathbf{C})} \right) \right], \tag{19.1}$$

where $g_m(\mathbf{C})$ is the geometric mean of the vector components given by

$$g_m(\mathbf{C}) = \left(\prod_{i=1}^{D} c_i \right)^{\frac{1}{D}}, \tag{19.2}$$

and log is the natural logarithm. Note that the dependence of individual compositional entries on each other is an important consequence of the constrained Simplex space. If in the course of an alteration process a subset of elements is variable and is added or lost from a given rock volume, the concentrations of all components, the variable ones but also the immobile ones will change merely due to the closure condition. To overcome such problems Aitchison (1997) proposed a so-called perturbation mechanism to model open system reactions such as alteration. Equation (19.3) shows the application of an alteration vector $\mathbf{C}^*$ on a chemical composition $\mathbf{C}$ for yielding a composition C^{**} (Aitchison 1997)

$$C^{**} = \mathbf{C} \oplus \mathbf{C}^* = \left(\frac{c_1 c_1^*}{\sum_{i=1}^{D} c_l c_i^*}, \frac{c_2 c_2^*}{\sum_{i=1}^{D} c_l c_i^*}, \ldots, \frac{c_D c_D^*}{\sum_{i=1}^{D} c_l c_i^*} \right). \tag{19.3}$$

In the analysis of Kolb *et al.* (2006) the chemical subcompositions of 13 Martian surface elements in wt% (Na, Mg, Al, Si, P, S, Cl, K, Ca, Ti, Cr, Mn, Fe) were selected from the literature. The elements were taken from different landing sides and geographical locations by different Mars rovers: Mars Pathfinder (Brückner *et al.* 2003; Foley *et al.* 2003); MER-A (Gellert *et al.* 2004); and MER-B (Rieder *et al.* 2004).

Kolb *et al.* (2006) did neither consider Viking samples in their study because five elements (Na, P, K, Cr, Mn) were not analyzed nor volatile components and O, Ni, Zn, and Br, where the concentrations are only available for a few mission samples (Kolb *et al.* 2006). Because in some Pathfinder samples Cr and Mn contents are below the detection limits of about 0.1 wt% (Brückner *et al.* 2003) and for meeting the requirements of strictly positive entries for clr transformation, missing data of these minerals were replaced by 0.065 wt%, i.e. 65% of the detection limit, according to Martín-Fernández *et al.* (2003). The clr transformation in Kolb *et al.* (2006) was applied prior to factor determination to account for the closed nature of chemical data and to prevent bias on the factor loadings obtained (Aitchison *et al.* 2005).

Figure 19.2 shows the results illustrated in a clr biplot according to Mars Pathfinder and Mars Exploration Rover data, which describes 74% of the variance. The vectors indicate the distribution of clr-transformed variables across the chemical variability of Mars' surface materials, where the origin of the vectors represent the origin in the clr-transformed Euclidean space. Five classes of sample suites can be identified in Figure 19.2: Domain of soil, MER-B evaporites, MER-B basalts, MER-A basalts and Mars Pathfinder basalt-andesites. The soils plot close to the origin, while the source rocks are located in different positions of the biplot (Kolb *et al.* 2006).

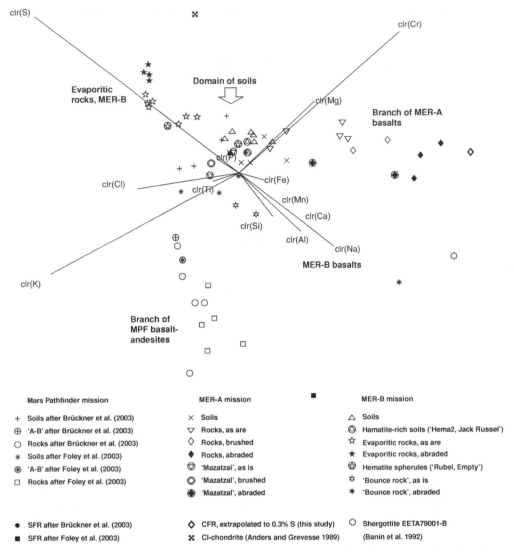

Figure 19.2 clr biplot for gross classification of Martian soil and rock samples based on 13 elements (Na, Mg, Al, Si, P, S, Cl, K, Ca, Ti, Cr, Mn, Fe); soil free rock (SFR) and crust free rock (CFR). After Kolb *et al.* (2006).

This behaviour represents the chemical uniformity among Martian soils, which is in contrast to the rather heterogeneous nature of rock compositions. The trends shown in Figure 19.2 allow discrimination between coated and abraded rocks which are located more proximal and more distal to the region of soils. The rock Mazatzal represents an exception and does not fall on the basaltic branch (Figure 19.2). Bishop *et al.* (2002) developed an airborne dust model on rock surfaces. By considering modification by dust storms most likely the veneer of this rock is cemented dust as a product of chemical interactions, which is also in agreement with microscopic information obtained from Mazatzal (McSween *et al.* 2004).

Magmatic rocks are closer located on the side of clr(Si) and the evaporites plot in the area which is spanned by clr(S), clr(Cl), and clr(Mg). MER-A basalts plot close to clr(Mg) and clr(Cr); basalt-andesites moved closer toward the majority of clr(K) and clr(Si) and basalts investigated by MER-B are more of an intermediary character. The shift between brushed and abraded basalts observed by MER-A indicates the formation of chemical weathering crusts different from fresh rocks. The overall analysis of Kolb *et al.* (2006) concludes that these observations are consistent with petrological models of Mars rocks.

As pointed out in Kolb *et al.* (2006), chemical weathering may have imposed also the incorporation of volcanic gases into the soil. This process could lead to the formation of secondary sulfates and chlorides some gigayears ago. One can assume that fresh rocks represent reservoirs for soil formation aside of volcanic gases, because as shown in Figure 19.2 weathering rinds represent most likely intermediate stages between source rock and soil.

Kolb *et al.* (2006) applied also linear least squares unmixing to recast the soil composition at Mars in terms of anticipated end-member components. These authors considered the following seven end members: two samples of soil free rock, crust free rock, shergottites, volcanic gas, meteorites, ferric oxides, where each component reflects an influence from chemical or physical weathering. The elemental values of these end members are given in percentage in Table 3 of Kolb *et al.* (2006).

Figure 19.3(a) shows the difference vectors based on mix 2 (Kolb *et al.* 2006, Table 3) between target soil composition in percentage and recalculated soil composition from least squares unmixing in comparison with the uniform composition (horizontal line) as a reference. The vertical bars indicate relative uncertainties inherent in the measurements (Brückner *et al.* 2003; Foley *et al.* 2003; Gellert *et al.* 2004; Rieder *et al.* 2004). The unmixing calculation with soil free rock after Brückner *et al.* (2003) provides the best results by means of omission of ferric oxides as denoted end member (Kolb *et al.* 2006).

The compositional analysis of the Martian surface by Kolb *et al.* (2006) shows evidence from the alteration vector of MER-A basalts of the release of Mg and Fe species upon attack

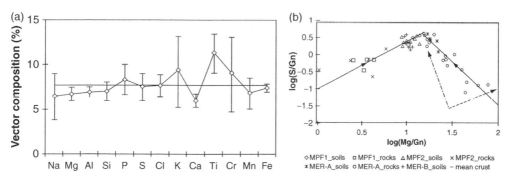

Figure 19.3   (a) Difference vectors between target soil composition and calculated soil composition towards the target composition from least squares unmixing in comparison with the uniform composition which is shown by the horizontal line. The vertical bars indicate relative uncertainties inherent in the measurements. (b) Log-ratio contrast of Mg versus S. The trend lines toward soil for basalt-andesitic and MER-A basaltic branches are separated, indicating external source of S and tendency of compositional symmetry among the rock branches concerning variability of Mg. After Kolb *et al.* (2006).

of volcanic gases and acidic fluids, which is consistent with models and experiments on acid-sulfate weathering of olivine-bearing basaltic rocks.

Figure 19.3(b) shows the log-ratio contrast between Mg and S for all Martian samples except Viking samples and MER-B rock samples. The mean crust was taken from least squares calculation of Table 5 in Kolb *et al.* (2006). Addition of $MgSO_4$ to mean crust illustrated by the dashed arrow does not coincide with the soil formation vector which is marked by the chain dotted arrow. This implies that the soil can not be explained by simple addition of $MgSO_4$ phases. One can also see that the vector toward $MgSO_4$-rich compositions shown by the dashed arrow is parallel to the vector of the basalt-andesitic branch, which means that from the view of basalt-andesites soils are formed by addition of $MgSO_4$ phases, but the basaltic branch may serve as a main source of $MgSO_4$ via remote weathering (Kolb *et al.* 2006).

From the meteoritic contribution in soils one can enable the deduction of minimum formation rates and maximum deposition soil ages. The influence of short-term meteoritic accretion rates are consistent with lunar research and sedimentation rates on the order of metres per gigayear which are also consistent with estimations of dust mantling on Mars (Kolb *et al.* 2006).

Mass balance calculation between escape to space (Lammer *et al.* 2003) and on the incorporation of atmospheric oxygen into Martian soil can be used for the estimation of the oxidation state of S-bearing volcanic gases. Irreversible bonding of atmospheric oxygen takes mainly place via oxidation of Fe and S species at the surface. Atmospheric oxygen is produced by dissociation of H_2O molecules at the upper atmosphere related to H and H_2 escape to space. Oxygen which cannot escape to space may react with the surface (Lammer *et al.* 2003; Kolb *et al.* 2006). The overall amount of sequestered oxygen could be equivalent to an ocean of H_2O with 40 cm depth, which results in an oxidized soil layer of 4–14 m, depending on the oxidation state of S-bearing volcanic gases. Consequently, hydrogen sulfide may represent an important component in volcanic gas released. However, recent nonthermal escape calculations of hot O atoms by Krestyanikova and Shematovich (2005) and Filzmoser and Hron (2009) may reduce the amount of oxygen which could have been available for the oxidation of the Martian surface.

As one can see, the alteration of the surface composition of Mars is mainly produced by chemical and physical weathering which is strongly connected to the planet's early volcanic outgassing, atmosphere, climate and geophysical processes. As a contrast we discuss the recent application of the same compositional modelling method to the innermost planet, Mercury, which represents a body without significant atmosphere.

19.3   Compositional analysis of Mercury's surface

Mercury is the closest planet in the solar system to our Sun. The only information obtained from spacecraft comes from three fly-bys performed in 1974 by the Mariner 10 spacecraft and three recent fly-bys in 2008 and 2009 by the MESSENGER spacecraft. Ground-based observations are difficult to obtain due to Mercury's location close to the Sun. One Mercury solar day corresponds to about 176 Earth days. The surface temperature at the dayside ranges from 700 K at the subsolar point up to 90 K in the aphelion midnight point. The planetary radius is about 2440 km which is about 2.6 times smaller than that of the Earth and about 1.39 times smaller than Mars. Because of a large iron core Mercury's density of about 5.3 g cm^{-3}

is the highest for a solar system planet (Lewis 1988). The Mariner 10 fly-bys discovered the existence of an intrinsic magnetic field (Ness *et al*. 1976) with field strengths between 200 nT and 400 nT at the equator.

Mariner 10 reported also no observational evidence of a proper atmosphere, however, a thin gaseous envelope around the planet was observed (Broadfoot *et al*. 1976). This thin gaseous envelope is an exosphere where the neutral atoms and molecules move in a collision-less medium along ballistic trajectories (Wurz and Lammer 2003; Killen *et al*. 2007). As illustrated in Figure 19.1, Mercury's surface interacts in a complex manner with the particle flux and radiation environment of the Sun. The related weathering processes are different compared with the chemical weathering on the Martian surface. Observations on board Mariner 10 with the UV spectrograph established the presence of H, He, and O in Mercury's exosphere (Broadfoot *et al*. 1976; Shemansky and Broadfoot 1977), while ground based observations established the presence of Na, K and Ca (Potter and Morgan 1985, 1986; Bida *et al*. 2000; Killen *et al*. 2007, and references therein). MESSENGER observed during the first fly-by the tailward distributions of exospheric Na and Ca (McClintock *et al*. 2008a) and during the second fly-by Na, Ca and Mg (McClintock *et al*. 2009).

Compositional modelling and related studies are relevant on Mercury or similar airless bodies because a detailed study of the various exospheric species can provide important information about the planet's surface composition and its evolution. In preparation for measurements with the SERENA instrument on board BepiColombo MPO (Milillo *et al*. 2010) it is important to obtain knowledge of the planet's surface composition because the exosphere is continuously eroded and refilled by interaction processes coupled to the solar wind plasma and solar radiation, so that one can consider the planet's environment as a *magnetosphere–surface–exosphere* system.

In a recent study, Wurz *et al*. (2010) investigated the presently available information on the mineralogical surface composition of Mercury and developed a mineralogical model that can serve as a basis for future exosphere models. As for Mars the compositional data analysis approach based on the method of Aitchison (1986) was applied for the analysis of Mercury's surface composition which represents a kind of mineralogical surface map.

The present knowledge of the mineralogical composition of Mercury's surface has been discussed in previous works (Sprague *et al*. 2007, and references therein). Although there are no direct measurements of Mercury's surface composition from the Mariner 10 fly-bys, there are disk-averaged spectra available as well as some spatially resolved observations from the MASCS instrument from the first MESSENGER fly-by (McClintock *et al*. 2008b). The main knowledge of Mercury's mineralogical composition has been derived from ground-based observations in the visible and infrared spectral regions (Warell *et al*. 2006; Sprague *et al*. 2009), but the spatial resolution is limited to 200–300 km at best because of disturbances in the Earth's atmosphere (Sprague *et al*. 2007, 2009). Therefore, the results from these ground-based observations shown in Table 19.1, which were developed by Wurz *et al*. (2010), have to be interpreted as global averages or at least as averages over a large surface area.

The small size of the regolith grains complicates the spectroscopic identification of min-erals on the surface. From laboratory studies it is known that grain sizes of about 30 μm fit best with the observed spectra (Warell and Blewett 2004). The observed spectral red-dening of the surface by space weathering can only be reproduced if the surface soils contain at least a few wt% of FeO in the bulk material (Hapke 2001; Burbine *et al*. 2002). Additionally mid-infrared spectral studies of Mercury's surface indicate also sodium-rich

Table 19.1 Expected mineralogical composition of Mercury's surface developed by Wurz *et al.* (2010). The last column shows the range of mineral abundances.

Mineral	Composition	Abundance (mol %)	Range (mol %)
Iron nickel metal	FeNi	0.07	0.04–0.15
Troilite	FeS	0.15	0–0.5
Daubreelite	$FeCr_2S_4$	0.15	0–0.3
Oldhamite	CaS	0.15	0–0.3
Sphalerite	ZnS	0.58	0–1
Feldspar group			
Albite	$NaAlSi_3O_8$	17.44	13–21
K-feldspar	$KAlSi_3O_8$	0.39	0.2–0.7
Anorthite	$CaAl_2Si_2O_8$	8.72	6.7–11
Ilmenite	$FeTiO_3$	0.07	0.02–0.15
Apatite	$Ca_5(PO_4)_3OH$	1.45	0–2
Pyroxene group			
Wollastonite	$CaSiO_3$	2.91	2.4–3.5
Ferrosilite	$Fe_2Si_2O_6$	0.36	0.1–0.5
Enstatite	$Mg_2Si_2O_6$	29.06	24–34
Olivine group			
Fayalite	Fe_2SiO_4	2.18	1.7–2.7
Fosterite	Mg_2SiO_4	36.33	31–41

feldspars and pyroxene (Sprague and Roush 1998) and alkali basalts (Sprague *et al.* 1994), as well as clino-pyroxene (Sprague *et al.* 2002).

We apply constraints related to the mineralogical composition which arise from exosphere observations, in particular for Na, K, and Ca and used a mixture of end members related to three mineralogical groups (feldspar, pyroxene, olivine) shown in Table 19.1. One can see from these minerals that they are connected to 15 species: O, Na, Mg, Al, Si, P, S, K, Ca, Ti, Cr, Fe, Ni, Zn, OH (Wurz *et al.* 2010).

To account for the relative nature of compositional data, ratios among entries are considered rather than absolute values (Aitchison 1986). As discussed above, the log-ratio methodology for compositional data analysis has been successfully applied for the analysis of chemical data of the Martian surface (Kolb *et al.* 2006).

The calculation of intermediates between two compositions by using the perturbation mechanism given in Equation (19.3) can be generalized to the case of the intermediates between n compositions $C_1, C_2, \ldots, C_n$. Wurz *et al.* (2010) applied this strategy to calculate intermediate compositions for the chemical data. Chemical subcompositions of Mercury's minerals of the above-mentioned 15 elements were produced. By establishing the intermediate compositions, those elements which are not simultaneously present in some compositions appear as a '0'. To avoid the '0'-effect in the perturbation mechanism procedure, the intermediate subcomposition is calculated and then the rest of elements can be imputed. By calculating the intermediate composition in, e.g., the Pyroxene group which is related to the minerals Ca, SiO_3, $Fe_2Si_2O_6$, and $Mg_2Si_2O_6$, the compositions are formed by Si, O, Ca, Fe, and Mg. For

Table 19.2 Comparison between the additive and multiplicative surface composition model results for Mercury's surface elemental abundance (Wurz *et al.*, 2010).

Model	O	Na	Mg	Al	Si	P	S	
Additive	58.61	1.34	16.2	2.71	17.4	0.208	0.519	
Multi. Ca fixed	58.8	1.34	16.1	2.64	17.4	0.208	0.529	
Multi. Ca = 1.67%	59.42	1.32	15.8	2.62	17.3	0.268	0.591	
Model	K	Ca	Ti	Cr	Fe	Ni	Zn	OH
Additive	0.030	1.67	0.015	0.042	0.872	0.004	0.291	0.069
Multi. Ca fixed	0.03	1.269	0.015	0.042	1.279	0.004	0.291	0.069
Multi. Ca = 1.67%	0.03	1.67	0.014	0.041	0.611	0.004	0.285	0.069

this mineral group, the intermediate composition is produced by applying the perturbation mechanism to the subcomposition (Si, O) and then, imputing the rest of the elements to obtain the full composition (Mateu-Figueras *et al.* 2003). This method of imputation guarantees that the ratios between the elements are preserved. Following this strategy, the surface composition is obtained and can be compared with simple additive correlation methods used in previous studies. Table 19.2 shows the abundance of the expect surface elements for the modelled additive and multiplicative methods in units of at% when Ca is fixed to a concentration of 1.67 at%.

One can see from Table 19.2 that in the case of Mercury the absolute differences between the results produced by the additive method and the multiplicative method are generally small. On the other hand there are relative differences between some trace elements which can be large (e.g. S when one fixes the Ca element). These effects can be explained because the most abundant minerals on Mercury's surfaceshown in Table 19.1 are Albite, Enstatite and Fosterite, which are mainly composed of the chemical elements O, Mg and Si.

One can see that the main difference between the composition modelled with the additive and the multiplicative methods is related to the amount of Fe. When the multiplicative method is applied, the ratio of Na/K is kept constant to a value of about 44 in all cases. Ca is most likely sputtered from the surface (Bida *et al.* 2000) and an amount of 1.67 at% reproduces the observed exosphere column density very well (Wurz and Lammer 2003); we fixed this value in one calculation. If we assume that Ca = 1.67%, the multiplicative method gives only slightly lower Fe contents compared with the additive method. If one takes Ca not as a constant, its composition is lower and the amount of Fe is much higher.

In addition, the different effect of both methods is distributed among the 15 elements in the composition not causing large absolute differences in the trace elements. One should note that the subcomposition formed by the trace elements in some minerals like Troilite, Oldhamite and Sphalerite (see Table 19.1) is close to the barycentre of the simplex ($e = [1/D, \ldots, 1/D]$). As stated by Martín-Fernández and Bren (2001), when the data are near to the barycentre both Euclidean and log-ratio tools produce approximate results.

The derived surface compositions can be used for the calculation of the release of surface minerals due to solar wind sputtering and related refilling of the planet's exospheric densities in a self-consistent way.

19.4   Conclusion

It was shown that compositional data analysis is an important tool in planetology for analysing the surface mineralogical composition of different planets. On Mars compositional modelling was performed on surface materials in order to shed some light on chemical weathering scenarios and the role of meteoritic accumulation which occurred over the history of the red planet. The main results are in agreement with discoveries from experimental acid-sulfate weathering on olivine-bearing basalts and the persistence of secondary silica in evaporitic rocks. In the case of Mercury, optical observations in the visible and infrared, as well as the exosphere composition indicate that the planet's upper-most surface layer has undergone solar-induced space weathering over several gigayears. Thus, mineralogical models reflect the present space-weathered surface. By applying a mineralogical model which is based on end-member minerals we compared the results obtained from the simple additive composition modelling method with a complex but more realistic multiplicative approach and found that the subcomposition formed by the trace elements in some minerals are near the barycentre of the simplex space geometry, so that in this particular case and contrary to Mars both methods lead to comparable results.

Acknowledgement

H. Lammer, J.A. Martín-Fernández and H. I. M. Lichtenegger acknowledge support from the *Büro für Akademische Kooperation und Mobilität* of the Austrian Academic Exchange Service under the ÖAD-Acciones Integradas project ES 14/2010 *Surface-exosphere composition modelling of airless bodies in the Solar System.*

References

Aitchison J 1986 The Statistical Analysis of Compositional Data Monographs on Statistics and Applied Probability. Chapman and Hall Ltd (reprinted 2003 with additional material by The Blackburn Press), London (UK). 416 p.

Aitchison J 1997 The one-hour course in compositional data analysis or compositional data analysis is simple. In *Proceedings of IAMG' 97 – The III Annual Conference of the International Association for Mathematical Geology* (ed. Pawlowsky-Glahn V), vol. I, II and addendum. International Center for Numerical Methods in Engineering (CIMNE), Barcelona (Spain). pp. 3–35.

Aitchison J, Kay JW and Lauder IJ 2005 *Statistical Concepts and Applications in Clinical Medicine.* Chapman and Hall/CRC, Boca Raton, FL (USA). 339 p.

Anders E and Grevesse N 1989 Abundances of the elements – meteoritic and solar. *Geochimica et Cosmochimica Acta* **53**, 197–214.

Banerdt WB, Golombek MP and Tanaka KL 1992 Stress and tectonics on Mars. In *Mars* (ed. Kieffer H, Jakosky B, Snyder C and Matthews M) University of Arizona Press, Tucson, AZ (USA). pp. 249–297.

Banin A, Han FX, Kan I and Cicelsky A 1997 Acidic volatiles and the mars soil. *Journal of Geophysical Research* **102**, 13341–13356.

Bibring JP, Langevin Y, Gendrin A, Gondet B, Poulet F, Berthè M, Soufflot A, Arvidson R, Mangold N, Mustard J and Drossart P 2005 Mars surface diversity as revealed by the omega/mars express observations. *Science* **307**, 1576–1581.

Bida TA, Killen RM and Morgan TH 2000 Discovery of calcium in mercury's atmosphere. *Nature* **404**, 159–161.

Bishop JL, Murchie SL, Pieters CM and Zent AP 2002 A model of dust, soil, and rock coatings on Mars Physical and chemical processes on the Martian surface. *Journal of Geophysical Research* **107(E11)**, 7.1–7.17.

Broadfoot AL, Shemansky DE and Kumar S 1976 Mariner 10: Mercury atmosphere. *Geophysical Research Letters* **3**, 577–580.

Brückner J, Dreibus G, Rieder R and Wänke H 2003 Refined data of alpha proton x-ray spectrometer analyses of soils and rocks at the mars pathfinder site: Implications for surface chemistry. *Journal of Geophysical Research* **108(E12)**, 8094.

Bullock MA and Moore JM 2007 Atmospheric conditions on early mars and the missing layered carbonates. *Geophysical Research Letters* **34(19)**, L19201.

Burbine TH, McCoy TJ, Nittler LR, Benedix GK, Cloutis EA and Dickinson TL 2002 Spectra of extremely reduced assemblages: Implications for Mercury. *Meteorites & Planetary Science* **37**, 1233–1244.

Christensen PR, Bandfield JL, Clark RN, Edgett KS, Hamilton VE, Hoefen T, Kieffer HH, Kuzmin RO, Lane MD, Malin MC, Morris RV, Pearl JC, Pearson R, Roush TL, Ruff SW and Smith MD 2000 Detection of crystalline hematite mineralization on mars by the thermal emission spectrometer: Evidence for near-surface water. *Journal of Geophysical Research* **105(E4)**, 9623–9642.

Christensen PR, Wyatt MB, Glotch TD, Rogers AD, Anwar S, Arvidson RE, Bandfield JL, Blaney DL, Budney C, Calvin WM, Fallacaro A, Fergason RL, Gorelick N, Graff TG, Hamilton VE, Hayes AG, Johnson JR, Knudson AT, McSween JHY, Mchall GL, Mchall LK, Moersch JE, Morris RV, Smith MD, Squyres SW, Ruff SW and Wolff MJ 2004 Mineralogy at meridiani planum from the mini-tes experiment on the opportunity rover. *Science* **306**, 1733–1739.

Clark BC 1993 Geochemical components in martian soil. *Geochimica et Cosmochimica Acta* **57**, 4575–4581.

Feldman WC, Boynton WV, Tokar RL, Prettyman TH, Gasnault O, Squyres SW, Elphic RC, Lawrence DJ, Lawson SL, Maurice S, McKinney G, Moore KR and Reedy RC 2002 Global distribution of neutrons from mars: Results from Mars Odyssey. *Science* **297**, 75–78.

Filzmoser P and Hron K 2009 Correlation analysis for compositional data. *Mathematical Geosciences* **41**, 905–919.

Foley CN, Economou T and Clayton RN 2003 Final chemical results from the mars pathfinder alpha proton x-ray spectrometer. *Journal of Geophysical Research* **108(E12)**, 8096.

Gellert R, Rieder R, Anderson RC, Brückner J, Clark BC, Dreibus G, Economou T, Klingelhöfer G, Lugmair GW, Ming DW, Squyres SW, d'Uston C, Wänke H, Yen A and Zipfel J 2004 Chemistry of rocks and soils in gusev crater from the alpha particle x-ray spectrometer. *Science* **305**, 829–833.

Hapke B 2001 Space weathering from Mercury to the asteroid belt. *Journal of Geophysical Research* **106(E5)**, 10.039–10.074.

Hodges RRJ 1980 Methods for monte carlo simulation of the exospheres of the moon and mercury. *Journal of Geophysical Research* **85**, 164–170.

Killen R, Cremonese G, Lammer H, Orsini S, Potter AE, Sprague AL, Wurz P, Khodachenko ML, Lichtenegger HIM, Milillo A and Mura A 2007 Processes that promote and deplete the exosphere of mercury. *Space Science Reviews* **132**, 433–509.

Kolb C, Martín-Fernández JA, Abart A and Lammer H 2006 The chemical variability at the surface of mars: implication for sediment formation and rock weathering. *Icarus* **183**, 10–29.

Krestyanikova MA and Shematovich VI 2005 Stochastic models of hot planetary and satellite coronas: a photochemical source of hot oxygen in the upper atmosphere of mars. *Solar System Research* **39**, 22–32.

Lammer H, Lichtenegger HIM, Kolb C, Ribas I, Guinan EF, Abart R and Bauer SJ 2003 Loss of water from mars: Implications for the oxidation of the soil. *Icarus* **165**, 9–25.

Lewis J 1988 Origin and composition of Mercury In *Mercury* (ed. Vilas F, Chapman C and Matthews M). University of Arizona Press, Tucson, AZ (USA). pp. 651–666.

Martín-Fernández JA and Bren M 2001 Some practical aspects on multidimensional scaling of compositional data. In *Proceedings of the Annual Conference of the International Association for Mathematical Geology* (ed. Sharp E) Cancun (Mexico). p. 16.

Martín-Fernández JA, Barceló-Vidal C and Pawlowsky-Glahn V 2003 Dealing with zeros and missing values in compositional data sets using nonparametric imputation. *Mathematical Geology* **35** (3), 253–278.

Mateu-Figueras G, Pawlowsky-Glahn V and Barceló-Vidal C 2003 Distributions on the simplex. In *Proceedings of CoDaWork'03, The 1st Compositional Data Analysis Workshop* (ed. Thió-Henestrosa S and Martín-Fernández JA). http://ima.udg.es/Activitats/CoDaWork03/. University of Girona, Girona (Spain). CD-ROM.

McClintock WE, Bradley ET, Vervack RJ, Killen RM, Sprague AL, Izenberg NR and Solomon SC 2008a Mercury's exosphere: Observations during messenger's first mercury flyby. *Science* **321**, 92–95.

McClintock WE, Izenberg NR, Holsclaw GM, Blewett DT, Domingue DL, Head JW, Helbert J, McCoy TJ, Murchie SL, Robinson MS, Solomon SC, Sprague AL and Vilas F 2008b Spectroscopic observations of Mercury's surface reflectance during MESSENGER's first Mercury flyby. *Science* **321**, 62–65.

McClintock WE, Vervack RJ, Bradley ET, Killen RM, Mouawad N, Sprague AL, Burger MH, Solomon SC and Izenberg NR 2009 Messenger observations of Mercury's exosphere: Detection of magnesium and distribution of constituents. *Science* **324**, 610–613.

McSween JHY, Arvidson RE, Bell JF, Blaney D, Cabrol NA, Christensen PR, Clark BC, Crisp JA, Crumpler LS, Marais DJD, Farmer JD, Gellert R, Ghosh A, Gorevan S, Graff T, Grant J, Haskin LA, Herkenhoff KE, Johnson JR, Jolliff BL, Klingelhoefer G, Knudson AT, McLennan S, Milam KA, Moersch JE, Morris RV, Rieder R, Ruff SW, de Souza PA, Squyres SW, Wnke H, Wang A, Wyatt MB, Yen A and Zipfel J 2004 Basaltic rocks analyzed by the Spirit Rover in Gusev crater. *Science* **305**, 842–845.

Milillo A, Fujimoto M, Kallio E, Kameda S, Leblanc F, Narita Y, Cremonese G, Laakso H, Laurenza M, Massetti S, McKenna-Lawlor S, Mura A, Nakamura R, Omura Y, Rothery D, Seki K, Storini M, Wurz P, Baumjohann W, Bunce E, Kasaba Y, Helbert J and Sprague A 2010 The BepiColombo mission: an outstanding tool for investigating the Hermean environment. *Planetary and Space Science* **58**, 40–60.

Milillo A, Wurz P, Orsini S, Delcourt D, Kallio E, Killen RM, Lammer H, Massetti S, Mura A, Barabash S, Cremonese G, Daglis IE, DeAngelis E, Di Lellis AM, Livi S, Mangano V and Torkar K 2005 Surface-exosphere-magnetosphere system of mercury. *Space Science Reviews* **117**, 397–443.

Morris RV, Golden DC, Bell III JF, Shelfer TD, Scheinost AC, Hinman NW, Furniss G, Mertzman SA, Bishop JL, Ming DW, Allen CC and Britt DT 2000 Mineralogy, composition, and alteration of mars pathfinder rocksand soils: Evidence from multispectral, elemental, and magnetic data on terrestrial analogue, snc meteorite, and pathfinder samples. *Journal of Geophysical Research* **105**, 1757–1817.

Ness NF, Behannon KW, Lepping RP, Whang YC and Schatten KH 1976 Observations of Mercury's magnetic field. *Icarus* **28**, 479–488.

Potter AE and Morgan TH 1985 Discovery of sodium in the atmosphere of Mercury. *Science* **229**, 651–653.

Potter AE and Morgan TH 1986 Potassium in the atmosphere of Mercury. *Icarus* **67**, 336–340.

Rieder R, Gellert R, Anderson RC, Brückner J, Clark BC, Dreibus G, Economou T, Klingelhöfer G, Lugmair GW, Ming DW, Squyres SW, d'Uston C, Wänke H, Yen A and Zipfel J 2004 Chemistry of rocks and soils at meridiani planum from the alpha particle x-ray spectrometer. *Science* **306**, 1746–1749.

Shemansky DE and Broadfoot AL. Interaction of the surface of the Moon and Mercury with their exospheric atomospheres. *Reviews of Geophysics and Space Physics* **15**, 491–499.

Sprague A, Donaldson-Hanna KL, Kozlowski RWH, Helbert J, Maturilli A, Warell JB and Hora JL 2009 Spectral emissivity measurements of mercuy's surface indicate Mg- and Ca-rich mineralogy, k-spar, na-rich plagioclase, rutile, with possible perovskite, and garnet. *Planetary and Space Science* **57**, 364–383.

Sprague A, Emery JP, Donaldson KL, Russel RW, Lynch DK and Mazuk AL 2002 Mercury: Mid-infrared (3-13.5 μm) observations show heterogeneous composition, presence of intermediate and basic soil types, and pyroxene. *Meteorites Planetary Science* **37**, 1255–1268.

Sprague A, Warell J, Cremonese G, Langevin Y, Helbert J, Wurz P, Veselovsky I, Orsini S and Milillo A 2007 Mercury's surface composition and character as measured by ground-based observations. *Space Science Reviews* **132**, 399–431.

Sprague AL and Roush TL 1998 Comparison of laboratory emission spectra with Mercury telescopic data. *Icarus* **133**, 174–183.

Sprague AL, Kozlowski RWH, Witteborn FC, Cruikshank DP and Wooden DH 1994 Mercury: Mid-infrared (7.3–13.5 μm). spectroscopic observations showing features characteristic of plagioclase. *Icarus* **109**, 156–167.

Toulmin P, Baird AK, Clark BC, Keil K, Rose JHJ, Christian RP, Evans PH and Kelliher WC 1977 Geochemical and mineralogical interpretation of the viking inorganic chemical results. *Journal of Geophysical Research* **82**, 4625–4634.

Warell J 2003 Properties of the hermean regolith: III. Disk-resolved vis-nir reflectance spectra and implications for the abundance of iron. *Icarus* **161**, 199–222.

Warell J and Blewett DT 2004 Properties of the hermean regolith: V. New optical reflectance spectra comparison with lunar anorthosites, and mineralogical modelling. *Icarus* **168**, 257–276.

Warell J, Sprague AL, Emery JP, Kozlowski RWH and A. L 2006 The 0.7–5.3 μm IR spectra of Mercury and the Moon: Evidence for high clinopyroxene on Mercury. *Icarus* **180**, 281–291.

Wurz P and Lammer H 2003 Monte-Carlo simulation of Mercury's exosphere. *Icarus* **164**, 1–13.

Wurz P, Whitby JA, Rohner U, Martín-Fernández JA, Lammer H and Kolb C 2010 Self-consistent modelling of mercury's exosphere by sputtering, micro-meteorite impact and photon-stimulated desorption. *Planetary and Space Science* **58**, 1599–1616.

Ziegler JF 2004 Srim-2003. *Nuclear Instruments & Methods in Physics Research Section B* **219**, 1027–1036.

20

Spectral analysis of compositional data in cyclostratigraphy

Eulogio Pardo-Igúzquiza and Javier Heredia
Instituto Geológico y Minero de España (IGME), Geological Survey of Spain, Madrid, Spain

20.1 Introduction

Many sequences of data (time series and one-dimensional spatial sequences) are compositional data; as, for example, any variable measured as a percentage. Because the addition of all the percentages gives a constant value of 100% and because the domain of definition of each percentage is restricted to the open interval (0, 100%), special statistical techniques must be developed for their analysis, using the *stay-in-the-simplex approach* (see Chapter 3) or some transformation must be applied to the compositional data in order to allow use of classical statistical techniques (Aitchison, 1986) (see also Chapter 2). Although the recognition of these kinds of problems dates back to Pearson (1897), it was Aitchison (1982) who developed a sound statistical theory for the analysis of compositional data and decisively contributed to the diffusion of the awareness that the analysis of compositional data requires a special treatment. Professor Aitchison has also been the leader and the inspiration for the spreading of the interest in statistical research on compositional data analysis techniques (Aitchison and Egozcue 2005) (see also Chapter 1).

A good deal of applications of compositional data analysis have used geochemical-like data, where a number of variables are measured on different rock fragments or soil samples. Then one tries to solve some problem using multivariate statistics applied to the transformed compositional data in order to avoid spurious correlations between the variables. Although different studies have been done in forecasting compositional time series (Mills 2009), less

Compositional Data Analysis: Theory and Applications, First Edition. Edited by Vera Pawlowsky-Glahn and Antonella Buccianti.
© 2011 John Wiley & Sons, Ltd. Published 2011 by John Wiley & Sons, Ltd.

work has been done in the spectral analysis of compositional time series. Nevertheless, Erba *et al.* (1992) recognize the compositional character of cyclostratigraphic time series and apply the spectral analysis to the logit transformed original variables. In the present chapter there is an extension of the methodology of Erba *et al.* (1992) in order to show the importance of recognizing the compositional character of the data when performing the spectral and cross-spectral analysis between pairs of time series.

20.2   The method

The basic idea of spectral analysis of compositional data is that the simplex geometry modifies the original spectral content of a cyclic component. A transformation of the compositional variable to the unrestricted real space must be done in order to allow the spectral analysis to reveal the true spectral content of the data sequence. Spectral analysis essentially consists of estimating a spreading function (the spectral density or power spectrum), which distributes the variance of the process over a frequency axis. When the process is compositional, its total variance is the variance of its coordinate in the two-part compositional time series or, more generally, the trace of the variance-covariance matrix of all the coordinates (Pawlowsky-Glahn and Egozcue 2001) (see also Chapter 3).

   Let us imagine that a physical process results in a perfectly cyclic wave behaviour that can be modelled by a single sinusoid. In the spectral analysis it should have power only for the frequency of the sinusoid. However, many times what is measured from the stratigraphic sequence is a proxy of the physical process and because the way in which the measurements are obtained, the experimental values have a compositional character. For example, a core recovered from a drill hole is sampled and for each sample a percentage is calculated. This results in a two-part composition, with the variable of interest and its complement to adding to 100%. The two-part simplex is the segment (0, 100), where the compositional value is plotted. This two-part simplex is represented for different times giving a compositional time series. A two-part compositional sinusoid can be defined as that whose only coordinate is $y(t) = B + A\sin(\omega t + \varphi)$, where B is the mean value, A is the amplitude, ω is the angular frequency and φ a phase angle and t can be discrete or continuous time or space. The compositional time series is obtained using the ilr^{-1} transformation (Egozcue *et al.* 2003) (see also Chapter 2), which in the two-part case is proportional to the inverse logit transformation. The logit transformation of $\mathbf{x} = (x_1, x_2)$, $x_1 + x_2 = 100$ (using percentages) is

$$y = \text{logit}(\mathbf{x}) = \ln\frac{x_1}{x_2} = \ln\frac{x_1(\%)}{100 - x_1(\%)}, \quad \mathbf{x} = \text{logit}^{-1}(y) = \left(\frac{100\exp(y)}{1 + \exp(y)}, \frac{100}{1 + \exp(y)}\right).$$

(20.1)

The sinusoid $y(t)$ is transformed into a compositional sinusoid using logit^{-1}, i.e. $\mathbf{x}(t) = \text{logit}^{-1}(y(t))$. Figure 20.1 shows three compositional sinusoids obtained by applying the inverse logit transformation to a perfect sinusoidal coordinate.

   Figure 20.1 shows that when the proxy variable is measured on the simplex implies that a perfectly symmetrical sinusoidal wave (in coordinates) may become asymmetrical in the simplex if the variable approaches the border of the simplex. When the power spectrum of the first part of the compositional sinusoid, $x_1(t)$, is estimated, the asymmetry of the sinusoidal wave implies that a number of harmonics of the original single frequency must be invoked.

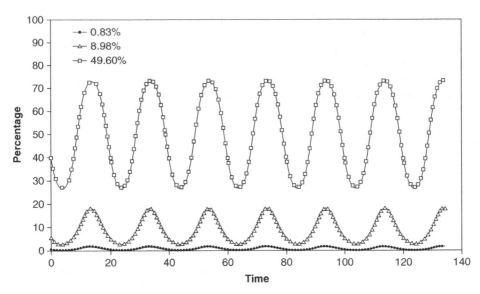

Figure 20.1   Three two-part compositional sinusoids. Parts x_1, x_2 are the ordinate from the curve to 0 and to 100, respectively. The center values are $\text{logit}^{-1}(B)$.

The new spectral peaks, although they may appear as statistically significant, are not related to the original physical cause but constitute an artifact induced by the compositional character of the data. Figure 20.2(a) shows the power spectrum of the time series of Figure 20.1 with a mean (called center in compositional analysis) of 49.8%, that is with variations around the center of the simplex. It may be seen that this does not produce significant harmonics. However, as may be seen in Figure 20.2(b), the power spectrum of the compositional time series shown in Figure 20.1 with a mean of 8.98%, shows important harmonics of the fundamental frequency. These harmonics are a mere artifact introduced by the geometry of the simplex. By applying the logit transformation to the percentage variable $\mathbf{x}(t)$ the constrained simplex is transformed into the unconstrained real line (coordinate). Spectral analysis can be

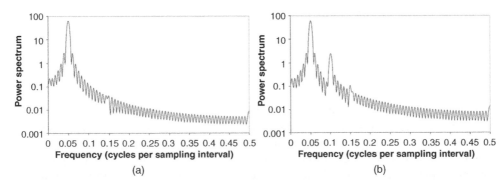

Figure 20.2   Power spectrum of the compositional time series shown in Figure 20.1. (a) Compositional time series with center 49.6%. (b) Compositional time series with center 8.98%.

applied to the unrestricted coordinate time series. The obtained results are easier to interpret (i.e. the effect of spurious harmonics is attenuated). Furthermore, in cross-spectral analysis of three-part compositional time series, the two functions of interest in cyclostratigraphy are the coherency and the phase spectrum. Simulation results have shown that the coherency of the three-part time series is best estimated from the two-coordinate time series obtained using the isometric log ratio (ilr) transform of the three-part composition. However the phase is best estimated from the phase spectrum between the two logit-transformed time series. This is illustrated in the next section.

20.3   Case study

We have used the data set provided by Erba *et al.* (1992), who study fluctuations in abundance of different species of calcareous nannofossils measured at cores recovered from drill holes at two locations in the Albian Gault Clay Formation in southern England. Although Erba *et al.* (1992) consider the compositional character of their data, the analysis has been extended by considering the cross-spectral analysis between two time series of abundances. Erba *et al.* (1992) considered samples at 5 cm along drill holes randomly taking 300 calcareous nannofossils. They identified the different species and then they calculated the percentages of abundance for each species. Only the species *Watznaueria barnesae* and *Repabulum parvidentatum* from the Sevenoaks location are considered here. Figure 20.3(a) shows the compositional time series of *W. barnesae* (with mean of 6.03%) and *R. parvidentatum* with a mean of 10.27%, while Figure 20.3(b) shows the logit-transformed time series after subtracting a linear trend. The smoothed Lomb–Scargle power spectrum of the logit-transformed time series is shown in Figure 20.3(a) while the achieved confidence level is given in Figure 20.3(b). The Lomb–Scargle periodogram deals with irregular time series without the need of interpolation for having a constant sampling interval. From Figure 20.4(a) it may be seen how the time variation of *W. barnesae* reveals significant spectral peaks with periods of 400 ky and 100 ky which are in perfect correspondence with the long and short eccentricity Milankovitch cycles. Although the 23 ky cycle related to the precession Milankovitch cycle is also present, its confidence level is 85% while only peaks with confidence level higher than 90% have been considered as significant in Figure 20.4(b). With respect to the time variation of abundance of

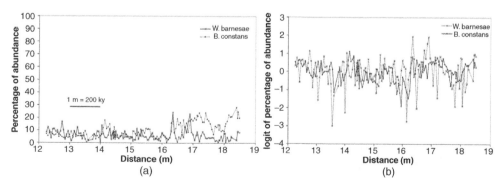

Figure 20.3   (a) Compositional time series of percentage of *W. barnesae* and *R. parvidenta-tum*. (b) Detrended logit transform of the compositional time series shown in (a).

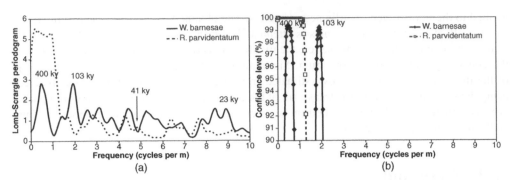

Figure 20.4 (a) Lomb–Scargle power spectrum of the compositional time series given in Figure 20.3(a). (b) Achieved confidence level of the power spectrum given in (a). For *W. barnesae*, the significant spectral peaks have periods of 400 ky and 100 ky, which have a clear correspondence with the long and short eccentricity periods of Milankovitch cycles. However for *R. parvidentatum*, there is only high power content around the long eccentricity period.

R. parvidentatum, Figure 20.3(a) shows much of the power is concentrated around the long eccentricity Milankovitch cycle and the evidence for any other Milankovitch cycle is not significant (only 28% confidence level for the obliquity 41 ky Milankovitch cycle). Cross-spectral analysis will be used in order to detect the spectral content of this species and its relationship with the time abundance variation of *W. barnesae*.

Cross-spectral analysis gives information about the frequency relationship of the joint temporal variation of abundance of the two species. In cyclostratigraphy, the most interesting functions of cross-spectral analysis are the coherency and the phase spectrum (Weedon 2003). Because there are two compositional time series there are different posibilities of cross-spectral analysis. Intensive computer simulation using synthetic compositional time series has shown that the best strategy for estimating the coherency is to consider the two time series of coordinates from the ilr transform applied to the three-part composition (x_1, x_2, x_3) for percentages of *W. barnesae*, *R. parvidentatum* and percentage of the rest of species, respectively. The ilr transformation gives two orthonormal components called balances (Egozcue *et al.* 2003; Egozcue and Pawlowsky-Glahn 2005, 2006):

$$b_1 = \sqrt{\frac{2}{3}} \log \frac{\sqrt{x_1 x_2}}{x_3} = \sqrt{\frac{1}{6}} (\log(x_1) + \log(x_2) - 2\log(x_3)), \qquad (20.2)$$

$$b_2 = \sqrt{\frac{1}{2}} \log \frac{x_1}{x_2} = \sqrt{\frac{1}{2}} (\log(x_1) - \log(x_2)). \qquad (20.3)$$

Figure 20.5(a) shows the coherency between the two time series of coordinates of the (ilr) transform. The spectral features that appear as significant [Figure 20.5(b)], are the cycles with periods of 51 ky and 23 ky. Then, both time series are highly cross-correlated at the level of the precession cycle and what could be a harmonic of the short eccentricity Milankovitch cycle. The phase of coherency is shown in Figure 20.6. The phase from the (ilr) transform Figure 20.6(a) is not as convenient as the phase from the logit-transformed time series which

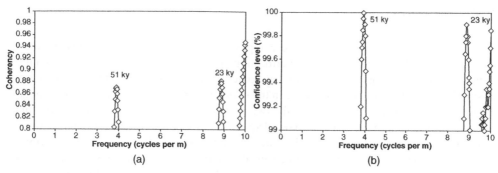

Figure 20.5 (a) Coherency (the squared correlation coefficient between two time series as a function of the frequency content of both sequences) calculated from coordinates of the (ilr) transform. (b) Achieved confidence level of the coherency.

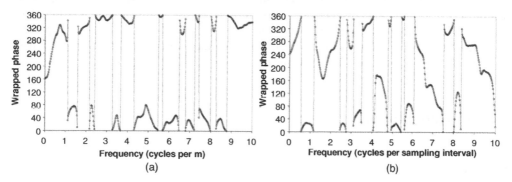

Figure 20.6 (a) Wrapped phase spectrum from coordinates of the ilr transform. (b) Wrapped phase spectrum from the logit-transformed time series.

gives estimates of 341 and 270° as phase difference for the 51 and 23 ky (highly coherent) periods, respectively. The interpretations of these results will be given in the next section.

20.4   Discussion

According to paleoecological and paleogeographic studies, the calcareous nannofossils were sensitive to physical and chemical changes in surface waters masses of Mesozoic oceans. While *R. parvidentatum* is considered as a cool water indicator because of the affinity of this species with high latitudes, *W. barnesae* is a nonfertility index in relation to low seasonality, thus low gradients in the ocean which implies less nutrients and less fertility (abundance of the species). Seasonality, related with eccentricity and precession, is revealed by the direct spectral analysis of *W. barnesae*. With respect to coherency, both species are coherent for the short cycles: the 51 ky with almost in phase variation and the 23 ky with variation not in phase. This result indicates that a maximum in the relative abundance of *R. parvidentatum* is coincidental with a minimum of *W. barnesae* and vice versa (Figure 20.3).

20.5 Conclusions

Many sequences in cyclostratigraphy are compositional. It has been shown how ignoring this compositional characteristic when performing spectral analysis can lead to misleading results because the appearance of spectral harmonics which are spurious. This is so because a sinusoid in real space will be asymmetric in the two-part simplex (for example a percentage) with a flatter part close to the border of the two-part simplex. When analysing a single time series, the logit transformation seems to be adequate to transform the compositional sequence from its restricted two-part simplex to the unrestricted real space. However, when performing cross-spectral analysis between two sequences, the more adequate strategy has proved to be the estimation of coherency from the two time series of coordinates that are obtained from the isometric log-ratio transformation of the three-part series formed by the two compositional time series and the complementary time series. Additionally, in this strategy, the phase of the highly coherent spectral components is best estimated from the phase spectrum obtained by using the logit transformation of each of the two compositional time series. The spectral features obtained from the coherency have proved to be highly informative in revealing the Milankovitch connection of a species when the spectral analysis of the logit of the species failed to show such evidence. A case study using percentages of calcareous nannofossils has proved to be a good example to illustrate the previous points.

Acknowledgement

The authors thank Prof. Dr J. J. Egozcue for his critical comments which were helpful for an effective improvement of this chapter.

References

Aitchison J 1982 The statistical analysis of compositional data (with discussion). *Journal of the Royal Statistical Society, Series B (Statistical Methodology)* **44**(2), 139–177.

Aitchison J 1986 *The Statistical Analysis of Compositional Data*. Monographs on Statistics and Applied Probability. Chapman and Hall Ltd (reprinted 2003 with additional material by The Blackburn Press), London (UK). 416 p.

Aitchison J and Egozcue JJ 2005 Compositional data analysis: where are we and where should we be heading? *Mathematical Geology* **37**(7), 829–850.

Egozcue JJ and Pawlowsky-Glahn V 2005 Groups of parts and their balances in compositional data analysis. *Mathematical Geology* **37**(7), 795–828.

Egozcue JJ and Pawlowsky-Glahn V 2006 Simplicial geometry for compositional data. In *Compositional Data Analysis: From Theory to Practice* (ed. Buccianti A, Mateu-Figueras G and Pawlowsky-Glahn V). The Geological Society, London (UK). pp. 145–160.

Egozcue JJ, Pawlowsky-Glahn V, Mateu-Figueras G and Barceló-Vidal C 2003 Isometric logratio transformations for compositional data analysis. *Mathematical Geology* **35**(3), 279–300.

Erba E, Castradori D and Ripepe M 1992 Calcareous nannofossils and milankovitch cycles: the example of albian gault clay formation (southern england). *Paleogeography, Paleoclimatology, Paleoecology* **93**, 47–69.

Mills T 2009 Forecasting obesity trends in England. *Journal of the Royal Statistics Society, Series A* **172**(1), 107–17.

Pawlowsky-Glahn V and Egozcue JJ 2001 Geometric approach to statistical analysis on the simplex. *Stochastic Environmental Research and Risk Assessment (SERRA)* **15**(5), 384–398.

Pearson K 1897 Mathematical contributions to the theory of evolution. On a form of spurious correlation which may arise when indices are used in the measurement of organs. *Proceedings of the Royal Society of London* **LX**, 489–502.

Weedon G 2003 *Time-Series Analysis and Cyclostratigraphy: Examining Stratigraphic Records of Environmental Cycles*. Cambridge University Press, Cambridge (UK). 259 p.

21

Multivariate geochemical data analysis in physical geography

Jennifer McKinley and Christopher David Lloyd

School of Geography, Archaeology and Palaeoecology, Queen's University, Belfast, UK

21.1 Introduction

Geographical analysis for environmental, geological or economic applications frequently comprises exploration of correlations between variables in multivariate compositional data sets [see Evans and Jones (1981), for a review of compositional data analysis in a geographical context]. However, most analyses of compositional data by geographers do not recognise the special characteristics of compositional data. There is increasing recognition in geography, as other disciplines, of the need to deal properly with compositional data (Lloyd, 2010) and the current chapter is intended to contribute in this respect.

Geographic attributes are rarely Gaussian and often exhibit special characteristics associated with compositional data, this can cause problems when interpreting correlations between attributes in that skewness and outliers in the data will influence the correlation measures (Filzmoser and Hron 2009). Therefore transformation of the multivariate data sets becomes an integral part of any approach. Soil geochemistry data are typically used for mapping geology, soils and natural resources but care needs to be taken to honour inherent constrained behaviour to ensure integrity of the data is maintained if the decisions about the development of natural resources are based on the interpretation of the data.

A geochemical survey over the land surface of Northern Ireland was undertaken by the Geological Survey of Northern Ireland between 2004 and 2006 as part of the Tellus project

Compositional Data Analysis: Theory and Applications, First Edition. Edited by Vera Pawlowsky-Glahn and Antonella Buccianti.

funded by the Department of Enterprise, Trade and Investment of Northern Ireland and by the Rural Development Programme through the Northern Ireland Programme for Building Sustainable Prosperity. This comprised the collection of 22 000 samples, of soils, stream sediments and stream waters. The geochemical data comprise X-ray fluorescence (XRF) analyses of rural soils sampled at 20 cm depths collected on a grid at one site per 2 km². Each soil sample is a composite of five soil auger samples, collected from the centre and each corner of a 20 × 20 m square. The soil geochemical survey reflects changes in underlying bedrock, soil type, anthropogenic influences and anomalies associated with precious and base metal mineralisation. Previous work on the geochemical data from the Tellus survey has investigated the appropriate use of transformation methods such as log-ratios and stepwise conditional transform (McKinley and Leuanthong 2010) to reproduce the inherently complex multivariate relations.

This chapter examines the implications of the compositional nature of the soil geochemical data and the reliability of the results and interpretations based upon the relationship between different geochemical elements. XRF analyses of rural soils provide concentrations of the chemical elements per sample parts per million (ppm). Balances (Egozcue and Pawlowsky-Glahn 2005) provide the basis of the analysis. The compositional biplot is used in combination with geological interpretation (geological and structural controls on soil geochemistry) to determine the balance partitioning scheme.

21.2   Context

Figure 21.1 shows a simplified solid geology map of Northern Ireland. The diversity of geology found in Northern Ireland ranges from 600 million year old metamorphosed sedimentary and volcanic rocks to basaltic lavas and lacustrine sedimentary rocks formed between ca. 55 and 62 million years ago. At least 80% of the bedrock is covered by superficial deposits (glacial till and post-glacial peat) formed as a result of the advance of ice sheets and their meltwaters. For this research, a local scale study area (Figure 21.1) comprising an area, within the county of Fermanagh and the southern part of County Tyrone, was selected [comparable with the study by McKinley and Leuanthong (2010)] to investigate the use of balances in computing correlations between the multivariate geochemical elements. The county of Fermanagh and the southern part of County Tyrone comprises shallow water Carboniferous limestone sediments with interbedded sandstones (Mitchell 2004). The Carboniferous includes the unfossiliferous Ballyness Formation and the Clogher Valley Formation of the Tyrone group and the Kilskerry group of the Fintona block. The Ballyness Formation comprises red sandstones and conglomerates. The Clogher Valley Formation is divided into micrite, siltstones, sandstones, evaporate beds and grey mudstone. The Kilskerry Group comprises green mudstones and interbedded sandstones and conglomerates. Devonian strata in the Fintona Block, the Shanmullagh Formation consist of pebbly sandstones, purplish grey mudstones and mudstone. There are several structural controls apparent in the solid geology of the area; north-east–south-west trending terrane fault boundaries, the Killadeas–Seskinore fault and the Omagh thrust, define the margins of the Devonian Shanmullagh Formation. A more northerly trending alignment defines the line of the Tempo–Sixmilecross Fault and outlines the Carboniferous Kilskerry Group of the Fintona Block. Further south a north-east–south-west boundary follows the trace of the Clogher Valley Fault and defines the Carboniferous Ballyness and Clogher Valley formations.

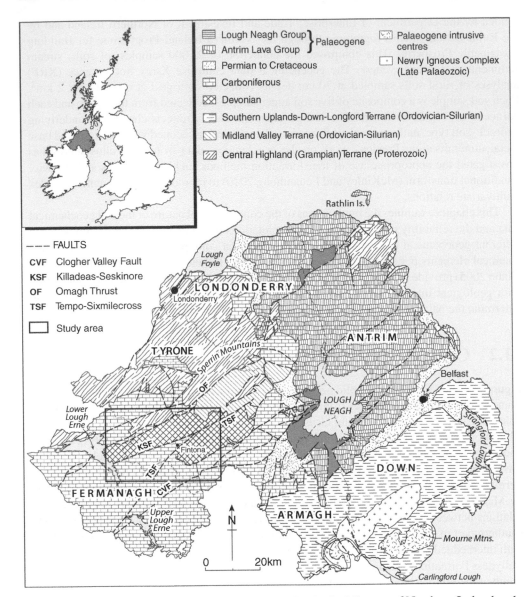

Figure 21.1 Based on mapping published by the Geological Survey of Northern Ireland and reproduced with permission.

Evidence of north–south trending basement lineaments in the area may indicate a continuation of the Omagh lineament. This area has been the focus of mineral prospecting (Arthurs and Earls 2004) for base metals from the eighteenth to early twentieth century. Anomalies of lead, barium and silver have been found (Smith *et al.* 2004) in the faulted sedimentary rocks. The geological environment of Lower Carboniferous shallow water sediments near a basin margin fault, is typical of many of the major lead-zinc deposits of the Irish Midlands (e.g. Navan, County Meath, Republic of Ireland). However, the thick drift cover over the area hinders detection of the subtle geochemical anomalies. Therefore correct interpretation of soil geochemical data from the Tellus survey is important to accurately assess the economic potential of the area.

Knowledge of the geological and structural controls of the area can be used to inform the choice of balances for the correlation analysis. Namely that:

- There are lithostratigraphically distinctive Carboniferous and Devonian rocks.

- The stratigraphy is controlled by basement tectonics.

- Previous research indicates an inferred relationship between base metals and basin faulting (Mitchell 2004).

21.3 Data

Aitchison (2003) introduced the additive log-ratio (alr) and the centred log-ratio (clr), while Egozcue *et al.* (2003) developed the isometric log-ratio (ilr) transform [see Egozcue and Pawlowsky-Glahn (2006), for a summary account] for the transformation of compositional data. Outputs from the alr and clr transforms are subject to some restrictions in their treatment by standard methods, whereas ilr-transformed data can be analysed directly using standard univariate or multivariate statistical methods. Egozcue and Pawlowsky-Glahn (2005) have developed a form of ilr coordinates called balances. Balances represent the relative variation in two groups of parts and they may have straightforward interpretations. Balances, like ilr coordinates more generally, can be analysed using standard multivariate statistical approaches. Balances offer a means of analysing simultaneously variation within groups of parts and between groups of parts (Egozcue and Pawlowsky-Glahn 2005). The following analysis is based on balances.

The geochemical data were grouped into four classes, as detailed in Table 21.1. These were based on prior knowledge of the dominant geological formations and the structural controls in the study area. Fault boundaries and nomenclature are provided in Figure 21.1. Figure 21.2 shows ordinary kriged (OK) mapped outputs for selected elements within each of the four group assignments. The spatial variability of elements, as demonstrated by the examples shown, is markedly different. A north–south alignment is apparent for the map of silver (Ag) whereas the distribution of copper (Cu) appears to follow an inclined orientation (similar to the alignment of the fault-bounded Devonian Shanmullagh sandstone and mudstone formation).

Table 21.1 Group assignments.

Group	Elements
North–south basement lineament	Ag, Pb, Sb
TSF defined Kilskerry Group	Se, U
CVF defined Clogher Valley and Ballyness formations	Sr, Cd, Sn
OF and KSF defined Shanmullagh Formation	As, Cu, Ni, Co, Zn, Cr, Th

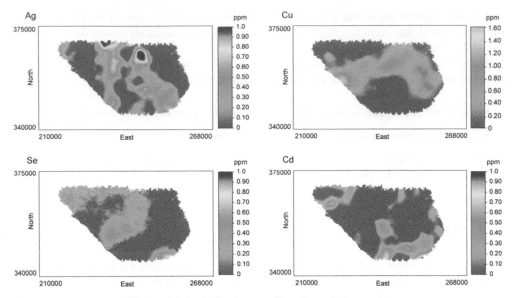

Figure 21.2   Kriged maps for selected elements.

Biplots (Gabriel 1971) allow the simultaneous visualisation of information on variables and observations of a data matrix. The software CODAPACK includes functionality for construction of compositional biplots (Aitchison and Greenacre 2002) and this is made use of in the present analysis. The compositional biplot is shown in Figure 21.3. The lines projecting from the centre are termed rays and the observations are indicated by the points. The different directions for the different rays indicate the relative importance of these geochemical elements in distinguishing samples. The ray for Cu can be very clearly distinguished from those for all other groups.

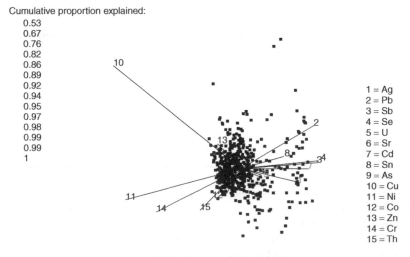

Figure 21.3   Compositional biplot.

Table 21.2 Partitions.

	Ag	Pb	Sb	Se	U	Sr	Cd	Sn	As	Cu	Ni	Co	Zn	Cr	Th
1	1	1	1	-1	-1	-1	-1	-1	-1	-1	-1	-1	-1	-1	-1
2	1	1	-1												
3	1	-1													
4				1	1	-1	-1	-1	-1	-1	-1	-1	-1	-1	-1
5				1	-1										
6						1	1	1	-1	-1	-1	-1	-1	-1	-1
7						1	1	-1							
8						1	-1								
9									1	1	1	1	1	1	-1
10									1	1	1	1	1	-1	
11									1	1	1	1	-1		
12									1	1	1	-1			
13									1	1	-1				
14									1	-1					

Given this information, one possible partitioning scheme (Egozcue and Pawlowsky-Glahn 2005), which starts with a split between Ag, Pb, Sb and all other elements is given in Table 21.2. The '1' entries indicate the numerators and the '−1' entries indicate denominators of the log-ratios. Egozcue and Pawlowsky-Glahn (2005) provide full details of the derivation of balances and Filzmoser and Hron (2009) provide a further account in the context of correlation analysis. The following section outlines some analyses of the balances based on this partitioning scheme.

21.4 Analysis

One way of representing the partitions, along with statistical summaries, is to compute a balance dendrogram (Thió-Henestrosa *et al.* 2008). Figure 21.4 shows the balance dendrogram for the present case. As discussed by Thió-Henestrosa *et al.* (2008), the balance dendrogram shows the partition of the composition using vertical and horizontal links. The sample mean or centre with respect to each balance is the point where the vertical bars end. The length of the vertical bars represents the proportion of the sample total variance which corresponds to each balance. The sum of these lengths represents the total variance (Thió-Henestrosa *et al.* 2008). For each balance, summary statistics are represented by the box plots, comprising the percentiles 5, 25, 50, 75 and 90, on each horizontal bar. The balance dendrogram provides a convenient graphical summary of a composition given a particular partitioning scheme.

The exploration of relations between balances is a key next step and the correlation matrix for the balances is given in Table 21.3. The largest absolute value (0.84) is for B1 against B6. Since balances can be analysed using standard multivariate statistical procedures, it was possible to explore trends in the balances through the use of principal components analysis (PCA). Table 21.4 summarises the components and the proportion of variance explained in each case for PCA based on the correlations. The variances of the log-ratios differ markedly

Table 21.3   Balances: correlation matrix.

	B1	B2	B3	B4	B5	B6	B7	B8	B9	B10	B11	B12	B13	B14
B1	1.00	-0.18	-0.32	0.73	0.64	0.84	-0.33	-0.72	-0.15	0.38	-0.40	-0.43	0.54	0.60
B2	-0.18	1.00	-0.40	-0.20	-0.06	-0.14	-0.03	0.15	0.11	0.06	-0.17	0.05	-0.16	-0.17
B3	-0.32	-0.40	1.00	-0.04	-0.21	-0.08	0.48	0.15	-0.29	-0.48	0.33	-0.16	-0.22	0.08
B4	0.73	-0.20	-0.04	1.00	0.40	0.79	-0.11	-0.67	-0.28	0.25	-0.30	-0.42	0.48	0.53
B5	0.64	-0.06	-0.21	0.40	1.00	0.52	-0.19	-0.48	0.13	0.37	-0.16	-0.19	0.39	0.43
B6	0.84	-0.14	-0.08	0.79	0.52	1.00	-0.15	-0.56	-0.30	0.12	-0.36	-0.48	0.48	0.54
B7	-0.33	-0.03	0.48	-0.11	-0.19	-0.15	1.00	0.37	-0.10	-0.36	0.15	-0.02	-0.26	-0.09
B8	-0.72	0.15	0.15	-0.67	-0.48	-0.56	0.37	1.00	0.01	-0.47	0.32	0.32	-0.47	-0.48
B9	-0.15	0.11	-0.29	-0.28	0.13	-0.30	-0.10	0.01	1.00	0.49	-0.16	0.50	0.06	-0.39
B10	0.38	0.06	-0.48	0.25	0.37	0.12	-0.36	-0.47	0.49	1.00	-0.42	0.11	0.48	0.11
B11	-0.40	-0.17	0.33	-0.30	-0.16	-0.36	0.15	0.32	-0.16	-0.42	1.00	0.34	-0.08	-0.24
B12	-0.43	0.05	-0.16	-0.42	-0.19	-0.48	-0.02	0.32	0.50	0.11	0.34	1.00	0.23	-0.69
B13	0.54	-0.16	-0.22	0.48	0.39	0.48	-0.26	-0.47	0.06	0.48	-0.08	0.23	1.00	0.04
B14	0.60	-0.17	0.08	0.53	0.43	0.54	-0.09	-0.48	-0.39	0.11	-0.24	-0.69	0.04	1.00

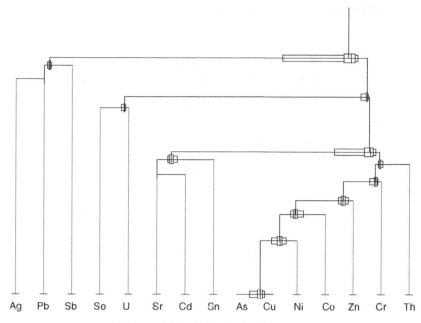

Figure 21.4   Balance dendrogram.

because the ranges of values on which they are based vary. This is one reason for using correlations rather than covariances. Nearly 80% of the variation is explained by the first five components. Table 21.5 shows the correlations of each balance with each component. The largest correlation with component 1 is, perhaps predictably, with B1.

Table 21.4   Balances: PCA summary 1.

Component	Total	% of Variance	Cumulative %
1	5.11	36.51	36.51
2	2.72	19.43	55.94
3	1.50	10.71	66.65
4	0.98	7.01	73.66
5	0.87	6.18	79.84
6	0.76	5.46	85.30
7	0.52	3.71	89.01
8	0.41	2.94	91.96
9	0.32	2.31	94.27
10	0.23	1.66	95.93
11	0.21	1.51	97.44
12	0.19	1.37	98.81
13	0.11	0.77	99.58
14	0.06	0.42	100.00

Table 21.5   Balances: PCA summary 2.

	1	2	3	4	5
B1	0.94	0.00	0.03	−0.07	0.03
B2	−0.15	0.32	−0.65	−0.25	0.45
B3	−0.27	−0.69	0.38	0.33	−0.04
B4	0.82	−0.22	0.14	−0.02	0.22
B5	0.66	0.15	0.12	0.14	0.05
B6	0.85	−0.24	0.05	−0.09	0.27
B7	−0.37	−0.44	0.08	0.50	0.57
B8	−0.82	−0.09	−0.11	−0.05	0.17
B9	−0.15	0.76	0.07	0.49	−0.07
B10	0.48	0.68	0.01	0.29	−0.11
B11	−0.48	−0.22	0.54	−0.38	−0.09
B12	−0.49	0.62	0.48	−0.10	0.15
B13	0.56	0.36	0.55	−0.19	0.29
B14	0.67	−0.45	−0.20	0.11	−0.24

Variogram analysis was used to examine the spatial variability in the balances. The coefficients of the variograms were used for spatial prediction (OK) to investigate the relationship between balances and the geologically and structurally defining groups. The modelled variogram and the kriged output estimated for B1 are shown in Figures 21.5 and 21.6, respectively. The modelled variogram indicates a bounded structure for B1. The kriged output shows a distinct spatial distribution of high and low balances. Areas where the balance values are close to zero, indicate locations where all elements are present in similar abundance, areas with

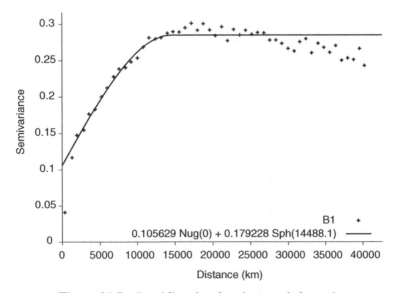

Figure 21.5   Omnidirectional variogram: balance 1.

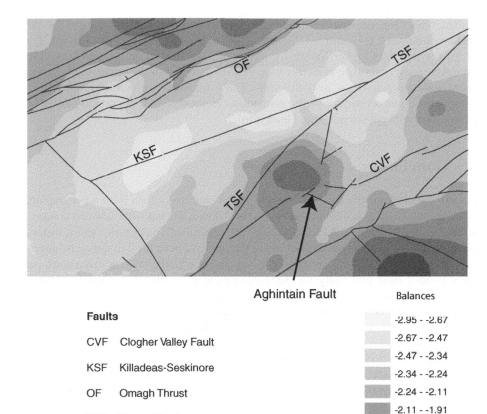

Faults

CVF   Clogher Valley Fault

KSF   Killadeas-Seskinore

OF   Omagh Thrust

TSF   Tempo-Sixmilecross

Balances

-2.95 - -2.67
-2.67 - -2.47
-2.47 - -2.34
-2.34 - -2.24
-2.24 - -2.11
-2.11 - -1.91
-1.91 - -1.63
-1.63 - -1.23
-1.23 - -0.65
-0.65 - 0.17

Figure 21.6   Kriged map of balance 1.

highly negative balance values, signify locations where the elements other than Ag, Pb and Sb are most dominant.

21.5   Discussion

The area where the balance values are highly negative (-2 and smaller) relates to the Devonian Shanmullagh Formation (red sandstones, siltstone and mudstones) along with Carboniferous mudtones, sandstones and conglomerates. These stratigraphic units have small amounts of Ag, Pb and Sb but greatest relative abundance of elements As, Cu, Ni, Co, Zn, Cr and Th. An interesting finding, observable from the kriged map estimated for B1, is that the KSF fault does not appear to influence significantly the spatial distribution of these elements. Conversely, the area of high negative balances does appear to be fault-bounded (by the OF and TSF). This suggests that a sound partitioning cannot solely be derived from a graphic representation such as the biplot but requires geological and geochemical background knowledge. Areas with balances between 0.17 and −1.23 correspond to an area north of the Omagh Thrust and an area within the Clogher Valley, focused along the trace of an inferred fault (Figure 21.6; the Aghintain fault). The elements (Ag, Pb, Sb) assigned to the north–south basement lineament are most significant in these areas. These findings reinforce the inference of a north–south basement control for these geochemical elements [most importantly silver (Ag), often associated with gold deposits and lead (Pb)].

21.6   Conclusion

The chapter details the application of balances in the analysis of data from a geographically extensive ground-based geochemical survey. A local area study has been selected comprising lithostratigraphically distinctive rocks, tectonically controlled stratigraphy and a history of economically significant mineralization. These factors have been used to define balances for the correlation analysis. Results from the analyses concur with the inference of a north–south basement control most especially for the geochemical elements of Ag and Pb. However, the results indicate that the spatial distribution of geochemical elements is related to both stratigraphy and fault-bounded controls, and that background knowledge of these is required to derive a robust analysis based on balances. The findings suggest possibilities for further analysis including incorporating distance from key fault lines and a separate evaluation of outliers (in this case Cu) indicated from the dendrogram.

Balances are not subject to limitations in terms of their treatment by standard multivariate statistical methods and they provide a powerful means of working with compositional data. In other words, the use of balances is a statistically valid framework for the analysis of compositions. Therefore, this approach will help ensure a more accurate interpretation of the nature of the geochemical variability and consequently any geological, environmental and economic inferences from the Tellus project.

Acknowledgement

Geological Survey of Northern Ireland (GSNI) is thanked for provision of the Tellus survey data.

References

Aitchison J 2003 *The Statistical Analysis of Compositional Data*. The Blackburn Press, Caldwell, NJ (USA). 435 p.

Aitchison J and Greenacre M 2002 Biplots for compositional data. *Applied Statistics* **51**(4), 375–392.

Arthurs JW and Earls G 2004 Mineral. In *Geology of Northern Ireland – Our Natural Foundation* (ed. Mitchell W). Geological Survey of Northern Ireland, Belfast (UK).

Egozcue JJ and Pawlowsky-Glahn V 2005 Groups of parts and their balances in compositional data analysis. *Mathematical Geology* **37**(7), 795–828.

Egozcue JJ and Pawlowsky-Glahn V 2006 Simplicial geometry for compositional data. *Compositional Data Analysis in the Geosciences: From Theory to Practice* (ed. Buccianti A, Mateu-Figueras G and Pawlowsky-Glahn V). Geological Society, London (UK). pp. 145–159.

Egozcue JJ, Pawlowsky-Glahn V, Mateu-Figueras G and Barceló-Vidal C 2003 Isometric logratio transformations for compositional data analysis. *Mathematical Geology* **35**(3), 279–300.

Evans I and Jones K 1981 Ratios and closed number systems. In *Quantitative Geography: A British View* (ed. Wrigley N and Bennett R). Routledge and Kegan Paul, London (UK). pp. 123–134.

Filzmoser P and Hron K 2009 Correlation analysis for compositional data. *Mathematical Geosciences* **41**, 905–919.

Gabriel KR 1971 The biplot – graphic display of matrices with application to principal component analysis. *Biometrika* **58**(3), 453–467.

Lloyd CD 2010 Exploring population spatial concentrations in Northern Ireland by community background and other characteristics: an application of geographically weighted spatial statistics. *International Journal of Geographical Information Science* **24**, 1193–1221.

McKinley J and Leuanthong O 2010 An examination of transformation techniques to investigate and interpret multivariate geochemical data analysis – the Tellus case study. In *GeoENV VII: Geostatistics for Environmental Applications* (ed. Atkinson PM and Lloyd CD). Springer, Dordrecht (Germany). pp. 231–242.

Mitchell W (ed.) 2004 *Geology of Northern Ireland – Our Natural Foundation*. Geological Survey of Northern Ireland, Belfast (UK).

Smith R, Smith C and Legg I 2004 *Mineral Exploration in the Clogher Valley area, Co. Tyrone. Part 1: Follow-up Investigations*. Geological Survey of Northern Ireland, Belfast (UK). Geological Survey of Northern Ireland Technical Report GSNI/96/3.

Thió-Henestrosa S, Egozcue JJ, Pawlowsky-Glahn V, Kovács LO and Kovács G 2008 Balance-dendrogram, a new routine of codapack. *Computer and Geosciences* **34**(12), 1682–1696.

Combining isotopic and compositional data: a discrimination of regions prone to nitrate pollution

Roger Puig[1], Raimon Tolosana-Delgado[2], Neus Otero[1] and Albert Folch[3]

[1] *Faculty of Geology, University of Barcelona (UB), Spain*
[2] *Maritime Engineering Laboratory, Technical University of Catalonia, Spain*
[3] *Department of Geology, Autonomous University of Barcelona, Spain*

22.1   Introduction

In the last few decades, nitrate pollution has become a major threat to groundwater quality, as the threshold value for drinking water [50 mg l^{-1}, Directive 98/83/EC; EC (1998)] is achieved in most of the local and regional aquifers in Europe. High nitrate levels in drinking water poses a health risk, because the ingestion of high nitrate concentration can cause methahemoglobinaemia in children and babies (Magee and Barnes 1956), and some authors pointed out that nitrogen compounds can act as human cancer promoters (Ward *et al.* 2005; Volkmer *et al.* 2005). Nitrate pollution is linked to the intensive use of synthetic and organic fertilizers, as well as to septic systems effluents. In Catalunya (north-east Spain), according to the nitrate directive (91/767/EU), twelve areas have been declared as vulnerable to nitrate

Compositional Data Analysis: Theory and Applications, First Edition. Edited by Vera Pawlowsky-Glahn and Antonella Buccianti.
© 2011 John Wiley & Sons, Ltd. Published 2011 by John Wiley & Sons, Ltd.

pollution from agricultural sources (Decrets 283/1998, 436/2004 and 136/2009), covering more than one third of the territory.

To improve water management in these areas, it is essential to determine the origin of pollution and the evolution of nitrogen compounds. Nitrate isotopes are a unique tool for this purpose. Nitrogen and oxygen isotopes of dissolved nitrate can be used as tracers of nitrate pollution, distinguishing between chemical fertilizers and manure/sewage (Wassenaar 1995; Kendall and McDonnell 1998). However, in order to use the isotopic composition of dissolved nitrate as a tracer of nitrate origin, one must bear in mind that several processes (e.g. volatilization, nitrification and denitrification) change the isotopic composition of the sources, leading to overlapping isotopic signatures for different nitrate sources. For example, ammonia volatilization results in an increase of the $\delta^{15}N$ residual ammonium (Letolle 1980). In denitrification processes, nitrate is reduced to N_2, increasing the isotopic composition ($\delta^{15}N$ and $\delta^{18}O$) of the remaining nitrate in waters (Böttcher *et al.* 1990). This fact, considered a drawback in the application of nitrate isotopes as tracers of sources, can be applied to trace processes themselves, e.g. the isotopic composition of dissolved nitrate is used to distinguish between dilution and denitrification in groundwater samples in areas where a diminution in nitrate concentration is observed (Grischek *et al.* 1998; Cey *et al.* 1999; Mengis *et al.* 1999). In this sense, this approach towards the identification of processes could be later complemented with geochemical and biogeochemical reaction modelling applied to all chemical and isotopic variables (Bethke 2008). A further step in the investigation of denitrification processes is to determine the factors controlling the reaction. This has been done coupling chemical data with the $\delta^{15}N$ and/or $\delta^{18}O$ of dissolved nitrate and the isotopic composition of the ions involved in denitrification reactions, as $\delta^{34}S$ and $\delta^{18}O$ of dissolved sulfate, and/or $\delta^{13}C$ of dissolved inorganic carbon (Aravena and Robertson 1998; Pauwels *et al.* 2000). This approach was proposed in an ongoing project performed in several areas classified as vulnerable to nitrate pollution in Catalunya. Five of the vulnerable areas (Maresme, Osona, Lluçanès, Empordà and Selva) have been studied coupling classical hydrochemistry data with a comprehensive isotopic characterization, including $\delta^{15}N$ and $\delta^{18}O$ of dissolved nitrate, $\delta^{34}S$ and $\delta^{18}O$ of dissolved sulfate, $\delta^{13}C$ of dissolved inorganic carbon, and δD and $\delta^{18}O$ of water (Vitòria *et al.* 2005, 2008; Puig *et al.* 2007; Otero *et al.* 2009). The key goals of this project were (i) to identify the main sources of nitrate pollution in the areas, (ii) to verify if denitrification (natural attenuation) processes were taking place, and (iii) to determine the factors controlling the denitrification reactions. In this framework, the present chapter aims to put forward a statistical methodology to integrate isotope data together with geochemical data. This methodology will be applied to discriminate sample groups affected by different nitrate pollution sources.

22.2 Study area

The studied areas are located in Barcelona and Girona provinces, four of them belong to the Ter river basin and one is located along the coast (Figure 22.1). The Ter river is a major stream in the area, it crosses the Osona, Lluçanès, Empordà and Selva studied areas. Its mean discharge in the Osona area is 509 hm^3 year^{-1} (at the gauging station of Roda de Ter), in the Empordà area 1026 hm^3 year^{-1} (at the gauging station of Colomers), and in the Selva area 261 hm^3 year^{-1} (at the gauging station of Cellera de Ter). The studied areas have a sub-Mediterranean climate with mean rainfall between 550 and 850 mm year^{-1} for all the areas. The potential evapotranspiration, calculated by the Thornwaite method, is in the

Figure 22.1 Map of Catalunya showing in grey the areas classified as vulnerable to nitrate pollution from agricultural sources. The studied areas are indicated by the first letter of their names (see text).

range of rainfall (570–910 mm year^{-1}). The following sections describe the main geological, hydrogeological and land use characteristics of each studied area.

22.2.1 Maresme

In the Maresme vulnerable zone the studied area is 3 km^2. The geology of the area consists of Holocene alluvial deposits of coarse sands derived from the weathered granodiorite that forms the Catalan Coastal Range. The main hydrogeological units are an unconfined sandy aquifer underlaid by an aquitard composed of silts and clays and a confined sandy aquifer. The thickness of these units is 5–40, 5–15 and 15–20 m, respectively. The unconfined aquifer is the only one affected by groundwater extractions and its water table varies between 4 and 30 m in depth. The Maresme area is characterized by intensive agricultural activity. Flowers, fruit and vegetable crops are the main agricultural products, and about half of them grow under greenhouse conditions. Fertilization is carried out with inorganic fertilizers usually injected through trickle irrigation systems that use groundwater extracted from partially penetrating wells (5–40 m deep). The soil type in this area is usually coarse sand with a low organic matter content (<3% of dry soil) and a low C:N ratio of approximately 1:2 (Guimerà *et al.* 1995). This low natural fertility, together with the low water-holding capacity of the soil, requires high fertilizer and irrigation applications, resulting in high nutrient leaching in the upper zone of the aquifer affected by groundwater withdrawals. Recirculation of the shallow groundwater by the irrigation system causes high concentrations of nitrates (up to 300 mg^{-1}) in the groundwater (ACA 2009). This area, where only chemical fertilizers are used, is considered as representative of nitrate pollution from inorganic fertilizers, as shown by Vitòria *et al.* (2005) using an isotope approach.

22.2.2   Osona

The study performed in the Osona area covers 600 km^2. From a geological perspective, the area is constituted of Paleogene sedimentary materials overlaying hercynian crystalline (igneous and metamorphic) rocks. The stratigraphic sequence primarily consists of carbonate formations, with an alternation of calcareous, marl and carbonate sandstone layers. It is worth noting the presence of disseminated pyrite in marls. These formations show a quite uniform dipping of about 7–10° to the west. The area is hydrogeologically constituted by a series of confined aquifers located in the carbonate and carbonate–sandstone layers. Marl strata act as confining layers. In this area the porosity is mainly related to the fracture network. Main production wells for agriculture and farm demand usually reach depths of more than 100 m, searching for the most productive confined aquifers. Alluvial aquifers are scarce and generally nonproductive in the area; except those located at the Ter river terraces. In the central part of Osona nitrate pollution is widely extended, with a median concentration above 100 mg l^{-1} during the last 5 years. In this region, of 1263.8 km^2, there are more than 1000 pig farms, with 990 000 pigs, 110 000 cows and 67 000 sheep (IDESCAT 1999). This intensive farming activity produces huge amounts of organic residues, 10 900 t year^{-1} of nitrogen. Fertilizers are also applied, but only in the surroundings of the villages, as the use of pig manure close to urban areas is forbidden. 93% of the municipalities are connected to the sewage network; therefore the contribution of sewage to groundwater nitrate pollution is expected to have negligible influence, compared with agricultural sources. In this area, although chemical fertilizers are also applied, the main contribution to nitrate pollution is linked to an excess of manure application as fertilizer or in uncontrolled dumps, as demonstrated by Vitòria *et al.* (2008) and Otero *et al.* (2009) using a multi-isotopic approach. Hence this area is considered as representative of manure nitrate pollution.

22.2.3   Lluçanès

The Lluçanès area is located in the north-west of the Osona area, and the studied area covers 400 km^2. The geology of the area is constituted by Paleogene sedimentary materials, which include continental detritic facies (conglomerates, sandstones and clays) and marine facies (silts, marls and limestones). The lithostratigraphic units configure an upper and lower deltaic complexes whose clastic contributions are from the north. A monoclinal structure shows a regional dip <4° to the south–south-west–west. The main hydrogeological unit in the Lluçanès area consists of sandstones and conglomerates of the upper deltaic complex (north–north-east area), and is a high productivity confined aquifer. Silt and marl interspersed in sandy levels work as an aquitard, and lutites, sandstones and conglomerates give rise to local aquifer levels with low capacity and limited recharge. The regional flow direction is north–south for all the hydrogeological units of the Tertiary materials. The Lluçanès land uses distribution comprises forest (57%), crops (25%), fields (15%) and urban areas (3%). Since agricultural areas are mostly dry-farmed crops, water resources in the studied area are not as affected by agriculture water demand as they are by the application of manure onto fields as fertilizer. In the studied area the intensive livestock activity is predominantly pig raising: there are 366 farms with around 70 000 animals that produce more than 2500 t year^{-1} of nitrogen as organic residues. On the other hand, the influence of sewage to nitrate contamination is not discarded, because some villages dump to surface waters. Hence, in the Lluçanès area, mainly there is a surplus application of nitrogen organic compounds from agricultural and livestock activities.

22.2.4   Empordà

In the nitrogen vulnerable zone of the Empordà the studied area is located in the south and covers 200 km^2. The geology of the Empordà area consists of Paleogene detritic and carbonate sedimentary rocks (in the east and south of the studied area), Neogene clay facies (in the west), and Holocene alluvial deposits of Ter river and its tributary. These Tertiary and Quaternary materials lay on Paleozoic discordant bedrock. Paleozoic outcrops are located in the western and southern parts of the watershed. This area presents a complex distribution of hydrogeological units due to the high lithological diversity (Puig *et al.* 2007). The main units are: (i) an unconfined aquifer with sand and gravel, and some clay in the matrix (mainly from Quaternary); and (ii) a confined and sometimes unconfined fractured aquifer with thickness discontinuity (mainly from Tertiary). This area has a notable agricultural activity (mainly maize, sunflower and fruit crops) which uses synthetic fertilizers, and the water demand for irrigation is remarkably increased in summer months. Organic fertilizers are also applied as a consequence of an intensive pig farming activity (462 pigs km^{-2}) which produces large amounts of organic residues. Thus, the Empordà area is considered a mixed area where both fertilizers and animal manure are used.

22.2.5   Selva

In the Selva vulnerable zone the studied area is 350 km^2. The area belongs to a tectonic basin surrounded by three ranges with 1000 m altitude above sea level. This basin was created during the distensive periods after the Alpine orogenesis, and it has a Neogene sedimentary poorly consolidated and volcanic filling. The surrounding ranges consist of Paleozoic igneous and metamorphic rocks, and pre-Alpine Paleogene sedimentary rocks (mainly limestone and sandstone). From a hydrogeological perspective in the Selva area we can differentiate a regional and a local flow system. Thus, four geological domains can be hydrogeologically distinguished: (i) the granitic materials of the surrounding ranges, which act as the main recharge area of the granitic basement of the depression; (ii) the Neogene materials of the sedimentary basin, whose local flows originate from the range areas as a lateral recharge, and from the uppermost parts of the basement; (iii) the main faults oriented north-west–south-east and north-east–south-west behaving as an independent hydrogeological unit connected with the rest of the units (fractures responsible for the thermal springs occurring within the basin); and (iv) the upper alluvial formations with two alluvial aquifers associated with two streams. This area is also considered a mixed area with regards to land use and nitrate pollution sources.

22.3   Analytical methods

All the sampling surveys were conducted on production wells. In the Maresme area, two sampling surveys in a small area (3 km^2) were performed: the first of 8 samples and the second of 23 samples. In the Osona area several sampling surveys were carried out: the first, in a reduced area of 36 km^2 with 38 samples, and three more surveys, in a larger area of 600 km^2, with 59, 58 and 32 samples, respectively. In the Lluçanès area one field sampling was done in an area of 400 km^2, with 30 samples. In the Empordà area two surveys were conducted in a 200 km^2 area: the first survey of 24 samples and the second of 40 samples. In the Selva area only one survey was executed, with 38 samples in a 350 km^2 area. Physicochemical parameters (pH, temperature, electrical conductivity, dissolved O_2 and Eh) were measured *in situ*, using a flow cell to avoid contact with the atmosphere. Samples were stored at 4 °C and in a dark environment. Chemical parameters were determined by standard analytical techniques; the chemical characterization comprises major ions (Cl^-, SO_4^{2-}, HCO_3^-, Na^+, Ca^{2+}, Mg^{2+}, K^+), nitrogen compounds (NO_2^-, NO_3^-, NH_4^+), Fe and Mn. The δD and $\delta^{18}O$ of water were obtained by means of isotope ratio mass spectrometer (IRMS) with a Delta S Finnigan Mat, following the methodological approaches of Friedman (1953) and Epstein and Mayeda (1953), respectively. For $\delta^{15}N_{NO_3}$ and $\delta^{18}O_{NO_3}$ analysis, dissolved nitrate was concentrated using anion-exchange columns Bio Rad® AG 1-X8(Cl^-) 100–200 mesh resin, after extracting sulfates and phosphates by precipitation with $BaCl_2$ and filtration (Mayer *et al.* 2001). Afterwards dissolved nitrate was eluted with HCl and converted to $AgNO_3$ by the addition of silver oxide. The silver nitrate solution was then freeze-dried obtaining the pure $AgNO_3$ for analysis usign a method modified from Silva *et al.* (2000). Two surveys of the Osona area were analysed for nitrogen and oxygen isotopes of dissolved nitrate following the methods of Sigman *et al.* (2001) and Casciotti *et al.* (2002). For sulfur and oxygen isotopic analysis, the dissolved sulfate was precipitated as $BaSO_4$ by the addition of $BaCl_2 \cdot 2H_2O$, after acidifying the sample with HCl and boiling it. For $\delta^{13}C$ analysis unfiltered splits of samples were treated with $NaOH-BaCl_2$ solution to precipitate carbonates, and then filtered at 3 μm. The sulfur, nitrogen and carbon isotopic composition was determined with an Elemental Analyser (Carlo Erba 1108) coupled with an IRMS (Delta C Finnigan Mat). The oxygen isotopic composition of nitrate and sulfate was analysed in duplicate with a Thermo-Chemical Elemental Analyser (TC/EA Thermo-Quest Finnigan) coupled with an IRMS (Delta C Finnigan Mat). Results are expressed in terms of δ per mil relative to the following international standards: Vienna Standard Mean Ocean Water (V-SMOW) for δD and $\delta^{18}O$, atmospheric N_2 (AIR) for N isotopes, Vienna Canyon Diablo Troilite (V-CDT) for S isotopes, and Vienna Pecdee Belemnite (V-PDB) for C isotopes. The isotope ratios were checked using international and internal laboratory standards. Reproducibility precision ($=1\sigma$) of the samples calculated from standards systematically interspersed in the analytical batches is $\pm 1.5°/_{oo}$ for δD, $\pm 0.2°/_{oo}$ for $\delta^{18}O_{H_2O}$, $\pm 0.3°/_{oo}$ for $\delta^{15}N_{NO_3}$, $\pm 0.2°/_{oo}$ for $\delta^{34}S$, $\pm 0.5°/_{oo}$ for both $\delta^{18}O_{NO_3}$ and $\delta^{18}O_{SO_4}$, and $\pm 0.2°/_{oo}$ for $\delta^{13}C_{HCO_3}$. Isotopic samples were prepared in the laboratory of the Applied Mineralogy and Environment Research Group and analysed at the Scientific-Technical Services of the University of Barcelona, except the isotopic composition of dissolved nitrates of two surveys in the Osona area, which were analysed at the Woods Hole Oceanographic Institution.

22.4   Statistical treatment

22.4.1   Data scaling

Before any statistical analysis can be done, we should scale the data set in an adequate way. This means transforming our variables in such a way that the variations in each variable (or set of variables) are comparable between them. For this particular study, this implies the following issues.

- **Removing the local mean water line from the isotopic composition.** As is well-known, the largest variation observed in δD and $\delta^{18}O$ in meteoric water is due to the altitude/continentality effects (Figure 22.2). As this has already a well-known

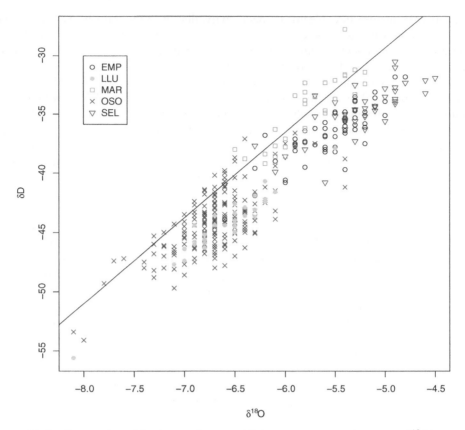

Figure 22.2 Scatterplot of hydrogen isotope (δD) versus oxygen istotope (δ^{18}O) compositions in water, with indication of the local meteoric water line (LMWL) calculated with data from the Global Network of Isotopes in Precipitation (GNIP) from stations 0818001 and 0818002 located in Barcelona (IAEA/WMO 2004).

explanation, this effect should be removed before any statistical analysis of isotopes can be aplied. Otherwise, it would mask any other effect. To do so, we can project the measured δD and δ^{18}O$_{H_2O}$ onto the LMWL (IAEA/WMO 2004),

$$\delta D = 7.32 + 7.29\, \delta^{18}O - H_2O$$

and extract the orthogonal deviations of each data point with respect to this line. These deviations are going to enter the analysis of isotopes instead of δD and δ^{18}O.

- **Log-ratio transforming the compositional variables.** If we have $\mathbf{Z}$ a compositional data set, then we must apply our statistical analyses to the clr-transformed data set, $\mathbf{Y} = \mathrm{clr}(\mathbf{Z})$, or to the ilr-transformed one, $\mathbf{X} = \mathrm{ilr}(\mathbf{Z})$. Both are related through $\mathbf{X} = \mathbf{Y} \cdot \mathbf{V}$ and $\mathbf{Y} = \mathbf{X} \cdot \mathbf{V}^t$, where $\mathbf{V}$ is the $D \times (D-1)$-matrix of definition of the used ilr basis. In the following sections we will either use clr- or ilr- transformed compositions wherever a real-valued data set is needed.

- **Scaling both subsets to be comparable.** This contribution uses a combined data set, with a compositional $\mathbf{Z}$ and a real $\mathbf{X}^r$ subset of variables, actually an isotopic composition array. Tolosana-Delgado *et al.* (2005) showed that the isotopic delta ratios are an excellent first-order approximation to a full log-ratio treatment with isotopes differentiated as extra geochemical component. This is due to the extremely low variability that isotopes show in comparison with the common geochemical variation. In order to adequately combine these two sources of information, we will first construct a coordinate data set, joining the ilr-transformed composition with the isotopic deltas. However, the variability of each of the two parts is not going to be comparable. For instance, in the present data set the geochemical composition has a *metric variance* (a single-number measure of the total variability of a subset of samples) (Pawlowsky-Glahn and Egozcue 2001) of 4.93, whereas the metric variance of the isotopic delta variables (once the LMWL was removed) is almost 95. Thus, if we simply combine them, the delta variability is going to mask the compositional one, being around 19 times larger. One should therefore either *downweight* the isotopic delta set, or increase the importance of the geochemical part by multiplying each subset with a constant:

$$\mathbf{X} = \left[\alpha \mathrm{ilr}(\mathbf{Z}); \beta \mathbf{X}^r \right] . \tag{22.1}$$

Lacking any way of deciding the values of α and β, it is reasonable to scale each of these two data sets by the inverse of the metric variance, i.e. to take

$$\alpha^{-1} = \sqrt{\mathrm{Tr}[\mathrm{Var}[\mathrm{ilr}(\mathbf{Z})]]}, \quad \beta^{-1} = \sqrt{\mathrm{Tr}[\mathrm{Var}[\mathbf{X}^r]]} .$$

In this way, both isotopic and geochemical parts contribute equally to the total variability.

22.4.2   Linear discriminant analysis

Let $\mathbf{X}$ be a data set, of P real variables and N individuals (either previously ilr-transformed compositions, isotopic delta variables or a mix of both types). Assume that this set is split into K groups, each of N_k samples, and denote by $\mathbf{m}_k$ and $\mathbf{S}_k$ the empirical centre and variance matrix of each group. The goal of linear discriminant analysis (LDA) is to find $K - 1$ directions of the P-dimensional real space, $\mathbb{R}^P$, where the separation between the groups is optimal (e.g. Mardia *et al.* 1979; Fahrmeir and Hamerle 1984; Krzanowski 1988; Krzanowski and Marriott 1994). Consider that each group has a prior likelihood of p_k^0, either chosen by the analyst or estimated as $\hat{p}_k^0 = N_k/N$. In a parametric framework, LDA hypotheses lead a sample $\mathbf{x}$ to have a posterior probability to belong to group k proportional to

$$p_k(\mathbf{x}) \propto p_k^0 \exp \left[-\frac{1}{2} d_{\mathrm{Mah}}^2(\mathbf{x}, \mathbf{m}_k | \mathbf{W}) \right],$$

where $d_{\mathrm{Mah}}^2(\mathbf{x}, \mathbf{m}_k | \mathbf{W})$ is the squared Mahalanobis distance from the sample to the centre of the group, with regard to the pooled *within-groups variance* matrix

$$\mathbf{W} = \sum_{k=1}^{K} p_k^0 \mathbf{S}_k .$$

If we define the global centre as $\mathbf{m} = \sum_{k=1}^{K} p_k^0 \mathbf{m}_k$, and the *between-groups variance* as

$$\mathbf{B} = \sum_{k=1}^{K} p_k^0 \cdot (\mathbf{m}_k - \mathbf{m}) \cdot (\mathbf{m}_k - \mathbf{m})^t,$$

then the solution can be found as the eigen-decomposition of the matrix $\mathbf{Q} = (\mathbf{W}^{-1} \cdot \mathbf{B})$. Its first $K - 1$ eigenvectors are the sought directions of optimal separations between groups, whereas the corresponding eigenvalues represent the ratios of the between- and within-group variances projected onto these directions, i.e. the discriminating power of each eigenvector.

If we are dealing with a D-part compositional data set, then $P = D - 1$. The Mahalanobis distance can be taken as $d_{\text{Mah}}^2(\mathbf{x}, \mathbf{m}_k | \mathbf{W}) = \text{ilr}(\mathbf{x} \ominus \mathbf{m}_k) \cdot \mathbf{W}^{-1} \cdot \text{ilr}^t(\mathbf{x} \ominus \mathbf{m}_k)$, and the between-groups variance may be obtained with $\mathbf{B} = \sum_{k=1}^{K} p_k^0 \cdot \text{ilr}(\mathbf{m}_k - \mathbf{m}) \cdot \text{ilr}^t(\mathbf{m}_k - \mathbf{m})$, where $\text{ilr}(\mathbf{m}) = \sum_{k=1}^{K} p_k^0 \text{ilr}(\mathbf{m}_k)$ is the global centre. All these elements may also be computed with compositions and variance matrices expressed as alr coordinates or even with clr coefficients, if the within-groups variance matrix is inverted with the Moore–Penrose generalized inversion.

When dealing with a mixed data set, it is safer to keep all computations in ilr coordinates, previously normalized to have unit metric variance, as shown in Equation (22.1).

22.4.3   Discriminant biplots

Classical biplots are bad tools for displaying the discrimination between groups $\mathbf{Q}$, being optimized to display the global variance $\mathbf{S} = \mathbf{W} + \mathbf{B}$. However, we can construct a biplot (a joint graphical representation of variables and observations) devised to display differences between groups. Following Gabriel (1971), a biplot is constructed from the decomposition of a centred data matrix $\mathbf{X}^* = \mathbf{X} - \mathbf{1}_N^t \cdot \mathbf{m}$ in a couple of matrices

$$\mathbf{X}^* = \mathbf{F} \cdot \mathbf{H}^t,$$

where $\mathbf{F}$ has N rows and P columns and $\mathbf{H}$ is a square matrix in which P columns represent orthogonal directions of $\mathbb{R}^P$. The two-dimensional graphical representation is obtained plotting the first two columns of $\mathbf{F}$ as dots (one for each individual) and the first two columns of $\mathbf{H}$ as rays (one for each part). In a variance biplot, $\mathbf{H}$ is chosen as the matrix of eigenvectors of $\mathbf{S}$ scaled by the square roots of their eigenvalues, and $\mathbf{F}$ is estimated with generalized inversion (Gower and Hand 1996),

$$\hat{\mathbf{F}} = (\mathbf{X}^* \cdot \mathbf{H}) \cdot (\mathbf{H}^t \cdot \mathbf{H})^{-1}. \tag{22.2}$$

In the same way, we propose to obtain a discriminant biplot by defining

$$\mathbf{H} = \mathbf{E} \cdot \mathbf{D}, \tag{22.3}$$

where matrix $\mathbf{E}$ is the eigenvectors of $\mathbf{Q}$ stored in the columns, and the diagonal matrix $\mathbf{D}$ contains the square roots of their eigenvalues (in the right order), and estimating $\mathbf{F}$ with Equation (22.2).

In the case of having a compositional data set, the eigendecomposition of $\mathbf{Q}$ is actually done on the ilr scores, but we would like the biplot $\mathbf{H}$ matrix to have a row for each part in the composition, so that we can draw its ray in the biplot. This is obtained simply as

$$\mathbf{H}_{clr} = \mathbf{V} \cdot \mathbf{E} \cdot \mathbf{D}. \qquad (22.4)$$

The individual matrix $\mathbf{F}$ is estimated equally with Equation (22.2), using matrix $\mathbf{H}$ obtained from Equation (22.3). Remember that the link between two rays represents the log-ratio of the two involved parts, as in a classical compositional covariance biplot.

When using a combined data set, we compute the eigendecomposition of $\mathbf{Q}$ with the data set of Equation (22.1). The first $D - 1$ coordinates of each eigenvector can be applied to Equation (22.4) to obtain the positions of the clr-transformed parts in the joint biplot. The remaining coordinates of each eigenvector are associated with isotopic variables, and need only be scaled by Equation (22.3).

22.5   Results and discussion

The representation of the association between variables and sample groups is intended to (i) discriminate the five sampled zones, (ii) observe which variables or combination of variables condition this discrimination, and (iii) determine whether the associations showed by the plot are related either to the anthropogenic sources of pollution or to the geological background. In this sense the effect of altitude/continentality, that could mask a possible discrimination controlled by the origin of pollution, has been removed by using the deviations of δD and $\delta^{18}O_{H_2O}$ with regards to the local meteoric water line, instead of using the isotopic composition of water. In the following discriminant biplots, only geochemical data (Figure 22.3), only isotope data (Figure 22.4) and both data sets together (Figure 22.5) have been used.

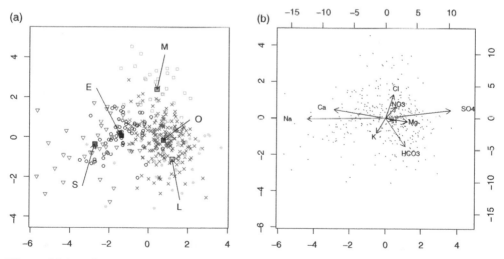

Figure 22.3 Discriminant plot (a) with indication of group means (plot centred using the global mean), and biplot (b), where arrows represent the explanatory variables using the geochemical data set. A clr transformation was used.

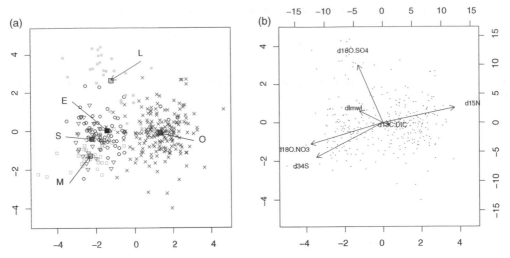

Figure 22.4 Discriminant plot (a) with indication of group means (plot centred using the global mean), and biplot (b), where arrows represent the explanatory variables using the isotopic data set.

In the discriminant plot using only geochemical data (Figure 22.3), we can discriminate Osona-Lluçanès, from Empordà-Selva, and from Maresme, but the sample groups are not perfectly split up. The explanatory variables are the couples of NO_3^--Cl^-, Ca^{2+}-Na^+ and Mg^{2+}-HCO_3^- contents. For instance, the Maresme area presents some of the highest NO_3^- concentrations due to the application of synthetic fertilizers, and is separated from the rest

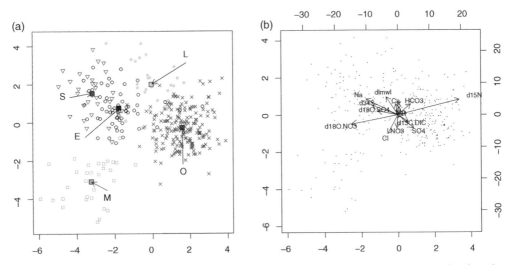

Figure 22.5 Discriminant plot (a) with indication of group means (plot centred using the global mean), and biplot (b), where arrows represent the explanatory variables, using both the geochemical (clr-transformed) and isotopic data sets.

of the zones in the direction of the NO_3^--Cl^- arrows, as mineral fertilizers have also high contents of chloride (Otero et $al.$ 2005). On the other hand, Empordà and Osona-Lluçanès areas, which also present high NO_3^- contents (specially Osona), seem to be better explained by the Ca^{2+}-Na^+ and Mg^{2+}-HCO_3^- arrows, respectively. The Selva area appears to be discriminated in the direction of Ca^{2+}-Na^+ contents, which agrees with the hydrochemical facies of local (Ca^{2+}-HCO_3^-) and regional (Na^+-HCO_3^-) flow systems, and with the Ca^{2+}-Na^+ cation exchange process that is also occurring in this zone. SO_4^{2-} content cannot be considered an explanatory variable because it is not able to separate the Maresme and Osona-Lluçanès areas, which present high concentrations of this anion. With regard to SO_4^{2-}, it only can be concluded that the Na^+/SO_4^{2-} ratio in the Empordà-Selva areas is lower than the average for all zones. In this discriminant plot some of the studied areas are discriminated by NO_3^- contents, but the main nitrate sources that contribute to nitrate contamination, fertilizers and pig manure, represented by the Maresme and the Osona areas, respectively, are not in extreme positions. The separation of Osona-Lluçanès from Empordà-Selva is due to bedrock signature. Therefore, the discriminant plot entering only geochemical data does not allow an easy distinction of the five vulnerable zones according to the main nitrate source.

In the discriminant plot using only isotope data (Figure 22.4), Osona and Lluçanès samples are well separated from Maresme, Empordà and Selva samples. The explanatory variables are $\delta^{15}N$, $\delta^{18}O_{SO_4}$ and $\delta^{34}S$-$\delta^{18}O_{NO_3}$, indicating the combined influence of nitrate sources and the processes undergone by nitrate (mainly nitrate reduction favoured by sulfide and/or organic matter oxidation). In the link formed by $\delta^{15}N$ and $\delta^{34}S$-$\delta^{18}O_{NO_3}$ (which means using the difference between $\delta^{15}N$ and the sum of $\delta^{34}S$ and $\delta^{18}O_{NO_3}$), Osona and Maresme areas are in extreme positions, which agrees with their different nitrate contamination origin. Osona presents higher $\delta^{15}N$ values and lower $\delta^{34}S$ and $\delta^{18}O_{NO_3}$ values, because the main nitrate source in this area is pig manure and denitrification processes are occurring linked to sulfide oxidation. Maresme presents lower $\delta^{15}N$ and higher $\delta^{18}O_{NO_3}$ values because the main nitrate source is synthetic fertilizers and the reduction of nitrate is not detected. Anyhow, although Empordà and Selva samples plot next to the Maresme ones, they are not only affected by mineral fertilizers, but a nitrate contribution of pig manure is known in these areas. The Lluçanès sample group is remarkably well separated in the direction of the $\delta^{18}O_{SO_4}$, as the main contributions of SO_4^{2-} in this area are fertilizers and the presence of evaporites. If processes were not involved, from this discriminant plot we could relate pollution origin in the mixed areas mainly to fertilizers, with minor contribution of pig manure. However, we must take into account whether denitrification is occurring, and how it is occurring, because the $\delta^{34}S$ and $\delta^{18}O_{SO_4}$ variables can discriminate areas with the same nitrate source, depending on which is the main reaction that is controlling nitrate reduction. For instance, the $\delta^{34}S$ values in Osona, linked to the presence of pyrite and so to denitrification by pyrite oxidation, give rise to a clear separation of this sample group from the rest. The $\delta^{13}C_{DIC}$ and the deviation with respect to the local meteoric water line (dlmwl) do not exert an important role in discriminating the different areas. Thus, the use of only isotope data allows to distinguish the zones in a clearer way by the nitrate source influence, but some difficulties arise interpreting the mixed areas.

Using both data sets, in the discriminant plot obtained (Figure 22.5), the different sample groups are better separated. We have a clear discrimination of samples with the ratio ($\delta^{15}N$-$\delta^{18}O_{NO_3}$) + log (HCO_3^-/Cl^-): whereas the Maresme sample group is in one extreme and the Osona sample group in the opposite, which are considered as source end members, those samples from the mixed areas (Empordà and Selva) are in between. Taking in account the ratio

Table 22.1 LDA reclassification table obtained with each set of data, with percentage of correct reclassification.

		Geochemistry				Isotopes					Both				
Predicted group	E	L	M	O	S	E	L	M	O	S	E	L	M	O	S
True group	E	49	0	0	12	3	43	2	1	4	8	52	0	0	4	2
	L	1	14	0	13	2	4	21	0	4	0	2	21	0	6	0
	M	1	0	22	8	0	1	0	30	0	0	1	0	30	0	0
	O	6	8	6	176	0	6	1	2	180	0	3	1	0	185	0
	S	4	0	1	3	31	9	0	1	0	20	5	0	0	0	25
% good		81.111					87.24					92.88				

$\delta^{13}C_{DIC}$ - ($\delta^{34}S$-$\delta^{18}O_{SO_4}$), the sample group distribution can give us an idea of how denitrification processes are occurring: low $\delta^{13}C_{DIC}$ values (together with an increase of $\delta^{15}N$ and $\delta^{18}O_{NO_3}$) mean that organic matter oxidation is the reaction linked to denitrification processes; and low $\delta^{34}S$ and $\delta^{18}O_{SO_4}$ values imply that the reaction involved in natural attenuation of nitrate is sulfide oxidation. We must also bear in mind that the length of the arrows is related to the discriminant power of the variables that they represent, so the short length of the $\delta^{13}C_{DIC}$ arrow can be interpreted as if organic matter oxidation is taking place in all the areas where denitrification processes are occurring.

The discriminant biplots with only geochemical data, only isotope data and both data subsets separate the sample groups according to the following percentages of reclassification: 81, 87 and 93% (Table 22.1). As could be expected, the best discrimination is obtained when using both data subsets, but the discriminant biplot with only isotope data is useful enough to separate sample groups affected by different nitrate sources. Thus, the isotope data set is a powerful tool by itself, though some Empordà and Selva samples are missclassified.

22.6   Conclusions

A statistical methodology has been applied to the geochemical and isotope data set of five vulnerable zones. This procedure consists of a linear discriminant analysis of sample groups and the corresponding discriminant biplot, where the explanatory variables are plotted depending on their discriminant power. In order to implement this statistical methodology that integrates isotope with geochemical data together, both data subsets have been scaled so that their variations are comparable, and geochemical data have been transformed and treated as compositional data. A discriminant biplot has been generated by means of the eigendecomposition of **Q** matrix, that is defined as it follows: $\mathbf{Q} = (\mathbf{W}^{-1} \cdot \mathbf{B})$, where **W** and **B** are the *within-groups* and *between-groups variance matrices*, respectively. Note discrimination power when using only the isotope data set, although the optimum separation of sample groups is achieved using both geochemical and isotope data subsets. Moreover, in this case, in the direction defined by the variables $\delta^{13}C_{DIC}$, $\delta^{34}S$ and $\delta^{18}O_{SO_4}$, a separation of sample groups depending on the reactions associated with denitrification processes is suggested. Further research is needed focusing on the assessment of natural attenuation of nitrate and the reactions involved, applying statistical methods to compositional data sets.

Acknowledgements

This study was funded by CICYT projects CGL-2008-06373-CO3-01/03-BTE from the Spanish Government, project 2009 SGR 103 from the Catalan Government, and the I3P Programme funded by the EU. Funding is also acknowledged from the Spanish Ministry of Science and Innovation through a *Juan de la Cierva* subprogramme, supported by the European Social Fund. The authors would like to thank the Serveis Cientifico-Tècnics of the University of Barcelona for their services.

References

ACA 2009 Data for groundwater quality in Catalunya http://aca-web.gencat.cat/aca/appmanager/aca/aca/. Accessed December 2009.

Aravena R , Robertson WD 1998 Use of multiple isotopc tracers to evaluate denitrification in ground water: Study of nitrate from a large-flux septic system plume. *Ground Water* **36**(6), 975–981.

Bethke C 2008 *Geochemical and Biogeochemical Reaction Modelling*, 2nd edition Cambridge University Press, Cambridge (UK). 564 p.

Böttcher J, Strebel O, Voerkelius S and Schmidt HL 1990 Using isotope fractionation of nitrate-nitrogen and nitrate-oxygen for evaluation of microbial denitrification in sandy aquifer. *Journal of Hydrology* **114**, 413–424.

Casciotti KL, Sigman DM, Galanter M, Bölkhe JK and Hilkert A 2002 Measurement of the oxygen isotopic composition of nitrate in seawater and freshwater using the denitrifier method. *Analytical Chemistry* **74**, 4905–4912.

Cey E, Rudolph D, Aravena R and Parkin G 1999 Role of the riparian zone in controlling the distribution and fate of agricultural nitrogen near a small stream in southern Ontario. *Journal of Contaminant Hydrology* **37**, 45 67.

EC 1998 Council directive 98/83/EC, of 3 November 1998, on the quality of water intended for human consumption. Official Journal of the European Communities, L 330, of 5.12.1998, Brussels. http://eur-lex.europa.eu/LexUriServ/LexUriServ.do?uri=OJ:L:1998:330:0032:0054:EN:PDF.   Accessed October 2004.

Epstein S and Mayeda T 1953 Variation of ^{18}O content of waters from natural sources. *Geochimica et Cosmochimica Acta* **4**, 213–224.

Fahrmeir L and Hamerle A 1984 *Multivariate Statistische Verfahren*. Walter de Gruyter, Berlin (Germany). 796 p.

Friedman I 1953 Deuterium content of natural waters and other substances. *Geochimica et Cosmochimica Acta* **4**, 89–103.

Gabriel KR 1971 The biplot – graphic display of matrices with application to principal component analysis. *Biometrika* **58** (3), 453–467.

Gower JC and Hand DJ 1996 *Biplots*. Chapman and Hall Ltd, London (UK). 277 p.

Grischek T , Hiscock KM , Metschies T 1998 Dennis PF W. N, Factors affecting denitrification during infiltration of river water into a sand and gravel aquifer in Saxony, Germany. *Water Research* **32** (2), 450–460.

Guimerà J, Marfà O, Candela L and Serrano L 1995 Nitrate leaching and strawberry production under drip irrigation management. *Agriculture, Ecosystems and Environment* **56**, 121–135.

IAEA/WMO 2004 Global network of isotopes in precipitation. The GNIP Database. http://www-naweb.iaea.org/napc/ih/IHS resources_gnip.html. Accessed October 2004.

IDESCAT 1999 Cens agrari del banc d'estadístiques de municipis i comarques http://www.idescat.cat/. Accessed October 2004.

Kendall C and McDonnell JJ 1998 *Isotope Tracers in Catchment Hydrology*. Elsevier Science BV, Amsterdam (The Netherlands). 839 p.

Krzanowski WJ 1988 *Principles of Multivariate Analysis: A User's Perspective*. Clarendon Press, Oxford (UK). 563 p.

Krzanowski WJ and Marriott FHC 1994 *Multivariate Analysis, Part 2 - Classification, Covariance Structures and Repeated Measurements*. Edward Arnold, London (UK). 280 p.

Letolle R 1980 Nitrogen-15 in the natural environment. In *Handbook of Environmental Isotope Geochemistry, Vol. 1. The Terrestrial Environment* (ed. Fritz P and Fontes JC). Elsevier, Amsterdam (The Netherlands). pp. 407–433.

Magee PN and Barnes JM 1956 The production of malignant primary hepatic tumors in the rat by feeding dimethylnitrosamine. *British Journal of Cancer* **10**, 114–122.

Mardia KV, Kent JT and Bibby JM 1979 *Multivariate Analysis*. Academic Press, London (UK). 518 p.

Mayer B, Bollwerk SM, Mansfeldt T, Hütter B, Vezier J 2001 The oxygen isotopic composition of nitrate generated by nitrification in acid forest floors. *Geochimica et Cosmochimica Acta* **65**(16), 2743–2756.

Mengis M, Schiff SL, Harris M, English MC, Aravena R, Elgood RJ and MacLean A 1999 Multiple geochemical and isotopic approaches for assessing ground water NO_3^- elimination in a riparian zone. *Ground Water* **37**(3), 448–457.

Otero N, Torrentó C, Soler A, Menció A and Mas-Pla J 2009 Monitoring groundwater nitrate attenuation in a regional system coupling hydrogeology with multi-isotopic methods: The case of Plana de Vic (Osona, Spain). *Agriculture, Ecosystems and Environment* **133**, 103–113.

Otero N, Vitòria L, Soler A and Canals A 2005 Fertilizer characterization: major, trace and rare earth elements. *Applied Geochemistry* **20**(8), 1473–1488.

Pauwels H, Foucher JC and Kloppmann W 2000 Denitrification and mixing in a schist aquifer: Influence on water chemistry isotopes. *Chemical Geology* **168**, 307–324.

Pawlowsky-Glahn V, Egozcue JJ 2001 Geometric approach to statistical analysis on the simplex. *Stochastic Environmental Research and Risk Assessment (SERRA)* **15**(5), 384–398.

Puig R, Soler A and Mas-Pla J 2007 Determination of the sources of nitrate pollution and evaluation of natural attenuation processes using multi-isotopic methods in the Baix Empordà basin (NE Spain). In *Water Pollution in Natural Porous Media at Different Scales. Assessment of Fate, Impact and Indicators* (ed. Candela L, Vadillo I, Aagaard P, Bedbur E, Trevisan M, Vanclooster M, Viotti P and López-Geta JA). Instituto Geológico y Minero de España, Madrid (Spain). pp. 239–245.

Sigman DM, Casciotti KL, Andreani M, Bradford C, Galanter M and Bölkhe JK 2001 A bacterial method for the nitrogen isotopic analysis of nitrate in seawater and freshwater. *Analytical Chemistry* **73**, 4145–4153.

Silva SR, Kendall C, Wilkison DH, Zieglerc AC, Chang CCY and Avanzino RJ 2000 A new method for collection of nitrate from fresh water and the analysis of nitrogen and oxygen isotope ratios. *Journal of Hydrology* **228**, 22–36.

Tolosana-Delgado R, Otero N and Soler A 2005 A compositional approach to stable isotope data analysis. In *Proceedings of CoDaWork'05, The 2nd Compositional Data Analysis Workshop* (ed. Mateu-Figueres G and Barceló-Vidal C). University of Girona, Girona (Spain).

Vitòria L, Soler A, Aravena R and Canals A 2005 Multi-isotopic approach (15N, 13C, 34S, 18O and D) for tracing agriculture contamination in groundwater. In *Environmental Chemistry: Green Chemistry and Pollutants in Ecosystems* (ed. Lichtfouse E, Schwartzbauer J and Robert D). Springer-Verlag, Berlin (Germany). pp. 43–56.

Vitòria L, Soler A, Canals A and Otero N 2008 Environmental isotopes (N, S, C, O, D) to determine natural attenuation processes in nitrate contaminated water: Example of Osona (NE Spain). *Applied Geochemistry* **23**, 3597–3611.

Volkmer BG, Ernst B, Simon J, Kuefer R, Bartsch GJ, Bach D and Gschwend JE 2005 Influence of nitrate levels in drinking water on urological malignancies: a community-based cohort study. *British Journal of Urology International* **95**(7), 972.

Ward MH, DeKok TM, Levallois P, Brender J, Gulis G, Nolan BT and VanDerslice J 2005 Workgroup report: Drinking-water nitrate and health-recent findings and research needs. *Environmental Health Perspectives* **113**(11), 1607–1614.

Wassenaar L 1995 Evaluation of the origin and fate of nitrate in Abbotsford Aquifer using the isotopes of ^{15}N and ^{18}O in NO_3^-. *Applied Geochemistry* **10**, 391–405.

23

Applications in economics

Tim Fry

School of Economics, Finance & Marketing, RMIT University, Melbourne, Australia

23.1   Introduction

There are many examples in economics where the focus is on shares of total or compositional data. It is well known that there are three approaches to analysing compositional data. Researchers can ignore the compositional nature of their data and apply traditional multivariate statistical techniques; they can transform the data from the unit simplex to real Euclidean space, apply traditional multivariate statistical techniques and then apply an inverse transformation back to the unit simplex or they can work directly in the unit simplex. It is extremely rare that economists apply the tools of compositional data analysis to their problems. In economics, researchers typically ignore the compositional nature of their data.

This chapter looks at the areas in economics where compositional data analysis techniques have been used and how the data have been analysed. Of the two approaches: transformation and working in the simplex, the most common approach to dealing with compositional data in economics has been the use of log-ratio transformations. Both the additive log-ratio (alr) and the centred log-ratio (clr) transformations have been used with the alr being the more popular. As the focus in economics is on the estimation of model parameters of economic interest the transformation approach is applied to yield an estimating model with desirable properties. Thus, it is less common to see researchers applying the inverse transformation back to the unit simplex.

The most dominant area of application has been that of consumer demand systems. However, there are a limited number of applications in other areas. The next section discusses the use of compositional data analysis techniques in consumer demand systems. This is

Compositional Data Analysis: Theory and Applications, First Edition. Edited by Vera Pawlowsky-Glahn and Antonella Buccianti.
© 2011 John Wiley & Sons, Ltd. Published 2011 by John Wiley & Sons, Ltd.

followed by a look at a range of empirical studies in other areas of economics that employ compositional data analysis techniques; some applications of time series analysis techniques to economic data; some suggested new directions for researchers and the chapter ends with some concluding remarks.

23.2   Consumer demand systems

The analysis of data on consumer demand has a long heritage in economics. In such research the focus is typically on understanding how the consumer allocates their budget among the commodities available. We begin by reviewing the traditional approach to consumer demand modelling (a more detailed exposition may be found in Pollak and Wales) (1992). Economic theory starts with a consumer who chooses their consumption quantities as if they maximise their utility subject to a budget constraint. Formally, we have:

$$\max_{\mathbf{q}}(U(\mathbf{q}) : \mathbf{p}'\mathbf{q} \leq m, \ \mathbf{q} \geq 0),$$

where $\mathbf{q}$ is a $D \times 1$ vector of quantities with prices $\mathbf{p}$, m is total expenditure (the budget) and $U(\mathbf{q})$ is a non-decreasing quasi-concave utility function that reflects the consumer's preferences. The solution of this maximisation problem yields a system of demand equations:

$$q_i = Q_i(\mathbf{p}, m, \boldsymbol{\beta}), \quad i = 1, \ldots, D,$$

with $\boldsymbol{\beta}$ the vector of parameters of the utility function. Because these demand systems are the solution to the above maximisation problem, they will by construction satisfy the following regularity conditions: adding up; homogeneity of degree zero in prices and income; and symmetry and negative semi-definiteness of the implied Slutsky matrix.

Economists do not typically estimate the system of equations for quantities. Rather the analysis looks at the determinants of the budget shares for the commodities. The system of share equations to be estimated is specified as:

$$s_i = S_i(\mathbf{p}, m, \boldsymbol{\beta}) + u_i, \quad i = 1, \ldots, D. \tag{23.1}$$

Thus the observed quantities are transformed to expenditure shares $s_i = p_i q_i / m$ and these shares are then taken to be equal to the corresponding functional form from economic theory $S_i(\mathbf{p}, m, \boldsymbol{\beta}) = p_i Q_i(\mathbf{p}, m, \boldsymbol{\beta})/m = S_i(\mathbf{Z}, \boldsymbol{\beta})$ plus a multivariate normal error term, u_i. This transformation ensures that the error term in the share system will possess a homoscedastic variance matrix and removes dependence upon the numeraire.

Since the budget shares add to one (termed *adding up*) the estimation of this nonlinear system of equations is not straightforward. In economics the solution to the adding up constraint has been to delete an equation from the system, estimate the remaining $D-1$ equations and to obtain estimates of the excluded (deleted) equation and variance-covariance parameters through parameter restrictions imposed by the adding up constraint. We will term this the traditional approach and point out that it relies upon a result in Barten (1969) who shows invariance of the maximum likelihood estimation procedure to the choice of equation to delete.

Whilst economists pay attention to choosing utility functions, or using a 'dual' approach, indirect utility functions, that satisfy a number of properties this, at best, only results in share specifications $S_i(\mathbf{Z}, \boldsymbol{\beta})$ that are bounded to the unit simplex. The statistical model in Equation (23.1) is not correctly bounded. The problem arises from the appending of an error term to the theoretical model for shares to form the statistical model. Economists have been slow to recognise or deal with this problem. Indeed the overwhelming majority of empirical work with demand share data has adhered to the traditional approach.

There have been instances of economists seeking a statistical model which combines the functional forms from economic theory with a stochastic specification that will appropriately bound to the unit simplex. One of the earliest papers to address this issue was by Woodland (1979) who suggested the use of the Dirichlet distribution for the stochastic component, u_i. His paper considers the estimation of shares from the Dirichlet based model and, with both empirical examples and a simulation study, compares the model to the traditional approach assuming multivariate normality. Despite recognising that the Dirichlet represents *an attractive alternative to the Normal model* he suggests that whilst not being theoretically valid the normality based model–the traditional approach–may still yield valid results.

Whilst the Dirichlet model represents an example of working directly in the simplex, the few economists applying compositional data analysis rely on the use of log-ratio transforms. This approach was first suggested by Ronning (1992) but is primarily associated with the work by Fry *et al.* (1996, 2000, 2001). Modelling involves transforming the shares using the alr transformation and modelling the log-ratio transformed data, $\mathbf{y}$, in terms of $\boldsymbol{\mu}$ and $\boldsymbol{\Sigma}$. In particular, we may parameterise the mean, $\boldsymbol{\mu}$, to depend upon a set of variables, $\mathbf{Z}$, and a set of parameters, $\boldsymbol{\beta}$, according to a multivariate regression model:

$$y_i = \ln\left(\frac{s_i}{s_D}\right) = \mu_i(\mathbf{Z}, \boldsymbol{\beta}) + v_i \,,$$

where $\mathbf{v}$ is a stochastic term which is distributed as multivariate normal $(\mathbf{0}, \boldsymbol{\Sigma})$. The advantage of this model is that, within this framework, the shares are distributed as additive logistic normal and the basis–in this application the expenditures on the commodities–as multivariate log-normal. The use of a similar additive logistic normal approach in the context of modelling vote shares in multiparty elections has also been suggested by Katz and King (1999). Later work (Honaker *et al.* 2002) has shown that such electoral data (and potentially economic data on consumer demand?) is better modelled using an additive logistic t distribution.

The remaining issue is the specification of the functional form for the $\mu_i(\mathbf{Z}, \boldsymbol{\beta})$. By analogy with the arguments in Fry *et al.* (1996) the parameterisation chosen should retain any parameter interpretations from the underlying economic theory and, further, it should retain the logical consistency argument that shares from the model are restricted to the unit simplex. Such a parameterisation is given by:

$$y_i = \ln\left(\frac{S_i(\mathbf{Z}, \boldsymbol{\beta})}{S_D(\mathbf{Z}, \boldsymbol{\beta})}\right) + v_i \,, \tag{23.2}$$

where $S_i(\mathbf{Z}, \boldsymbol{\beta})$ is the theoretical specification for the share of i which retains the logical consistency requirement. Estimation of the model is discussed in Fry *et al.* (1996) and for the case with autocorrelated stochastic terms in McLaren (1996).

The specification given by Equation (23.2) has been applied in a number of empirical studies. McLaren *et al.* (1995) show that the use of the alr transformation provides a convenient statistical test for the application of the widely used Almost Ideal Demand System. This test is obtained via a parameter restriction within the Modified Almost Ideal Demand System. In their application with annual Australian expenditure data on four commodities from 1953/54 to 1992/93 they find that the Modified Almost Ideal Demand System is the preferred specification. Conniffe and Eakins (2003) apply the methodology to annual Irish expenditure data from 1979 to 1999 for five commodities. They conclude that whilst theoretically more appropriate, the compositional data analysis approach does not yield parameter estimates that are markedly different to the traditional approach. Thus, for aggregate applications with time series data they suggest that the stochastic approach taken is comparatively unimportant.

The use of log-ratio transformations to estimate systems of demand equations has been independently proposed by economists working with addilog and indirect addilog functional forms (Chavas and Segerson 1987) for the theoretical share specification. These functional forms are generalisations of the multinomial logit (MNL) form first proposed by Theil (1969). These models are unusual in that they are typically estimated using the clr or alr transformations as the use of one of these transformations yields estimating equations that are log-linear. Addilog and generalised addilog models have been applied in empirical studies (Bewley 1982a,b, 1986).

The use of log-ratio transformations like the alr is appropriate when the observed share data does not include zeros. In applications with aggregate time series data on shares or with aggregate commodity groupings such as 'food' such zero shares are unlikely to be observed. However, the use of survey data and/or disaggregate categories is highly likely to yield zero share observations. In such cases we cannot rely on the approach specified above. The question of how to deal with zero values for the observed share data is discussed in Fry *et al.* (2000). They argue strongly that a zero replacement technique, that is ratio preserving, is simple to implement, easy to work with, has simple rationale and gives sensible results, should be used. Their paper proposes such a technique that they term *modified Aitchison*. They apply the modified Aitchison procedure with data from the 1988/89 Australian household expenditure survey and demonstrate the success of the procedure. Fry *et al.* (2001) further show how the modified Aitchison procedure can be motivated by duality theory to provide a model specification that is a consistent theoretical and stochastic specification for the occurrence of zero demand over a range of expenditures and/or prices. They use this approach to further analyse data from the 1988/89 Australian household expenditure survey.

The requirement that a zero replacement procedure for compositional data analysis retains the share ratios for the nonzero components means that those share ratios are, by construction invariant to the addition of components to the composition. This is exactly the property of Independence of Irrelevant Alternatives (IIA) that is inherent in the multinomial logit (MNL) model for discrete choice. Namely, that the odds ratio is invariant to additions or deletions to the choice set. Thus, the modified Aitchison zero replacement procedure, that has become the default procedure, is consistent with the MNL theoretical specification for $S_i(\mathbf{Z}, \boldsymbol{\beta})$. IIA in the discrete choice literature is viewed as an extremely restrictive property for models to have. This has led to the development of a range of alternative model specifications such as the nested logit that do not embody IIA but which can allow for tests for IIA.

Nested logit forms can be estimated in a straightforward sequential manner using alr transformations and multivariate linear regression models (Bechtel 1990) and allow for a simple statistical test for the multinomial logit form. Building upon this observation (Fry and Chong 2005) derive a zero replacement procedure that is consistent with a more general nested logit form for $S_i(\mathbf{Z}, \boldsymbol{\beta})$ and that allows for a simple statistical test to determine the form of share specification and zero replacement procedure that is consistent with the data. As yet, this procedure has not been applied in practice.

Several researchers (Fry *et al.* 1996, 2000, 2001; Conniffe and Eakins 2003) have argued that it is with the use of survey data and/or disaggregate categories that any differences between the compositional data analysis and the traditional approaches will become evident. Surprisingly, however, there have been few empirical applications of the compositional data analysis approach with such micro data. Fry *et al.* (2000) use Australian data to illustrate the robustness of their modified Aitchison zero replacement technique when estimating a system of Engel Curves. Fry *et al.* (2001) provide an economic theory rationale for the use of modified Aitchison and show the properties of Engel Curves in such situations. They again illustrate their results using Australian data to estimate systems of Engel Curves. Koch (2007) represents the only other application of the modified Aitchison combined with the use of the alr transformation. He uses South African survey data for 2000 to estimate systems of Engel Curves and expenditure elasticities for both four and six commodity groupings. Finally, we note that Kim and Min (2007) apply the modified Aitchison procedure to deal with zero share observations, but combine it not with a compositional data analysis technique but with a semiparametric procedure.

23.3   Miscellaneous applications

This section looks at a small number of studies that have used compositional data analysis techniques in areas other than consumer demand. Anyadike-Danes (2004, 2007) has studied movements in employment, unemployment and inactivity rates over time by region and by gender in the UK. He uses the alr transformation to model the movements in the labour market shares and then the inverse transformation to produce predictions for the labour market shares. This work provides an understanding of regional and gender divides in labour market outcomes.

Compositional data analysis techniques have also found application in studies of income distribution. Longford and Pittau (2006) and Pittau and Zelli (2006) use household survey data to model income distribution. Their work involves fitting mixture distributions to the data to yield the probabilities that a household with a given income comes from a certain component of the mixture. In Longford and Pittau (2006) these probabilities are then summarised, using ternary plots and logistic regression and comparisons made across European countries. Pittau and Zelli (2006) take the mixture probabilities by income level and, using a clr transformation, model them as a function of a number of covariates describing the household.

Two applications that work directly in the simplex are by van der Ark *et al.* (1999) and Desarbo *et al.* (1995). Desarbo *et al.* (1995) are concerned with determining the impact of covariates on a constant sum criterion at the segment level. They propose the use of the Dirichlet distribution to deal with the constant sum (compositional) feature of their data and illustrate their approach with an empirical example. van der Ark *et al.* (1999) recognise that the latent budget model is a reduced rank model for compositional data. They then exploit the geometry of the simplex to determine a criterion by which the model can be identified. The new procedure is then applied and compared with correspondence analysis.

Finally, Arauzo *et al.* (2007) use panel (longitudinal) survey data to study the sectoral and regional entry and exit rates of Spanish firms. A compositional data analysis approach is not taken in the study. However, use is made of the modified Aitchison procedure to replace zero observations on the rates prior to the application of a multivariate regression model for the log entry and exit rates.

23.4   Compositional time series

Most of the work in this field has been focused upon methodological developments. However, research with time series compositional data has often involved application to data from economics. More detail on compositional time series can be found in Chapter 7. Once again when the compositional nature of the data is recognised there have been two approaches: to work directly in the simplex or to transform to Euclidean space and use multivariate time-series analysis techniques. The stay in the simplex research is not as common and has relied upon the use of the Dirichlet distribution within a Bayesian state space framework. Grunwald *et al.* (1993) develop this methodology and apply it to data on world car production.

The transformation approach is more popular and use has been made of both the alr and the clr transformations. Of the two transformations the clr was introduced by Quintana and West (1988) but has not been used as often as the alr. Quintana and West (1988) use a dynamic regression model for the clr-transformed data that allows for interventions, and apply their model to data on Mexican imports. Larrossa (2003) uses the clr transformation and the modified Aitchison approach to zeros to study data on capital stock from a number of countries using biplots.

The alr transformation has proved more popular (Brunsdon and Smith 1998; Silva and Smith 2001) and the transformed data has been modelled using multivariate time series models. These have included vector autoregressive moving average models, with or without covariates, with and without differencing for stationarity and additionally including a variance component (generalised autoregressive conditional heteroscedasticity). Applications have been to UK voting intention data, to Australian labour force status data and to Indian microfinance data.

One last transformation approach that has been used in this field is that of Wang *et al.* (2007). They propose a forecasting model that involves a nonlinear transformation of the compositional data onto a hypersphere. The model is applied to forecasting the output shares of the primary, secondary and tertiary production sectors in China.

23.5   New directions

We note that many other areas of economics involve modelling compositional data, but typically ignore the nature of the data. Occasionally researchers [Morana (2007) in the context of factor demand modelling and McKitrick (1998) in the context of computable general equilibrium modelling] have advocated the use of compositional data techniques, but this suggestion has not, as yet, been followed. Interestingly, although as described above compositional data techniques have been used in modelling income distribution shares, they have not been used to understand income inequality. However, common measures of income inequality such as Gini coefficients are constructed using income share (compositional) data and perhaps new measures of income inequality might be derived using compositional data techniques.

One area in economics that has large potential for compositional data techniques to be applied is in the measurement of market concentration and contestability. Given data on the shares of the market an index of concentration is computed and interpreted. Several indices exist, however, the Herfindahl–Hirschman index (Herfindahl 1950; Hirschman 1964) is probably the most widely used of these. The Herfindahl–Hirschman index of concentration for a market is given by the sum of the squared market shares. That is, $\text{HHI} = \sum_{i=1}^{D} s_i^2$. The inverse of the index given by $N = \text{HHI}^{-1} = 1/\text{HHI}$ is less often used but has a convenient interpretation. N gives us the equivalent number of equally sized entities (companies, products) that could contest the market. The lower bound of HHI occurs at equal shares for all D entities and is equal to $1/D$. HHI attains its upper bound of 1 when $D = 1$. That is, when only one entity exists that has a share of one. In the economics and related literature HHI and N are treated as *descriptive*. However, can we look at their statistical properties and hence derive results that would allow us to conduct more rigorous analysis? In what follows three situations are of interest to us.

First, the Herfindahl–Hirschman index is used in competition law. The regulatory authorities (e.g. the US Department of Justice) use the values of the Herfindahl–Hirschman index before (HHI_P) and after (HHI_M) a proposed merger to decide whether any competition issues arise from the proposed merger. In the US case, the rules are as follows:

- If $\text{HHI}_M < 0.1$, no competition issues arise.

- If $0.1 \leq \text{HHI}_M < 0.18$ and

  - $\text{HHI}_M - \text{HHI}_P \leq 0.01$, no competition issues arise;

  - $\text{HHI}_M - \text{HHI}_P > 0.01$, anticompetitive issues arise.

- If $\text{HHI}_M \geq 0.18$ and

  - $\text{HHI}_M - \text{HHI}_P < 0.005$, no competition issues arise;

  - $0.005 \leq \text{HHI}_M - \text{HHI}_P < 0.01$, anticompetitive issues arise.

- If $\text{HHI}_M - \text{HHI}_P > 0.01$, then serious anticompetitive issues arise.

In the European Union, the situation is simpler: if $\text{HHI}_P \geq 0.1$ and $\text{HHI}_M - \text{HHI}_P > 0.025$, then anticompetitive issues arise.

Since a merger is the amalgamation of two parts of our composition, say parts 1 and 2 (firms one and two) then, using $(s_1 + s_2)^2 = s_1^2 + s_2^2 + 2s_1 s_2$, the difference in the index after the merger is simply $\text{HHI}_M - \text{HHI}_P = 2s_1 s_2$. Thus, all that we need to know is the distribution of $s_1 s_2$ to assign probabilities to the outcomes in the rules.

The second area concerns comparing market concentration in two separate markets. For example, if HHI is computed from two sets of market shares, say from regions A and B, can we test equality of HHI from the two markets A and B? The last area of interest concerns modelling the determinants of the concentration of markets. Thus, if we have data on $t = 1, \ldots, T$ compositions (e.g. vote shares for D political parties in each of T electoral divisions) could a regression model be fitted to HHI_t? The answer to these last two questions depends upon us utilising compositional data analysis techniques to determine the distribution of HHI.

23.6   Conclusion

There are many examples in economics where the focus is on shares of total or compositional data. However, it is extremely rare that economists apply the tools of compositional data analysis to their problems. In other words, in economics researchers ignore the compositional nature of their data. This chapter has highlighted the areas in which compositional data techniques have been applied and discussed which approaches (transformation or staying in the simplex) have been used. It is in the field of consumer demand that most of the work has appeared and the alr transformation is the approach of choice. Indeed, it was to deal with zero shares in consumer demand that researchers developed the modified Aitchison approach. Given the scarcity of application of compositional data techniques some new directions or areas in economics, such as measuring concentration, where such techniques might be applied, have been suggested.

References

Anyadike-Danes M 2004 The real north-south divide? regional gradients in UK male non-employment. *Regional Studies* **38**(1), 85–95.

Anyadike-Danes M 2007 How well are women doing? Female non-employment across UK regions. *Applied Economics* **39**(14), 1843–1854.

Arauzo JM, Manjon M, Martín M and Segarro A 2007 Regional and sector-specific determinants of industry dynamics and the displacement-replacement effects. *Empirica* **34**(2), 89–115.

Barten AP 1969 Maximum likelihood estimation of a complete system of demand equations. *European Economic Review* **1**(1), 7–73.

Bechtel GG 1990 Share-ratio estimation of the nested multinomial logit model. *Journal of Marketing Research* **27**(2), 232–237.

Bewley RA 1982a The generalised addilog demand system applied to Australian time series and cross section data. *Australian Economic Papers* **21**(38), 177–192.

Bewley RA 1982b On the functional form of Engel curves: The Australian household expenditure survey 1975-76. *Economic Record* **58**(1), 82–91.

Bewley RA 1986 *Allocation Models: Specification, Estimation and Applications*. Ballinger, Cambridge, MA (USA). 339 p.

Brunsdon TM and Smith TMF 1998 The time series analysis of compositional data. *Journal of Official Statistics* **14**(3), 237–253.

Chavas JP and Segerson K 1987 Stochastic specification and estimation of share equation systems. *Journal of Econometrics* **35**(2–3) 2–3.

Conniffe D and Eakins J 2003 Does the stochastic specification of the linear expenditure system matter? *Economic and Social Review* **31**(1), 23–32.

Desarbo W, Ramaswamy V and Chatterjee R 1995 Analyzing constant-sum multiple criterion data: A segment-level approach. *Journal of Marketing Research* **32**(2), 222–232.

Fry JM, Fry TRL and McLaren KR 1996 The stochastic specification of demand share equations: Restricting budget shares to the unit simplex. *Journal of Econometrics* **73**(2), 377–385.

Fry JM, Fry TRL and McLaren KR 2000 Compositional data analysis and zeros in micro data. *Applied Economics* **32**(8), 953–959.

Fry JM, Fry TRL, McLaren KR and Smith T 2001 Modelling zeroes in microdata. *Applied Economics* **33**(3), 383–392.

Fry TRL and Chong D 2005 A tale of two logits, compositional data analysis and zero observations. In *Proceedings of CoDaWork' 05, The 2nd Compositional Data Analysis Workshop* (ed. Mateu-Figueras G and Barceló-Vidal C). http://ima.udg.es/Activitats/CoDaWork05/. University of Girona, Girona (Spain).

Grunwald GK, Raftery AE and Guttorp P 1993 Time series of continuous proportions. *Journal of the Royal Statistical Society, Series B (Statistical Methodology)* **55**(1), 103–116.

Herfindahl O 1950 *Concentration in the US steel industry*. PhD thesis, Columbia University (USA).

Hirschman AO 1964 The paternity of an index. *American Economic Review* **54**(5), 761–762.

Honaker J, King G and Katz J 2002 A fast, easy, and efficient estimator for multiparty electoral data. *Political Analysis* **10**(1), 84–100.

Katz J and King G 1999 A statistical model for multiparty electoral data. *American Political Science Review* **93**(1), 15–32.

Kim I and Min I 2007 An alternative semiparametric estimate of the base-independence equivalence scale: An application to us consumer expenditure survey data. *Applied Economics* **39**(10), 1307–1314.

Koch S 2007 South African household expenditure shares: South African household data pitfalls. *Studies in Economics and Econometrics* **31**(1), 1–28.

Larrossa J 2003 A compositional statistical analysis of capital stock. In *Proceedings of CoDaWork' 03, The 1st Compositional Data Analysis Workshop* (ed. Thió-Henestrosa S and Martín-Fernández JA). http://ima.udg.es/Activitats/CoDaWork03/. University of Girona, Girona (Spain). CD-ROM.

Longford N and Pittau M 2006 Stability of household income in European countries in the 1990's. *Computational Statistics and Data Analysis* **51**(2), 1364–1383.

McKitrick R 1998 The econometric critique of computable general equilibrium modeling: The role of functional forms. *Economic Modelling* **15**(4), 543–573.

McLaren K 1996 Parsimonious autocorrelation corrections for singular demand systems. *Economics Letters* **53**(2), 115–121.

McLaren K, Fry J and Fry T 1995 A simple nested test of the almost ideal demand system. *Empirical Economics* **20**(1), 149–161.

Morana C 2007 Factor demand modelling: The theory and the practice. *Applied Mathematical Sciences* **1**(31), 1519–1549.

Pittau M and Zelli R 2006 Trends in income distribution in Italy: A non-parametric and a semi-parametric analysis. *Journal of Income Distribution* **15**, 90–118.

Pollak R and Wales T 1992 *Demand System Specification and Estimation*. Oxford University Press, New York, NY (USA). 218 p.

Quintana JM and West M 1988 Time series analysis of compositional data. In *Bayesian Statistics 3* (ed. Bernardo JM, DeGroot MH, Lindley DV and Smith AFM), Oxford University Press, New York, NY (USA). pp. 747–756.

Ronning G 1992 Share equations in econometrics: A story of repression, frustration and dead ends. *Statistical Papers* **33**(1), 307–334.

Silva D and Smith T 2001 Modelling compositional time series from repeated surveys. *Survey Methodology* **27**(2), 205–215.

Theil H 1969 A multinomial extension of the linear logit model. *International Economic Review* **10**(3), 251–259.

van der Ark LA, van der Heijden P and Sikkel D 1999 On the identifiability in the latent budget model. *Journal of Classification* **16**(1), 117–137.

Wang H, Liu Q, Mok H, Fu L and Tse W 2007 A hyperspherical transformation forecasting model for compositional data. *European Journal of Operational Research* **179**(2), 459–468.

Woodland A 1979 Stochastic specification and the estimation of share equations. *Journal of Econometrics* **10**(3), 361–383.

Part V

SOFTWARE

24

Exploratory analysis using CoDaPack 3D

Santiago Thió-Henestrosa and Josep Daunis-i-Estadella
Department of Computer Science and Applied Mathematics, University of Girona, Spain

24.1 CoDaPack 3D description

In 2001, the University of Girona research group on Compositional Data Analysis stated that the set of routines programmed by John Aitchison (Aitchison 1986) under the name of CODA (with Basic as the language) and NEWCODA (with Matlab 5) were difficult to use for scientists and other users with no programming skills. For this reason, the group began to translate the routines into a more accessible environment. Excel® was selected due to its wide distribution and its familiarity to almost everyone that uses computers. CoDaPack 3D is implemented as a set of menus from an Excel® datasheet and returns numerical results either on the same sheet or as graphical results in an independent window. Initially, the package contained the same routines created by Aitchison, but it quickly grew with other features. It is publicly available as freeware at http://ima.udg.edu/CoDaPack, and, to date, it has been proven to work under Excel® 2003 and Excel® 2007 with Windows XP® or Windows Vista®.

CoDaPack 3D is programmed in VisualBasic and OpenGL, but this is invisible to the user, who only sees menus inside Excel® (Figure 24.1). At this time, CoDaPack 3D has 9 menus with 40 routines:

- *Transformations*: routines that transform the data from the simplex to the real space or vice versa.

- *Operations*: routines that perform some operations on the data inside the simplex.

Compositional Data Analysis: Theory and Applications, First Edition. Edited by Vera Pawlowsky-Glahn and Antonella Buccianti.
© 2011 John Wiley & Sons, Ltd. Published 2011 by John Wiley & Sons, Ltd.

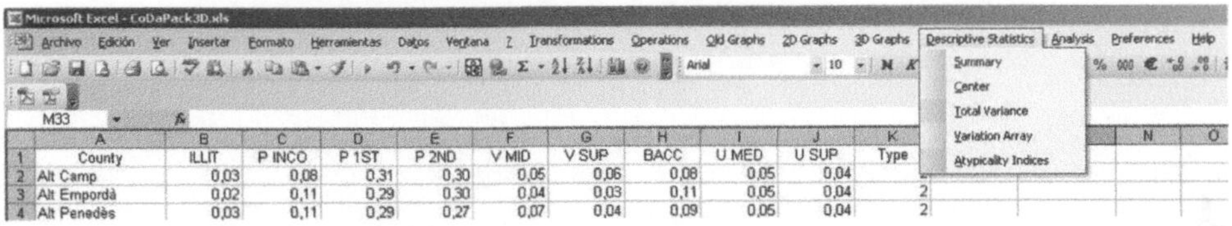

Figure 24.1 Menus of CoDaPack 3D integrated inside Excel® and a data structure. The Descriptive Statistics menu is extended showing its five routines.

- *Old Graphs*: routines from earlier versions of CoDaPack that generate low quality two-dimensional graphical representations in the simplex or in real space.

- *2D Graphs*: routines that generate two-dimensional graphical representations in the simplex or in real space.

- *3D Graphs*: routines that generate three-dimensional graphical representations in the simplex or in real space.

- *Descriptive Statistics*: routines that return descriptive statistics for a data set (Figure 24.1).

- *Analysis* that in the present version only performs the Logistic Normality test.

- *Preferences* that customise some general parameters of CoDaPack.

- *Help*.

To work with CoDaPack 3D, the data must be located on a worksheet in a copy of the file CoDaPack 3D.xls. The data are organised by columns where a column represents a part of a composition or a variable in the real space. CoDaPack 3D interprets the first row of every column as a label of this part or variable (Figure 24.1). For this reason, the first row should never contain data, just a label of the part or variable or may be left blank. The second and succeeding rows contain observations (cases).

The number of observations (cases) is defined by the occurrence of a blank row. If there are more data after this blank row, CoDaPack 3D ignores them. All parts or variables must have the same length, otherwise CoDaPack 3D halts execution with an error message. The same thing happens when cells containing characters or negative values are encountered. The maximum size of a Data Set in CoDaPack 3D is 200 parts and 6000 observations. Each routine selected from the pull-down menus appears as a new window (Figure 24.2). All CoDaPack 3D routines behave similarly. The left side of its window contains the *Select Columns* structure, and the middle of the window contains the *Inputs* structure and the *Store In (Initial Column)* box. Between the left and the middle parts, there are two arrows to pass information between them. When this window is opened, the *Select Columns* list contains the first row of each column of the Excel® sheet, and the letter that identifies the columns of any Excel® sheet. Also, if the routine has been executed before, this window contains in *Inputs, Store In (Initial Column)* and on the buttons the values used during the last execution. First, the user is required to select the parts to be used in the routine by marking a set of rows in the *Select Columns* list and then clicking on the arrow. Then, the name of the selected column appears in the middle, inside the *Inputs* structure. The user must repeat this operation to select each part to be used. If the user wants to unselect a part in the *Inputs* structure, the part inside the *Inputs* structure

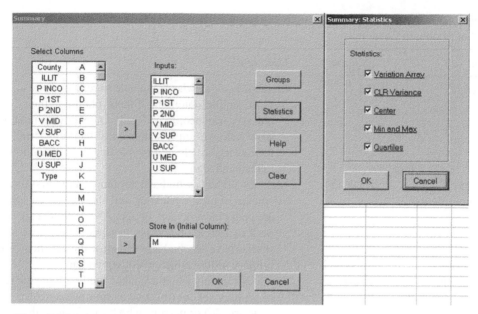

Figure 24.2   The main window of the Summary routine, and the Statistics window.

has to be marked and the arrow clicked, which now indicates the opposite direction. Double clicking on a single column name achieves the same outcome, namely, the part is selected or unselected as appropriate. Finally, the user should select the column where the numerical results are to be placed. On the right side, there are always some buttons: the *Clear* button clears all values in the *Inputs* and *Store In* columns. The *Help* button opens the CoDaPack 3D help directly for the routine. The *Cancel* button closes the window returning directly to Excel® without any execution. The *OK* button activates the routine with the parts and parameters selected. For some routines, there are specific buttons to specify some non-default values. For a more complete description of CoDaPack 3D, the reader is referred to Thió-Henestrosa and Martín-Fernández (2005, 2006) and Thió-Henestrosa *et al.* (2008), or to the manual included with the package.

24.2   Data set description

The data set used (Generalitat de Catalunya. Institut d'Estadística de Catalunya 2009) contains census data concerning the level of education in Catalonia, an autonomous community of Spain. Catalonia is situated in the north-east of Spain, and it is bounded by the Mediterranean Sea (from the south-west to the north-east) as well as France and Andorra (in the north). Its capital and most important city is Barcelona.

Catalonia is divided into 41 counties of different surface area (from 144.7 to 1784.1 km^2) and population (from 4171 to 2 213 701 inhabitants). For each of these counties, in the year 2001 the administration counted, in a census operation, how many people there were in each of the nine levels of instruction (Table 24.1), where the level of instruction refers to the highest educational level attained by the person in 2001. The 41 counties are classified into three groups: mainly rural (group 1), mainly urban (group 3) and part rural and part urban (group 2).

Table 24.1 A description of the variables used in the census data describing the educational level of people in Catalonia.

Title	Identifier	Description
Illiterate	ILLIT	Includes persons who are unable to read or write a short, simple statement of facts relating to their ordinary life, people who only know how to read or write a short, simple statement of facts concerning their everyday life and people that only know how to read and sign
Primary incomplete	P INCO	Includes students who have not finished the first five courses of primary school, and also all students that are under eleven years of age. Also included are persons who are no longer in school and whose studies have reached a level below the certificate of primary school education or equivalent
Primary first stage complete	P 1ST	Includes students who have taken the first five courses of primary school (must be a minimum age of 11 years old) or persons who have obtained the certificate of primary education or schooling
Primary second stage complete	P 2ND	Includes those who have completed eight courses of primary school (must be a minimum age of 14 years old) or who have obtained the level of primary graduate school. They also include people whose highest level of education is the old elementary high school. From the year 1996, this category includes persons who have obtained the secondary school certificate (must be a minimum age of 16 years old).
Vocational training middle grade or official	V MID	Includes persons who are in possession of the title of assistant technician, and, also, who studied to be an industrial officer
Vocational training superior grade or expertise	V SUP	Includes people who are in possession of the title of technical specialist, and, also, those with the level of industrial expertise
Baccalaureate	BACC	Includes people who have such studies as the highest degree
University medium grade	U MED	Includes persons who have received a university degree consisting of fewer than five courses
University superior grade	U SUP	Include persons who received a university degree consisting of five or more courses

24.3   Exploratory analysis

In every exploratory data analysis, the first step is to check for data errors, such as typing errors, unusual data values, etc. Usually, the way to do this checking is by means of the same tools of exploratory data analysis (numerical and graphical) described in this chapter. Compositional data techniques do not permit zero values because of the use of logarithms. In order to replace the zeros, CoDaPack 3D has a routine on the Operation menu. The data used are counts and the sum of the parts of the observations is not constant. This is not a problem in using CoDaPack 3D because in all routines the data are closed if necessary, that is, each part is divided by the sum of the parts for the observation.

24.3.1   Numerical analysis

The first approximation to the description of the data set is achieved by numerical results obtained with the Summary routine (Figure 24.2) that can be found in the Descriptive Statistics menu. This routine returns by default the Variation Array, the centred log-ratio (CLR) Variance, the Centre, the Min and Max and the quartiles. The centre (Figure 24.3) shows for each level of instruction the geometric mean of the percentages for the 41 counties. The main levels of instruction are Primary first and second stage completed, with centres of 0.298 and 0.277, respectively. Note that these percentages do not correspond to the percentage for the total population of Catalonia with this level of instruction because each observation pertains to a county and every county has a different population. Figure 24.3 demonstrates that the range, from the Max to the Min, of values of all parts is greater in those counties with a large mean.

One of the most informative statistics is the CLR variance that estimates the variability of each part. The Illiterate has the greatest variance, followed by University superior and Primary incomplete, while the Vocational training middle grade has the least variance. The

M	N	O	P	Q	R	S	T	U	V	W	X
Variation Array											
	ILLIT	P INCO	P 1ST	P 2ND	V MID	V SUP	BACC	U MED	U SUP		
ILLIT		0,110	0,141	0,108	0,132	0,170	0,162	0,170	0,186	**Variances**	
P INCO	-1,699		0,063	0,086	0,061	0,112	0,098	0,097	0,132		
P 1ST	-2,741	-1,042		0,028	0,046	0,073	0,096	0,082	0,142		
P 2ND	-2,671	-0,971	0,070		0,047	0,068	0,061	0,068	0,111		
V MID	-1,023	0,676	1,718	1,648		0,030	0,066	0,044	0,082		
V SUP	-0,932	0,767	1,809	1,739	0,091		0,070	0,037	0,074		
BACC	-1,606	0,093	1,135	1,065	-0,583	-0,674		0,021	0,026		
U MED	-1,056	0,644	1,685	1,615	-0,033	-0,124	0,550		0,019		
U SUP	-0,911	0,788	1,829	1,759	0,111	0,020	0,694	0,144			
	Means									**Tot var**	0,347
	ILLIT	P INCO	P 1ST	P 2ND	V MID	V SUP	BACC	U MED	U SUP		
CLR Variance	0,065	0,042	0,037	0,032	0,028	0,035	0,033	0,030	0,043		
Compositional Descriptive Statistics											
	ILLIT	P INCO	P 1ST	P 2ND	V MID	V SUP	BACC	U MED	U SUP		
Center	0,019	0,105	0,298	0,277	0,053	0,049	0,096	0,055	0,048		
Min	0,008	0,053	0,233	0,196	0,040	0,028	0,051	0,031	0,026		
Max	0,033	0,144	0,392	0,347	0,071	0,073	0,139	0,076	0,094		
Q25	0,016	0,094	0,268	0,261	0,048	0,043	0,083	0,048	0,039		
Median	0,021	0,104	0,289	0,276	0,052	0,048	0,095	0,053	0,044		
Q75	0,024	0,120	0,322	0,294	0,059	0,056	0,111	0,062	0,061		

Figure 24.3   Descriptive Statistics output.

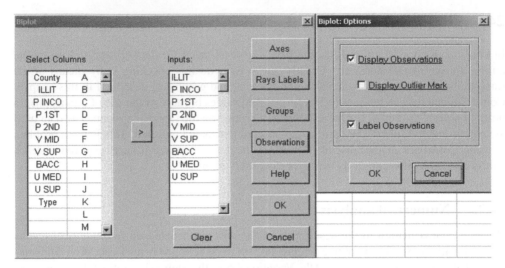

Figure 24.4   The Biplot menu.

CLR variance may be decomposed with the variances of the log-ratios that are estimated in the Variation Array. The sum of the variances of the log-ratios that include a part (in the numerator or denominator) is equal to four times the CLR variance of this part. The log-ratios with greatest variance are associated with the parts with the greatest CLR variance (Illiterate and University superior). It is noteworthy that the three log-ratios with the lowest CLR variance correspond to University superior versus University medium, University medium versus Baccalaureate and University superior versus Baccalaureate. Thus, there is low variability between these pairs throughout the 41 counties.

24.3.2   Biplot

The CLR variance may be illustrated by means of a compositional Biplot. The routine of CoDaPack 3D by default draws only the rays of a biplot, but it is also possible to display (Figure 24.4) observations, with or without labels. In our example (Figure 24.5) the biplot

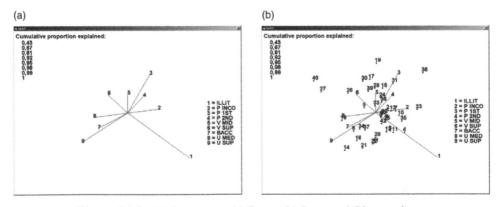

Figure 24.5   Biplot output. (a) Rays. (b) Rays and Observations.

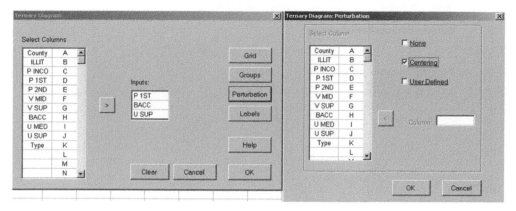

Figure 24.6   The Ternary Diagram menu with perturbation.

only represents 67% of the total variance, so not all the parts are represented well. For example the CLR variance of part 2, Primary incomplete, is similar to the CLR variance of part 9, University superior, but it is clearly shorter on the biplot. For this reason part 2 is not represented well by the biplot.

24.3.3   The ternary diagram

Compositional biplots have different rules of interpretation from standard biplots. These rules are described by Aitchison and Greenacre (2002) and Daunis-i-Estadella *et al.* (2006). By following one of these rules, it is possible to see a linear pattern between parts 3=P 1ST, 7=BACC and 9=U SUP. This three-part subcomposition could be drawn by means of a ternary diagram. The routine of CoDaPack 3D (Figure 24.6) that draws the ternary diagram permits centring the data by means of a perturbation (von Eynatten *et al.* 2002). Note that as only three parts are involved in a ternary diagram, such a diagram is limited to three variables at a time. Figure 24.7 shows ternary diagrams for the original and centred parts, and it is easy to see the linear pattern mentioned before.

(a)   (b)

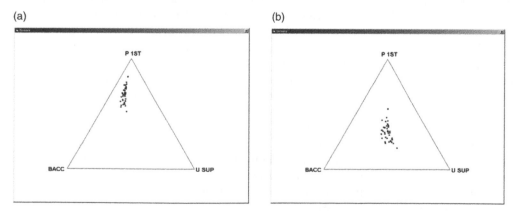

Figure 24.7   Ternary diagram output. (a) Without centring. (b) With centring.

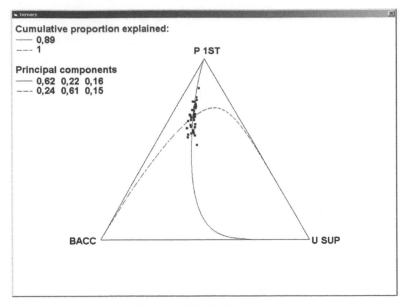

Figure 24.8 PCA output.

24.3.4 Principal component analysis

This linear pattern may be quantified with CoDaPack 3D by means of a Principal Component Analysis (PCA). Figure 24.8 shows this linear trend where the first component of PCA captures 89% of the total variability exhibited by the three parts.

24.3.5 Balance-dendrogram

Another powerful tool used to describe data is the Balance-Dendrogram (Egozcue and Pawlowsky-Glahn 2005a,b; Thió-Henestrosa *et al.* 2008). This tool describes the data set by means of a sequential binary partition of the parts. Figure 24.9 shows the main menu for the Balance-Dendrogram routine where the user chooses the parts to be employed, where to store the results, and the method used to obtain a partition. The partition could be obtained by default, read from a spreadsheet or defined by the user. Figure 24.10 shows the menu of CoDaPack 3D used to define a partition. In our example, the first step of the partition consists of separating the Illiterate from the other parts. The second step works on the nonilliterate parts and separates parts corresponding to primary studies from parts that involve higher levels of education. Inside this group, step 5 separates those having undergone vocational studies from those having baccalaureate and university degrees. Figure 24.11 shows the resulting dendrogram. As it has been shown previously, the main portion of the total variance is due to the Illiterate category. This fact could be shown on the dendrogram, where the sum of black vertical bars is proportional to the total variability. In fact, each vertical line corresponds to the variability of each balance and a subsequent cluster of parts. The boxplot marked as A in Figure 24.11, corresponding to the first partition (Illiterates versus the other parts), is not symmetric, and it shows a large skew to the right. Also, the boxplot marked as B, which corresponds to the partition between the vocational parts (V MID and V SUP) against the

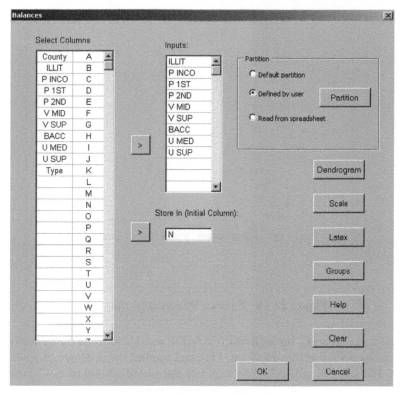

Figure 24.9   The Balance-Dendrogram menu.

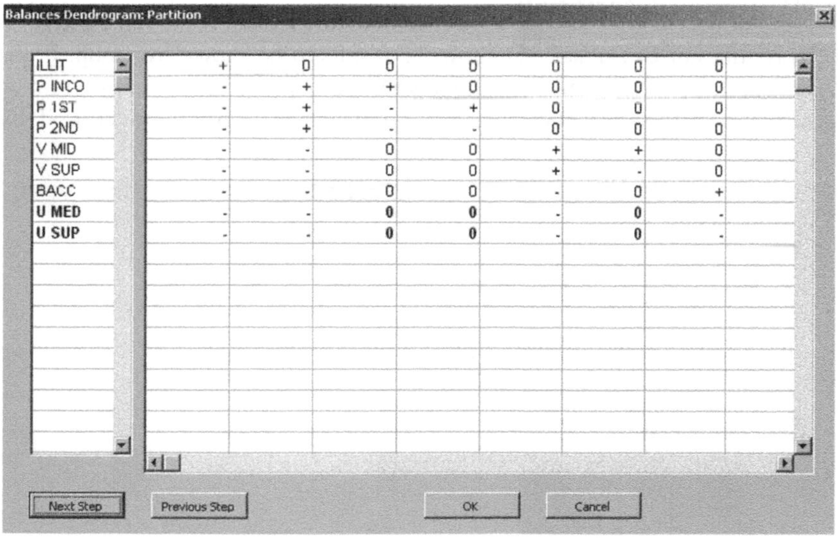

Balances Dendrogram: Partition

ILLIT	+	0	0	0	0	0	0
P INCO	-	+	+	0	0	0	0
P 1ST	-	+	-	+	0	0	0
P 2ND	-	+	-	-	0	0	0
V MID	-	-	0	0	+	+	0
V SUP	-	-	0	0	+	-	0
BACC	-	-	0	0	-	0	+
U MED	-	-	**0**	**0**	-	**0**	-
U SUP	-	-	**0**	**0**	-	**0**	-

Next Step   Previous Step   OK   Cancel

Figure 24.10   The Definition of a Partition form.

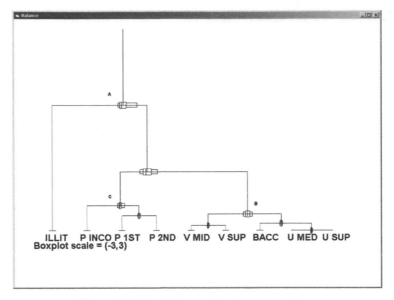

Figure 24.11 Balance-Dendrogram output.

baccalaureate and university parts (BACC, U MED and U SUP), is comparatively small relative to the variability of this partition. This fact implies the presence of some atypical observation. The same thing happens on step 3 of the partition, marked as C. Furthermore, the Balance-Dendrogram routine calculates the Isometric log-ratio coordinates that enable the user to use all standard statistical methods.

24.3.6   By groups description

Some of the routines of CoDaPack 3D allow the results to be shown by group. In the case of graphical routines, it is possible to distinguish the groups by means of different colours or symbols. Figure 24.12 shows how the groups are managed: if groups are desired, the user

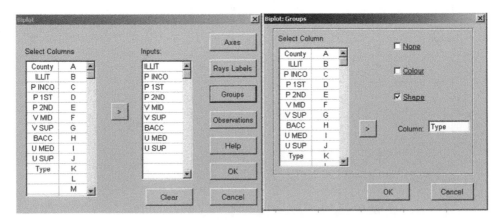

Figure 24.12 The menu to define the groups on a Biplot.

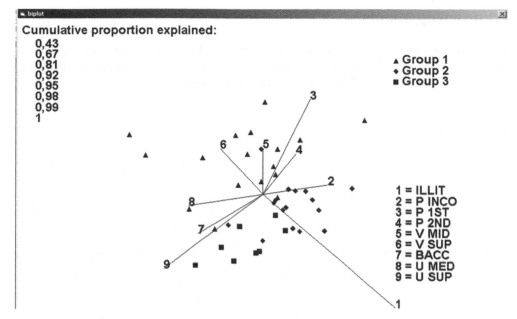

Figure 24.13   A Biplot displaying observations by groups.

should choose between Colour and Shape and should also indicate the column that contains the group variable. This column must be coded numerically from 1 to 10 (the maximum number of groups that can be accommodated). Figure 24.13 shows a biplot that displays the observations with different shapes depending on the group to which they belong. In this example, group 1 corresponds mainly to rural counties, group 3 to urban counties and group 2 to counties that are part rural and part urban. It is possible to see that observations belonging to group 3 appear near ray 9 [corresponding to CLR(U SUP)], while the observation of group 1 appears on the upper side of the biplot and is associated with rays 3, 4, 5 and 6 (corresponding to CLR of P 1ST, P 2ND, V MID and V SUP).

24.4   Summary and conclusions

The objective of this chapter is to demonstrate how CoDaPack 3D can be used for the exploratory data analysis of compositional data. Compositional data analysis, following Aitchison's approach, is mainly based on the study and interpretation of log-ratios. CoDaPack 3D provides a user-friendly freeware environment that performs most of the techniques of this approach. From an exploratory viewpoint, CoDaPack 3D provides numerical results through the Descriptive Statistics menu and graphical output with Biplot, Ternary Diagram, Principal Component Analysis and Balance-Dendrogram routines. Some of these routines allow grouping of the results by means of a previous classification, in which case the groups are represented by different colours or symbols. Other capabilities of CoDaPack 3D are three-dimensional graph production, transformations of the data to and from the real space, and operations inside the simplex such as centring, perturbation and amalgamation, among others.

Acknowledgements

This research has been supported by the Spanish Ministry of Science and Innovation (projects CSD2006-00032 and MTM2009-13272) and by the Agència de Gestió d'Ajuts Universitaris i de Recerca of the Generalitat de Catalunya (Ref. 2009SGR424).

References

Aitchison J 1986 *The Statistical Analysis of Compositional Data.* Monographs on Statistics and Applied Probability. Chapman and Hall Ltd (reprinted 2003 with additional material by The Blackburn Press), London (UK). 416 p.

Aitchison J and Greenacre M 2002 Biplots for compositional data. *Applied Statistics* **51**(4), 375–392.

Daunis-i-Estadella J, Barceló-Vidal J and Buccianti A 2006 Exploratory compositional data analysis. In *Compositional Data Analysis in the Geosciences: From Theory to Practice.* (ed. Buccianti A, Mateu-Figueras G and Pawlowsky-Glahn). Geological Society, London (UK). pp. 161–174.

Egozcue JJ and Pawlowsky-Glahn V 2005a Coda-dendrogram: a new exploratory tool. In *Proceedings of CoDaWork'05, The 2nd Compositional Data Analysis Workshop* (ed. Mateu-Figueras G and Barceló-Vidal C). University of Girona, http://ima.udg.es/Activitats/CoDaWork05/Girona (Spain).

Egozcue JJ and Pawlowsky-Glahn V 2005b Groups of parts and their balances in compositional data analysis. *Mathematical Geology* **37**(7), 795–828.

Generalitat de Catalunya. Institut d'Estadística de Catalunya 2009 Anuari estadístic de Catalunya 2009. http://www.idescat.cat/pub/?id=aecandn=765andlang=en.

Thió-Henestrosa S and Martín-Fernández JA 2005 Dealing with compositional data: the freeware codapack. *Mathematical Geology* **37**(7), 773–793.

Thió-Henestrosa S and Martín-Fernández JA 2006 Detailed guide to codapack: a freeware compositional software. In *Compositional Data Analysis in the Geosciences: From Theory to Practice* (ed. Buccianti A, Mateu-Figueras G and Pawlowsky-Glahn). Geological Society, London (UK). pp. 101–118.

Thió-Henestrosa S, Egozcue JJ, Pawlowsky-Glahn V, Kovács LO and Kovács G 2008 Balance-dendrogram. a new routine of codapack. *Computer and Geosciences* **34**(12), 1682–1696.

von Eynatten H, Pawlowsky-Glahn V and Egozcue JJ 2002 Understanding perturbation on the simplex: a simple method to better visualise and interpret compositional data in ternary diagrams. *Mathematical Geology* **34**(3), 249–257.

robCompositions: an R-package for robust statistical analysis of compositional data

Matthias Templ[1,2], Karel Hron[3] and Peter Filzmoser[1]
[1]*Department of Statistics and Probability Theory, Vienna University of Technology, Austria*
[2]*Methods Unit, Statistics Austria, Vienna, Austria*
[3]*Department of Mathematical Analysis and Applications of Mathematics, Palacký University, Czech Republic*

25.1 General information on the R-package `robCompositions`

The programming language and software environment R (R Development Core Team 2010) is nowadays one of the most widely used and most popular software tools for statistics and data analysis. It is free and open-source (GPL 2) and it can be downloaded for all computer platforms from the comprehensive R archive network (`http://cran.r-project.org`). It is enhanceable via *packages* which consist of code and structured standard documentation explaining the input and output arguments (and more) of each function including code application examples.

Two contributed packages for compositional data analysis are currently available with R version 2.10.1.: the package `compositions` (van den Boogaart *et al.* 2008), and the package `robCompositions` (Templ *et al.* 2010). While `compositions` is devoted in particular

Compositional Data Analysis: Theory and Applications, First Edition. Edited by Vera Pawlowsky-Glahn and Antonella Buccianti.
© 2011 John Wiley & Sons, Ltd. Published 2011 by John Wiley & Sons, Ltd.

to classical statistical procedures, `robCompositions` provides tools for a robust statistical analysis of compositional data together with corresponding graphical tools. A comprehensive overview is available using the command in R shown in Listing 25.1.

```
R> help(package ='robCompositions')
```

Listing 25.1 Information on the functions and data included in the R-package robCompositions.

The prefix `R>` in the code listings means that a command is applied in R. Text after the symbol # denotes comments.

25.1.1   Data sets included in the package

To explicitly display the data sets available one can use the commands shown in Listing 25.2 (the package has to be loaded first). Note that we do not intend to provide a comprehensive overview of the data sets included in the package. They are introduced only briefly in order to give the reader an idea about their meaning. Some of these data sets are used in papers on robust methods, and they can now be used to check the results presented there. All data sets from Aitchison (1986) on compositional data are available in the package `compositions`.

```
R> require(robCompostions)        # loads the package
R> data(package='robCompositions') # lists the data in the package

Data sets in package robCompositions:
arcticLake              Arctic lake sediment data
coffee                  Coffee data
expenditures            Household expenditures data
expendituresEU          Mean consumption expenditures data
haplogroups             Haplogroups data
machineOperators        Machine operators data set
phd                     PhD students in the EU
skyeLavas               Aphyric Skye lavas data
```

Listing 25.2 List of data sets in the R-package robCompositions.

Listing 25.2 outlines the names of the data sets available in the package robCompositions. By typing `help(rcdata)` in R a structured help file of the data set `rcdata` (or the named function) pops up. For example, `help(expendituresEU)` shows the help file of the `expendituresEU` data set. In the following we give a short description of the data included in the package:

- arcticLake: Sand, silt, clay compositions of 39 sediment samples at different water depths in an Arctic lake (Aitchison 1986, p. 359).

- coffee: A subset of three compositional parts of 27 commercially available coffee samples of different origins (Korhoňová *et al.* 2009).

- expenditures: The expenditure data set (Aitchison 1986, p. 395) contains household expenditures (in former Hong Kong dollars) on five commodity groups of 20 single men.

- expendituresEU: Mean consumption expenditures (in Euro) of households on 12 domestic year costs in all 27 member states of the European Union (Eurostat 2008).

- haplogroups: Distribution of European Y-chromosome DNA (Y-DNA) of 12 haplogroups by region in percentages (Eupedia 2010).

- machineOperators: Compositions of 8-h shifts of 27 machine operators, spent on four classified activities (Aitchison 1986, p. 382).

- phd: Classification of PhD students by different kinds of studies (in %) (Eurostat 2009).

- skyeLavas: AFM compositions of 23 Aphyric Skye lavas (Aitchison 1986, p. 360).

25.1.2   Design principles

Almost all functions in the package robCompositions make use of function overloading and the method dispatch of R. Each function returns an object of a certain class. Print, summary and plot methods are implemented for objects of these classes. The method dispatch of R has the advantage that the user may apply simple and standardized functions on the result objects, such as print, summary and plot. The corresponding method for printing and plotting is then selected automatically depending on the class of the object. This principle is shown below in practical applications.

The package robCompositions currently depends on five other packages, namely the packages utils, robustbase, rrcov, car and MASS from where few functions are imported. robCompositions makes use of the name space management system for R-packages, which does not allow to redefine or overwrite functions that are defined in robCompositions. This is desirable because the user can be sure that if another function (e.g. from a different package) with the same name as used in robCompositions is loaded beforehand, always the function from robCompositions will be used.

Compiled code is used for computationally intensive procedures such as calculating Aitchison distances between matrices. The compiled code has been made accessible by the R/C interface. This allows much faster computations than using interpreted R code.

25.2   Expressing compositional data in coordinates

Three different possibilities to express compositions in coordinates are implemented in the package robCompositions: the additive log-ratio transformation (alr), the centred log-ratio transformation (clr), both defined in Aitchison (1986), and the isometric log-ratio transformation (ilr) (Egozcue *et al.* 2003). Although these transformations are already implemented in the package compositions, our implementation differs because variable names and absolute values are preserved.

Note that in this package only one specific ilr transformation according to Hron *et al.* (2010) is available. However, this is fully sufficient for all methods described in this chapter and in related papers about robust compositional data analysis.

To show the usage of the corresponding R functions, the expenditures data set (Aitchison 1986) is loaded, and the first three observations are shown in Listing 25.3.

```
R>   data(expenditures)          # loads the data
R>   head(expenditures, 3)       # prints the first three observations
     housing foodstuff alcohol tobacco other
1        640       328     147     169   196
2       1800       484     515    2291   912
3       2085       445     725    8373  1732
```

Listing 25.3 Display of the first three observations of the expenditures data set.

The mentioned transformations can then be applied by using Listing 25.4.

```
R> alr(expenditures) # by default, the last col. is chosen as ratioing variable
R> clr(expenditures)
R> ilr(expenditures)
```

Listing 25.4 Applying log-ratio transformations with the R-package robCompositions.

Also the inverse transformations are implemented. Their use is shown in Listing 25.5 for the alr transformation.

```
R> a <- alr(expenditures, 3)      # choose 3rd column as ratioing variable
R> class(a)                       # display the class of object "a"
[1] "alr"                         # the class of object "a"
R> x <- invalr (a)                # inverse alr for object "a"
R> expenditures[1,1]/expenditures[1,2] # ratio of the first 2 parts,
[1] 1.951220                      # first composition
R> x[1,1]/x[1,2]                  # check
[1] 1.951220
R> head(x, 3)                     # check
   housing foodstuff alcohol tobacco other
1      640       328     147     169   196
2     1800       484     515    2291   912
3     2085       445     725    8373  1732
```

Listing 25.5 Transformation and back-transformation.

Listing 25.5 shows some special and user-friendly features which are provided by the package robCompositions. The first thing to note is that the invalr() needs no information about the chosen ratioing variable, needed to generate the object a, as well as no information about the original names of the expenditures data (expenditures has 5 column names, object a only 4) to exactly reproduce the original data set (see Listings 25.4 and 25.5). Also the absolute values are preserved after applying the inverse transformation. Note that the functions alr() and invalr() allow to set specific parameters so that only the transformed data is returned without additional information. This is especially useful for simulations because of a reduction of computational costs.

Note that the data used can be expressed in percentages by using the function shown in Listing 25.6 that normalizes the expenditures data to a chosen sum (the default is 1).

```
R> ConstSum (expenditures, const=100)
```

Listing 25.6   Data mapped to a constant sum.

25.3   Multivariate statistical methods for compositional data containing outliers

In the following, the application of some popular (robust) methods for a statistical analysis of compositions are described. The first two methods are devoted to outlier detection and principal component analysis, where the following functions are implemented:

outCoDa: used for outlier detection;

print.outCoDa: print method for objects of class 'outCoDa';

plot.outCoDa: plot method for objects of class 'outCoDa';

pcaCoDa: apply (robust) principal component analysis;

print.pcaCoDa: print method for objects of class 'pcaCoDa';

plot.pcaCoda: plot method for objects of class 'pcaCoDa' (compositional biplot).

In addition to code explanations, the functions are illustrated by the expendituresEU data set (Eurostat 2008).

25.3.1   Multivariate outlier detection

Potential multivariate data outliers are identified by using robust Mahalanobis distances with the function outCoDa(). The Mahalanobis distance is defined for regular $(D - 1)$-dimensional data $\mathbf{x}_i$, $i = 1, \ldots, n$, as

$$\text{MD}(\mathbf{x}_i) = \left[(\mathbf{x}_i - T)^\top C^{-1} (\mathbf{x}_i - T) \right]^{1/2}$$

and represents a popular tool for outlier detection (Maronna *et al.* 2006; Filzmoser and Hron 2008). Here, the estimated covariance structure is used to assign a distance to each observation indicating how far the observation is from the centre of the data cloud with respect to the covariance structure. The choice of the location estimator T and the scatter estimator C is crucial. In the case of multivariate normal distribution, the (squared) Mahalanobis distances based on the classical estimators arithmetic mean and sample covariance matrix follow approximately a χ^2 distribution with $D - 1$ degrees of freedom. In the presence of outliers, however, only robust estimators of T and C lead to a Mahalanobis distance being reliable for outlier detection. A popular choice for robust location and covariance estimation is the Minimum Covariance Determinant (MCD) estimator (Rousseeuw and van Driessen 1999). Usually, also in this case a χ^2 distribution with $D - 1$ degrees of freedom is used as an approximate distribution, and a certain quantile, like the quantile 0.975, is used as a cut-off value for outlier identification: observations with larger (squared) robust Mahalanobis distance are considered as potential outliers.

The procedure described above follows the concept of Filzmoser and Hron (2008), and the corresponding function outCoDa() uses the isometric log-ratio transformed compositions to search for outliers in real space. The function includes four arguments: the data x, the quantile quantile of the χ^2_{D-1} distribution, the method used (either 'standard' for classical estimation or 'robust' for MCD estimation – the latter is the default method), and h as the size of the subsets used for the MCD estimator (Filzmoser and Hron 2008). The latter three function arguments have sensible defaults, but they can be set by the user as well.

Outlier detection for the example data set expendituresEU is done with the commands shown in Listing 25.7. Since the parameters for quantile and method are the default parameters, the code in Listing 25.7 could be shortened to outCoDa (expendituresEU).

```
R> data(expendituresEU)
R> outRob <- outCoDa(expendituresEU, quantile=0.975, method ="robust")
R> outRob
--------------------
[1] "8 out of 27 observations are detected as outliers."
--------------------
```

Listing 25.7 Outlier detection using robust methods.

As mentioned above, almost all functions in the package robCompositions make use of function overloading and the method dispatch of R, which now can be easily illustrated considering the comments in Listing 25.7. The function outCoDa() returns an object of class 'outdect'. By typing the name of the result object (here: outRob) in the R console, the corresponding print method [print.outCoDa()] is selected automatically. Within our example, the print result simply reports that 8 out of 27 observations are detected as outliers.

Generally, the resulting Mahalanobis distances are stored in an object of class 'outCoDa' (mahalDist), displayed in Listing 25.8, where a logical vector indicating outliers and non-outliers (outlierIndex) is shown as well.

```
R> outRob $mahalDist
 [1]   1.672683  2.374380  2.622180   1.942029 2.109558  2.195739  2.357287
 [8]   2.117132  7.134956  2.059403 10 .310637 1.896691  2.446645  2.191185
[15]   8.872333  2.506627 14 .812581  2.317635 1.623112 11 .410675 2.275384
[22] 16 .414259 2.247400  1.487761   1.798884 23.614927  9.142826
R> outRob $outlierIndex
 [1] FALSE FALSE FALSE FALSE FALSE FALSE FALSE FALSE  TRUE FALSE  TRUE FALSE
[13] FALSE FALSE  TRUE FALSE  TRUE FALSE FALSE  TRUE FALSE  TRUE FALSE FALSE
[25] FALSE  TRUE  TRUE
R> which(outRob $outlierIndex) # index of outlying compositions
 [1]  9 11 15 17 20 22 26 27
```

Listing 25.8 Accessing list elements: robust Mahalanobis distances and index of outliers.

Note that by typing names (object) into the R console, information about the list of objects is shown, with object equal to outRob in our example shown in Listings 25.7 and 25.8.

The resulting Mahalanobis distances (ordered according to the index of the observations), together with the corresponding cut-off value, can be graphically presented using the plot function plot.outCoDa(). R first searches for objects of class 'outCoDa' if a function called `plot.outCoDa()` exists. Therefore, the user again only needs to know the name of the generic function, `plot()`, which is simple to keep in mind. The outliers (observations 9, 11, 15, 17, 20, 22, 26 and 27) are marked with the symbol '+' by default (see Listing 25.9).

```
R> outStand <- outCoDa (expendituresEU, quantile=0.975, method="standard")
R> plot(outStand) # plots classical estimates of Mahalanobis distance
R> plot(outRob)   # plots robust estimates of Mahalanobis distance
```

Listing 25.9   Diagnostic of objects from class 'outCoDa'

Using standard (classical) estimates for outlier detection (here stored in object `out-Stand`, see Listing 25.9), all the outliers are masked, and no observation is detected as outlier (see Listing 25.10). Two observations are placed very close to the cut-off value which equals the 97.5% quantile of the χ_d^2 distribution ($d = 1$ in our example). Observations with (squared) Mahalanobis distance above such a cut-off value are considered as potential outliers (Rousseeuw and Van Zomeren 1990). Therefore, using standard estimates for outlier detection, which are themselves driven from outliers, may report that no artifacts are included in the data. However, observations 9, 11, 15, 17, 20, 22, 26 and 27 may highly influence further estimations based on standard estimates.

```
R> outStand
-------------------
[1] "0 out of 27 observations are detected as outlier."
-------------------
```

Listing 25.10   Outlier masking when using standard estimators.

Figure 25.1 can be easily produced by the commands shown in Listing 25.9.

25.3.2   Principal component analysis and the robust compositional biplot

The function `pcaCoDa()` computes scores and loadings in the clr space. By setting the function parameter `method` one can choose between standard principal component analysis ('standard') or its robust counterpart ('robust' – the default method). Function `plot.pcaCoDa()` then creates a (robust) compositional biplot (see Listing 25.11) according to Aitchison and Greenacre (2002) and Filzmoser et al. (2009a). The results obtained by standard and robust estimates are shown in Figure 25.2.

```
R> resStand <- pcaCoDa (expendituresEU, method="standard")
R> resRob <- pcaCoDa (expendituresEU)
R> plot(resStand); plot(resRob)
```

Listing 25.11   Robust principal component analysis and the robust compositional biplot.

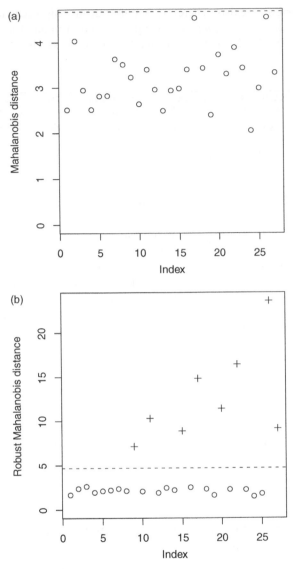

Figure 25.1 Graphical display of objects from class 'outCoDa'.

Outliers can play an important role whenever classical estimates are used. For PCA, outliers may affect the estimates of the correlation. This is no longer the case when robust PCA is applied to the transformed data [Figure 25.2(b)], because the correlations are estimated robustly. In this example, the observations (approximated by the scores) and the correlations between the variables do not differ drastically for the first two principal components. However, the classical (standard) version of the compositional biplot [Figure 25.2(a)] would indicate a different relation for example between the variables *food* and *health*.

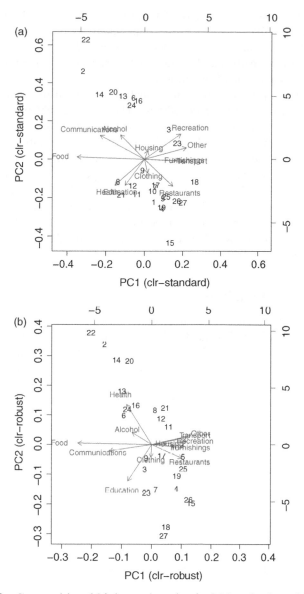

Figure 25.2 Compositional biplots using classical (a) and robust (b) estimates.

Although the robust compositional biplot [Figure 25.2(b)] (as well as the classical one) represents only rank-2 approximations of the multivariate data structure, there are also some artifacts according to outlier detection visible, e.g. positions of observations 15 (Luxembourg) and 18 (Netherlands). The top-to-bottom ordering of observations determines economic positions of the EU countries, starting with 22 (Romania) and 2 (Bulgaria). The direction of arrows shows that in poorer countries the variable *alcohol* plays an important role in the overall expenditures, contrary to higher expenditures on *clothing* and *restaurants* in countries with higher economic position. The interpretation of the compositional biplot according to Aitchison and Greenacre (2002) enables also to conclude that there are quite strong relations (stable ratios), e.g. between the variables (expenditures on) *education*, *clothing* and *housing*, *furnishings*, respectively. In particular the latter relation reflects a similar proportion between the corresponding costs in all the EU countries.

The same idea as for principal component analysis was used for robust factor analysis for compositional data (Filzmoser *et al.* 2009b). The package robCompositions provides the function `pfa()` for standard and robust factor analysis. The utility function `factanal.fit.principal1()` which is called by `pfa()` internally is doing the actual estimation with the parameters defined via `pfa()`. The main difference to usual implementations of factor analysis is that uniquenesses are no longer of diagonal form (for details, see Filzmoser *et al.* 2009b). This kind of factor analysis is designed for centred log-ratio transformed compositional data. A robust version where the covariance matrix is estimated from the isometric log-ratio transformed compositions could be chosen [for details, see the examples in the corresponding R-help file of the package robCompositions (Templ *et al.* 2010) or in Filzmoser *et al.* (2009b)].

25.3.3 Discriminant analysis

To demonstrate robust discriminant analysis with the package robCompositions, the `coffee` data set (Korhoňová *et al.* 2009) is loaded and robust Fisher discriminant analysis (Filzmoser *et al.* 2009c) is applied (Listing 25.12). Three compositional parts are available in the data set: 1-hydroxy-2-propanone, 2-methylpyrazine and 5-methylfurfural. As a result, two natural groups are formed by Arabica coffee (16 samples) and various blends of the Arabica and Robusta coffee (11 samples). The information about the sort is saved in the fourth variable (`sort`) of this data set (Listing 25.12).

```
R> data(coffee)
R> head(coffee, 2)              # display the first 2 compositions
   Metpyr 5-Met furfu    sort
1 12.50 8.51 6.2 arabica
2  5.33 11.80 17.8 arabica
R> dres <- daFisher (coffee [,1:3], grp=coffee [,4], method
="robust", plotScore=TRUE)
```

Listing 25.12   Discriminant analysis with standard and robust methods.

In the last line of the code shown in Listing 25.12 (robust) discriminant analysis is applied. By setting the function argument `plotScore` to `TRUE`, a plot showing the discriminant scores will be generated (see Figure 25.3). Note that in the two-group case, the second direction does not include information on the group separation (Filzmoser *et al.* 2009c).

The discriminant rules rely on the assumption of normal distribution of compositions (Mateu-Figueras and Pawlowsky-Glahn 2008). Therefore, it is necessary to have a possibility of a proper multivariate normality test. In the package, a battery of Anderson–Darling tests according to Aitchison (1986) for data in orthonormal coordinates is available via the function `adtestWrapper()`. Robust multivariate methods such as robust discriminant analysis are allowing for certain deviations from normality, and only the main bulk of the data needs

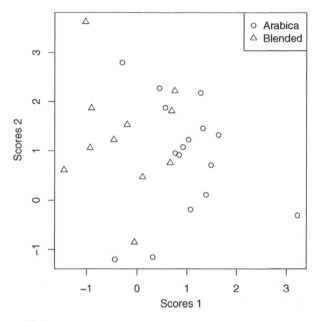

Figure 25.3   Fisher discriminant scores of the two sorts of coffee.

to follow multivariate normality. Therefore, a robust version based on the MCD estimator (Rousseeuw and van Driessen 1999) of the Anderson–Darling test for multivariate normality is implemented in the function `adtestWrapper()` [more information can be found in the manual by Templ *et al.* (2010)].

25.4   Robust imputation of missing values

Real-world compositional data sets often include missing values that need to be imputed before applying statistical methods. Since the relevant information of compositional data is contained in the ratios between the parts, special care for the imputation methodology is needed.

Two new imputation algorithms for estimating missing values in compositional data are introduced in Hron *et al.* (2010). The first proposal uses the *k*-nearest neighbour (*k*-nn) procedure based on the Aitchison distance. Hron *et al.* (2010) outlined that it is important to adjust the estimated missing values to the overall size of the compositional parts of the neighbours. As a second proposal, an iterative model-based imputation technique is introduced. This method initializes the missing values by the *k*-nearest neighbour procedure. The method is based on iterative and sequential regressions, hereby accounting for the whole multivariate data information. Sequentially means that in each step one variable is used as response while the other variables are used as predictors. Before a regression is applied, the data are expressed in orthonormal coordinates (ilr-transformed data) using a special basis in each step [for details, see Hron *et al.* (2010)]. The whole procedure is continued until the imputed values stabilize or a maximum number of iterations is reached.

```
R> expenditures[1,3]
147
R> expenditures[1,3] <- NA                  # set one value to NA
R> imputed1 <- impKNNa (expenditures)       # imputes the missing values using knn
R> imputed1
--------------------------------------
[1] "1 missing value was imputed"
--------------------------------------
R> imputed $xImp[1,3]
152 .1033
R> imputed2 <- impCoda (expenditures, method="ltsReg") # imputes the missing values
R> imputed2                                 # using model -based imputation

--------------------------------------
[1] "1 missing value was imputed"
[1] "2 iterations were needed"
[1] "the last change was 0"
--------------------------------------
R> imputed2 $xImp [1,3]
0.1016139
R> adjust(imputed2)$xImp [1,3]              # preserves absolute values

150 .7718
```

Listing 25.13   *k*-nn and model-based imputation.

In Listing 25.13 one value of the expenditures data is set to missing to demonstrate the imputation methods. The original value is 147 (see Listing 25.2) whereas the values imputed with *k*-nn and model-based imputation are 152.1033 and 150.7718, respectively.

Note that the imputation method from Palarea-Albaladejo and Martín-Fernández (2008), which is based on EM-based regression using the alr transformations, is reimplemented as well in the function `alrEM()`.

Also diagnostics for imputed data can be performed where the goal is to visualize the imputed values in order to graphically evaluate the quality of the imputation. The package provides three diagnostic plot methods. In Listing 25.14 the commands to create a multiple scatterplot and a ternary diagram are shown (Aitchison 1986).

```
R> plot(imputed2, which=1)
R> plot(imputed2, which=3, seg1=FALSE)
warning:
In plot.imp(xi, which = 3, seg1 = FALSE) :
ternary diagram is only visible for 3 variables, you have 5
only the first three variables are selected for plotting
```

Listing 25.14   Diagnostic plots for imputed data.

Figure 25.4 shows the results obtained by the code in Listing 25.14. The multiple scatterplot in Figure 25.4(a) shows pairwise bivariate scatterplots of the original compositions, where it is easy to see that the imputed data point fits well to the main part of the data, where the imputed values are highlighted by the symbol +. Note that this plot can also be generated for instance for data in coordinates, taking care of the compositional nature of the data.

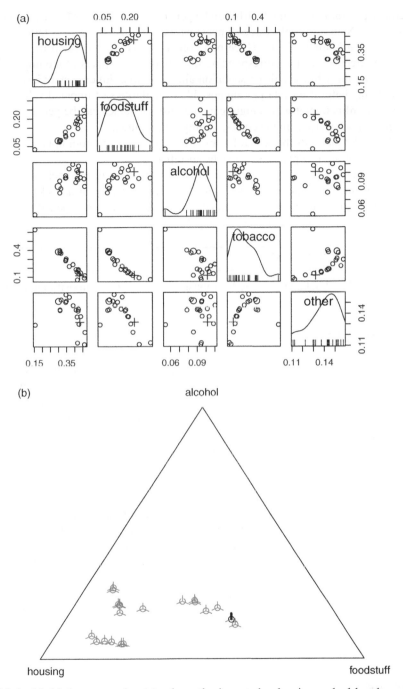

Figure 25.4 Multiple scatterplot (a) where the imputed value is marked by the symbol +
and ternary diagram (b) (with thick-lined spike for the missing value) for graphical evaluation
of the imputation results.

The ternary diagram in Figure 25.4(b) can be generated by `plot(xImp2, which=3, seg1=FALSE)`. Naturally, only three dimensions can be displayed in a ternary diagram (see the warning in Listing 25.14). Three-part compositions are presented by three spikes, pointing in the directions of the corresponding three parts. The spikes of the imputed values are highlighted by thick lines. This presentation allows a three-dimensional view of the data to be gained, being helpful for interpreting possible irregularities of imputed values.

A parallel coordinate plot which highlights the missing values can be generated with the function argument `which=2` in the corresponding plot method.

For all these plot methods the colour for highlighting imped and nonimputed values, their symbols and other graphical parameters can be specified.

Note that a package *vignette* for imputation is also included in the package. A package vignette is a document explaining parts or all of the functionality of a package in a more informal way than the strict format of reference help pages. It can be opened in the default pdf viewer of the operating system using Listing 25.15.

```
R> vignette ("imputation")
```

Listing 25.15   Package vignette for robust imputation of compositional data.

25.5   Summary

The design principles and the usage of the R-package robCompositions were shown using practical data examples. The examples are intended as a guidance for the user in order to successfully apply the functions included in the package robCompositions to their own data sets.

Compositional data, such as expenditures, income components, tax components, chemical concentrations, etc. virtually always contain outliers in the simplex. The robust methods included in the package fit the main part of the data minimizing the effect of data outliers and measurement errors. Most of the functions for multivariate analysis have an argument where classical (standard) methods can be compared with their robust counterparts. This allows possible differences to be seen which are due to outlying observations.

The package provides additional functionality for special tasks which were not shown in this chapter. For example, distances between compositions (or matrices) are provided by the function `aDist()`, where the function internally calls C-code by using the R-C interface. This is important whenever time complexity plays a role, i.e. when working with moderate or large data sets or when running simulations. Some other useful functions for a robust analysis of compositional data are provided, for example, for the estimation of a variation matrix using robust methods. The package will be extended with further functionality also in the future.

The package is under GPL 2. Therefore it can be used for free and the code is open source, whereas the intellectual rights on the code are preserved by this license.

References

Aitchison J 1986 *The Statistical Analysis of Compositional Data*. Monographs on Statistics and Applied Probability. Chapman and Hall Ltd (reprinted 2003 with additional material by The Blackburn Press), London (UK). 416 p.

Aitchison J and Greenacre M 2002 Biplots for compositional data. *Journal of the Royal Statistical Society, Series C (Applied Statistics)* **51**(4), 375–392.

Egozcue JJ, Pawlowsky-Glahn V, Mateu-Figueras G and Barceló-Vidal C 2003 Isometric logratio transformations for compositional data analysis. *Mathematical Geology* **35**(3), 279–300.

Eupedia 2010 Distribution of European Y-chromosome DNA (Y-DNA) haplogroups by region in percentage. http://www.eupedia.com/europe/european_y-dna_haplogroups.shtml.

Eurostat 2008 Mean consumption expenditures (in euro) of households on 12 domestic year costs in all 27 member states of the European Union (2005). http://epp.eurostat.ec.europa.eu/statistics_explained/index.php/Household_consumption_expenditure.

Eurostat 2009. *Europa in Zahlen - Eurostat Jahrbuch 2009 - Europäische Statistiken von A bis Z.* Eurostat Pressmitteilung 135/2009.

Filzmoser P and Hron K 2008 Outlier detection for compositional data using robust methods. *Mathematical Geosciences* **40**(3), 233–248.

Filzmoser P, Hron K and Reimann C 2009a Principal component analysis for compositional data with outliers. *Environmetrics* **20**, 621–632.

Filzmoser P, Hron K, Reimann C and Garrett R 2009b Robust factor analysis for compositional data. *Computers and Geosciences* **35**, 1854–1861.

Filzmoser P, Hron K and Templ K 2009c Discriminant analysis for compositional data and robust estimation. Technical Report SM-2009-3, Department of Statistics and Probability Theory, Vienna University of Technology, Austria. 27 p.

Hron K, Templ M and Filzmoser P 2010. Imputation of missing values for compositional data using classical and robust methods. *Computational Statistics and Data Analysis* **54**(12), 3095–3107.

Korhoňová M, Hron K, Klimčíková D. Müller L, Bednář P and Barták P 2009 Coffee aroma – statistical analysis of compositional data. *Talanta* **80**(82), 710–715.

Maronna R, Martin R and Yohai V 2006 *Robust Statistics: Theory and Methods.* John Wiley & Sons Ltd, New York, NY (USA). 436 p.

Mateu-Figueras G and Pawlowsky-Glahn V 2008 A critical approach to probability laws in geochemistry. *Mathematical Geosciences* **40**(5), 489–502.

Palarea-Albaladejo J and Martín-Fernández J 2008 A modified EM alr-algorithm for replacing rounded zeros in compositional data sets. *Computers & Geosciences* **34**(8), 902–917–1861.

R Development Core Team 2010 *R: A Language and Environment for Statistical Computing.* R Foundation for Statistical Computing, Vienna (Austria).

Rousseeuw P and van Driessen K 1999 A fast algorithm for the minimum covariance determinant estimator. *Technometrics* **41**, 212–223.

Rousseeuw P and Van Zomeren B 1990 Unmasking multivariate outliers and leverage points. *Journal of the American Statistical Association* **85**(411), 633–651.

Templ M, Hron K and Filzmoser P 2010 *robCompositions: Robust Estimation for Compositional Data.* Manual and package, version 1.4.1.

van den Boogaart G, Tolosana-Delgado R and Bren M 2008 *Compositions: Compositional Data Analysis.* R package version 1.01-1.

26

Linear models with compositions in R

Raimon Tolosana-Delgado[1] and Karl Gerald van den Boogaart[2]

[1]*Maritime Engineering Laboratory, Technical University of Catalonia, Spain*
[2]*Institute for Stochastic, Technical University Bergakademie Freiberg, Germany*

26.1   Introduction

One of the most common uses of statistics is the exploration and checking of *the existence* of some intrinsic relations between (sets of) variables, which may help in elucidating which processes are playing a role in the phenomenon under study. *Linear models* are the keystone of the methods applied in these cases: a linear model explains a *dependent variable* (or vector) as a *linear combination* of other variables (or sets of variables), called *explanatory variables* or *covariates*. When both the explained and explanatory variable sets are formed by measurable, continuous variables, then the method is typically called (multivariate or multiple) *linear regression*. When the explanatory variables are classifiable, categorical variables, then the method is called (multivariate) *analysis of the variance* (ANOVA), or *analysis of the covariance* (ANCOVA) when we have a mixture of categorical and continuous explanatory variables. In general, however, linear models are only applicable when the explained variable is a continuous one, because of an underlying, typically forgotten assumption: that the distribution of the explained variable *conditional* to the explanatory ones is a *normal distribution*. Obviously, in the case of having an explained vector (multiple regression, MANOVA, MANCOVA), this conditional distribution is a joint multivariate normal one. Moreover, the whole

Compositional Data Analysis: Theory and Applications, First Edition. Edited by Vera Pawlowsky-Glahn and Antonella Buccianti.
© 2011 John Wiley & Sons, Ltd. Published 2011 by John Wiley & Sons, Ltd.

structure of analysis is actually most often extended from the univariate to the multivariate case by simple *Cartesian product* of the separate, univariate variables of the explained set. In other words, apparently there is no real need in most multiple regression applications to attack the problem from a multivariate point of view: each explained variable could be studied separately as well.

Most of these issues fail when compositional variables are involved, especially when they play the role of explained variable set. First, the classical concept of linear combination has been shown to be meaningless on the simplex, as a classical linear combination may very well provide negative predictions (Aitchison 1986). Secondly, the conditional distribution of a composition cannot be a multivariate normal one, because of the well-known constraints of positivity and constant sum. Finally, and most importantly, a compositional vector is *not* a multivariate object because of our wish to treat the variables together: it is *intrinsically multivariate*, because each component cannot be interpreted without relating it to any of the other components.

However, these issues are all solved by simply posing the linear model on the coordinates of the observed compositional vectors with respect to any basis of the simplex, i.e. using log-ratio transformed compositions (e.g. ilr and alr) instead of raw ones. Now the linear combinations become meaningful with respect to the geometry of the simplex (Pawlowsky-Glahn and Egozcue 2001) (see also Chapter 11), the conditional distribution is thus a normal distribution on the simplex (or an additive logistic normal one; Aitchison 1986), and the multivariate nature of the data set is preserved, as long as the methods applied are linear.

Classical textbooks on linear models include Mardia *et al.* (1979, chapters 6 and 12), Fahrmeir and Hamerle (1984, chapters 4 and 5) and Krzanowski (1988, chapters 14 and 15) though mostly from the point of view that the multivariate variables are built through a Cartesian product of univariate variables. Only Eaton (1983) presents the subject in a general perspective, building upon an arbitrary Euclidean space structure for random vectors. For this reason, we mostly follow Eaton's (1983) approach.

This chapter shows how can we fit a linear model on the simplex using the package 'compositions' and the basic linear model functionality of R, the open source statistical software (R Development Core Team 2004). Its most relevant contribution is a graphical way of simultaneously checking several relevant null hypotheses about compositional parameters without using univariate tests on a log-ratio transformed data set. This is particularly important for data sets with many variables, when one wants to check which subcompositions may influence or be controlled by a covariable. The next section introduces a data set that will be used for illustration purposes throughout the chapter, and does a preliminary analysis of this data set. The following sections present ANOVA for a binary variable (Section 26.3), ANOVA for an ordered, categorical variable (Section 26.4) and regression with compositional response (Section 26.5). Finally, Section 26.6 introduces regression with an explanatory composition, where the geometry of the simplex plays an even more important role. Some methodological comments and conclusions regarding the data set close the chapter (Section 26.7).

26.2   The illustration data set

26.2.1   The data

Among the most important of the properties characterizing a sediment is to count their petrographic and/or chemical composition, and grain-size distribution. Grantham and Velbel

(1988) presented a data set of 72 samples of modern sands from low-order streams of Coweeta Basin (North Carolina, USA). The drainage basin is formed by Upper Precambrian meta-sedimentary bedrock. The southern part presents metamorphic/meta-sedimentary source rocks which are more mature (gneisses, meta-sandstones, quartzite, schists) compared with the northern part (meta-greywackes, mica schists, amphibolites). In spite of its small size ($\sim$16 km^2) the basin exhibits strongly contrasting relief (from less than 700 m up to 1600 m above sea level) and precipitation profiles (from 1700 to 2400 mm-year^{-1} approximately), thus having very different weathering conditions. Samples of coarse, medium and fine sands from eight watersheds were analysed for their content in rock fragments (Rf), polycrystalline quartz (Qp), monocrystalline quartz (Qm) and mica grains (M). We will focus our attention on the subcomposition formed by $\{Qm, Rf, M\}$. Polycrystalline rock fragments were discarded, to keep diagrams simpler. Thanks to the fact that log-ratio analysis is subcompositionally coherent (Aitchison 1986) (see also Chapter 11), we may be sure that results regarding the subcomposition $\{Qm, Rf, M\}$ will not be altered by this reduction of dimension. Moreover, the techniques explained in the following sections can be used with compositions of any number of parts: only their graphical representation becomes slightly harder. The data set and a script with the whole R code used in this chapter, can be downloaded from the book's website, at www.wiley.com/go/compositional_data_analysis.

26.2.2   Descriptive analysis of compositional characteristics

The preliminary steps of any analysis must be, obviously, to load the necessary packages and the data set:

```
> require("compositions")
> autoload("bpy.colors", "sp")
> require("lmtest")

> x = read.csv("GraVel.csv", header = TRUE)
> colnames(x)

[1] "watershed" "position" "CCWI" "discharge" "relief" "grain"
[7] "Qm"        "Qp"       "Rf"   "M"
```

With the command require one loads a package entirely, while with the command autoload we only prepare the system to load a particular function from a given package the first time we call it. The function bpy.colors will be used to obtain a nice, quasi-continuous colour palette that becomes a greyscale when printed in black and white. This is very useful when dealing with continuous variables. The data set is in this case stored in a csv file, and the first row contains the column names, therefore we use the function read.csv with the accessory argument header=TRUE. From the several variables available, shown by colnames, we select

```
> xc = acomp(x, c(7, 9, 10))
> colnames(xc)

[1] "Qm" "Rf" "M"

> gsize = factor(x[, 6], levels = c("c", "m", "f"), ordered = TRUE)
> loc = x[, 2]
```

excluding polycrystalline quartz, as mentioned before. Here, `loc` and `gsize` will be used as covariates, and `xc` contains a three-part composition, explicitly declared by the command `acomp` (from Aitchison composition). The first step should be an exploratory analysis of the composition and each covariate.

The command `summary` is very useful in this case,

```
> summary(xc)                              $mean.ratio
                                                  Qm        Rf        M
$mean                                      Qm 1.0000000 1.7824903 3.129611
        Qm        Rf        M             Rf 0.5610129 1.0000000 1.755752
0.5317617 0.2983252 0.1699131              M  0.3195286 0.5695566 1.000000
attr(,"class")
[1] acomp                                  $variation
                                                  Qm        Rf        M
Qm 0.0000000 1.071052 0.4693781            Qm 1.0000000 4.290899 3.224343
Rf 1.0710515 0.000000 1.6163752            Rf 1.5123732 1.000000 3.193372
M  0.4693781 1.616375 0.0000000            M  0.4302113 1.437290 1.000000
$expsd
        Qm        Rf        M              $max
Qm 1.000000 2.814870 1.983993                     Qm        Rf  M
Rf 2.814870 1.000000 3.565725             Qm 1.0000000 10.714286 48
M  1.983993 3.565725 1.000000             Rf 4.1214286  1.000000 70
                                           M  0.8983957  3.793651  1
$min
            Qm         Rf         M        $missingness
Qm 1.00000000 0.24263432 1.1130952        missingType
Rf 0.09333333 1.00000000 0.2635983        variable NMV BDT MAR MNAR SZ Err
M  0.02083333 0.01428571 1.0000000              Qm  72   0   0    0  0   0
                                                Rf  72   0   0    0  0   0
                                                 M  72   0   0    0  0   0
$q1
            Qm        Rf         M
Qm 1.0000000 0.6615676 2.3244391
Rf 0.2330629 1.0000000 0.6957672          attr(,"class")
M  0.3101810 0.3131733 1.0000000          [1] "summary.acomp"

$med                                       > summary(gsize)
            Qm        Rf         M
Qm 1.0000000 1.9992369 2.772034            c  m  f
Rf 0.5009178 1.0000000 1.453588           24 24 24
M  0.3607462 0.6887499 1.000000            > summary(loc)

$q3                                        north south
                                              36    36
```

A compositional summary contains the geometric centre and the variation matrix (Aitchison 1986) (see also Chapter 24). But, following the idea of variation matrix as an array of variances of all possible log-ratios, a compositional summary contains also equivalent arrays of some other classical descriptive statistics (van den Boogaart and Tolosana-Delgado 2008). Finally, it also prints a table reporting the missing values presence. Summaries for factors are just frequency tables. In this case, we see that the design is balanced, with the same number of coarse, medium and fine sand samples, and the same number of samples placed in the southern and northern parts of the basin. No data are missing.

26.3   Explanatory binary variable

To display the dependence of a composition on a factor (binary or multi-category), we can give the accessory arguments pch (symbols), col (foreground colour) or bg (background colour) values based on functions of the desired factor:

```
> plot(acomp(xc), pch = 20 + as.integer(loc), col = 1 + as.integer(loc))
```

The command plot applied to an acomp data set of three parts generates a ternary diagram. The preceding instruction displays both symbols and colours according to the binary factor loc [Figure 26.1(b)]. Figure 26.1 shows other elements that will be added in the next steps. In this step, it suffices to see that there seems to be a consistent difference between samples from the northern and southern sub-basins: the former appear poorer in Qm than the latter.

If we want to use a compositional linear model to model this effect, we state that

$$\mathbf{x} = \mathbf{a} \oplus l \odot \mathbf{b} \oplus \boldsymbol{\epsilon}, \tag{26.1}$$

or in coordinates

$$\mathrm{ilr}(\mathbf{x}) = \mathrm{ilr}(\mathbf{a}) + l \cdot \mathrm{ilr}(\mathbf{b}) + \mathrm{ilr}(\boldsymbol{\epsilon}),$$

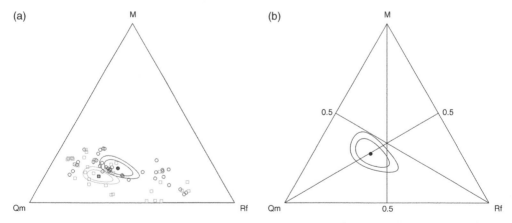

Figure 26.1   (a) Ternary diagram of the data, displaying the dependence of the composition on the location. Each sub-basin (north versus south) centre is represented with 90% and 99% confidence regions. (b) Ternary diagram of the **b** parameter, representing the perturbation leading from the mean composition at the northern sub-basin to the mean of the southern sub-basin. Three isoproportion lines of proportion 1:1 are shown for reference, equivalent to null hypotheses of no change within a subcomposition.

where $l = 0, 1$, respectively, for 'north' and 'south'. As usual, the error term ϵ is considered normally distributed: in this case it has a centred normal distribution on the simplex $\epsilon \sim \mathcal{N}_S^D(\mathbf{n}, \boldsymbol{\Sigma}_\epsilon)$, or in terms of the coordinates ilr$(\epsilon) \sim \mathcal{N}^{D-1}(\mathbf{0}, \boldsymbol{\Sigma}_\epsilon)$. Here $\mathbf{n} = [1, 1, \ldots, 1]/D$ is the neutral element of perturbation.

To build a model [Equation (26.1)] on R, we make use of the function $lm(y \; x)$, where x and y are, respectively, the explanatory and the explained variable sets. In the case that one of these is a compositional data set, we must introduce it as ilr-transformed, by virtue of the *principle of working on coordinates* (Pawlowsky-Glahn 2003) (see also Chapter 3). This is all done with

```
> resB = lm(ilr(xc) ~ loc)
> resB

Call:
lm(formula = ilr(xc) ~ loc)

Coefficients:
             [,1]      [,2]
(Intercept) -0.2500  -0.5909
locsouth    -0.3174  -0.2094
```

which fits a conventional linear model to each ilr coordinate. The result is the set of coordinates ilr($\mathbf{a}$) and ilr($\mathbf{b}$), which can be back transformed to obtain the compositions $\mathbf{a}$ intercept and $\mathbf{b}$ slope:

```
> compcoef = ilrInv(coef(resB), orig = xc)
> compcoef

              Qm        Rf        M
(Intercept) 0.4742517 0.3330141 0.1927342
locsouth    0.4431630 0.2828795 0.2739575
attr(,"class")
[1] acomp
```

With the model 'as is', we can also apply standard tests, like an analysis of the variance:

```
> anova(resB)

Analysis of Variance Table

            Df  Pillai approx F num Df den Df    Pr(>F)
(Intercept) 1 0.79685  135.324      2     69 < 2.2e-16 ***
loc         1 0.18144    7.647      2     69  0.001001 **
Residuals  70
---
Signif. codes: 0 '***' 0.001 '**' 0.01 '*' 0.05 '.' 0.1 ' ' 1
```

This shows that both coordinate vectors are significantly different from zero, ilr($\mathbf{a}$) $\neq 0 \neq$ ilr($\mathbf{b}$). Thus we may conclude that the difference between the northern and southern sub-basins displayed in Figure 26.1 is significant. To graphically represent this model we may recall first that **a** is the centre for $l = 0$ (northern sub-basin) and that **b** corresponds to the difference between the two centres:

```
> centreN = acomp(compcoef[1, ])
> centrediff = acomp(compcoef[2, ])
> centreS = centreN + centrediff
```

The function `vcov` is also useful to extract the estimation variance of the full vector of coefficients of the linear model, that we need to draw elliptic confidence regions around there model parameters. Since we only want the variance matrices of **a** and **b** separately, we must extract them:

```
> vaux = vcov(resB)
> colnames(vaux)

[1] ":(Intercept)" ":locsouth" ":(Intercept)" ":locsouth"

> varintercept = vaux[-2 * (1:2), -2 * (1:2)]
> varintercept = ilrvar2clr(varintercept)
> varslope = vaux[2 * (1:2), 2 * (1:2)]
> varslope = ilrvar2clr(varslope)
```

Finally, with these elememts we may plot the data set, the two centres of each sub-sample, and the slope composition **b**, accompanied with their respective 90 and 99% confidence regions (Figure 26.1). This is the generating code:

```
> par(mfrow = c(1, 2), mar = c(3, 3, 1, 1))
> plot(acomp(xc), pch = 20 + as.integer(loc), col = 1 + as.integer
  (loc))
> plot(acomp(centreN), pch = 21, bg = 2, add = TRUE)
> plot(acomp(centreS), pch = 22, bg = 3, add = TRUE)
> r90 = sqrt(qchisq(0.9, df = 2))
> r99 = sqrt(qchisq(0.99, df = 2))
> ellipses(acomp(centreN), varintercept, col = 2, r = r90)
> ellipses(acomp(centreS), varintercept, col = 3, r = r90)
> ellipses(acomp(centreN), varintercept, col = 2, r = r99)
> ellipses(acomp(centreS), varintercept, col = 3, r = r99)
> plot(centrediff, pch = 19)
> ellipses(centrediff, varslope, r = r)
> ellipses(centrediff, varslope, r = r99)
> isoProportionLines(by = 0.5)
```

Figure 26.1(b) shows an atypical display of the model: as stated, we plotted the slope composition **b** with some confidence regions, but we have also drawn three reference lines: along one of these lines, the proportion between two parts of the composition is constant and equal to 1:1. In other words, looking at the relative position of the ellipses with respect to these lines, we can graphically assess the hypothesis that **b** does not imply a change in a given subcomposition. In this particular case, we see that most surely the ratio M/Rf remains constant along **b**, i.e. that the main difference form the northern to the southern sub-basin is an enrichment in Qm. Note however that the line associated with the hypothesis 'Qm/Rf

is constant' also intersects the 90% confidence region, thus we could also consider this as reasonable. Both at the same time are nevertheless not credible, as the intersection point of the two lines $Qm/Rf = 1/1$ and $M/Rf = 1/1$ (the barycentre) is out of the confidence regions, even at 99% confidence.

26.4   Explanatory categorical variable

The same sort of analysis can be extended to factors (categorical variables) with more than two levels, illustrated using the factor gsize:

```
> resF = lm(ilr(xc) ~ gsize)
> anova(resF)

Analysis of Variance Table

              Df  Pillai approx F num Df den Df    Pr(>F)
(Intercept)   1 0.77624  117.945      2     68 < 2.2e-16   ***
gsize         2 0.82392   24.169      4    138 3.613e-15 ***
Residuals    69
---
Signif. codes: 0 '***' 0.001 '**' 0.01 '*' 0.05 '.' 0.1 ' ' 1
```

This ANOVA test tells us that the global model actually reproduces an effect on the data, but we do not know whether all levels are equally significant or whether some are significant and some not: e.g. coarse and medium sand could not differ between them, while fine sand could be different from them. We should then continue by testing some partial hypothesis. However, in multilevel factors the set of interesting hypotheses that may be tested are much more than simply 'the mean between the first and second groups has only differences in this or that subcomposition'. If we have K levels we could at least check the $K(K-1)/2$ pairwise differences between groups, but all these tests are redundant, not independent. R uses the concept (and the command) of contrasts, to decide which tests to use by default in an analysis with a factor. In our case, gsize has three levels, and the typical contrast matrices for this case are

```
> contrasts(factor(gsize, ordered = FALSE))
  m f
c 0 0
m 1 0
f 0 1

> contrasts(gsize)
                .L            .Q
[1,] -7.071068e-01  0.4082483
[2,]  3.925231e-17 -0.8164966
[3,]  7.071068e-01  0.4082483
```

The first one is valid when the factor has no order, and tells us that we evaluate how significant the differences are from coarse to medium sand average composition, and from coarse to fine sand average composition. In other words all levels are contrasted against the first level.

The second matrix is used for ordered factors (like in this case), and tells us that we evaluate a sort of 'linear' change between the three levels (in the form of the perturbation between the compositional means for the first and the last level) against a sort of 'quadratic' change (where the first and the third will be contrasted against the second, given a double weight). Note how similar these matrices are to the matrices of definition of ilr bases (see Chapter 11).

Because the categorical variable gsize is ordered, the coefficients of the fitted model are based on the second contrast matrix,

```
> coef(resF)
                    [,1]         [,2]
(Intercept) -0.4087158 -0.6955758
gsize.L      -1.1053227  0.8414134
gsize.Q       0.1710143 -0.1874837

> compcoef = ilrInv(coef(resF), orig = xc)
> compcoef
              Qm         Rf          M
(Intercept) 0.5317617 0.29832518 0.1699131
gsize.L     0.4012629 0.08405358 0.5146836
gsize.Q     0.3153969 0.40169073 0.2829124
attr(,"class")
[1] acomp
> centreM = acomp(compcoef[1, ])
> effectL = acomp(compcoef[2, ])
> effectQ = acomp(compcoef[3, ])
```

We have thus a global mean effect centreM, denoted as $\mathbf{x}_0$ and playing the role of the intercept, a linear effect effectL $\mathbf{x}_l$ and a quadratic effect effectQ $\mathbf{x}_q$,

$$\mathbf{x}(f) = \mathbf{x}_0 \oplus f \odot \mathbf{x}_l + f^2 \odot \mathbf{x}_q \oplus \epsilon, \qquad (26.2)$$

where in this case $f = -1, 0, 1$, respectively, for coarse, medium and fine sands. The current version of R (v. 2.6.2) does not include any command to directly test that any of these compositional coefficients is null (i.e. the vector $\mathbf{n} = [1/3, 1/3, 1/3]$). We either check the full model with anova and aov commands, or else use the coeftest command from the lmtest package to test the hypothesis over each coefficient independently, e.g. the first ilr coordinate of $\mathbf{x}_l$. None is useful for us, because of the arguments mentioned in Section 26.1 about the intrinsic multivariate nature of compositions. We therefore switch again to a graphical assessment of how important are the two effects, by using the error covariance matrix of the coefficient estimator, as in the preceding section:

```
> vaux = vcov(resF)
> rownames(vaux)

[1] ":(Intercept)" ":gsize.L" ":gsize.Q" ":(Intercept)" ":gsize.L"
[6] ":gsize.Q"

> varintercept = ilrvar2clr(vaux[3 * (0:1) + 1, 3 * (0:1) + 1])
> varL = ilrvar2clr(vaux[3 * (0:1) + 2, 3 * (0:1) + 2])
> varQ = ilrvar2clr(vaux[3 * (0:1) + 3, 3 * (0:1) + 3])
```

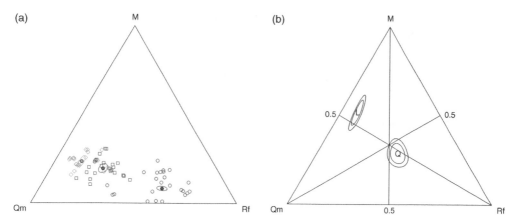

Figure 26.2 (a) Ternary diagram of the data, displaying the dependence of the composition on grain size. Each grain size subpopulation (coarse, medium and fine) centre is represented with 90% and 99% confidence regions. (b) Ternary diagram of the effects parameters, representing the perturbations associated with linear and quadratic effects between categories [Equation (26.2)]. Three isoproportion lines of proportion 1:1 are shown for reference, equivalent to null hypotheses of no change within a subcomposition. 'L' indicates the linear effect x_l whereas 'Q' shows the quadratic effect x_q.

Again, the data and the fitted model are compared in Figure 26.2. For the sake of saving space, the generating code is provided in the book's website. In Figure 26.2, we may find the compositional data set distinguished by sand size levels, each with 90% and 99% confidence regions for their means. But even more interesting for our purposes is Figure 26.2(b): the model coefficient representation as a composition, with the elliptic confidence regions and the same 90% and 99% confidence. Figure 26.2(b) shows that the linear effect x_l may well be on the isoproportion line $M/Qm = 1/1$, thus suggesting that this effect might be a pure depletion in Rf towards the finer sand fractions. In the same way, the quadratic effect x_q falls very near to the barycentre, and in fact the 99% confidence region includes it: we should thus conclude that the quadratic effect is barely significant. In other words, we may build a model where the change from coarse to medium sand is the same as the change from medium to fine sand, i.e. a linear model. This will be obtained in the next section.

26.5   Explanatory continuous variable

Grain sizes are measured in geology in ϕ units: if a grain has an average diameter d (in mm), then $\phi = -\log_2(d)$. Typically, a sand with $0 < \phi \leq 1$ is considered *coarse*, with $1 < \phi \leq 2$ is *medium*, and with $2 < \phi \leq 3$ is *fine*. These were the actual categories used by Grantham and Velbel (1988). Thus we may take gsize as a factor (as in the last section), or as a continuous variable (as we will do in this section). This line converts the factor into its equivalent numerical variable:

```
> gs = as.integer(gsize) - 1
```

using the lower limit of the intervals defined before to identify the grain sizes. We can now construct the model

$$\mathbf{x}(\phi) = \mathbf{a} \oplus \phi \odot \mathbf{b}_\phi \oplus \epsilon, \qquad (26.3)$$

where now $\mathbf{b}_\phi$ represents the perturbation occurring in the composition when ϕ changes one unit. This model is fit as before, with function lm,

```
> resC = lm(ilr(xc) ~ gs)
> anova(resC)

Analysis of Variance Table

            Df  Pillai approx F num Df den Df    Pr(>F)
(Intercept)  1 0.77595   119.48      2     69 < 2.2e-16 ***
gs           1 0.79822   136.48      2     69 < 2.2e-16 ***
Residuals   70
---
Signif. codes:  0 '***' 0.001 '**' 0.01 '*' 0.05 '.' 0.1 ' ' 1
```

Again, an analysis of the variance table shows that the slope composition $\mathbf{b}_\phi$ is significantly different from $\mathbf{n}$ the neutral element of perturbation. We can also extract the model coefficients and their estimation error covariance matrices,

```
> compcoef = ilrInv(coef(resC), orig = xc)
> intercept = acomp(compcoef[1, ])
> slope = acomp(compcoef[2, ])
> vaux = vcov(resC)
> varintercept = ilrvar2clr(vaux[2 * (0:1) + 1, 2 * (0:1) + 1])
> varslope = ilrvar2clr(vaux[2 * (0:1) + 2, 2 * (0:1) + 2])
```

and draw with them confidence regions for the intercept and slope compositions, and for the centres of coarse, medium and fine sand compositions. In this way we visually represent several issues (Figure 26.3). First, the intercept coefficient (and its confidence region) corresponds to the average coarse sand composition. Secondly, the slope coefficient may be considered (at 90% confidence) to lie on the line $Qm/M = 1{:}1$, thus the process here described may be a simple depletion on rock fragments towards finer grains without affecting the ratio Qm/M. Thirdly, if we compare the slope coefficient of the continuous model with the linear effect coefficient of the factor model (Section 4), we see that they are essentially the same (just a shorter effect in the continuous approach). This is a further hint that the model does not need the quadratic contribution: Equation (26.3) represents a model which is a valid simplification of the factor model of Equation (26.2).

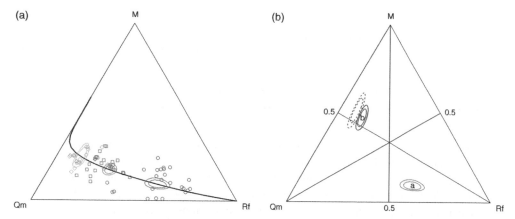

Figure 26.3   (a) Ternary diagram of the data, displaying the dependence of the composition on the grain size. Each grain size subpopulation (coarse, medium and fine) centre is represented with 90% and 99% confidence regions (Mardia *et al.* 1979). (b) Ternary diagram of the regression model parameters (solid lines), together with the linear effect parameter of the categorical model of Section 26.4. Three isoproportion lines of proportion 1:1 are shown for reference, equivalent to null hypotheses of no change within a subcomposition.

26.6   Explanatory composition

The final option within the scope of regression and linear models is to have a compositional data set playing the role of a set of covariates of a continuous, explained variable, simply as

$$y = y_0 + \sum_{i=1}^{D-1} \gamma_i \, \mathrm{ilr}_i(\mathbf{x}) + \epsilon,$$

or better, seen as an (affine) projection of the composition onto a given direction of the simplex,

$$y - y_0 + \langle \mathbf{g}, \mathbf{x} \rangle_a + \epsilon. \tag{26.4}$$

In this case, the error term has the same scale as y, which is considered a real variable. Thus, it is natural to assume $y \sim \mathcal{N}(0, \sigma_\epsilon^2)$ centred normally distributed. The preferential direction $\mathbf{g} = \mathrm{ilr}^{-1}(\gamma_1, \ldots, \gamma_{D-1})$ is the gradient of y with respect to $\mathbf{x}$ (see Chapter 11), i.e. the direction on the simplex of maximal variation of y. The model [Equation (26.4)] can be fitted and globally tested with

```
> res = lm(gs ilr(xc))
> anova(res)

Analysis of Variance Table
```

```
Response: gs
          Df Sum Sq Mean Sq F value      Pr(>F)
ilr(xc)    2 38.314 19.1572  136.48 < 2.2e-16 ***
Residuals 69  9.686  0.1404
---
Signif. codes: 0 '***' 0.001 '**' 0.01 '*' 0.05 '.' 0.1 ' ' 1
```

Note that the ANOVA test tell us that the model is globally significant. But it could happen that the relevant information for predicting y is contained just in a subcomposition, and this may become more important for a larger ($D > 3$) explanatory composition. It may be thus necessary to devise ways to test for subcompositional independence. Aitchison (1986) proposes an exhaustive exploration of all possible subcompositions, where the best one is chosen using a maximum-likelihood information criterion. Other options exist, like systematically exploring a set of ilr bases (Tolosana-Delgado and von Eynatten 2010). We propose here a graphical assessment, that may be useful to suggest which subcompositions deserve to be tested. As we did in the preceding sections, this is based on building confidence regions on the ternary diagram for the model parameters themselves,

```
> gs0 = coef(res)[1]
> gs0

(Intercept)
  0.8262851
> gradient = ilrInv(coef(res)[-1], orig = xc)
> gradorth = ilrInv(c(coef(res)[3], -coef(res)[2]))
> scalar(gradient, gradorth)

[1] 5.551115e-17
> vaux = vcov(res)
> vargs0 = vaux[1, 1]
> vargradient = ilrvar2clr(vaux[-1, -1])
```

These lines extract the null response y_0 (intercept, gs0) and the gradient $\mathbf{g}$, as well as their respective variances (the first a scalar, the second a matrix of compositional variance). We also build a vector orthogonal to the gradient (gradorth, denoted $\mathbf{g}^{\perp}$), that will be used for plotting the regression surface. Figure 26.4 shows all these elements. The generating code may be found in the book's website.

It is worth noting that the regression surface is a plane on the simplex, because its level curves are compositional lines and the compositional gradient is always the same vector. To draw this surface, we use the following facts:

- The mean composition gives the mean grain size, as usual in regression; thus $\bar{y} = y_0 + \langle \text{Cen}[\mathbf{x}], \mathbf{g} \rangle_a$.

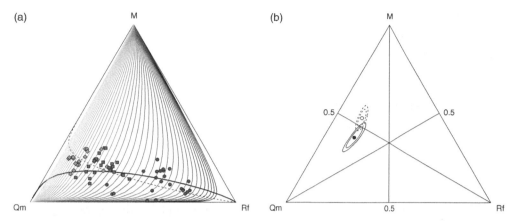

Figure 26.4 (a) Ternary diagram of the data, displaying the dependence of grain size on the composition. A family of iso-ϕ lines is displayed in a colour scale on the simplex, representing the fitted regression surface (in fact, a plane). The compositional gradient of this plane is shown as a thick line, orthogonal to the surface levels. The slope of the model in Section 26.5 is represented as a dashed line, for the sake of comparison. (b) Ternary diagram of the regression model parameters (solid lines), together with the slope parameter of the model in Section 26.5. Three isoproportion lines of proportion 1:1 are shown for reference, equivalent to null hypotheses of no change within a subcomposition, i.e. subcompositional independence.

- Changes on y occur only when the composition $\mathbf{x}$ moves along the direction of $\mathbf{g}$: because $\langle \mathbf{g}^{\perp}, \mathbf{g} \rangle_a = 0$, it is straightforward to verify that

$$
\begin{aligned}
y(\alpha, \beta) &= \langle \text{Cen}[\mathbf{x}] \oplus \alpha \odot \mathbf{g} \oplus \beta \odot \mathbf{g}^{\perp}, \mathbf{g} \rangle_a + y_0 \\
&= \langle \text{Cen}[\mathbf{x}], \mathbf{g} \rangle_a + \alpha \cdot \langle \mathbf{g}, \mathbf{g} \rangle_a + \beta \cdot \langle \mathbf{g}^{\perp}, \mathbf{g} \rangle_a + y_0 \\
&= \bar{y} + \alpha \cdot \|\mathbf{g}\|_a^2 + 0 = \bar{y} + \alpha \cdot \|\mathbf{g}\|_a^2
\end{aligned}
$$

from which we obtain that the line of level y must pass through the composition:

$$
\mathbf{x}(y) = \text{Cen}[\mathbf{x}] \oplus \alpha \odot \mathbf{g} = \text{Cen}[\mathbf{x}] \oplus \frac{y - \bar{y}}{\|\mathbf{g}\|_a^2} \mathbf{g}.
$$

Regarding Figure 26.4(b), we see that the gradient confidence region intersects the isoproportion line $Qm/M = 1/1$ (as occurred with the slopes obtained in the preceding sections). This suggests that we should check whether grain size is independent of the subcomposition $\{Qm, M\}$. We may do this by choosing an ilr basis based on a partition containing this subcomposition, and checking that its regression coefficient is zero:

```
> y = balance(xc, expr = ~ (Qm/M)/Rf)
> mydat = data.frame(y, gs)
> colnames(mydat)
```

```
[1] "QmM.Rf" "Qm.M" "gs"

> rescheck = lm(gs ~ QmM.Rf + Qm.M, mydat)
> coeftest(rescheck)

t test of coefficients:

            Estimate Std. Error t value  Pr(>|t|)
(Intercept) 0.826285   0.090907  9.0893 2.053e-13 ***
QmM.Rf      0.845908   0.052695 16.0528 < 2.2e-16 ***
Qm.M        0.208858   0.098354  2.1235 0.0373 *
---
Signif. codes:  0 '***' 0.001 '**' 0.01 '*' 0.05 '.' 0.1 ' ' 1
```

As we see from the coefficient test table, the coefficient multiplying the log-ratio Qm/M (corresponding to the last row) may be removed at a 1% significance level, whereas the log-ratio of $(Qm \cdot M)/Rf^2$ is highly significant. We may also see that the intercept ($gs0$) is the same now (see 'intercept estimate') as it was in the first, black-box model of this section. The procedure of building the basis within command balance and embedding the coordinates and the predicted variable in a single data frame is longer, especially when we have to write full formulae (the terms starting with ~) with many compositional parts. However, it is nowadays the easiest way to apply classical regression tests in R with compositions.

26.7   Conclusions

This chapter has shown how the package compositions can be used in conjunction with standard R routines to obtain linear models (regression and ANOVA) where compositional data are involved, either a response or as explanatory variables. The models themselves can be easily obtained plugging the functions ilr and lm, respectively, to compute ilr coordinates and fit linear models. The anova function (and eventually lmtest) may be applied directly to check the significance of the results. Using ilrInv we may transform the regression/ANOVA coefficients into compositional vectors that may be used to draw the fitted models (as lines or surfaces), and also to plot them and their confidence regions in a ternary diagram. This last representation is useful to visually check possible subcompositional independence hypotheses.

This methodology allowed us to explore the relationships between the composition of stream sands (in rock fragments, monocrystalline quartz and mica grains) and grain size, suggesting that in fact most of the relationship is concentrated in the rock fragment abundance (relative to the other two grain classes). This is an unsurprising result, as smaller grains are more easily monocrystalline (as quartz and mica are in this study) than polycrystalline (as a rock fragment must be). On the other hand, this data set was also split in to two sub-basins (one placed at the northern part of the study area, the other at the southern part). When exploring the differences between these two sub-basins, it has become clear than the southern sub-basin is essentially richer in monocrystalline quartz than the northern sub-basin. That was anticipated by Grantham and Velbel (1988) when presenting the data, who mentioned that the rocks producing these sands were more mature (i.e. richer in quartz) in the southern part than those from the northern part.

Acknowledgement

This research was funded by the Spanish Ministry of Science and Innovation through a 'Juan de la Cierva' programme (Ref. JCI-2008-1835), supported by the European Social Fund (ESF-FSE).

References

Aitchison J 1986 *The Statistical Analysis of Compositional Data*. Monographs on Statistics and Applied Probability. Chapman and Hall Ltd (reprinted 2003 with additional material by The Blackburn Press), London (UK). 416 p.

Eaton ML 1983 *Multivariate Statistics. A Vector Space Approach*. John Wiley & Sons, Ltd, New York, NY (USA). 512 p.

Fahrmeir L and Hamerle A (ed.) 1984 *Multivariate Statistische Verfahren*. Walter de Gruyter, Berlin (Germany). 796 p.

Grantham JH and Velbel MA 1988 The influence of climate and topography on rock-fragment abundance in modern fluvial sands of the southern Blue Ridge Mountains, North Carolina. *Journal of Sedimentary Petrology* **58**, 219–227.

Krzanowski WJ 1988 *Principles of Multivariate Analysis: A User's Perspective*. Clarendon Press, Oxford (UK). 563 p.

Mardia KV, Kent JT and Bibby JM 1979 *Multivariate Analysis*. Academic Press, London (UK). 518 p.

Pawlowsky-Glahn V 2003 Statistical modelling on coordinates. In *Proceedings of CoDaWork'03, The 1st Compositional Data Analysis Workshop* (ed. Thió-Henestrosa S and Martín-Fernández JA). http://ima.udg.es/Activitats/CoDaWork03/. University of Girona. Girona (Spain). CD-ROM.

Pawlowsky-Glahn V and Egozcue JJ 2001 Geometric approach to statistical analysis on the simplex. *Stochastic Environmental Research and Risk Assessment (SERRA)* **15**(5), 384–398.

R Development Core Team 2004 *R: A Language and Environment for Statistical Computing*. R Foundation for Statistical Computing, Vienna (Austria).

Tolosana-Delgado R and von Eynatten H 2010 Simplifying compositional multiple regression: Application to grain size controls on sediment geochemistry. *Computers & Geosciences* **36**(5), 577–589.

van den Boogaart KG and Tolosana-Delgado R 2008 "compositions": a unified R package to analyze compositional data. *Computers & Geosciences* **34**(4), 320–338.

Index

Compositional Data Analysis: Theory and Applications, First Edition. Edited by Vera Pawlowsky-Glahn and Antonella Buccianti.
© 2011 John Wiley & Sons, Ltd. Published 2011 by John Wiley & Sons, Ltd.

replacement
  Bayesian-multiplicative, 51, 56
  non-parametric, 45–46, 56
  parametric, 45–47
representant, *see* representative
representative, 14–15, 142, 264
resource selection, 222–223
robCompositions, 61–62, 341–342
robustness, 60, 70, 221, 322

sample membership indicator, 117–118, 123
sampling frame, 115–116, 122
SBP, *see* sequential binary partition
scaled Dirichlet distribution, 115–116, 122
scale-equivalent, 143
scale expansion, 177–178
scale-invariant, 177–179, 182
scale-invariant function, 177–179, 182
sequential binary partition (SBP), 6, 22, 26,
    35, 246, 336
simplex, 41, 43, 45, 51, 59
simplex-valued function of real variable,
    159
simplex-valued, *see* function
simplicial antiderivatives, 171–172
simplicial derivative, 163, 166–167, 170
single nucleotide polymorphism, 209, 220
singular value decomposition (SVD), 7, 22,
    23, 25, 104, 150
sinusoids, 284
size, 23
software
  CoDaPack, 294, 329–331, 333–335
  compositions, 8, 15
  robCompositions, 341–342
specialization (ecological), 228–229
spectral analysis, 283–284
spectral mapping, 105, 111
spectral peaks, 284–286
spurious correlation, 3–4, 14–15, 27
subcomposition, 15
subcompositional incoherence, 109–110,
    112, 119
surface
  composition, 269–270, 274–275, 277
  mineralogy, 270
  planetary, 267–269, 274

survey design, 114–115
SVD, *see* singular value decomposition

tangent, 168–169
Taylor linearization, 119, 121
Taylor polynomial, 170
ternary diagram, 36, 52, 61–62, 68, 70, 132,
    160–161
ternary plot, 210, 212, 214–215
test
  chi-square, 106, 111–112
  exact, 209–210
  likelihood ratio, 95, 209
theorem
  Bayes, 51
  fundamental of calculus, 159
  mean value, 174
  Pythagoras, 149
  Taylor, 170
  Weierstrass, 163
theropoda, 236, 238–239
theropods, 240–241
threshold, 44–46
time budgets, 229–230
time series
  stationary, 78, 92
  VARIMA, 87–89
  white noise, 92, 95
total variance, 22–27, 108, 205, 226
transcriptomics
  additive log-ratio (alr), 47, 59, 88, 257,
    293
  centered log-ratio, 20
  centered log-ratio (clr), 20
  isometric logratio, 6, 21, 35, 59
  isometric log-ratio (ilr), 149, 155, 183,
    198
  power, *see* powering
triangular inequality, 149

value below detection limit, 44
variation array, 23–24, 205, 333–334
variation matrix, 5, 7, 74, 76, 82
variation, multivariate coefficient of,
    225
variation-variogram, logratioVariogram
    (function), 82

Printed and bound by CPI Group (UK) Ltd, Croydon, CR0 4YY

27/10/2024

14580150-0005